Insect Pests

OF FARM, GARDEN, AND ORCHARD

seventh edition

Ralph H. Davidson
Professor Emeritus of Entomology
The Ohio State University

William F. Lyon
Associate Professor of Entomology
The Ohio State University

JOHN WILEY & SONS
New York Chichester Brisbane Toronto

Library of Congress Cataloging in Publication Data

Davidson, Ralph Howard.
Insect pests of farm, garden, and orchard.

Fourth-6th ed. by L. M. Peaires and R. H. Davidson.
Includes bibliographical and references and index.
1. Insects, Injurious and beneficial.
2. Insects, Injurious and beneficial—North America.
3. Insect control. I. Peaires, Leonard Marion, 1886–
Insect pests of farm, garden, and orchard.
II. Lyon, William Francis, 1937– joint author.
III. Title.
SB931.P34 1979 632'.7'097 78-31366
ISBN 0-471-03538-6

Printed in the United States of America

10 9 8 7 6 5 4 3 2 1

In Memory
of

E. Dwight Sanderson, author of the first edition and
co-author of the second and third editions,

and

Leonard M. Peairs, author of the fourth edition and
co-author of the fifth and sixth editions,
this book is sincerely dedicated

Preface

The seventh edition of *Insect Pests of Farm, Garden, and Orchard* has been written to keep abreast of the many changes that have occurred in this field during the past decade.

Major changes in this edition include: complete updating of references to literature and biological information; altering and rearranging of topics, incorporating the suggestions of users of previous editions; adding important insect and mite pests; deleting some pests no longer of much importance; omitting dilution and conversion tables and substituting conversion factors for changing from English to the metric system and vice versa; deleting many illustrations and substituting new ones where appropriate, with magnifications given for all illustrations; reducing the section that deals with chemicals and emphasizing control measures other than chemical; and adding a new chapter on environmental management.

The new chapter introduces the student to a developing multidimensional area of agricultural science. Although pest management is the popular terminology, integrated environmental management seems far more descriptive of what actually takes place.

Common and scientific names of pests, for the most part, are those approved by the Entomological Society of America.

Detailed chemical names and structural formulae may be obtained from special publications on the subject or provided by the instructor.

The history of entomology has been intentionally omitted from this book. Those desirous of such coverage should read references on the subject listed at the end of Chapter 1.

Because some pests may attack well over 250 hosts, it is inadvisable to list every host of every pest and it is difficult to group all pests into categories that will be satisfactory to everyone. The host groupings used have met with wide approval in past editions; only slight changes have been made in this edition.

In so doing we have kept overlapping information at a minimum. Competent teachers know the pest problems of a particular host in their area and can adapt the text material to their needs.

For those using the book in a single quarter or semester course we suggest that after covering the first nine chapters a detailed study be made only of the pests in a given geographical region. Mastery of the subject of economic entomology requires years of continued study and research.

Besides serving as a textbook for beginning college courses in applied or economic entomology, this book can be a valuable reference for more advanced or specialized courses. In addition, it continues to serve as a source of current information and an aid in diagnosing pest problems by research and extension entomologists, county agricultural agents, vocational agriculture teachers, consulting entomologists, farm managers, and pest control operators.

The most reliable information on controlling pests has been assembled and *any user of this information does so at his or her own risk.* Control measures for individual pests are not given in detail because they vary for any given state, province, region, or country. Directions for the use of suggested pesticides are given on the container label by the manufacturer. Readers with difficult pest problems are urged to consult specialists at local agricultural colleges or agricultural experiment stations for more detailed information.

Although the title of the book indicates that only insects are discussed, other pest problems often confronting entomologists—such as mites, spiders, snails, slugs, nematodes, symphylids, centipedes, millipedes, and sowbugs—are also included.

Our thanks are extended to the staff of John Wiley & Sons whose expert advice and assistance made the production of this edition possible.

The senior author welcomes a colleague as co-author of this edition.

Columbus, Ohio
December, 1978

Ralph H. Davidson
William F. Lyon

Acknowledgments

A great many people made valuable contributions and criticisms or furnished illustrations in each edition of this book, beginning with the first edition in 1912. We believe credit and acknowledgment of their help will be informative to the readers of the seventh edition.

First Edition:

J. A. Arnold
E. A. Back
W. E. Britton
F. H. Chittenden
C. S. Crandall
S. W. Fletcher
S. A. Forbes
H. Garman
C. P. Gillette
H. A. Gossard
G. W. Herrick
L. O. Howard
S. J. Hunter
C. F. Jackson
T. C. Johnson
J. C. Kendall
C. L. Marlatt
F. B. Mumford
P. J. Parrott
R. H. Pettit
A. L. Quaintance
C. V. Riley
P. H. Rolfs
W. E. Rumsey
M. V. Slingerland
J. B. Smith
R. I. Smith
H. E. Summers
T. B. Symons
R. W. Thatcher
F. L. Washburn
F. M. Webster

Second Edition:

J. A. Arnold
E. W. Berger
F. C. Bishopp
W. E. Britton
C. S. Crandall
S. W. Fletcher
S. A. Forbes
H. Garman
H. A. Gossard
G. W. Herrick
L. O. Howard
S. J. Hunter
T. C. Johnson
J. C. Kendall
Don C. Mote
F. B. Mumford
Herbert Osborn
P. J. Parrott
R. H. Pettit
A. L. Quaintance
P. H. Rolfs
W. E. Rumsey
J. B. Smith
R. I. Smith

H. E. Summers
T. B. Symons
R. W. Thatcher
W. R. Walton

Third Edition: The list was extensive so no one was listed according to the preface.

Fourth Edition:

F. S. Arant
W. J. Baerg
D. J. Caffrey
O. L. Cartwright
E. N. Cory
G. A. Dean
C. J. Drake
E. H. Dusham
Philip Garman
Arthur Gibson
Grace Griswold
H. M. Harris
T. J. Headlee
G. W. Herrick
W. S. Hough
J. S. Houser
Ray Hutson
G. F. Knowlton
G. M. List
Simon Marcovitch
Eugenia McDaniel
W. F. Morofsky
J. A. Munro
C. E. Palm
P. J. Parrott
B. A. Porter
H. J. Reinhard
W. J. Schoene
H. C. Severin
L. A. Strong
D. L. VanDine
B. W. Walden
W. H. White
A. M. Woodside

Fifth Edition:

D. D. Ahmed
E. W. Baker
D. J. Borror
C. R. Cutright
R. J. Daum
F. W. Fisk
D. L. Goleman
D. G. Hall
D. W. Jenkins
H. H. Jewett
J. N. Knull
D. J. Knull
F. R. Koutz
M. C. Lane
Robert Matheson
Mina Maxwell
D. F. Miller
R. H. Painter
T. H. Parks
Alvah Peterson
A. E. Pritchard
S. S. Ristich
R. E. Sailer
F. F. Smith
R. M. Thomas
W. L. Thompson
L. H. Townsend
J. A. Wilcox

Sixth Edition:

B. D. Blair
R. F. Brooks
W. A. Connell
A. S. Deal
R. P. Holdsworth Jr.
G. F. Knowlton
A. E. Michelbacher
R. B. Neiswander
Ray F. Smith

Seventh Edition:

L. D. Anderson
J. E. Appleby
F. S. Arant
E. L. Atkins
F. J. Benci
T. L. Bissell
B. D. Blair
D. J. Borror
N. W. Britt
O. L. Cartwright
W. J. Collins
T. B. Davich
A. E. Davidson
E. W. Davidson
A. S. Deal

J. K. Flessel
E. H. Glass
D. L. Goleman
W. D. Guthrie
G. E. Guyer
R. N. Hofmaster
R. P. Holdsworth Jr.
R. N. Jefferson
J. M. Kingsolver
G. F. Knowlton
Michael Kosztarab
J. R. Leeper
Kay Lindsay
Marion Meredith
D. E. Mullins
L. R. Nault
M. W. Neilson
D. G. Nielsen
H. D. Niemczyk
Wayne Parrish
R. L. Pienkowski
K. P. Pruess
Roy W. Rings
H. H. Ross
C. W. Sabrosky
G. R. Stairs
A. L. Steinhauer
G. C. Steyskal
C. G. Summers
D. W. S. Sutherland
A. N. Tissot
R. E. Treece
C. A. Triplehorn
R. E. White
F. E. Wood

Contents

1

IMPORTANCE OF INSECTS TO HUMANS

One of the highly fascinating biological sciencies is entomology, the study of insects. Insects are the most abundant form of animal life on the earth. They are found nearly everywhere in the world except in the open seas and some parts of the polar regions. They have been on the earth over 250 million years and seem destined to remain. Whether they are considered helpful, harmful, or neutral to humans depends very largely on whether man is cooperating, competing, or indifferent to their presence.

Beneficial Effects

Insects may be helpful to humans by producing, directly or indirectly, materials of economic value, such as silk, honey, beeswax, shellac, cochineal dye, cantharidin, and galls used in making tannic acid, permanent inks and dyes; by aiding in the production of fruits, vegetables, flowers, and seeds, because of pollenizing activity; by serving as food for fishes, birds, other wildlife, and humans; by destroying injurious insects and mites either as predators, parasites, or parasitoids; by serving as research subjects in the study of toxicology, physiology, genetics, and related fields; by acting as scavengers, attacking and destroying dead plants and animals; by destroying noxious plants; by their medicinal value, particularly honey bee venom for treating arthritics; and by serving as objects in art, ornamentation, and in other aesthetic ways.

Harmful Effects

Insects may be harmful to humans and cause great economic loss by damaging or destroying agricultural crops and other valuable plants; by aiding in the spread and development of bacteria, fungi, protozoa, helminths, rickettsia, and viruses that produce disease and sometimes death in man and domestic animals; by aiding in the spread and development of bacteria, fungi, mycoplasmas, and viruses that produce diseases of plants; by annoying people or animals in various ways; and by destroying or lowering the value of stored foods, other products, and possessions.

The helpful and neutral groups of insects are by far the larger since less than 1% of all insect species are considered harmful to man, but this harmful group alone causes losses averaging from 5 to 15% of the annual agricultural production. These losses have been estimated to have a monetary value of over 7 billion dollars each year. In addition there are the annual expenditures for pesticides estimated to be $4.5 billion worldwide, and for spraying and dusting equipment, over $100 million. Labor costs for applying the pesticides would have to be included in the total and, if inflation continues, these will escalate each year.

Attempts have been made to estimate the monetary losses sustained from insect and mite attacks to crops, livestock, and man. Such estimates always result in figures of outstanding magnitude and because of the many factors involved they always vary with the authorities making them. The significance of such figures might be expressed in another way. For example, if we estimate a 10% annual loss to a corn crop because of insect attack, this means that without these pests the farmer could produce on 9 acres the same amount of corn that now requires 10 acres; that one day out of every ten spent in production of the crop could be used in other ways; that the equipment and land representing capital outlay could be reduced. A similar parallel could be drawn for other crops and livestock.

No matter how carefully conceived the plan for estimating insect losses, it is difficult to achieve accuracy. Nevertheless it can be safely stated that, next to the vagaries of climate, insects and mites combined are one of the farmer's greatest problems the world over. People in all lines of endeavor suffer some direct loss and inconvenience and indirectly must assume their share of the losses that are felt directly by the farmer.

The need for control measures is based on many factors. Discussion of these is given in the chapters on applied control and management operations. Population levels thought to be necessary before any control operation is initiated are given in the discussions of individual pests.

Sources of Information in Applied Entomology

Entomologists and others concerned with studies of insect control must first have basic information. This is supplied by several standard textbooks and reference books, a few of which are included at the end of this chapter. Basic information must constantly be supplemented by progress reports on new observations and investigations in this country and abroad. Such information is commonly published in the periodicals devoted to entomology, several of which are included in the list mentioned, and in bulletins and reports issued by many agencies. The most prolific single source of such information in this country is the Science and Education Administration of the United States Department of Agriculture. Several series of publications are issued by the United States Department of Agriculture and in them are found results of research conducted in various divisions of the Service. Farmers' bulletins, technical bulletins, circulars, agriculture handbooks, yearbooks of agriculture, and variously numbered mimeographed series are some of the better known publications. The state agricultural experiment stations, the agricultural colleges, the extension service, and endowed research institutions also issue bulletins and circulars of various kinds. Insecticide manufacturers issue much valuable information in regard to new products and discoveries. Current publications from all these sources are usually furnished free or at a nominal charge to persons in a position to make use of the information they contain.

The following list includes only a few of the many books and periodicals available for those interested professionally, or as laymen, in the field of entomology. Other publications, perhaps equally valuable, are necessarily omitted. More specialized publications are listed in the appropriate chapters.

REFERENCES

Anderson, R. F., *Forest and Shade Tree Entomology,* John Wiley and Sons, New York, 1960.

Annals of the Entomological Society of America, published by the Entomological Society of America, 4603 Calvert Rd., College Park, Md.

Annual Review of Entomology, published by the Entomological Society of America, 4603 Calvert Rd., College Park, Md.

Baker, W. L., *Eastern Forest Insects,* USDA Misc. Pub. 1175, 1972.

Canadian Entomologist, published by the Entomological Society of Canada, Ottawa.

Carter, Walter, *Insects in Relation to Plant Disease,* 2nd ed., John Wiley and Sons, New York, 1973.

Chandler, A. C., and C. P. Reed, *Introduction to Parasitology,* 10th ed., John Wiley and Sons, New York, 1961.

Craighead, F. C., "Insect Enemies of Eastern Forests," USDA Misc. Pub., 657, 1950.

Essig, E. O., *Insects and Mites of Western North America,* Macmillan Co., New York, 1958.

Essig, E. O., *A History of Entomology,* Macmillan Co., New York, 1931.

Fichter, G. S., *Insect Pests,* Golden Press, New York, 1966.

Herms, W. B., and M. T. James, *Medical Entomology,* 6th ed., Macmillan Co., New York, 1969.

Howard, L. O., *A History of Applied Entomology,* Smithsonian Misc. Collections 84, 1930.

Johnson, W. T., and H. H. Lyon, *Insects That Feed on Trees and Shrubs,* Comstock Publishing Associates, Cornell Univ. Press, Ithaca, N.Y., 1976.

Journal of Economic Entomology, published by the Entomological Society of America, 4603 Calvert Rd., College Park, Md.

Little, V. A., *General and Applied Entomology,* 3rd ed., Harper and Row, New York, 1972.

Mallis, Arnold, *Handbook of Pest Control,* 5th ed., McNair-Dorland Co., New York, 1969.

Mallis, Arnold, *American Entomologists,* Rutgers Univ. Press, 1971.

Martin, Hubert, *The Scientific Principles of Crop Protection,* 5th ed., Edward Arnold Ltd., London, 1965.

Metcalf, C. L., W. P. Flint, and R. L. Metcalf, *Destructive and Useful Insects,* 4th ed., McGraw-Hill Book Co., New York, 1962.

Mittler, T. E., C. N. Smith, and R. F. Smith, *History of Entomology,* Entomological Society of America, College Park, Md., 1974.

Osborn, Herbert, "Fragments of Entomological History," published by the author, Columbus, Ohio, 1937.

Pfadt, R. E., *et al., Fundamentals of Applied Entomology,* 3rd ed., Macmillan Co., New York, 1978.

Truman, L. C., Gary Bennett, and W. L. Butts, *Scientific Guide to Pest Control Operations,* 3rd ed., Harvest Publishing Co., Cleveland, Ohio, 1976.

USDA Yearbooks of Agriculture, *Insects,* 1952; *Animal Diseases,* 1956; Supt. of Documents, Wash., D.C.

USDA Losses in Agriculture, Agr. Handbook 291, Supt. of Documents, Wash., D.C.

Westcott, C., *The Gardener's Bug Book,* 4th ed., Doubleday, New York, 1973.

Zim, H. S. and Clarence Cottam, *Insects, A Guide to Familiar American Insects,* Simon and Schuster, New York, 1956.

In addition to these the student will find the following abstracting journals helpful, especially for foreign literature: *Biological Abstracts, Chemical Abstracts, Review of Applied Entomology,* and *Bibliography of Agriculture*. The insect pest surveys of Canada and the USDA are helpful for determining distribution of insect pests.

Persons desiring additional information should get in touch with the county extention agent, the state agricultural college, or the state agricultural experiment station.

2

STRUCTURE, PHYSIOLOGY, AND METAMORPHOSIS

Recognition and control of insects are based on exact knowledge of their structure, physiology, and metamorphosis.

STRUCTURE AND PHYSIOLOGY

EXTERNAL STRUCTURE

Every insect is covered externally with a thin nonchitinous, waxy, acid-resistant layer called the *epicuticle,* which is composed of polymerized lipoprotein and which protects the insect from excessive dryness, humidity, and disease organisms; beneath this are two noncellular but porous layers, the *exocuticle* and *endocuticle,* containing *chitin,* a colorless polymerized glucosamine chemically related to cellulose, which resists action of ordinary corrosive chemicals; next is a continuous layer of living cells, the *epidermis,* which secrete the substances forming the previous layers; below the epidermis is a thin layer called the *basement membrane*. All these together comprise what is known as the *exoskeleton* or body wall of an insect (Fig. 1). The same layers

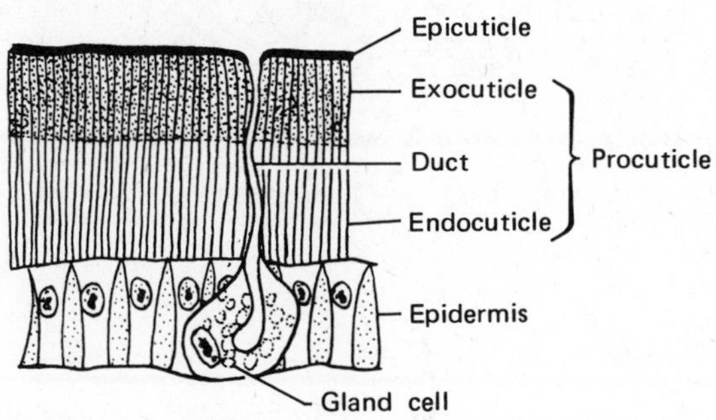

Figure 1. Diagrammatic section of body wall structure. (Adapted from Wigglesworth, Courtesy of H. H. Ross and John Wiley & Sons)

of the body wall line the fore-intestine, hind-intestine, and tracheae, since these structures are ectodermal in origin.

The insect skeleton is somewhat cylindrical and made up of a series of ring-like structures called segments. These are arranged in three groups: the head, consisting of 6 or 7 coalesced segments; the thorax, composed of 3 segments (*prothorax, mesothorax,* and *metathorax*) immediately following the head; and the abdomen, which numbers 11 segments in some primitive insects but is reduced in the majority of species either by fusion or telescoping of segments (Fig. 2). The three body regions are often covered by hairs (setae) and may have external protuberances such as horns, spines, or spurs. Some of these are hollow, some solid. Each seta has at its base a trichogen cell, and forming the socket, a tormogen cell.

Each segment is composed of several sclerotized or hardened plates, called

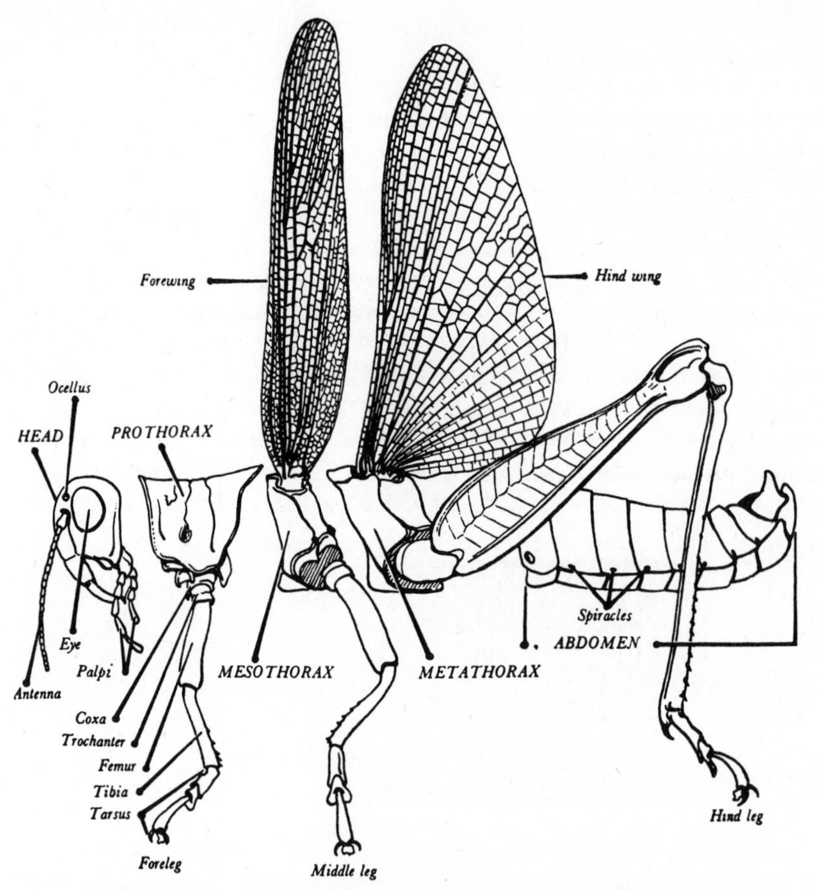

Figure 2. The external structure of a male grasshopper. (USDA)

sclerites, divided by juncture or union lines, called *sutures,* both named according to their location. The segments are connected by soft and flexible infolding intersegmental membranes. Some of the segments in most insects will clearly show a dorsal sclerite, called *notum, tergum,* or *tergite;* lateral sclerites, known as *pleura* (singular, *pleuron*) or *pleurites;* and a ventral sclerite called the *sternum* or *sternite.* The pleural sclerites are well developed on the thorax but are reduced or lacking in the abdominal region.

Structures on the head are the eyes, antennae, and mouthparts, this region functioning primarily as a sensory center and for the intake of food.

Most adult insects have a pair of lateral compound eyes, the surface of each composed of a few 100 to several 1000 usually hexagonal transparent areas called *facets;* beneath each facet is a cornea and visual unit called an *ommatidium,* which detects motion and is said to produce a composite or mosaiclike image. These eyes often occupy the greater portion of the insect head. Insects that see well have many *ommatidia,* viz., the dragonflies and house flies. Between the compound eyes, and also in some larvae and nymphs, there are two or three light sensitive simple eyes consisting of single facets. These are called dorsal *ocelli* (singular, *ocellus*). Some larvae possess from 1 to 8 lateral ocelli.

All adult insects and many immature stages have a pair of segmented antennae situated between the compound eyes or above the base of the mandibles. These are primarily sensory in function. Many modifications in form occur (Fig. 3), and these variations are of value in classification and determination of sex.

Attached to the fore or ventral part of the head are several appendages which, collectively, are called the mouthparts. These vary greatly in structure among different insects. Chewing mouthparts are considered the most primi-

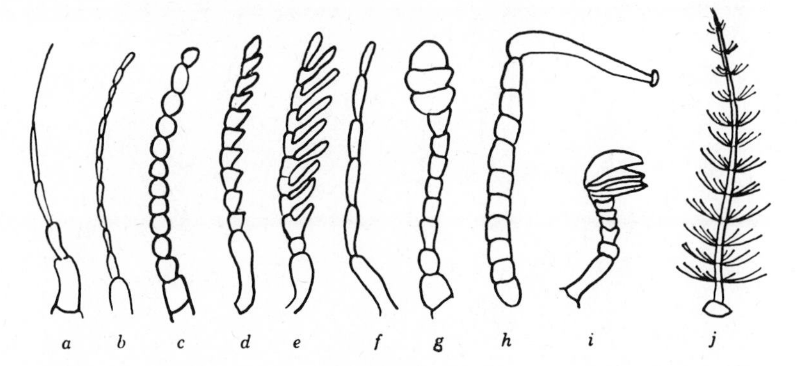

Figure 3. Types of insect antennae: *a,* setaceous or bristlelike; *b,* filiform or threadlike; *c,* moniliform or beadlike; *d,* serrate or sawlike; *e,* pectinate or comblike; *f,* filiform or threadlike; *g,* capitate or headlike; *h,* geniculate or elbowed; *i,* lamellate or platelike; *j,* plumose or plumed. (Redrawn from various sources.)

tive, a good example being those of a grasshopper (Fig. 4). Beginning anteriorly and going posteriorly the major structures are a flaplike upper lip or *labrum;* a pair of chewing jaws, called *mandibles;* a pair of *maxillae* or second pair of jaws, which bear organs of touch, smell, taste, and which also hold or cut tissues; a tonguelike ventral lobe called the *hypopharynx;* and a lower lip or *labium,* which functions in holding food and is also sensory. Each maxilla is made up of several parts; the jointed *palpus,* the *galea,* and the sharp distal structure, the *lacinia,* are the most obvious. The labium likewise is made up of several parts, the most conspicuous being a pair of segmented *palpi* and two distal lobes, often termed *ligula.* In insects that are adapted to feed on liquids, any of the above-named structures may be greatly modified, vestigial, or lacking. For example, the house fly can only suck up liquids; bed bugs and mosquitoes can pierce the skin and then suck blood; aphids, leafhoppers, and true bugs all pierce plant tissues and suck sap; butterflies and moths have a coiled tubelike structure that can be extended and used only in sucking up liquids (Figs. 5 and 6) but is incapable of piercing tissues; the hog louse has a piercing-sucking mouthpart, the piercing portion of which can be withdrawn into the head.

Some idea of the complexity of piercing–sucking mouthparts can be gained from the homopterous type, shown in Fig. 7, made from an electron microscope

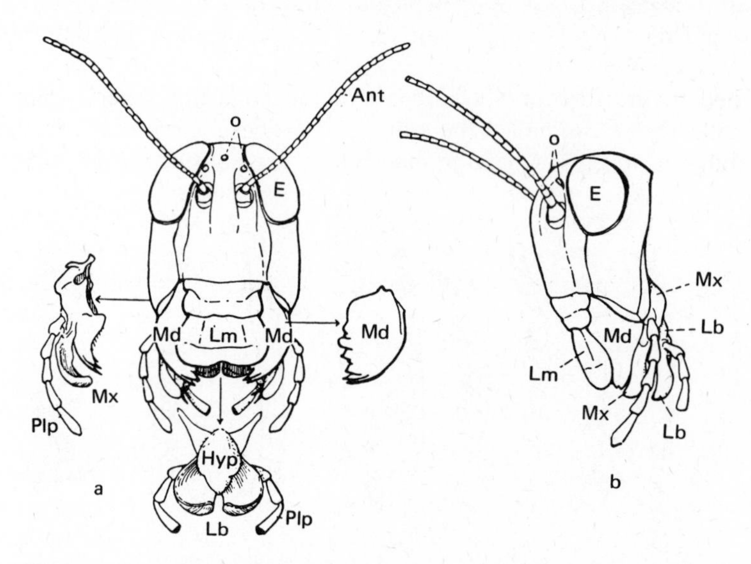

Figure 4. Head and mouthparts of a grasshopper: *a,* anterior aspect; *b,* lateral aspect. *Ant,* antenna; *E,* compound eye; *Hyp,* hypopharynx; *Lb,* labium; *Lm,* labrum; *Md,* mandible; *Mx,* maxilla; *O,* ocelli; *Plp,* palpus. (After Snodgrass, Courtesy of H. H. Ross and John Wiley & Sons)

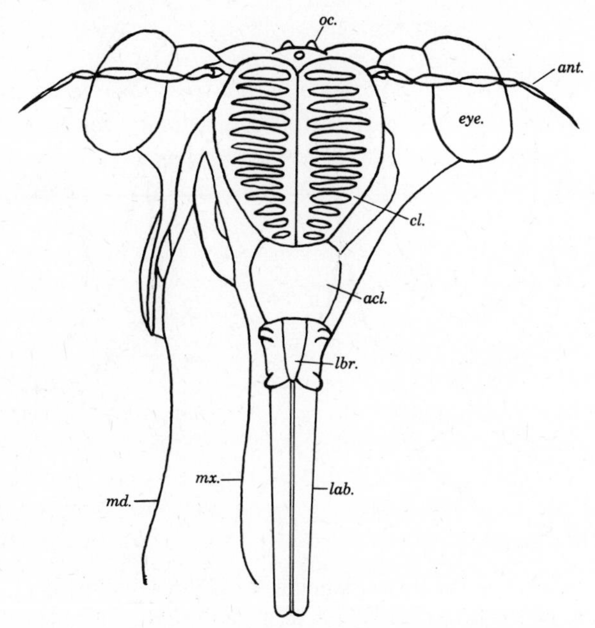

Figure 5. Front view of a cicada head showing the main structures of a piercing-sucking type of mouthpart; *ant.,* antenna; *acl.,* anteclypeus; *cl.,* clypeus; *eye,* compound eye; *lab.,* labium; *lbr.,* labrum; *md.,* mandible; *mx.,* maxilla; *oc.,* ocellus. Typical in the orders Homoptera and Hemiptera.

photograph of a cross section of the stylet bundle (the hairlike mandibles and maxillae within the labium) of a corn leaf aphid.*

Thrips have mouthparts intermediate between the chewing and piercing–sucking types; one mandible is missing and the one present functions as a rasping structure that breaks the plant cell walls and the sap is then sucked into the mouth. This type of mouthpart is called rasping–sucking.

A major function of the thorax is locomotion. Its appendages are the legs and wings. One pair of legs is attached to each of the segments of the thorax. Each leg is divided into several parts. The basal segment is the *coxa,* which fits into a cavity known as the *coxal cavity.* The next segment is small and is called the *trochanter.* Following this is the *femur,* often the largest segment of the leg, then the *tibia,* usually as long as the femur but much more slender. Attached to the tibia is the *tarsus* (plural, *tarsi*). This is composed of 1 to 5 segments, the terminal one generally bearing 1 or 2 claws and sometimes a

* For a detailed discussion see Wayne B. Parrish's paper, "The Origin, Morphology, and Innervation of Aphid Stylets (Homoptera)," *Ann. Ent. Soc. Amer.* 60:273–276, 1967.

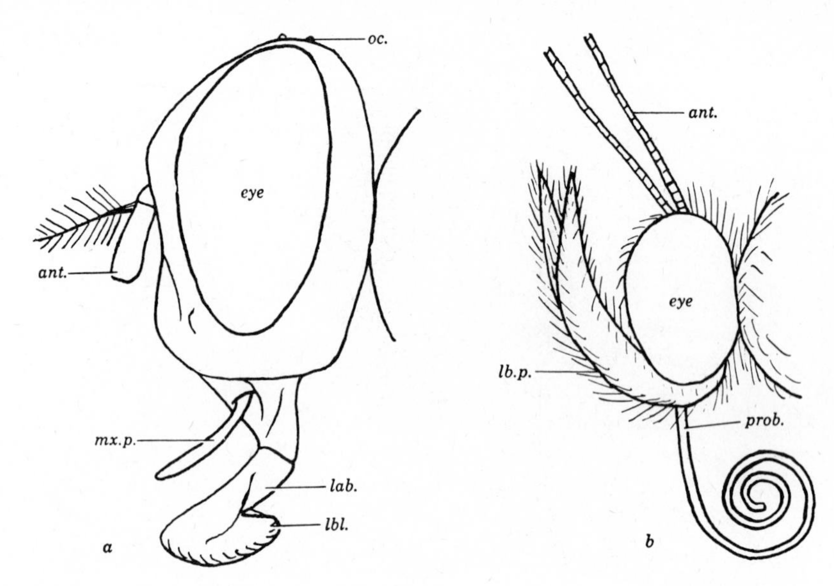

Figure 6. Insect mouthparts: *a,* lapping or sucking mouthparts of a house fly; *b,* siphoning or sucking mouthparts of a moth; *ant.*, antenna; *oc.*, ocellus; *eye,* compound eye; *mx.p.*, maxillary palpus; *lb.p.*, labial palpus; *prob.*, proboscis; *lab.*, labium; *lbl.*, labellum. The house fly antenna is of the aristate type.

padlike structure, the *pulvillus* or *arolium* (Fig. 2). Insect legs vary greatly, some common categories being running, jumping, grasping, digging, or swimming types. Variations in leg structure and tarsal segments are much used in the classification of insects.

Adult insects may be wingless or winged. If there is only one pair of wings, attachment is to the mesothorax, but if there are two pairs, attachment is to both meso- and metathorax (Fig. 2). Wings are usually membranous, often showing prominent thickened lines or veins that serve to strengthen the structure. In various groups the pattern formed by these veins is constant and frequently serves as a ready means of identification.

In many insects there are no obvious appendages on the abdomen. Some primitive forms show vestiges of legs. Others possess a pair of *cerci* at or near the tip of the abdomen. These may be very short or very long depending on the kind of insect. Several groups have in addition long segmented *anal filaments*. Females of some insects have a prominent structure that functions in the deposition of eggs. This is called an *ovipositor*. In the bees, wasps, and ants, the ovipositor is modified into a stinging organ which is retractile. The *spiracles* or external openings to the respiratory system are always present on the abdomen and 1 or 2 pairs are also found on the thorax, depending on the

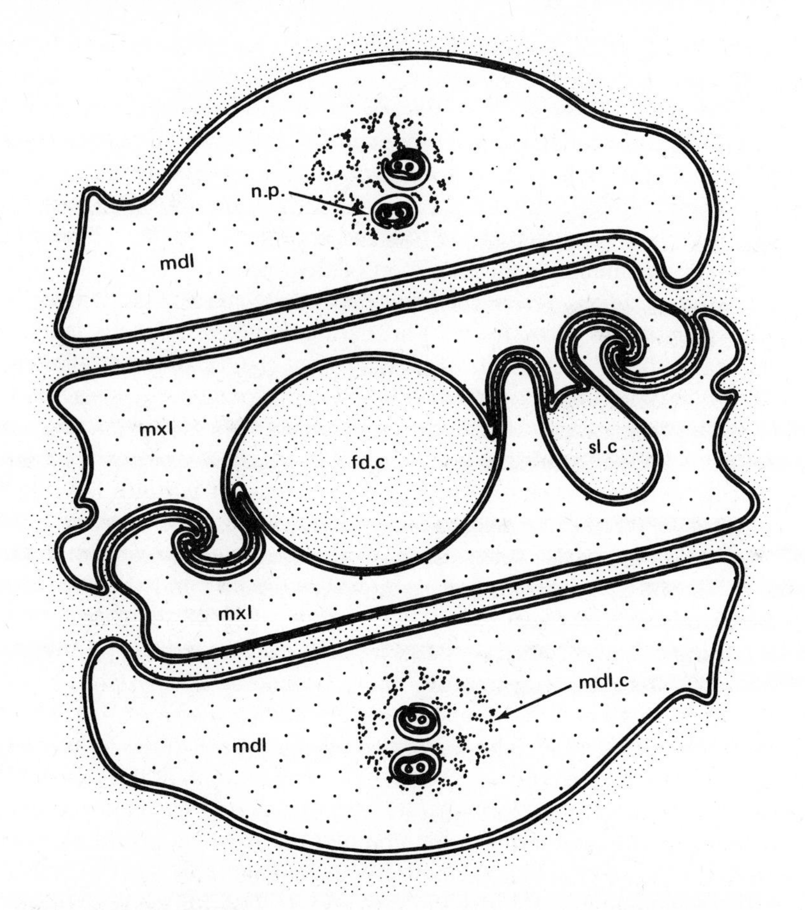

Figure 7. Diagrammatic cross section made from an electron microscope photograph of the stylet bundle of the corn leaf aphid, showing the outer pair of mandibular stylets (mdl) and the inner pair of maxillary stylets (mxl). The maxillae, by virtue of their coaptation, form the food canal (fd.c) and salivary canal (sl.c). Each mandible contains a mandibular canal (mdl.c) with two nerve processes (n.p) Scale: 20 mm = 0.5 μ. (Courtesy of Wayne B. Parrish and Kay Lindsay)

species of insect. Some of the major functions of the abdomen are digestion, respiration, excretion, and reproduction.

INTERNAL STRUCTURE AND PHYSIOLOGY

The internal structures of insects are highly developed and are comparable, in their differentiation into organs and systems of organs, to the structures of the vertebrates. The functions of the organs, with certain important exceptions, closely parallel those of the vertebrates.

The insect *skeletal system* has already been discussed because it seemed logical to do so in order to properly present the information on external anatomy. It is, of course, a most obvious contrast to the vertebrate condition which is truly an endoskeleton. However, in insects, there are rigid skeletal invaginations of the body wall, called *apodemes,* which serve for muscle attachment and are usually termed the endoskeleton of the insect.

The insect *muscular system* consists of many strong, segmentally arranged, cross-striated muscles. Grasshoppers are said to have over 900 distinct muscles and some caterpillars over 4000. Although insect muscles are quite small, they are very strong, and are often capable of extremely rapid contraction.

The insect *digestive system* consists of a tube, varying in length that extends from the mouth to the anal aperture. It is often called the alimentary canal. In some insects it is almost straight (Fig. 8); in others it is quite long and convoluted. It is divided into three principal parts: the fore-, mid-, and hind-intestine. In some insects the fore-intestine (stomodeum) can be further divided into the *pharynx, esophagus, crop,* and *gizzard,* and the hind-intestine (proctodeum) into the *ileum, colon,* and *rectum.* The mid-intestine is usually called the *stomach* (mesenteron) and anteriorly may sometimes possess pouch-like structures called *gastric caecae.* There is generally a *cardiac valve* at the juncture of the fore- and mid-intestine, and a *pyloric valve* at the juncture of the mid- and hind-intestine. The *salivary glands* connect to the mouth near the base of the hypopharynx and are considered evaginations of the fore-intestine. Near the anterior end of the hind-intestine are attached the slender *Malpighian tubules* that function primarily in the elimination of waste products of metabolism from the blood. Their distal ends are closed and their number varies from 1 to over 150, depending on the kind of insect. The Malpighian tubules are considered to be a part of the insect *excretory system.* The entire alimentary canal functions as the site for digestion and assimilation but the mid-intestine is the principal region involved. Digestive enzymes in insects are grouped into carbohydrases, proteases, lipases, and esterases, and function in the breakdown of starches, sugars, proteins, fats, and their esters, respectively.

The *circulatory system* of the insect is an open type. It consists of a *heart,* with lateral openings or *ostia* situated in the dorsal part of the abdomen, and the *aorta* which extends forward from the heart to the head. Flow is usually anterior, the blood bathing all the organs of the body. Its function is simply to transport nutritive materials to the tissues and to carry away certain wastes. With few exceptions the blood of insects contains no red corpuscles and plays no part in respiration as does vertebrate blood. Its pH range is 6 to 8.

The insect *respiratory system* consists of a series of slender branching tubes, called *tracheae,* which divide and subdivide into very tiny tubes, called *tracheoles.* These ramify throughout the body and ultimately reach cell groups

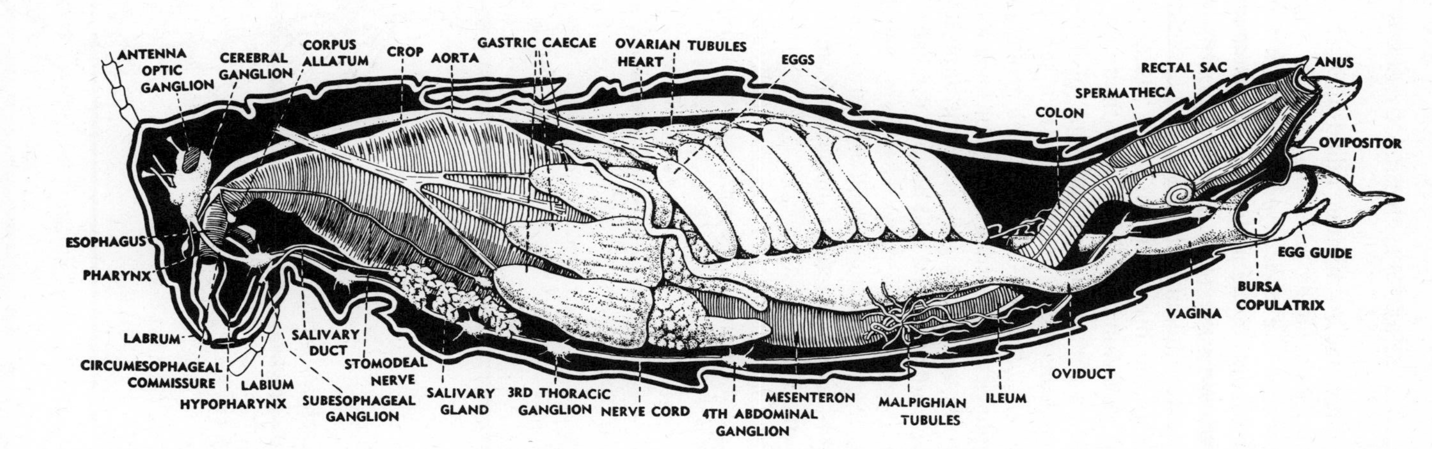

Figure 8. A diagrammatic view of the internal organs of a grasshopper with all the major structures labelled. The *fore-intestine* extends from the mouth to the bases of the gastric caecae; the *mid-intestine* from the gastric caecae to the Malpighian tubules; the *hind-intestine* from the Malpighian tubules to the anal aperture. (Drawing by Gerberg. Reproduced through courtesy of Robert Matheson and Comstock Publishing Assoc.)

or individual cells where they end in minute, liquid-filled sacs, through the walls of which the respiratory exchange takes place (Figs. 2 and 9). The external openings to the tracheae are called *spiracles,* with typically 2 pairs located on the thorax and 8 pairs on the abdomen. Movement of oxygen and carbon dioxide through the tracheal system is commonly by diffusion, but in some insects body movements undoubtedly serve to ventilate the tracheae. These movements may be called breathing. In some insects a considerable amount of respiration is by diffusion directly through the body wall.

The central *nervous system* consists of a large mass of nerve tissue in the dorsal part of the head called the *brain,* and a ladderlike series of paired segmental ganglia forming the *ventral nerve cord* which lies beneath the alimentary canal; just below the esophagus is the subesophageal ganglion composed of three fused ganglia (Fig. 8). In the more highly specialized insects many of the ganglia of the ventral nerve cord are fused and have shifted anteriorly. Insects also have what is termed a sympathetic nervous system, which functions in controlling the heart, digestive, respiratory, and possibly other systems, and a peripheral nervous system, which functions primarily from sensory stimulation by the external environment.

The *reproductive system* of insects (Fig. 8) has essentially the same arrangement found in other animals, the females having a pair of *ovaries* and a pair of

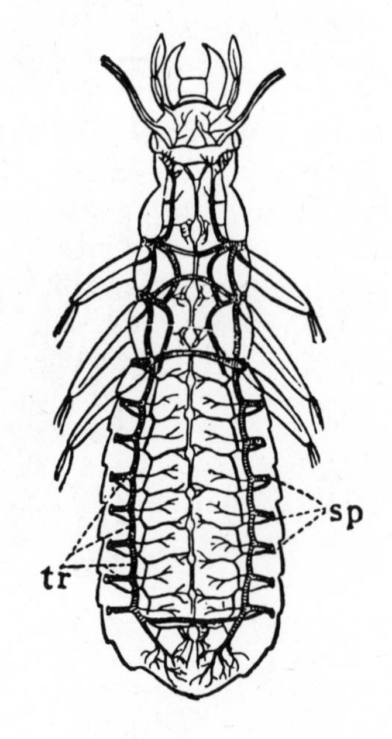

Figure 9. The tracheal respiratory system of an insect: *tr.,* tracheae; *sp.,* spiracles; the ventral nerve cord and brain are also illustrated. (Kolbe.)

oviducts which unite to form a common duct leading to the *vagina,* and the males having a pair of *testes,* each attached to a seminal duct or *vas deferens* which unite to form the ejaculatory duct leading to the copulatory organ or *aedeagus* (penis). In addition females may have a sperm storage pouch, the *spermatheca,* and a pair of *accessory glands* which secrete the material that covers egg masses or glues them to objects. The seminal vesicles of the male are enlarged or dilated portions of the vas deferens.

If greater detail on the physiology of insects is desired, consult a copy of Patton.*

DEVELOPMENT AND METAMORPHOSIS

Most insect reproduction is sexual, an egg cell developing only after union with a sperm cell from the male. The females of many kinds of insects lay eggs and are said to be *oviparous*. Some insects also have special modes of reproduction: development from unfertilized eggs is called *parthenogenesis;* development of many embryos from a single egg is termed *polyembryony;* nourishment of the young within the female body until development is well advanced is called *viviparous* reproduction; eggs hatching within the female body, the young being born without nourishment except from the yolk, is called *ovoviviparous* reproduction. Development within the egg is embryonic and after hatching or birth it is termed postembryonic.

Eggs of insects are of the greatest variety. Some notion may be gained of the common types by observing the many illustrations of pest species throughout the book. The number of eggs produced varies from one for some plant lice to many thousands for some social insects.

The newly hatched or born insect differs in size and often in form from the parent. The change that must take place before the young assumes the adult condition is called *metamorphosis*. The degree of change varies widely in different insects. In some it is slight and gradual; in others it is rather abrupt and complete. These variations have led to the classification of metamorphosis. If the changes are slight and gradual, if the young or *nymphs* resemble the adults, feed in the same habitat, and if wing development is external (for winged insects), the metamorphosis is described as *gradual* (Figs. 10 and 11). If the changes are very marked, the young or *larvae* not resembling the adults, the mouthparts, legs, antennae, and habitat often differing, if wing development is internal (for winged insects), and if there is an additional life stage, the *pupa,* the metamorphosis is described as *complete* (Fig. 12). In the mayflies, dragonflies, damselflies, and stoneflies, development is gradual, but the young

* Patton, R. L., *Introductory Insect Physiology,* W. B. Saunders Co., Philadelphia, 1963.

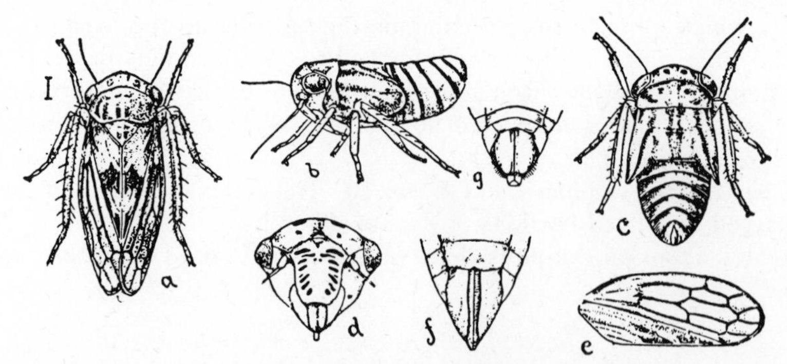

Figure 10. Gradual metamorphosis in Homoptera: The clover leafhopper, *Aceratagallia sanguinolenta* (Prov.). *a,* adult; *b,* nymph, side view; *c,* nymph, dorsal view; *d,* face; *e,* elytron; *f,* female genitalia; *g,* male genitalia. All enlarged. Straight line (top, left) indicates actual size of insect. (Osborn and Ball, USDA)

or *naiads* are aquatic and possess gill structures that are lost when the winged adult stage is reached. This has been termed *incomplete* metamorphosis by some entomologists (Fig. 13). Others group gradual and incomplete metamorphosis together and call them both incomplete. Some entomologists also have adopted the categories of simple and complex metamorphosis.* Regardless of the terms describing these variations, it is important for the student to have some knowledge of the complicated processes involved.

Growth begins when the sperm cell enters the micropyle of the egg. Most obvious growth is indicated by the various sizes of the developmental stages of an insect, whether it be a nymph, naiad, or larva. Since the exoskeleton cannot expand sufficiently to accommodate this increase in size, it is cast off during the process of *molting* or *ecdysis*. Layers of the exoskeleton shed during ecdysis are the epi- and procuticle, but only after new layers are formed beneath by the epidermal cells (see Fig. 1). This process is initiated by hormones produced in the head and prothoracic regions. The brain hormone is produced by the *corpus cardiacum,* a paired structure situated just posterior to the subesophageal ganglia. It liberates the hormone into the blood stream, activating the prothoracic gland; this gland then secretes the hormone *ecdysone* which controls metamorphosis. In some way environmental factors stimulate the corpus cardiacum to produce the brain hormone. Closely associated with the corpus cardiacum is another paired structure, the *corpus allatum,* that secretes the juvenile hormone *neotinin,* which in turn suppresses the action of

* R. E. Snodgrass applies the terms ametabolous (none), paurometabolous (gradual), hemimetabolous (incomplete), and holometabolous (complete) to describe the kinds of metamorphosis. For a detailed discussion of metamorphosis see his paper "Insect Metamorphosis," *Smithsonian Misc. Coll.,* 122 (9), 1–125, 1954.

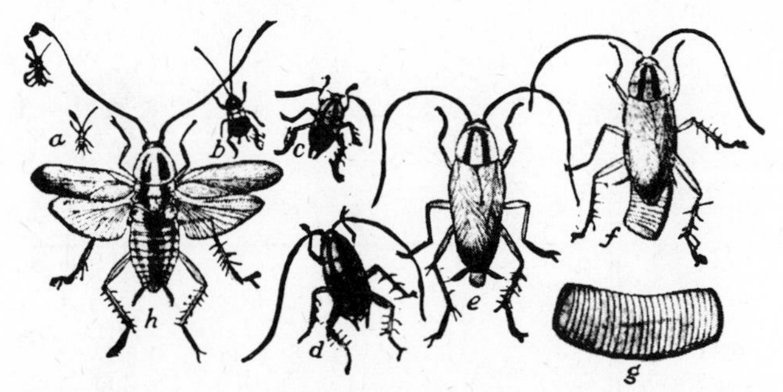

Figure 11. Gradual metamorphosis in Orthoptera: The German roach, *Blattella germanica* (L.). *a,* first instar; *b,* second instar; *c,* third instar; *d,* fourth instar; *e,* adult; *f,* adult female with egg case; *g,* egg case, enlarged; *h,* adult with wings spread. All natural size except *g.* (Riley.)

ecdysone. Apparently neotinin has a balancing function in the insect endocrine system, for when the last molt is reached it is no longer secreted.

Most insects molt four times but some molt less and others molt many more. The form of an insect between successive molts is called an *instar*. The *pupa* is a nonfeeding stage during which the larval tissues are transformed into adult characters. *Chrysalid* is a term that denotes the pupa of a butterfly. Some pupae are enclosed in *cocoons* formed of silk spun from the modified salivary glands of the larvae. In many flies the pupa is enclosed in the next-to-last larval molt skin or exuvium and is known as a *puparium*.

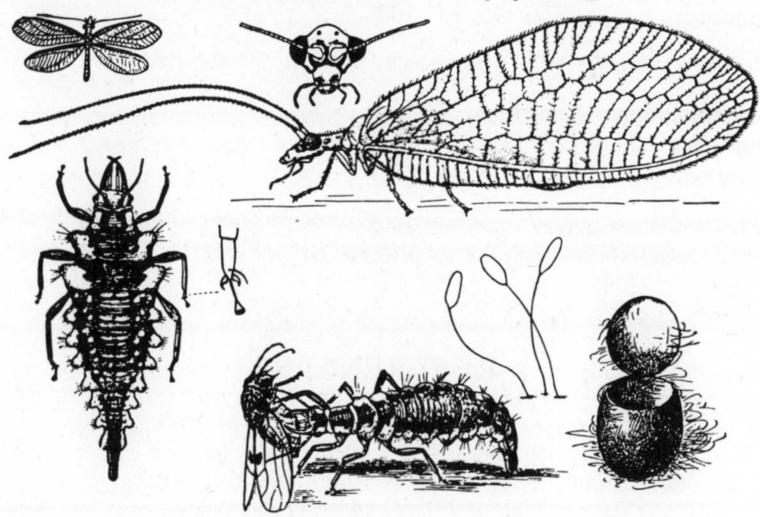

Figure 12. Complete metamorphosis in Neuroptera. Eggs, larvae, cocoon, and adults of a lacewing. Top left figure natural size, others enlarged. (USDA)

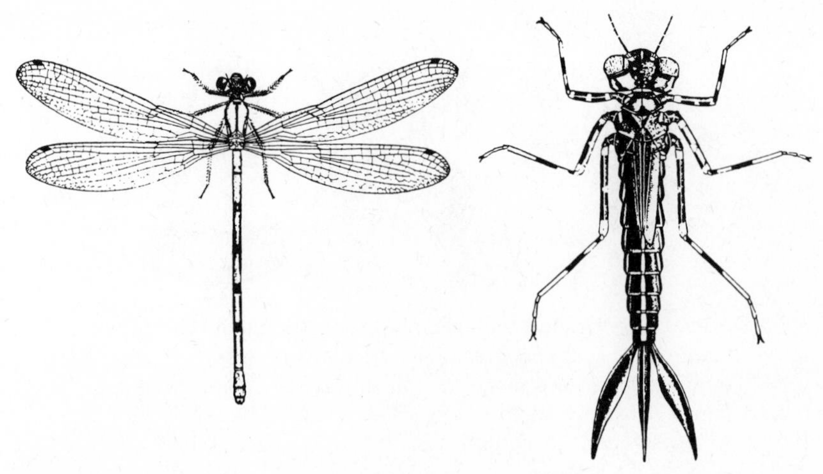

Figure 13. Incomplete metamorphosis of a damselfly, order Odonata; adult and naiad (or nymphal) stages. (Kennedy.)

Insect larvae are quite variable in appearance and are often classified into the following types:

campodeiform—resemble bristletails (Figs. 12, 49, 58)
elateriform—resemble wireworms (Figs. 34, 72, 186)
eruciform—caterpillars (Figs. 38, 83, 85, 171)
vermiform—maggots (Figs. 50, 192)
scarabaeiform—grubs (Figs. 75, 81, 100)

Pupae likewise have been categorized into the following types:

exarate—appendages free (Figs. 34, 75)
coarctate—puparia of flies (Figs. 130, 206)
obtect—appendages fused to body (Figs. 38, 171, 172)

Insects having complete metamorphosis but radically different larval instars are said to undergo hypermetamorphosis. Blister beetles are examples (see pp. 40, 265).

REFERENCES

Beck, S. D., *Insect Photoperiodism,* Academic Press, New York, 1968.

Borror, D. J., D. M. DeLong, and C. A. Triplehorn, *An Introduction to the Study of Insects,* 4th ed., Holt, Rinehart and Winston, New York, 1976.

Dethier, V. G., *The Physiology of Insect Senses,* John Wiley and Sons, New York, 1963.

Matheson, Robert, *Entomology for Introductory Courses,* 2nd ed., Comstock Publishing Associates, Ithaca, N.Y., 1951.

Peterson, Alvah, *Larvae of Insects,* Edwards Brothers, Inc., Ann Arbor, Mich., Part I, 1973; Part II, 1973.

Ross, H. H., *A Textbook of Entomology,* 3rd ed., John Wiley and Sons, New York, 1965.

Wigglesworth, V. B., *Principles of Insect Physiology,* 6th ed., rev., E. P. Dutton and Co., New York, 1965.

Wigglesworth, V. B., *Insect Hormones,* W. H. Freeman and Co., San Francisco, 1970.

3

CLASSIFICATION

Insects comprise nearly 80% of the million or more known species in the animal kingdom. They are truly the greatest variety show on earth. To facilitate learning about this enormous number of animals it is important that they be grouped or classified in some logical manner. The classifying can be done in various ways but the system usually followed by zoologists is based on structural characters. These include the number of cells, the type of symmetry, the number and character of the appendages, the internal and external arrangement of the body organs, and the nature of the skeletal system. On this basis the animal kingdom is divided into various categories called *phyla* (singular, *phylum*). Some of the major phyla arranged in order from the simplest to the most complex forms, with a few common examples, are as follows:

Phylum Protozoa: amoeba, paramecium, euglena.
Phylum Porifera: sponges.
Phylum Coelenterata: jellyfish, hydra, corals.
Phylum Platyhelminthes: flatworms, tapeworms, flukes.
Phylum Nemathelminthes: roundworms, nematodes.
Phylum Trochelminthes: rotifers.
Phylum Brachiopoda: lamp shells.
Phylum Bryozoa: moss animals.
Phylum Mollusca: snails, slugs, oysters.
Phylum Echinodermata: starfish, sea urchins.
Phylum Annelida: earthworms, leeches.
Phylum Arthropoda: crayfish, spiders, mites, insects.
Phylum Chordata: fishes, amphibians, reptiles, birds, mammals.

Insects belong to the phylum Arthropoda, along with several other related forms, all of which are characterized by an external skeleton of chitin, bilateral symmetry, a segmented body, paired and segmented appendages, a ventral nerve cord, and a dorsal heart.

All animals of a given phylum are further divided, according to similarities in structure, into groups called *classes*. For example, insects constitute the class Hexapoda or Insecta, and are characterized as adults by 1 pair of antennae, tracheal respiration, 3 pairs of true legs, and 3 body regions (head, thorax, and abdomen). The class Arachnida includes spiders, mites, ticks, harvestmen, scorpions, and tarantulas (Figs. 14, 136, 216, 292, 403, 428, 429). All these animals are characterized by 2 body regions (cephalothorax and abdomen), 4 pairs of legs (except some mites and ticks), absence of antennae, and respiration by means of tracheae, book lungs, or diffusion through the body wall. Centipedes, class Chilopoda, are elongate forms having 15 or more pairs of legs with only 1 pair occurring on each body segment (Fig. 15). Millipedes, in the class Diplopoda, have enlongated bodies with many legs, usually 2 pairs per body segment except for the first few back of the head (Fig. 16). The class Symphyla includes tiny centipedelike animals having 11 or 12 pairs of short legs (Fig. 232). Crayfish, sowbugs, water fleas, and lobsters, as well as many smaller forms, usually aquatic, often possessing gills, and having 5 or more pairs of legs, make up the class Crustacea (Fig. 17).

All the animals of each class are further divided into groups called *orders*, and this subdivision continues, with the orders being divided into *families*, the families into *genera* (singular, *genus*), and the genera into *species*. With each division the characters for separation become more minute and more detailed, species determination frequently being based on the internal genital structures of the male and/or female. Family names end in *idae;* superfamilies in *oidea*.

The generic and specific names assigned a particular animal constitute its scientific name. This is followed by the name (or its abbreviation) of the person who described and named the species. For example, the scientific name of the house fly is *Musca domestica* Linnaeus. If the name of the describer following a scientific name is enclosed in parentheses, it indicates that the species has

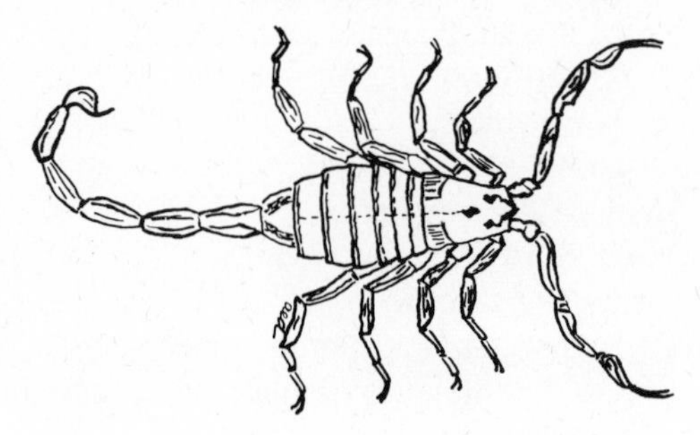

Figure 14. Class Arachnida, a scorpion, about natural size. (Drawing by Ann E. Davidson)

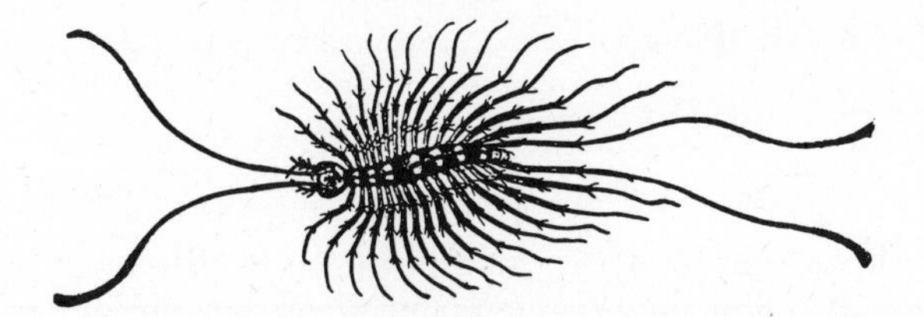

Figure 15. Class Chilopada: the house centipede, *Scutigera coleoptrata* (L.), a minor household nuisance, predatory in habit. (Marlatt, USDA)

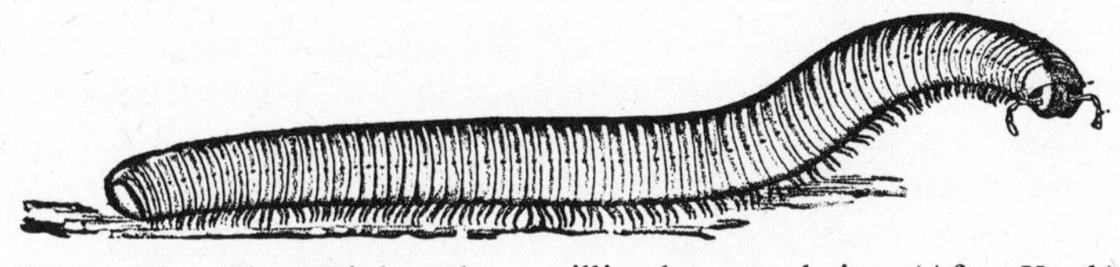

Figure 16. Class Diplopoda: a millipede, actual size. (After Koch)

been shifted to a genus other than the one in which it was placed when described. This method of naming animals originated in 1758 with the publication of *Systema Naturae,* 10th ed., by Linnaeus, and is called the binomial system of nomenclature. At times a subspecific name is used in addition to the specific name. These are called trinomials.

The class Hexapoda or Insecta is divided into orders on the basis of such characteristics as the presence or absence of wings, the number, wing texture, venation, the type of mouthparts, and type of metamorphosis.

All entomologists do not agree as to the exact limits of an order; consequently there is a variation in the number in different publications. Of the 24 orders listed in this chapter, 17 are generally considered of minor importance and 7 of major importance. A brief summary of the differentiating characteristics with examples will be given for all the orders, and descriptive features of some families of economic importance in all the orders will be presented.

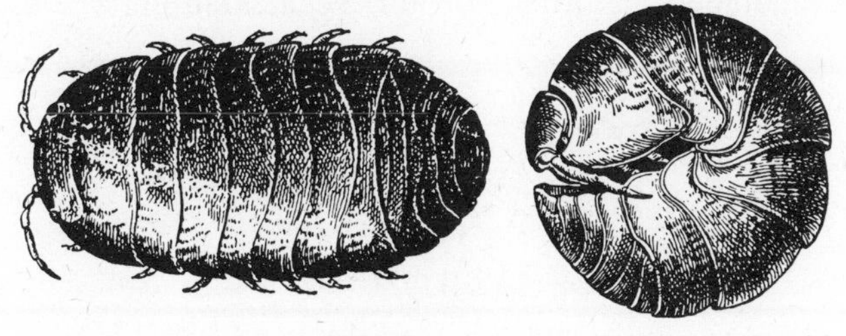

Figure 17. Class Crustacea: sowbugs, minor pests in glasshouses; left, extended; right, rolled into a ball and known as a pillbug; 4×. (Popenoe)

THE ORDERS OF INSECTS

ORDER PROTURA, the Proturans

These are tiny elongated wingless insects with gradual or no metamorphosis and no eyes or antennae. The first pair of legs might be mistaken for antennae because they are carried in an elevated position. Proturans are found in moist habitats, principally decaying organic material. No species are considered to be pests.

ORDER THYSANURA, the Bristletails

These are primitive wingless soft-bodied insects with gradual or no metamorphosis, eyes present or absent, vestigial abdominal legs, chewing mouthparts, long, segmented antennae, definite cerci, and a long segmented caudal filament present on some forms. Common species are about 1.2 cm in body length. Firebrats and silverfish are two species of importance as pests in the household (p. 515). They are placed in the family Lepismatidae.

ORDER COLLEMBOLA, the Springtails

These small-to-minute wingless insects have chewing-type mouthparts and gradual or no metamorphosis. No true compound eyes are present, and many species possess a taillike structure that folds beneath the body and functions as an organ of locomotion (Fig. 18). Commonly found in moist or damp soil rich in organic matter, the springtails are seldom considered as major pests, but one species, *Onychiurus armatus* (Tullberg), is known to feed on seedling plants and mushrooms and a few others have similar habits. Very little is known about the life cycles of these universally distributed insects. Common families are Poduridae, Entomobryidae, and Sminthuridae.

ORDER EPHEMEROPTERA, the Mayflies

Adults are delicate slender insects, varying from small to rather large size (35 mm). They have four membranous wings with many veins, the front pair being much larger than the hind pair. In some genera the hind wings are vestigial. The antennae are inconspicuous, the compound eyes large, the mouthparts vestigial; the abdomen has a pair of long segmented cerci and often a long

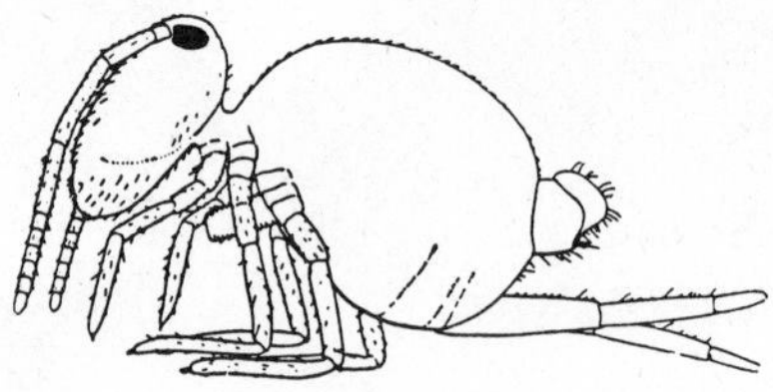

Figure 18. A springtail; order Collembola; greatly enlarged. (Lacroix, Conn. Agr. Exp. Sta.)

median caudal filament (Fig. 19). Metamorphosis is incomplete, the aquatic nymphs or naiads have chewing mouthparts and little resemblance to the adults. Members of the order are often extremely numerous in the vicinity of streams and lakes. Except for the fact that they are a source of food for fishes, mayflies are of no consequence to man. Common families are Ephemeridae, Heptageniidae, and Baetidae.

ORDER ODONATA, the Dragonflies and Damselflies

These are medium-to-large insects, some reaching a wingspread of over 10 cm. They have chewing mouthparts, bristlelike antennae, long slender bodies, and four long, narrow, net-veined wings of equal or subequal size. Metamorphosis is incomplete, the nymphs or naiads being aquatic and bearing little resemblance to the adults. Dragonflies are noted for their strong, graceful flight; damselflies have similar characteristics but are much smaller and more delicate creatures (Fig. 13). Both nymphs and adults are predatory in habit, the nymphs devouring mosquito larvae and other aquatic life, the adults catching various insects on the wing. The predatory habits of the adults are not considered of great value in the control of pest species. Members of this order are probably better known than most of the other minor groups because of their large size and their diurnal activity. Common families are Aeschnidae, Gomphidae, Corduliidae, Libellulidae, Agrionidae, Coenagrionidae, Lestidae, Cordulegastridae, and Petaluridae.

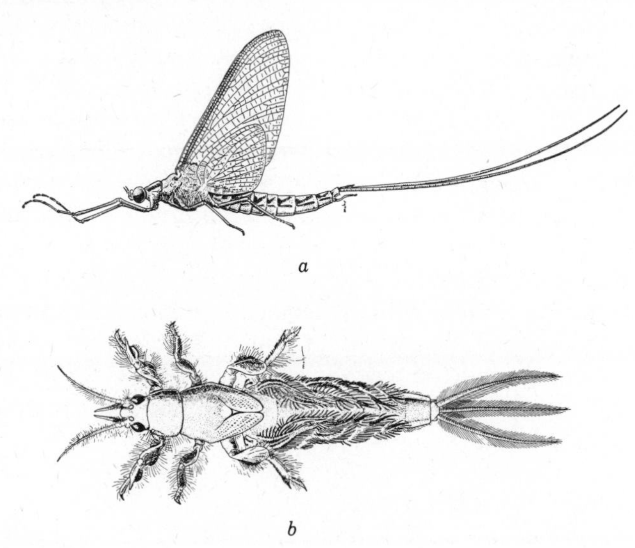

Figure 19. A mayfly, *Hexagenia limbata* (Serville): (*a*) adult, (*b*) naiad. Order Ephemeroptera. About natural size. (From B. D. Burks, Ill. Nat. Hist. Survey, Courtesy of N. W. Britt)

ORDER ORTHOPTERA

This group of terrestrial insects is the first of the so-called major orders; its members are variable in form and there is no common name that applies to all of them. Orthopterous insects have strong mandibulate mouthparts; their front wings are usually thickened and tend to be elongate and narrow in shape; the rear wings are membranous, broad, partly rounded in form, and folded fanlike when at rest. In the resting position the front wings cover and protect the rear wings. Some insects in this order are wingless. Sound is produced by the males of most species possessing jumping legs. Metamorphosis is gradual or simple. With very few exceptions the Orthoptera includes comparatively large insects. The number of families is small and all major ones will be mentioned.

Family Acrididae, the Short-Horned Grasshoppers or Locusts (this family is also called Locustidae).
Grasshoppers or locusts constitute one of the most important groups of destructive insects in the world. They are jumping insects with antennae usually shorter than half the length of the body; the auditory organs are found on the first abdominal segment, and the tarsi are 3-segmented. A discussion of the important species is given in a subsequent chapter (p. 117).

Family Tettigoniidae, the Long-Horned Grasshoppers
This family includes jumping forms with antennae as long as or longer than the body, which is always strongly arched, never flattened on top. A few species are wingless and are often mistaken for crickets, but members of this family possess 4 tarsal segments, whereas the true crickets have 3. Katydids, meadow grasshoppers, cave and Mormon "crickets" are typical members of this group. Cave "crickets" sometimes are troublesome in glasshouses, and Mormon "crickets" are at times very destructive to crops in Utah and nearby states (p. 119). Except for the latter insect the economic importance of this family is not great. Cave "crickets" are sometimes placed in the Gryllacrididae.

Family Gryllidae, the Crickets
Crickets are jumping insects with rather heavy bodies. The tarsi are 3-segmented and the ovipositor spearshaped. Common species represented are field, house, mole (Fig. 20), and tree crickets. Occasionally some of these species are quite destructive to crops. See pp. 120, 317. Some place the mole crickets in Gryllotalpidae.

Family Phasmatidae, the Walkingsticks
The American species are large, commonly 7.5 to 15 cm in length, nearly all wingless, very narrow elongate-bodied, with long slender legs. Walkingsticks devour the foliage of trees, usually oak, and occasionally cause serious damage. The females oviposit while in the trees, the eggs dropping to the ground. One

Figure 20. The northern mole cricket. (USDA)

generation develops each year and eggs are the overwintering stage. Adequate moisture is needed for hatching and escape from the egg. There are usually 5 nymphal instars.

Family Mantidae, the Praying-Mantids
A mantid is a very striking insect of elongate form with the front legs modified into prominent grasping organs that catch and hold the prey of this beneficial predatory insect. Both native and introduced species are quite large, some over 10 cm in length. Winter is passed as masses of eggs glued to twigs or other objects. One generation develops each season.

Family Blattidae, the Cockroaches
These are broad oval much flattened insects with well-developed running legs. Several species are dwelling pests (Fig. 371); others occur outdoors in rotting logs and stumps. Nearly 70 species are found in North America and over 2000 are known in the world. The species troublesome to man are discussed under household insects (p. 513). Authorities differ on number of families; some are "splitters," others "lumpers." Blattellidae is the other family of common pest species.

ORDER DERMAPTERA, the Earwigs
These are elongate slender insects, some reaching a length of over 13 mm. They have chewing mouthparts, gradual metamorphosis, and a pair of pincerlike cerci. When winged, the front pair are thickened and very short, the hind pair membranous and much folded. They closely resemble the rove beetles. A few species are minor pests of garden crops, and one, the European earwig (Fig. 228), sometimes becomes a nuisance in houses. Only a half dozen species are common in this country. The common family is Forficulidae.

ORDER ISOPTERA, the Termites
This group of small chewing insects includes many species in the world, but in North America the most destructive forms are called subterranean termites (Fig. 21). They live together in colonies under a caste system which includes workers, soldiers, supplementary reproductives, a king and a queen. Wingless and winged individuals develop, the winged forms possessing two pairs of

Figure 21. Subterranean termites; order Isoptera, greatly enlarged. (McDaniel, Mich. Agr. Exp. Sta.)

equal size and of reticulated membranous texture. Metamorphosis is usually described as gradual. A detailed description of the subterranean species is given under household insects (p. 509). Families represented in the United States are Rhinotermitidae, Kalotermitidae, and Hodotermitidae.

ORDER PLECOPTERA, the Stoneflies

Stoneflies vary in color, form, and size, some reaching a length of 5.5 cm. They have chewing mouthparts, long filiform antennae, a pair of cerci, and two pairs of membranous many-veined wings, the front ones narrow and elongate, the hind ones broad and folded fanlike under the front wings when the insect is not in flight. Metamorphosis is considered incomplete, the nymphs or naiads being aquatic and generally found in streams clinging to stones. In the immature stages they undoubtedly serve as food for other aquatic insects and fishes. Perlidae is the largest family of stoneflies.

ORDER PSOCOPTERA, the Psocids and Booklice

These are small-to-minute (less than 5 mm) insects with chewing mouthparts and gradual metamorphosis. Both winged and wingless forms occur. Winged forms have two pairs of membranous wings with simple venation, the second pair usually much shorter. They often are found on the bark of trees and are sometimes known as barklice (family Psocidae). The wingless forms, booklice (family Liposcelidae), are numerous at times and may become pests in libraries and homes. They are minute brown with relatively long legs and antennae (Fig. 22). They feed on glue, paste, dried insects, fungi, grain, and many milled food products as well as other organic substances. Booklice are easily controlled by methoxychlor, propoxur, diazinon, synergized pyrethrins, or by fumigation.

ORDER MALLOPHAGA, the Chewing Lice

Tiny flattened wingless insects with gradual metamorphosis and chewing mouthparts, make up this order of ectoparasites of birds and mammals. Most

Figure 22. A booklouse or cereal psocid, *Liposcelis* spp., order Psocoptera, greatly enlarged. (From Cir. 498 Calif. Agr. Ext. Service)

species attack birds and domesticated fowls; none is known to attack man. All life stages are spent on the host and spread is by contact. A detailed discussion of the important species in this order including illustrations is given in Chapter 24.

ORDER ANOPLURA, the Sucking Lice

This group includes all the blood-sucking lice ectoparasitic on mammals. They are small wingless flattened insects with gradual metamorphosis and piercing-sucking mouthparts. They have more pointed and elongate heads than the biting lice. These insects are irritating and annoying, greatly lowering the vitality of livestock. Species attacking man are vectors of some dreaded diseases. A detailed discussion of the important species along with illustrations is given in Chapter 24.

ORDER THYSANOPTERA, the Thrips

Small-to-minute slender insects constitute this order. Both wingless and winged individuals are found, the latter having four very narrow wings fringed with long hairs (Fig. 23). Metamorphosis is gradual but an incipient pupal stage is indicated. The mouthparts are classified as rasping-sucking. The order includes many plant-feeding as well as predaceous species which may occur in vast numbers. A more detailed description of these may be found elsewhere in this book (pp. 210, 311). Species of economic importance are in the family Thripidae.

Figure 23. The flower thrips, *Frankliniella tritici* (Fitch); enlarged about 23×. (Fla. Agr. Exp. Sta.)

ORDER HEMIPTERA, the True Bugs

Insects of this order may be wingless; when wings are present the first pair is usually thickened and leathery at the base and membranous at the tips, and the second pair is entirely membranous. At rest, the wings cover the abdomen with the membranous tips of the first pair overlapping. The mouthparts are piercing-sucking and metamorphosis is gradual. There are many families of true bugs but only a few of considerable importance will be named.

Family Cimicidae, the Bed Bugs

A description of the one common species of these wingless, blood-sucking ectoparasites of man and other animals is sufficient for the entire family (see Chapter 24). Other species are mentioned.

Family Anthocoridae, the Flower or Minute Pirate Bugs

This family of tiny bugs (3–5 mm) includes the important species *Orius tristicolor* (White) (Fig. 24). All species attack and destroy all stages of mites, scales, aphids, and leafhoppers.

Family Miridae, the Leaf or Plant Bugs

This group of rather small bugs (8–10 mm) includes a number of very destructive pests of agricultural crops. The antennae and proboscis are 4-segmented, and the membrane of the fore wing usually has 2 closed cells at its base. Important species are discussed and illustrated in succeeding chapters of this book.

Family Nabidae, the Damsel Bugs

This is a group of bugs mentioned only because it includes important predaceous species that destroy great numbers of destructive insects each year (Fig. 25).

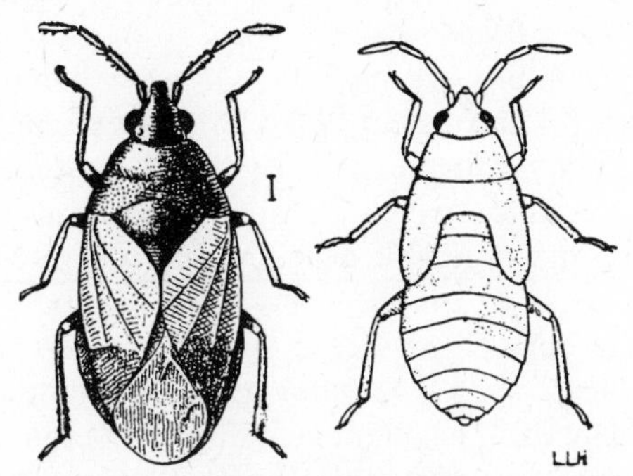

Figure 24. Adult and nymph of *Orius tristicolor* (White), a minute pirate bug that attacks eggs, mites, and small insects; family Anthocoridae; 18×. (Ewing, Ore. Agr. Exp. Sta.)

Family Reduviidae, the Assassin Bugs

Medium-to-large predatory bugs belong to this family, some reaching almost 5 cm in length. They have 4-segmented antennae, no ocelli, and a 3-segmented beak which fits into a prosternal groove. Several species are valuable predators and a few are likely to attack man but these are classed as being of minor importance.

Family Phymatidae, the Ambush Bugs

This family is important because it includes predatory species that are recognized by their greatly thickened front legs.

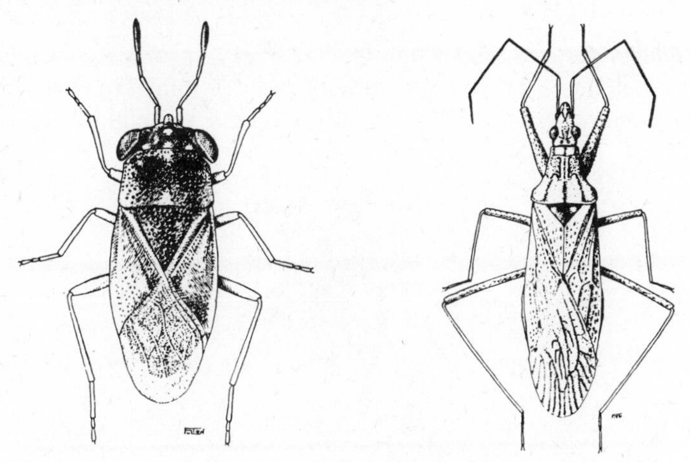

Figure 25. Geocoris decoratus Uhler, family Lygaeidae; 8× (left), and *Nabis alternatus* Parshley, family Nabidae; 5× (right); both predaceous insects. (Knowlton, Utah Agr. Exp. Sta.)

Family Tingidae, the Lace Bugs

All the members of this family are small (3–6 mm) with many wing veins, giving a lacelike appearance to the insect (Fig. 26). Some species have expanded lobed areas that extend over the head. The beak and antennae are 4-segmented, the tarsi are 2-segmented, and ocelli are absent. Several species are important pests of trees, shrubs, and vegetable crops, viz., sycamore lace bug, *Corythuca ciliata* (Say), oak lace bug, *C. arcuata* (Say), elm lace bug, *C. ulmi* Osborn and Drake, hawthorn lace bug, *C. cydoniae* (Fitch), cotton lace bug, *C. gossypii* (Fabricius), azalea lace bug, *Stephanitis pyrioides* Scott, rhododendron lace bug, *S. rhododendri* Horvath, eggplant lace bug, *Gargaphia solani* Heidemann. Most species overwinter as adults and pesticides are applied when nymphs are abundant, if warranted.

Family Lygaeidae, the Lygaeids

Lygaeids have 4-segmented beaks and antennae but only 5 or 6 veins in the wing membrane. They vary considerably in size, some being quite small, others nearly 2 cm in length. The chinch bug is considered the most injurious species in the family (see p. 163). All are plant-feeding species except those in the genus *Geocorus,* which are predaceous on other insects (Fig. 25).

Family Coreidae, the Coreids

Coreids are often large bugs, but there are some small species. They are elongate and robust in form; their beaks and antennae are 4-segmented, and the membranous portion of the front wings has many veins. Both plant-feeding and predaceous species are found. The squash bug (Fig. 188) is the best known of the pests belonging to this family, but there are other destructive species.

Family Pentatomidae, the Stink Bugs

This group includes a number of rather large (15 mm) broad, shield-shaped bugs, with 5-segmented antennae and a prominent scutellum (Fig. 27). Many

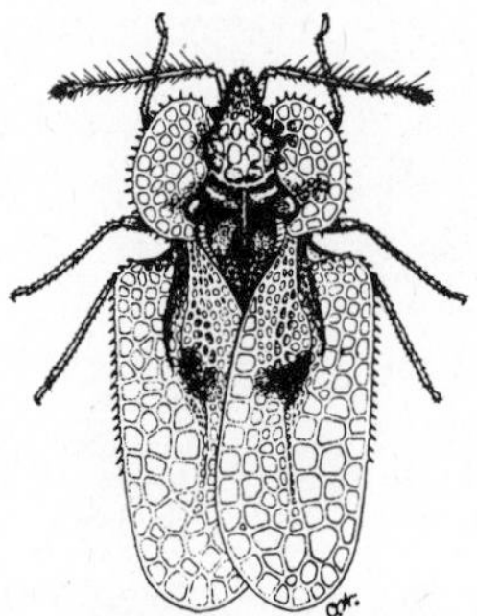

Figure 26. Adult sycamore lace bug, 7×; family Tingidae. (Wade.)

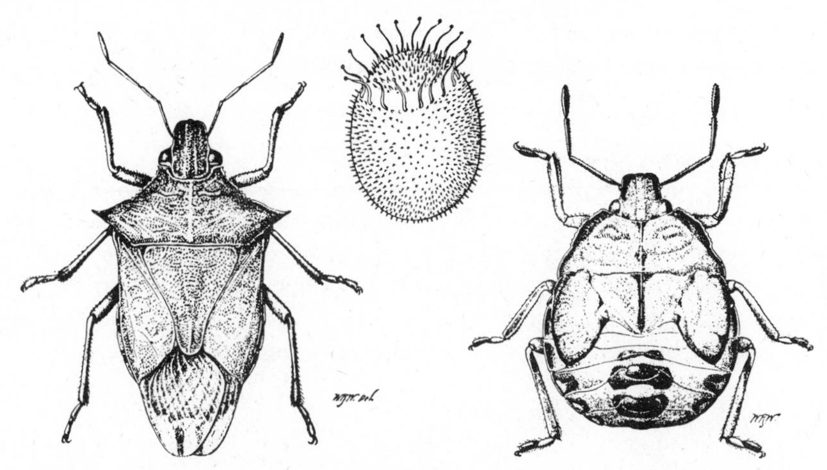

Figure 27. The spined soldier bug, *Podisus maculiventris* (Say), family Pentatomidae. From left to right: adult, egg, nymph; 4×. (Walton, USDA)

species are predatory and beneficial; several others are important plant pests. This family includes most of the bugs possessing stink glands.

In addition to the previously named families, the order includes several groups of common aquatic insects, many of which are predaceous, as well as a number of small families of terrestrial species, most of which are not sufficiently destructive to attract notice.

ORDER HOMOPTERA

There is no one common name generally acceptable to indicate all members of this order, which is sometimes considered a suborder of Hemiptera. Homopterous insects have piercing-sucking mouthparts similar in structure to those of the true bugs but attached at the base of the head, often appearing to arise from the thorax between the coxae. Wingless and winged individuals are found. Those with wings commonly have 2 pairs, both membranous; some families are characterized by species with the front pair slightly thickened. Male scale insects only have 1 pair of wings. Metamorphosis varies from gradual to one that approaches completeness. Members of this order feed only on plant juices, and many of them are quite injurious. Some transmit viruses and mycoplasmas causing serious diseases of plants. Size, with few exceptions, is medium to small or even minute. The number of families is not great, and the most common ones are characterized briefly in this summary. Detailed information about many species in this order is given in subsequent chapters.

Family Cicadidae, the Cicadas

Cicadas are often mistakenly called locusts. They are large insects, the common species being 3.8–6.5 cm in length, far exceeding all other members of the order (in the United States) in size. The so-called 17-year locust or periodical cicada is a destructive species; the common dog-day harvestflies or annual cicadas are of lesser importance (see p. 346).

Family Membracidae, the Treehoppers

This family includes medium to rather small insects with the pronotum or top of the prothorax greatly enlarged. The pronotum usually covers the head from above and projects backward, often hiding much of the wings and abdomen (Fig. 28). It may be strongly angulate with the angles sometimes produced into hornlike processes. Treehoppers feed on trees, shrubs, and legumes, and do further damage by their habit of ovipositing in woody plants (see p. 349).

Family Cicadellidae, the Leafhoppers

Leafhoppers are among the most abundant of insects. Many species are to be found in any locality. A few are of tiny size up to 15 mm in length, but most of the destructive species are quite small (Fig. 29). They are usually slender and elongate with rounded or pointed heads and a double row of spines extending the entire length of the hind tibiae. A considerable number of important species will be discussed in following chapters.

Family Cercopidae, the Spittlebugs

These insects resemble robust leafhoppers but can be distinguished from them by the circlet of spines at the base of the hind tibiae. The hind tibiae also bear

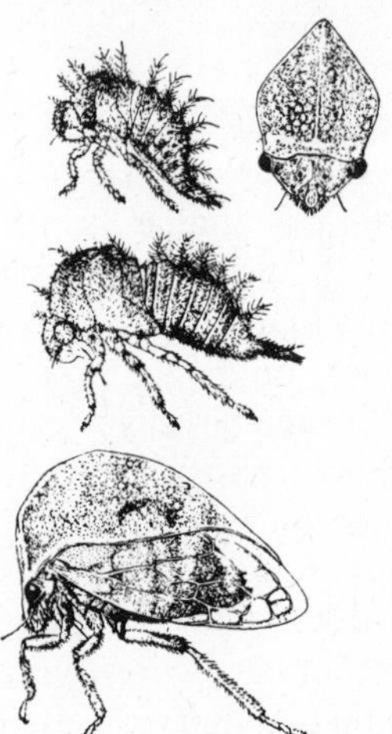

Figure 28. Nymphs and front and side views of *Stictocephala inermis* F. adults, a treehopper, family Membracidae; 2×. (Yothers, USDA)

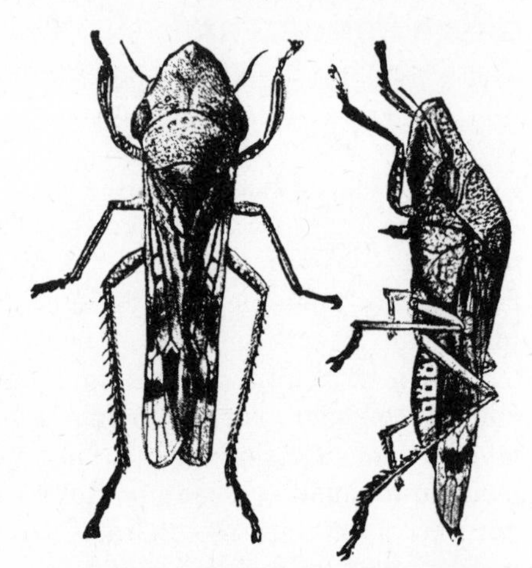

Figure 29. Dorsal and side views of the cotton sharpshooter or leafhopper, *Homalodisca triquetra* (Fabr.), family Cicadellidae; 2×. (Riley and Howard, *Insect Life*.)

a few heavy spurs, whereas a double row of spines is found on leafhoppers. The nymphal instars develop only inside foamy spittle masses which are easily detected on the host plants. Any perennial plant may be damaged, but legumes, strawberries, and many nursery crops are often seriously injured. The common species is discussed in detail under the section dealing with insects attacking legumes (Chapter 12).

Superfamily Fulgoroidea, the Planthoppers

This family includes a wide variety of forms, which can be distinguished from spittlebugs and leafhoppers by the antennae that are attached to the head below the compound eyes. The antennae are thickened at their base and terminate in a bristlelike process. Members of this group have rarely attracted attention by their feeding habits, but many of them are rather conspicuous and interesting insects. The ''splitters'' have divided this group into several families, the most important being Fulgoridae, Delphacidae, Issidae, and Flatidae.

Family Psyllidae, the Jumping Plantlice or Psyllids

Members of this family resemble tiny (3 mm) cicadas or winged aphids. The adults are more active than aphids, having hind legs fitted for jumping. Many species attack woody plants, which they injure in a variety of ways, but only a few species are classed as important pests, among them the pear psylla, boxwood psyllid, and potato psyllid (see pp. 267, 321). Some species cause gall formations.

Family Aleyrodidae, the Whiteflies

Whiteflies are minute insects with broad wings which are covered with fine, snow-white, waxy powder. In the North only a species found in glasshouses is injurious although a few others occur. In the South several species are seriously injurious to citrus and other plants. The immature stages resemble those of scale insects. Illustrations and further discussion are given in Chapters 16 and 22.

Family Aphididae, the Aphids or Plantlice

Aphids are small, sluggish, soft-bodied insects, usually green, but sometimes brown, red, black, or purple, and winged or wingless, which feed in groups on any part of the host. In most species a pair of dorsal abdominal projections near the posterior part of the body, called *cornicles,* are of value in identification, as are the antennal characters. Since there are many illustrations throughout this book, detailed description here is unnecessary. Aphids are among the most important plant pests and few crops are free from attack by one or more species.

Family Phylloxeridae, the Phylloxerans

This group of insects resembles aphids in appearance and in complexity of the life cycle. The best example of a pest species of this family is grape phylloxera (p. 438).

Family Diaspididae, the Armored Scales or Coccoids

In this family the soft body of the insect can be removed from under the hardened scale covering which is composed of two exuviae and a waxy secretion. Except in the first instar, scale insects have little or no resemblance to insects. They commonly appear as excrescences on the plants on which they feed. There are numerous species, many of which attack woody plants (Fig. 30) (see Chapters 17 and 22).

Family Coccidae, the Soft or Unarmored Scales or Coccoids

In this group of scales, the waxy covering over the insect is actually the body wall and it cannot be separated from the insect itself. Some species are large and rounded, others broad and flattened, but all species are not as hard as the armored scales. Soft scales, often excreting copious quantities of honeydew, are associated with ants. Pest species are covered in detail in Chapter 17 and 22.

Family Pseudococcidae, the Mealybugs

These are oval sluggish insects with short spines on the body margin and a covering of mealy white powder. As the period of egg-laying approaches for oviparous species, they are often buried in a mass of cottony fiber. A great many species occur both inside and outside glasshouses. Important ones are discussed in detail in Chapter 22.

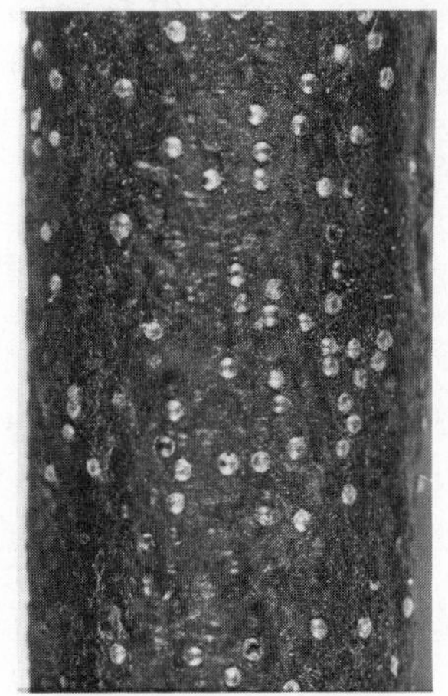

Figure 30. Pine needle scale, family Diaspididae (left); golden oak scale, family Asterolecaniidae (right), both about natural size. (Courtesy of Neiswander and Treece, OARDC.) At times both species are serious pests of their respective hosts.

Other important families of scales or coccoids are: Aclerdidae, Asterolecaniidae, Conchaspididae, Dactylopiidae, Kermesidae, Lacciferidae, Margarodidae, Ortheziidae, and Eriococcidae.

ORDER NEUROPTERA, the Nerve-Winged Insects

Neuropterans are usually medium-to-large insects, but several very small species occur in the order. They have four elongate, membranous wings of nearly equal size with numerous longitudinal veins and many cross veins. Bodies are slender, the antennae prominent, the mouthparts chewing, and the metamorphosis complete. The larval stages are predatory, and some species are valuable aids to humans in the natural control of small insects. The aphidlions or lacewings (Chrysopidae) (Fig. 12) are important as predators of insects such as plantlice, scales, and thrips. ''Doodlebugs'' are larvae of the antlions (Myrmeleontidae) which are predaceous on ants. The mantispids (Mantispidae) resemble praying mantids, the larvae attacking spider egg sacs and the nests of certain wasps. The dobsonflies, fishflies, and alderflies are often placed in the suborder Megaloptera. They are variable in size, some of the dobsonflies being more than 10 cm long with a wingspread of 12.5 cm. Their larvae are found principally in lakes and streams, the best-known being the hellgrammite or larvae of the dobsonfly. All these larvae are predaceous on other aquatic insects and serve as food for fishes or as fish bait. The alderflies are placed in the Sialidae, and the dobsonflies and fishflies in the Corydalidae.

ORDER COLEOPTERA, the Beetles

This is the largest order of insects, approximately 40% of all species being beetles. Its members are easily recognized by their hardened, opaque front wings, called elytra, which meet in a straight line down the back. The second pair of wings are membranous and folded beneath the first pair. They have chewing mouthparts, their metamorphosis is complete, and their antennae are of various types. Larvae have well-developed heads and jaws, usually three pairs of true legs but no prolegs, with a wide diversity of form and habits. The larvae of snout beetles or weevils are entirely footless. Pupae are usually naked and the appendages are free. The order includes a large number of families many of which have considerable economic importance. Representative families are briefly discussed.

Family Cicindelidae, the Tiger Beetles

These long-legged beetles are brightly colored, often metallic, with 5 tarsal segments on all legs. The antennae are filiform and the head is usually wider than the prothorax. They are predatory in both the larval and adult stages.

Family Carabidae, the Ground Beetles

These active beetles are usually black, sometimes brightly colored, nocturnal, and predatory, although some species are plant feeders. They have five tarsal segments on all legs, and the head is narrower than the prothorax. Antennae are filiform. The family is a large one, and the aggregate effect of the activities of the many predators is certainly great, though not accurately known.

Family Staphylinidae, the Rove Beetles

These black or brown beetles are all slender and elongate, with very short elytra. They are usually associated with decaying animal or vegetable matter, but recent studies show them to be predaceous on the insects occurring in such materials. Nearly 3000 species occur in the United States.

Family Silphidae, the Carrion Beetles

These beetles, being scavengers, are important to man in that they help eliminate dead animals. Broad, drab oval species, and red and black elongate forms are most commonly found.

Family Scarabaeidae, the Scarab Beetles

This is a large family of beetles varying considerably in size, form, color, and habits. Some species are scavengers, others are destructive crop pests. Some of the best-known forms are the May beetles, Japanese beetles, rose chafers, and dung beetles. They are all robust beetles with inconspicuous lamellate antennae and usually with five tarsal segments on all legs (Fig. 75). Important species are discussed in detail in succeeding chapters.

Family Buprestidae, the Metallic Wood Borers

These dark metallic beetles are extremely hard-bodied; some narrow and elongate, others broad and flattened, all tapering posteriorly. Anterior segments of the larvae are often expanded and flattened; hence the common name "flatheaded borers" is used to describe them (Fig. 268). Larvae usually feed under the bark and in dead or dying trees, although some species attack newly transplanted or unthrifty trees. Other woody plants are also attacked by some species.

Family Elateridae, the Click Beetles

This family consists of elongate, hard-bodied beetles, with serrate antennae, that are capable of flipping themselves into the air when accidentally turned on their backs. This flipping continues until they land upright on their feet. Their slender, cylindrical, brown-to-yellow larvae are called wireworms and are important crop pests feeding on roots and tubers (see Figs. 71, 72).

Family Lampyridae, the Fireflies

This is an interesting group of beetles, some of which have light-producing organs in both the larval and adult stages. The familiar "lightningbugs" of summer constitute one of the most common species. They are predaceous on other insects and are therefore beneficial. The larvae are called "glowworms." Firefly light is produced by two chemicals—luciferin and luciferinase—located in the posterior part of the abdomen, which are activated by adenosine triphosphate (ATP), an energy-storage compound present in all living cells. Because luciferin and luciferinase are now used in medical and biochemical research, firefly collecting has become a source of income, especially for children.

Family Cantharidae, the Soldier Beetles

Elongate, soft-bodied, brown or black beetles, ranging in size up to nearly 2 cm in length, comprise this family of predaceous insects. The soldier beetle resembles the firefly, but it does not have light-producing organs and its head is not concealed by the pronotum.

Family Dermestidae, the Skin Beetles

Beetles in this family are usually quite small and are scavengers as well as pests in the household. Included in the group are the carpet beetles, larder beetles, and hide beetles. Some are black, others mottled with gray. The larvae are all quite hairy (Fig. 31). Adults feign death when disturbed.

Family Nitidulidae, the Sap Beetles

Most of the members of this family are small beetles that feed on the sap from tree wounds, damaged fruits and vegetables, and fermenting plant fluids of decaying fruits and vegetables; a few are found in flowers or with fungi, and some are associated with dead animals. They are often black; a few are spotted. The elytra are short, exposing the terminal segments of the abdomen. Pest species are discussed in detail on p. 165.

Family Coccinellidae, the Lady Beetles

All are nearly hemispherical, often brightly colored, marked with spots, with

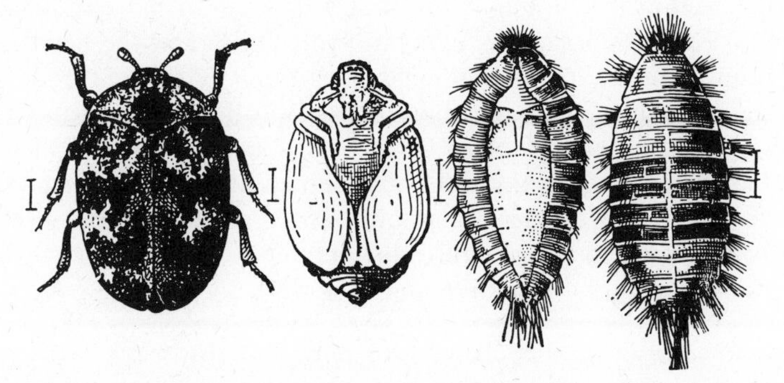

Figure 31. Adult, pupa, and larva, of the carpet beetle, *Anthrenus scrophulariae* (L.), family Dermestidae; 10×. (Riley, USDA)

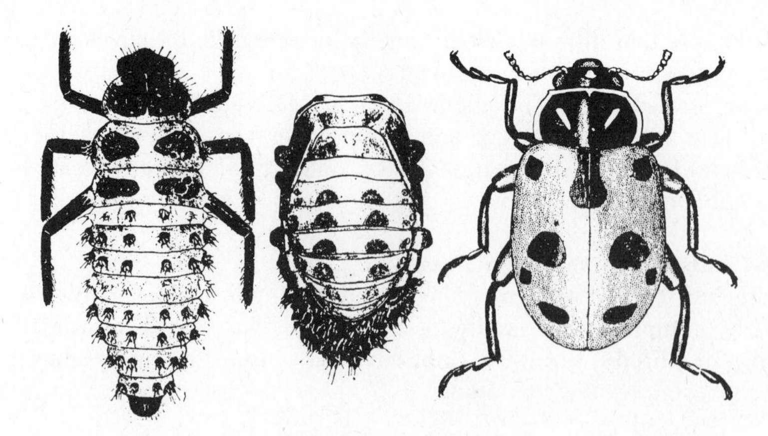

Figure 32. Larva, pupa, and adult of the sinuate lady beetle, *Hippodamia sinuata* Mulsant, family Coccinellidae; 7×. (Smith, Idaho Agr. Exp. Sta.)

three tarsal segments on all legs (Fig. 32). Many are very small and little noticed; others are larger, reaching nearly 1 cm in length. All but two species, the Mexican bean beetle and squash beetle, are predatory, and the family is justly regarded as one of highest value to humans. The predators feed on aphids, mealybugs, scale insects, and mites.

Family Tenebrionidae, the Darkling Beetles

Tenebrionids resemble ground beetles but have thickened antennae and only four segments in the hind tarsi. Some species are scavengers, others destroy stored products, especially grains. The larvae of a few species resemble wireworms and are called false wireworms (Fig. 74).

Family Meloidae, the Blister Beetles

Most of the beetles in this family are around 2.5 cm or more in length, black, gray, or striped, with a narrow, elongate, soft body, and the head wider than the pronotum. The elytra are rounded over their cylindrical bodies. Larvae are predators, feeding on grasshopper eggs and other soil insects. The adults are foliage feeders, and are often quite destructive (Fig. 33). Blister beetles contain a substance known as cantharidin, which may blister human skin when the beetles are handled. Hypermetamorphosis is also characteristic of the group, which means that 2 or more distinct types of larvae occur in the life cycle, these generally being active and sedentary forms.

Family Cerambycidae, the Long-Horned Beetles

Larvae of these rather large beetles are called roundheaded borers (Fig. 269). The adults are elongate, having unusually long antennae, with 5 tarsal segments

Figure 33. The margined blister beetle, *Epicauta pestifera* Werner. (USDA)

that appear as 4 because the fourth is extremely small and concealed. Most species are over 13 mm in length; some are nearly 10 cm. The adults feed on foliage and pollen, and the larvae tunnel the heartwood of trees and other woody plants. One member, the old house borer, *Hylotrupes bajulus* (L.), attacks wooden structures. Other species of importance are: locust borer, *Megacyllene robiniae* (Förster), redheaded ash borer, *Neoclytus acuminatus* (Fabr.), and pine sawyers, *Monochamus* species.

Family Chrysomelidae, the Leaf Beetles

A great variety of species occur in this family, some brightly colored, others drab; some are nearly 12 mm in length, others about 1 mm. They are usually oval in form with 4-segmented tarsi on all legs, the third segment bilobed. Both adults and larvae of some species attack foliage; others feed only on the roots in the larval stage. The corn rootworms, flea beetles, and cucumber beetles are common representatives (Figs. 34, 105, 106, 108, 109).

Family Bruchidae, the Seed Beetles

Small insects infesting the stored seeds of leguminous plants are members of this family. Some species begin their work in the developing seeds in the field. They are short stout beetles with truncated elytra, and the head is prolonged into a short, broad snout. Most species are less than 5 mm long.

Family Curculionidae, the Weevils

Many of our most serious crop pests belong to this family. They are also known as snout beetles (Figs. 110, 111) or curculios because of the elongate curved snouts on the head, at the tip of which are the chewing mouthparts. Adults are

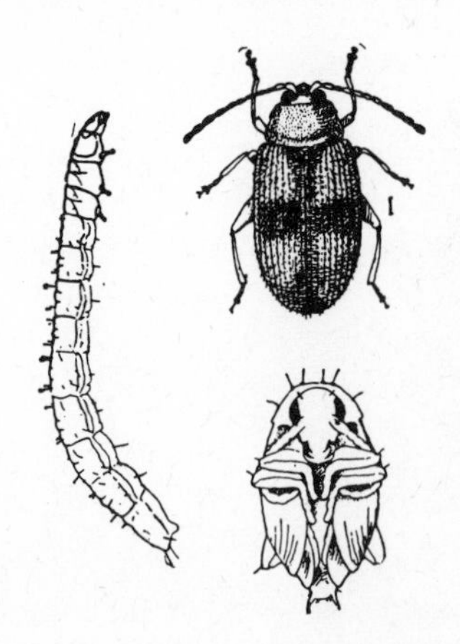

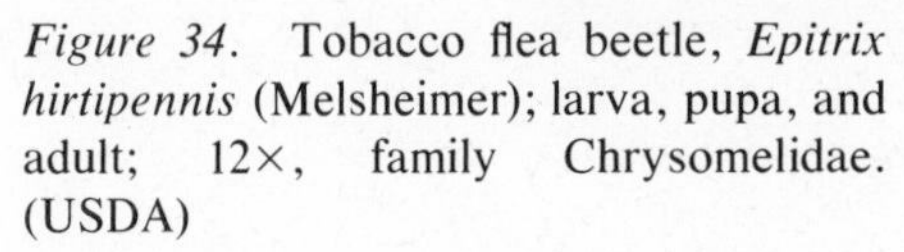

Figure 34. Tobacco flea beetle, *Epitrix hirtipennis* (Melsheimer); larva, pupa, and adult; 12×, family Chrysomelidae. (USDA)

Figure 35. Mountain pine beetle, *Dendroctonus ponderosae* Hopkins, adult; 8×, family Scolytidae. (from Swaine, Canadian Bark Beetles)

hard-bodied, usually dark colored, and feign death when disturbed. Larvae are white, thick-bodied, footless grubs. Both beetles and grubs may attack almost any part of the host, but the grubs more commonly feed within the plant whereas the beetles work externally. Many destructive species are discussed in later chapters.

Family Scolytidae, the Bark and Ambrosia Beetles

This group of rather small beetles with stubby snouts generally attacks woody plants, often boring beneath the bark in the cambium region. Both adults and larvae do damage, and death of the host usually results from girdling. Many important pests of forest, shade, and fruit trees belong to this family. Some common species discussed later are the shot-hole borer, peach bark beetle, and elm bark beetle. Ambrosia beetles tunnel deeply into the heartwood of trees. Their feeding is on fungi, which they cultivate in their galleries.

The following bark beetles are considered to be most destructive to forest trees in North America: southern pine beetle, *Dendroctonus frontalis* Zimmerman; western pine beetle, *D. brevicomis* LeConte; Jeffrey pine beetle, *D. jeffreyi* Hopkins; mountain pine beetle, *D. ponderosae* Hopkins (Fig. 35); Douglas-fir beetle, *D. pseudosugae* Hopkins; Mexican pine beetle, *D. approximatus* Dietz; lodgepole pine beetle, *D. murrayanae* Hopkins; eastern larch beetle, *D. simplex* LeConte; spruce beetle, *D. rufipennis* (Kirby); and red turpentine beetle, *D. valens* Lec.

ORDER STREPSIPTERA, the Twisted-Winged Parasites

These minute insects are mostly endoparasitic on other insects. Adult males are free-living and winged; the front wings are reduced to paddle-shaped appendages; the hind wings are greatly expanded, fan-shaped with few veins and, when at rest, are folded lengthwise. Mouthparts are vestigial; the antennae are enlarged and flabellate. Mature females are wingless with chewing mouthparts, often legless, and in the parasitic species they never leave their hosts, usually lack eyes, antennae, and legs. Metamorphosis is complete with hypermetamorphosis common. Principal hosts are in the orders Hymenoptera, Homoptera, Orthoptera, Thysanura, and Hemiptera. Members of four families are known to occur in the United States.

ORDER MECOPTERA, the Scorpionflies

The common species in this order are slender-bodied, some nearly 25 mm or more in length, with two pairs of narrow elongate wings of almost equal size, and having a relatively small number of veins. The legs are long—the insects resemble crane flies. The chewing mouthparts of these predatory insects occur at the tip of a downward prolongation of the head. Metamorphosis is complete but the larvae are rarely encountered. There are some very small species that are apterous. No economic importance is ascribed to the scorpionflies. The common name of these insects is derived from the scorpionlike resemblance of the recurved tip of the abdomen, which is evident in males of the family Panorpidae. The hanging scorpionflies are members of the family Bittacidae.

ORDER TRICHOPTERA, the Caddisflies

These insects resemble moths, having four wings sometimes with scales but always with numerous hairs. When not in flight the wings are held rooflike over the body. A few species are apterous. The adults are variable in size, some being quite small, others reaching 38 mm in length. Adults have chewing or vestigial mouthparts, and the larvae have well-developed chewing mouthparts. Metamorphosis is complete, the aquatic larvae constructing silken cocoons or cases, often of very interesting form, found attached to rocks or floating free in the water. Caddisflies are said to constitute the largest group of predominantly aquatic insects; they serve as food for fishes and other aquatic animals.

ORDER LEPIDOPTERA, the Moths, Skippers, and Butterflies*

Adults in this order are usually recognized by the dense covering of tiny scales and hairs on their wings and bodies, but there are a few species with wings devoid of scales. Their mouthparts consist of a long flexible tube formed for sucking or siphoning (Fig. 6), the palpi are often prominent, the antennae are of various types, and the 2 pairs of wings are membranous. Some wingless females occur in a few species. Metamorphosis is complete, and the chewing larvae are called caterpillars. These are elongate, cylindrical, wormlike crea-

* For those particularly interested in the Lepidoptera, obtain a copy of *Butterflies and Moths*, by R. T. Mitchell and H. S. Zim, Golden Press, New York. There are 432 illustrations in color.

tures with 3 pairs of thoracic legs and fleshy abdominal prolegs. There are normally 5 pairs of prolegs, but in the looping species this number is reduced. Many caterpillars spin elaborate silken cocoons in which pupation occurs. Commercial silk comes from the salivary glands of the silkworm, *Bombyx mori* (L.).

The order is a large one, the great majority of the species being moths. Since the larvae of most of these are phytophagous, it is not surprising that a great number of injurious species are to be found among them. Butterflies and skippers are relatively unimportant. They are diurnal and comprise only a few of the many families in the order. Only the more conspicuous and important families will be mentioned.

Family Pyralidae, the Pyralids

This is the third largest family in the order and includes many destructive species, some important ones being the European corn borer, sod webworms, pickleworm, melonworm, garden webworm, and the meal moths. Most of the members of this family are small brown moths with darker markings, often with gray mottling (Fig. 36). The destructive species are illustrated and described in succeeding chapters.

Family Olethreutidae, the Olethreutid Moths

Over 700 species in this large family of moths occur in the United States. Three of the very important pest species are the codling moth, oriental fruit moth, and grape berry moth. These are discussed thoroughly in the appropriate chapters.

Family Tortricidae, the Leaf Rollers

These are small moths, the larvae of which roll or fold the leaves of plants, tying them with silk and feeding on the inside. Several important species (Fig. 220) are discussed in the book.

Figure 36. The beet webworm, *Loxostege sticticalis* (L.), ♂ and ♀; slightly enlarged, family Pyralidae. (USDA)

Family Cossidae, the Carpenterworm Moths*

These large heavy-bodied gray moths with speckled or mottled wings have wood-boring larvae. The adult carpenterworm, *Prionoxystus robiniae* (Peck), has a wingspread of 75 mm; the fully grown larvae also approaches 75 mm in length. The leopard moth, *Zeuzera pyrina* (L.), is smaller; the larvae reach 50 mm in length and the moths about 65 mm in wingspread. Both species feed on deciduous hardwoods. Adults appear in late spring, lay eggs, and the hatching larvae tunnel the trunk or branches. One generation is completed in 2 to 3 years for the leopard moth and in 3 to 4 for the carpenterworm moth.

Family Sesiidae, the Clear-Winged Moths

All species in this family have much of the membranous wing area devoid of scales. Their wings are narrow and their bodies slender. Many adults resemble wasps and are diurnal in habit. The larvae are borers, usually in woody plants where damage is often quite serious. The peach tree borers, squash vine borer, lilac borer, and dogwood borer, are examples of destructive species (see Figs. 187, 303).

Family Psychidae, the Bagworm Moths

The larvae of members of this family construct large cases about themselves in which they feed on the foliage of trees and shrubs. The adult male has wings with very few scales; the female is wingless (see p. 350).

Family Tineidae, the Tineids

Some very small buff moths occur in this family. Common species known to most people are the clothes moths (p. 516).

Family Geometridae, the Measuring Worms

Adults of this family are medium-to-small moths, often delicate in appearance. Larvae are rather easily recognized by their looping or measuring method of crawling (Fig. 245). A few species will sometimes completely destroy the foliage of woody plants.

Family Arctiidae, the Tiger Moths

Arctiids may have a wingspread of 5 cm or less. They are often brightly colored, but several are entirely white. Others have black and orange patterns, suggesting the common name of the family. Larvae are hairy, many of our common "woolly worms" or "woolly bears" belonging to this group. Although a few species are pests, none is of great importance. See fall webworm, p. 359.

* Conn. Agr. Exp. Sta. Bul. 169, 1911; N.D. Agr. Exp. Sta. Bul 278, 1934.

Family Noctuidae, the Noctuids
This family includes about a third of all species of moths. Although many are medium-sized, having a wingspread of 5 cm or less, some reach a wingspread of nearly 10 cm. Dull brown-to-black with gray mottling is the prevailing coloration, but the second pair of wings is colorful in some species. Many of our serious crop pests are in this family, cutworms, armyworms, bollworms, fruitworms, corn earworm, and cabbage looper being common examples. Discussions of these, along with illustrations, will be given in later chapters.

Family Notodontidae, the Prominents
This group closely resembles the noctuids. Probably the best species to exemplify the family are the yellow-necked, walnut, and red-humped caterpillars which feed on the foliage of woody plants. They are included in discussions in Chapter 17.

Family Lymantriidae, the Tussock Moths
This is not an extensive group, but a few of its members are very destructive. The gypsy, brown-tail, and white-marked tussock moths are typical important species that are discussed in detail in Chapter 17.

Family Bombycidae, the Silkworm Moths
This family contains the species, *Bombyx mori* (Linnaeus), the true silkworm of Asia. It is reared artificially on mulberry leaves for the purpose of producing silk commercially. Culturing these caterpillars is still done in a number of countries, and the silk produced is said to have a commercial value of $200 to $500 million annually.

Family Lasiocampidae, the Tent Caterpillars
The important members of this family devour the foliage of trees. They are given special consideration in a later discussion (see pp. 356–358).

Family Saturniidae, the Giant Silkworm Moths
The largest of our moths belong to this family, some having a wingspread of 25 cm. Both adults and larvae attract attention because of their bright coloration and large size. The larvae feed on the foliage of trees and shrubs, but this is not considered important because of their small numbers. Some of the better-known species are the *cecropia, luna, polyphemus, promethea,* and *io* moths.

Family Citheroniidae, the Royal Moths
This family includes two very large species and several smaller ones. Possibly the best-known form is the larva of the regal moth, which is called the hickory horned devil. This harmless caterpillar reaches a length of 12.5 cm and has several prominent curved hornlike processes on the thoracic segments, giving

it a ferocious appearance. The other large species is the imperial moth, the larva of which feeds on foliage of trees and shrubs. Neither species is important as a pest.

Family Sphingidae, the Sphinx or Hawk Moths

Hawk moths vary considerably in size and coloration, some having a wing-spread of less than 5 cm, others over 15 cm. They are heavy-bodied with large narrowed fore wings and much smaller rounded hind wings. They usually fly at dusk and are able to hover in the air over flowers from which they suck nectar. Large species are often mistaken for hummingbirds. The larvae are called hornworms, and several species are important defoliators of tobacco (Fig. 37), tomato, grape (Fig. 38), and catalpa. The last, called the catalpa sphinx, *Ceratomia catalpae* (Boisduval), is the major pest of that host. Its large yellow-green and black caterpillars are often used as fish bait.

Family Hesperiidae, the Skippers

Skippers are dark or tawny yellow and brown, of medium-to-small size, usually with recurved or hooked antennae. Their larvae are peculiar in that they have a necklike constriction just back of the head. The diurnal skippers make short flights, hence the common name. A few species sometimes become destructive to crops.

Family Lycaenidae, the Gossamer-Winged Butterflies

These are usually very small butterflies, frequently colored blue, slaty, brown, or copper, the latter forms having a metallic sheen. They are of very little importance as pests.

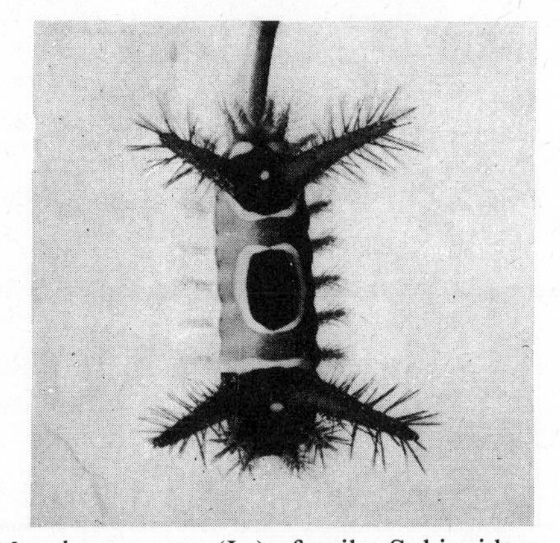

Figure 37. (Left) the tobacco hornworm, *Manduca sexta* (L.), family Sphingidae; note the long sucking mouthpart in extended position; ½×. (Conn. Agr. Exp. Sta.). (Right) the saddleback caterpillar, *Sibine stimulea* (Clemens), slightly enlarged; family Limacodidae. This larva is armed with venomous setae. (USDA)

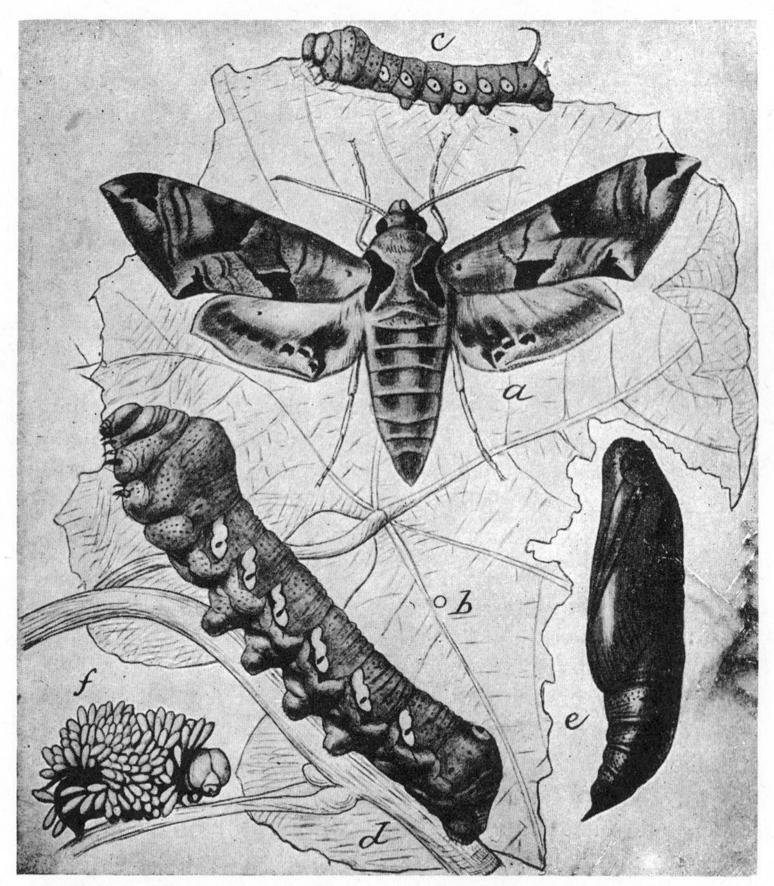

Figure 38. The achemon sphinx, *Eumorpha achemon* (Drury): *a,* moth; *b,* egg; *c,* young larva; *d,* mature larva; *e,* pupa; *f,* parasitized larva; all natural size; family Sphingidae. A minor pest of grapes. (USDA)

Family Pieridae, the White and Yellow Butterflies

There are many species in this family and the coloration is predominantly white, yellow, and orange, or combinations of these with some black spotting The imported cabbage butterfly and the alfalfa butterfly are two of the common destructive species; otherwise the members of the family are innocuous.

Family Papilionidae, the Swallowtail Butterflies

Large butterflies, many with taillike projections on the hind wings, are too well known to require further description. A pest species is the celery butterfly.

Family Nymphalidae, the Brushfooted Butterflies

This is the largest family of butterflies, but some taxonomists have been dividing it into several others. Nymphalids vary from small to quite large forms with a great variety of color combinations. They are usually recognized by the reduced size of the front legs and they walk only on the hind two pairs. Only a few pests occur in the family and they are of minor importance. Some common species are the monarch, viceroy, red admiral, mourningcloak, and anglewings.

ORDER DIPTERA, the Flies

Distinguishing characters for the flies are the following: two membranous wings or none, a pair of halteres, complete metamorphosis, and a variety of antennal types and mouthparts. The larvae are footless, with the head and mouthparts greatly reduced in most species. Both large and minute species are represented in this extensive order. Their habits are quite variable, some being predators or parasites of insects, some attacking man and domestic animals, some being vectors of disease-producing organisms, some scavengers, and some important plant pests. The number of families is large, and only a few can be mentioned.

Family Tipulidae, the Crane Flies

These are large, long-legged flies, resembling giant mosquitoes. They develop in damp habitats high in organic matter. The larval stage is dark and leatherlike, and often called "leatherjackets." Over 1450 species occur in North America; some damage plants, a few are predaceous.

Family Psychodidae, the Moth and Sand Flies

Tiny hairy flies about 5 mm or less in length that hold their pointed tipped wings rooflike over their body when at rest are moth flies; they do not bite. Pest species that frequent drains and sewers, the larvae feeding on decaying organic material, are often called drain flies or sewer gnats. Sand flies in the genus *Phlebotomus* are bloodsucking as adults; they are also vectors of organisms causing the following diseases in various parts of the world: kala-azar, pappataci fever, oriental sore, oroya fever, and espundia. Sand flies occur in southern United States and other tropical areas.

Family Culicidae, the Mosquitoes

This group includes many small, slender-bodied, long-legged flies with plumose antennae. The males have much larger and more densely plumed antennae. Mosquitoes are separated from similar flies by the tiny scales attached to the wing margins and to each wing vein (Fig. 39). The females are bloodsucking, and many species are vectors of the organisms causing dreaded diseases, namely, dengue, malaria, yellow fever, encephalitis, and filariasis. The larvae are aquatic and serve as food for other aquatic animals. Pest species are discussed in detail in Chapter 24.

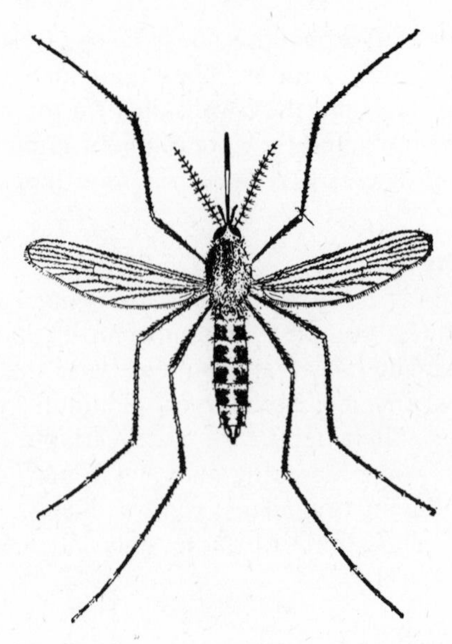

Figure 39. A salt-marsh mosquito, *Aedes sollicitans* (Walker); 3×. family Culicidae. (Smith, N. J. Agr. Exp. Sta.)

Family Ceratopogonidae, the Biting Midges

These tiny bloodsucking flies are often pests along seashores or shores of lakes and rivers. Common species that attack man are in the genus *Culicoides. Culicoides furens* (Poey) and *C. melleus* (Coquillett) are widely distributed species in tropical America from Brazil and the Mexican coast to the West Indies. Both occur along the coast of the United States from Massachusetts to Texas. In Florida they often appear in huge numbers and are the most abundant and troublesome species. Many other species in this same genus are also found in this and other parts of the world. The heaviest breeding of *C. furens* is in dense mangrove salt marshes where the soil is well-shaded, contains much organic matter, and is continually wet. Marshes that are subject to daily floodings by high tides are particularly favorable for larval development. The flies deposit their eggs in the mud near the water. Hatching occurs in a few days, and the slender larvae burrow in the sand or swim about in the water, continue development for a period of six to eight months, then change to pupae. Adult emergence takes place about a week later.

Family Chironomidae, the Midges

This family consists of many species of small mosquitolike flies. They often hover in swarms at dusk and generally occur near swampy areas. The larvae of most midges are aquatic and serve as food for other aquatic life.

Family Simuliidae, the Black Flies

Small, dark hump-backed flies, with scant venation in the wings and a vicious bite, make up this family (Fig. 40). The larvae develop in flowing streams. In Mexico and Central America they are vectors of a filarial worm that causes onchocerciasis in humans (see p. 570).

Family Cecidomyiidae, the Gall Gnats

These are also small flies (4 mm or less) resembling mosquitoes. They have long moniliform antennae, reduced wing venation and relatively long legs. Some species are predaceous, some scavengers, and some destructive plant pests, many of which cause the formation of galls. Some common species are the wheat midge, rose midge, boxwood leafminer, sorghum midge, chrysanthemum midge, and hessian fly.

Family Tabanidae, the Horse and Deer Flies

The females of these flies are vicious biters, feeding on the blood of livestock and man. Horse flies are large, some being 38 mm in length; deer flies are usually near 13 mm in length. Both species have semiaquatic larvae, and development takes place therefore in swamps, marshes, and along streams.

Family Asilidae, the Robber Flies

These rather large flies are predaceous on insects in both the larval and adult stages. The top of the head is hollowed out between the large compound eyes,

Figure 40. A black fly, *Simulium vittatum* Zett.; 15×. family Simuliidae. (Knowlton and Rowe, Utah Agr. Exp. Sta.)

and the abdomen is often long and tapered to a point. Many of these flies are quite hairy, with long legs equipped with spines and bristles (see Fig. 51*a*).

Family Bombyliidae, the Bee Flies

These are parasitic flies, the larvae feeding on various caterpillars and grasshoppers. The adults resemble bees and are often found on flowers.

Family Syrphidae, the Syrphid or Flower Flies

This family includes both small and large species, some reaching over 25 mm in length. They are generally black or brown with yellow stripes. Many resemble wasps or honey bees. The larvae are important predators of aphids and other small insects; the adults commonly are found hovering about flowers. A few species are scavengers (e.g., *Eristalis tenax* (L.), or drone fly) and some are plant pests (e.g., *Lampetia equestris* (Fabr.), narcissus bulb fly). Adults are distinguished by the spurious vein between the radial and medial veins. Adult syrphid flies do not bite.

Family Tephritidae, the Fruit Flies

These flies are all about the size of the house fly, and are recognized by the dark bands or spots on their wings. The larvae or maggots are the destructive stage. Some common species are the apple maggot, cherry maggot, and the Mediterranean fruit fly (check the index for these species).

Family Drosophilidae, the Vinegar or Pomace Flies

These tiny (3 mm) flies are troublesome around well-ripened fruits, in the field, in homes, or food processing plants. It is a large group; the members have short life cycles and are easily reared in laboratories. Hence they have been useful in genetic studies. See p. 505.

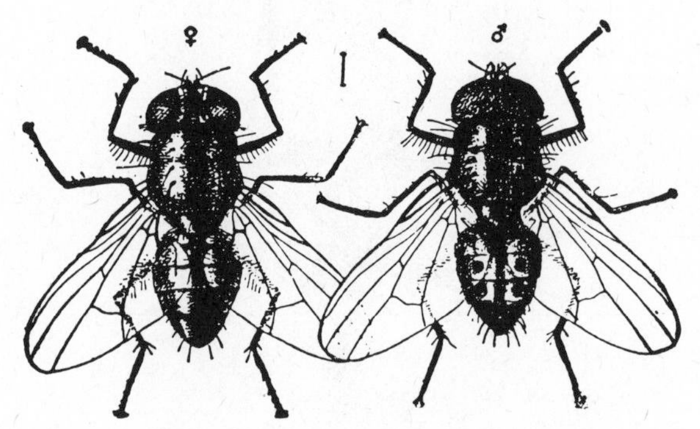

Figure 41. The little house fly, *Fannia canicularis* (L.); 10×, family Muscidae. Female left, male right. (USDA)

Family Agromyzidae, the Leafminer Flies

These tiny grayish to shiny black flies, usually less than 3 mm in length, are pests of various plants in the larval stage. They make winding mines between the leaf surfaces which increase in width as the larvae grow; some species mine stems, seeds, or stalks. Some common species are the holly leafminer, *Phytomyza ilicis* Curtis; chrysanthemum leafminer, *P. syngenesiae* (Hardy), columbine leafminer, *P. aquilegivora* Spencer; larkspur leafminer, *Phytomyza* sp. complex; corn blotch leafminer, *Agromyza parvicornis* Loew; alfalfa blotch leafminer, *A. frontella* (Rondani); pea leafminer, *Liriomyza huidobrensis* (Blanchard); vegetable leafminer, *L. sativae* Blanchard; serpentine leafminer, *L. brassicae* (Riley); and asparagus miner, *Ophiomyia simplex* (Loew). See pp. 220, 318.

Family Anthomyiidae, the Anthomyiids

These flies resemble house flies but are slightly smaller. Several species are very important plant pests, some common examples being the cabbage maggot, seed corn maggot, onion maggot, and the beet leaf miner. These are discussed in detail in subsequent chapters. Other species are scavengers, some aquatic and some parasitic.

Family Muscidae, the Muscids

This family includes the face fly, house fly, horn fly, stable fly, tsetse fly, and many others (Fig. 41). The horn, stable, face, and house flies are well-known species in this country, and a detailed discussion of their life cycles and control is given in Chapter 24.

Family Gasterophilidae, the Horse Bot Flies

These flies resemble honey bees. Their abdomen is rather pointed and usually curves under. The larval stages are important endoparasites of the alimentary tract of horses. See livestock pests.

Family Hippoboscidae, the Louse Flies

Winged species attack birds; the common wingless species is the sheep ked, an important ectoparasite of sheep. Their body is brown, flattened, and leathery, the mouthparts piercing-sucking, and reproduction truly viviparous.

Family Calliphoridae, the Blow Flies

This group includes the metallic blue or green flies commonly associated with dead animals. Some species lay their eggs in wounds and subsequent development may result in death of the animal. The screwworm fly is an example of a species with this habit.

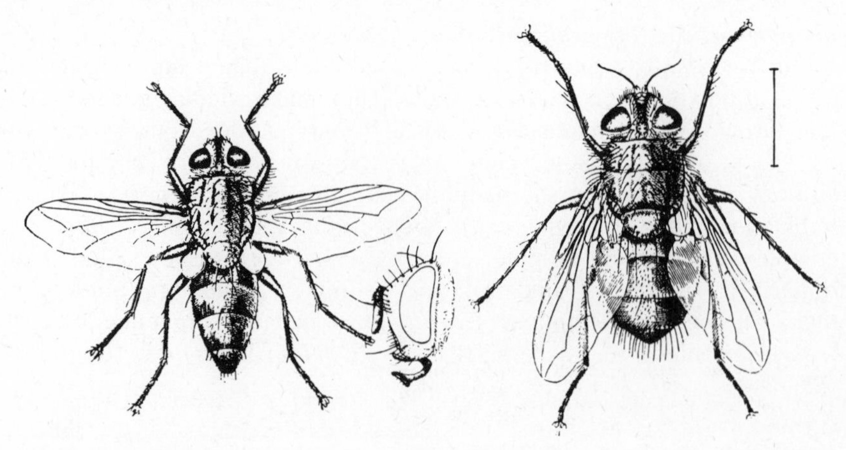

Figure 42. Winthemia leucanae (Kirk.), *left,* side view of head, *center,* and *Triachora unifasciata* (R.-D.), *right,* family Tachinidae, parasites of caterpillars; enlarged. (Riley.)

Family Tachinidae, the Tachinid Flies

These are small-to-rather-large hairy flies, usually with the arista bare (Fig. 42). They are parasitic on a wide variety of our pest insect species and undoubtedly play an important role in checking destructive outbreaks.

Family Oestridae, the Warble Flies

In this group are the important livestock pests, the sheep grub, and the ox warbles or cattle grubs. They are described in detail in Chapter 24.

ORDER SIPHONAPTERA, the Fleas

These are brown, laterally flattened, wingless insects with well-developed jumping legs and piercing-sucking mouthparts. The adults suck blood from birds and mammals, and some species are vectors of the organisms causing bubonic plague and endemic typhus. Metamorphosis is complete, the larvae developing in organic material consisting primarily of adult flea excrement, skin scales, and scabs of the host (see Chapter 24). A common family is Pulicidae.

ORDER HYMENOPTERA, the Bees, Wasps, and Ants

This large group of insects is characterized by having 4 membranous wings or none, complete metamorphosis, and chewing or chewing-sucking mouthparts. The larvae are legless and grublike except for some sawflies; their larvae resemble lepidopterous caterpillars but can be distinguished from them by the 6 to 8 pairs of abdominal prolegs without crochets. A variety of habits are represented in the order. Some attack plants, devouring foliage, mining leaves, boring in wood, or causing galls to form. Others are predaceous or parasitic on insects, pollinators of plants, or producers of honey and beeswax. Social behavior is highly developed in some groups of bees, wasps, and ants. The order is arranged into superfamilies for convenient discussion.

Superfamily Tenthredinoidea, the Sawflies

These insects have the abdomen broadly joined to the thorax, rather than constricted, the latter a characteristic of almost all the other members of the order. Their larvae also differ widely from the helpless young in other families. Sawfly larvae are caterpillarlike but have 6 to 8 pairs of abdominal prolegs in place of the 5 pairs possessed by true caterpillars. Prolegs of sawfly larvae have no crochets or hooks. Some species are small, a few are large, but more commonly the sawflies are of medium size. Defoliation and leafmining are common habits of the larvae (Figs. 270, 271). Common families are Argidae, Cimbicidae, Diprionidae, and Tenthredinidae.

Superfamily Siricoidea, the Horntails

This group of large insects (about 25–35 mm), usually brown or metallic blue-black with yellow, white, or reddish bands are borers, as larvae, in conifers and hardwoods. The adult female has prominent horny spinelike projections at the tip of the abdomen, the lower one being the ovipositor. A common species is the pigeon tremex, *Tremex columba* (L.), in the family Siricidae.

Superfamily Ichneumonoidea, the Ichneumon and Braconid Wasps

The largest of the common parasitic species belong to the family Ichneumonidae. It includes numerous species and is regarded as one of the most beneficial families of insects. Adults are variable in size from a few to over 40 mm in length, narrow-bodied, wasplike, with 16 or more segments in the antennae, 2 recurrent veins in the fore wings (Fig. 54), and a long prominent ovipositor on the females. The equally important family Braconidae comprises similar but usually smaller insects with one or no recurrent vein in the fore wings (Figs. 43 and 44). Pupation often takes place in silken cocoons outside the body of the host. Both ichneumons and braconids attack many caterpillars, beetle larvae, aphids, and spiders.

Superfamily Chalcidoidea, the Chalcid Wasps

Most chalcids are small; many very minute. Some are crop pests, but a great number are parasitic in the eggs and other developmental stages of moths, flies, beetles, and homopterous insects. They are frequently metallic black, broad-headed, with scarcely any wing venation (Fig. 45). This group ranks with the preceding one in importance to man. Some of the common families are Chalcididae, Trichogrammatidae, Eulophidae, Encyrtidae, Eupelmidae, Torymidae, Pteromalidae, and Eurytomidae.

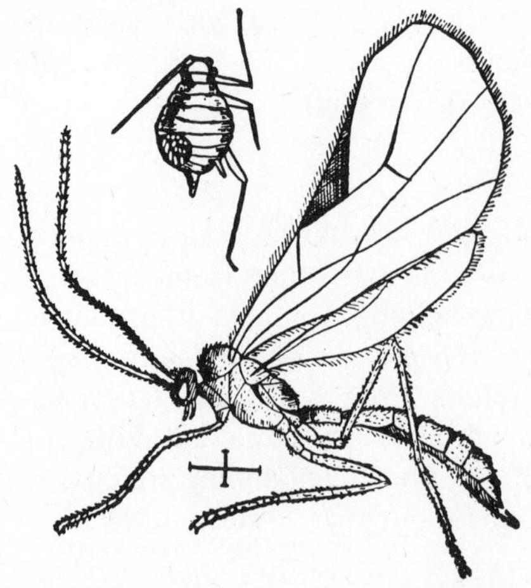

Figure 43. *Aphidius avenaphis* (Fitch), a parasite of aphids, and the body of the host showing the exit hole of the adult parasite; family Braconidae; enlarged. (J. B. Smith.)

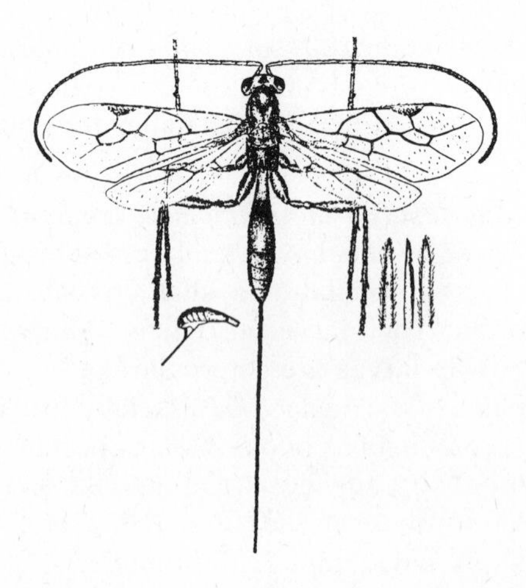

Figure 44. *Macrocentrus ancylivorus* Rohwer, a braconid parasite of several caterpillars. Lower left; side view of abdomen; all enlarged. (Fox, USDA)

Superfamily Proctotrupoidea, the Proctotrupids

This is the third major aggregation of parasitic insects. Most of the species are quite small and practically all are parasitic on other insects. The number that attack crop pests is enormous.

Species, among the small parasites, are usually identified only by specialists in the groups. Illustrations of parasites throughout this book will serve to acquaint the student with the more common species. In order to show the minute detail of these small insects the figures are greatly enlarged.

Superfamily Cynipoidea, the Gall Wasps

These are tiny black wasps with the abdomen laterally compressed and the posterior segments appearing telescoped. Some species are parasitic, but the majority attack plants, causing the development of tumorlike growths or galls. Oaks, rose, and blackberry are some of the common hosts. (Fig. 46).

Superfamily Scolioidea, the Ants and other Parasitic Wasps

This is a diverse group. Many parasitic forms attack the larvae of ground-nesting bees and wasps, white grubs, other beetle larvae, and crickets. A large group, the velvet ants (Mutillidae), occurs in arid areas of southwestern North America. The males are winged; the females are wingless and may inflict a painful sting. Perhaps the largest group of all are the ants (Formicidae), some being carnivorous, some scavengers, some plant-feeders. Several species listed as pests are illustrated and discussed in subsequent pages. It is possible the great-

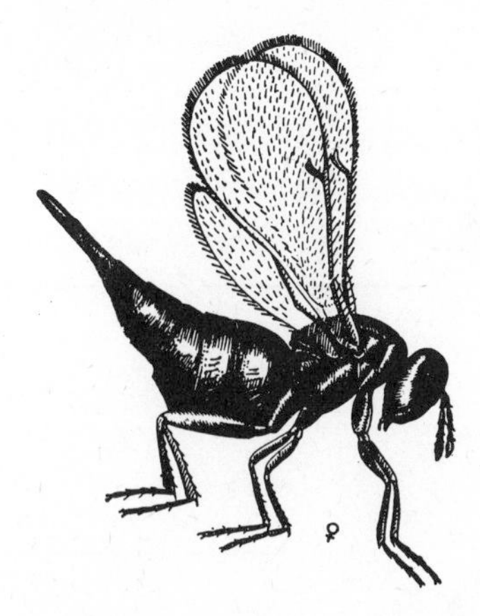

Figure 45. Aprostocetus diplosidis Crawford, a eulophid parasite of the sorghum midge; enlarged. (USDA)

est importance of ants is that they are predaceous on other insects. All ants are social and each colony contains a queen, males, and workers. Other families are Tiphiidae and Scoliidae.

Superfamily Vespoidea, the Social Wasps
Some of these wasps are predators and some are minor pests about dwellings. The family (Vespidae) can scarcely be considered as an economic group. Hornets, yellow jackets, spider wasps, and paper wasps are the best-known species. Sometimes they are a nuisance around homes or playgrounds.

Superfamily Sphecoidea, the Digger Wasps and Others
Quite a variety of wasps are ascribed to this group. All have unbranched hairs on the body. There are thread-waisted forms, cicada-killers, mud-daubers, sev-

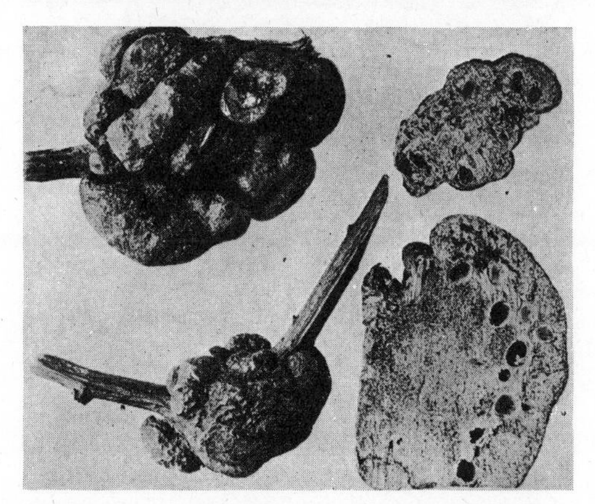

Figure 46. Galls of the blackberry gall maker, *Diastrophus turgidus* Bassett; natural size.

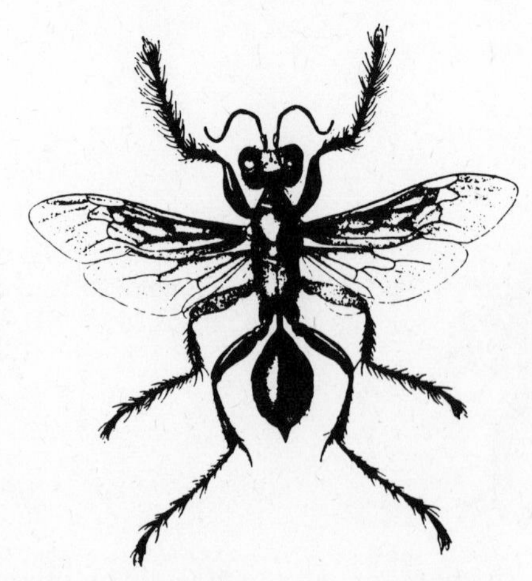

Figure 47. *Sphex atratus* Lepeletier, family Sphecidae, a digger wasp that provisions its nest with stung and stupefied grasshoppers; enlarged. (Walton, USDA)

Figure 48. Leafcutter bee damage to rose leaf; natural size. (Drawing by Ann E. Davidson)

eral mining species, and others. Many are predatory, capturing caterpillars, other insects, and spiders which serve as provision for their young (Fig. 47). A few species are parasitic on homopterous insects. They probably aid materially in the control of some pests. Sphecidae is the important family for this category.

Superfamily Apoidea, the Bees

Bees usually have modified hind tarsi and branched hairs on the body, but some species are less hairy on the abdomen. Females of most species have a well-developed stinger. The group is ranked as very important in the pollination of agricultural crops and the production of honey and beeswax. Some common species are honey bees, bumble bees, leafcutter bees, and carpenter bees. Leafcutter bees sometimes damage foliage (Fig. 48).

REFERENCES

Baker, E. W., and G. W. Wharton, *An Introduction to Acarology,* Macmillan Co., New York, 1952.

Borror, D. J., D. M. DeLong, and C. A. Triplehorn, *An Introduction to the Study of Insects,* 4th ed., Holt, Rinehart and Winston, New York, 1976.

Borror, D. J., and R. E. White, *A Field Guide to the Insects,* Houghton Mifflin Co., Boston, 1970.

Brues, C. T., A. L. Melander, and F. Carpenter, "Classification of Insects," *Harvard Museum of Comparative Anatomy Bul.,* 108, 1954.

Comstock, J. H., *An Introduction to Entomology,* 9th ed., Comstock Publishing Associates, Ithaca, N.Y., 1940.

Comstock, J. H., and W. J. Gertsch, *The Spider Book,* Doubleday, N.Y., 1940.

Jaques, H. E., *How to Know the Insects,* W. C. Brown Co., Dubuque, Iowa, 1947.

Kosztarab, M., The Armored Scale Insects of Ohio (Homoptera: Coccoidea: Diaspididae)," *Ohio Biological Survey Bul.* (n.s.) 2, 1964.

Levi, H. W., and L. R. Levi, *Spiders and Their Kin,* Golden Press, New York, 1968.

Muesebeck, C. F. W., K. V. Krombein, and H. K. Townes *et al., Hymenoptera of America North of Mexico,* USDA Agr. Monograph No. 2, 1956; 1st Supplement, 1958; 2nd Supplement, 1967; Government Printing Office, Wash., D.C.

Stone, Alan, C. W. Sabrosky, W. W. Wirth, R. H. Foote, and J. R. Coulson, *A Catalogue of the Diptera of America North of Mexico,* USDA Agr. Handbook No. 276, 1965; Government Printing Office, Wash., D.C.

Wharton, G. W., and H. S. Fuller, *A Manual of the Chiggers,* Entomological Society of Washington, D.C., 1952.

Williams, M. L. and M. Kosztarab, "Coccidae of Virginia with notes on Their Biology (Homoptera: Coccoidea)," Res. Bul. 74, Virginia Polytechnic Institute and State University, 1972, Blacksburg.

4

NATURAL CONTROL

The equilibrium that exists at any particular time among all the acting forces of nature has been termed the "balance of nature." These interacting forces can be grouped into two categories based on whether they promote or limit population growth.

The forces or factors that promote population growth of an insect, or any animal, are sometimes called its *biotic potential* or the ability to reproduce in the absence of destructive elements. The important factors are length of life cycle, number of generations per year, sex ratio, fecundity, and polyembryony. An insect that has a sex ratio of one, that is polyembryonic, that lays many eggs or gives birth to many young, and that has a short life cycle with repeated generations developing throughout the year possesses a high biotic potential. If such an insect reproduced uninhibited under ideal environmental conditions for an entire summer, the number of individuals at the end of that period would be enormous. This does not happen and the reason is that there are a number of factors acting in nature that limit or destroy animal populations. These factors combined have been called *environmental resistance*. They can be grouped into climatic, topographic, and biotic factors, and they are the subject of discussion in this chapter.

Natural control can be defined as any condition of the environment that checks insect populations and cannot be altered at will by man, at least on an extensive scale. Natural control encompasses all the factors mentioned under environmental resistance.

Climatic Factors

Climate is the long-term effect of the factors mentioned in this category; weather is the short-term or day-by-day effect. Temperature is perhaps the most important factor in climatic influence on insects. Its extremes prevent the existence of certain species in some regions or, in others, the insect is unable to maintain a population great enough for its activity to be noticed by man. The exposed life stages of insects are especially vulnerable. Prolonged periods of high (110–125°F; 43–52°C) or low (−20°F; −29°C) temperatures greatly reduce insect populations. Sudden changes in temperature also bring about control. For example, in the autumn a killing frost will eliminate the food plants of certain insects before they have reached the usual stage for overwintering and they therefore starve to death. Or a warm period in early spring stimulates hormonal activity of overwintering life stages to the point that they pass into the next stage, and these are then killed by a rapid temperature drop to freezing or below because the particular life stage is unable to withstand low temperatures. In general, insects, being cold-blooded, do have a wide range of acceptable temperatures.

Extremes of moisture, atmospheric humidity, soil moisture, and rainfall likewise act as checks or as favorable factors, depending on the insect species. Heavy rainfall may flood areas, killing many species or, accompanied by strong winds, may strike small, softbodied insects with such force that they will be washed from the plants and killed. The hibernating quarters of many species are greatly affected by an excess of moisture. A heavy snowfall acts as an insulating layer and actually protects life stages in hibernation, so that lack of snow in cold areas often results in lower populations. Humidity is a very important factor since the proper range (usually high) is necessary for the development of naturally occurring fungi, bacteria, and viruses that invade and kill insects. It is also essential for normal hatching, molting, and emergence.

Air currents or high winds may aid in the distribution of insects or be limiting factors in their numbers. Likewise sunlight and atmospheric pressures exert some influence on insects and their behavior and may be partly responsible for the size of a population in a given area.

Combinations of all these climatic factors operate to bring about natural control of insects.

Topographic Factors

These include the barriers that interfere with free migration from one region to another. Oceans, mountain ranges, deserts, broad rivers, and zones with prohibitively unfavorable climate make up the list of natural barriers. In addition, the physical and chemical nature of the soil has a direct bearing not only on the suitability of the environment for insects but also on their food supply.

Biotic Factors

All the factors of a biological nature that contribute to natural control of insects are usually placed in this category. They may be divided into the following groups: predators, parasites, natural resistance of plants or animals to attack by insects, and competition between the same or different species for a given food supply. *Always remember, in nature some must die that others might live.*

A *predator* may be defined as a living, active animal that catches and devours usually smaller or more helpless organisms or animals (called prey), killing them very soon after capture and requiring many individuals throughout its life cycle for proper nourishment and normal development. For example, during its life span, a lady beetle feeds on many less active smaller helpless creatures such as aphids or plant-lice in both the larval and adult life stages.

Birds are perhaps the most widely known of the vertebrate predators that contribute to the reduction of insect populations, but they are probably not the most important. The value of the group in nature, unaffected by the activities of man, is undoubtedly much greater than it usually is under conditions imposed by intensive agriculture. Nevertheless, many birds feed very largely upon insects, and they should be encouraged in every practical way. Some common species that are often considered important predators of insects are: swifts, wrens, warblers, nighthawks, gulls, killdeers, martins, woodpeckers, meadowlarks, orioles, starlings, blackbirds, quails, robins, and catbirds.

Many mammals such as hogs, bats, voles, moles, shrews, mice, anteaters, skunks, and even squirrels eat insect life stages. Even man may be considered a predator in those countries where certain insect species serve as food. For example, the eggs of May beetles and the larvae of a *Papilio* butterfly attacking the maguey plant in Mexico are often eaten by the natives. In some countries, grasshoppers, ants, termites, grubs, and caterpillars are a part of the diet of the people.*

In the amphibia we have the frogs, toads, and salamanders, which consume many insects. Usually the work of these animals is of importance only in small gardens, but occasionally toads, at least, have proved useful on a large scale.

A number of species of reptiles feed upon insects, but their actual influence on the populations of any important pest has not been demonstrated. Many fish feed almost exclusively upon insects, some of which are pest species like the immature stages of mosquitoes, and others feed on aquatic forms such as midge larvae, not considered pests, but valuable to man in that they serve as food for edible fish.

Invertebrates that attack insects are almost all in the phylum Arthropoda, except for hydra and planarians. Spiders are probably the most important of all noninsect predators of insects. The main disqualification of spiders as beneficial predators is that they feed on a wide variety of insect food and rarely concentrate enough on any one pest to accomplish a notable degree of control. Harvestmen, mites, and scorpions in the same class as the spiders, are with some exceptions predatory, and their importance as predators is unquestioned.

In the class Chilopoda, the centipedes are classified as predatory animals, and their food is largely insect, both beneficial and destructive species.

In the aggregate insects predaceous on insects become more numerous and

* Defoliart, G. R. "Insects as a source of Protein," *Bul. Ent. Soc. Amer.* 21:161–163, 1975.

are far more valuable than all the other groups of predators combined. Insect predators are found in many orders and are represented by a large number of families. Some families are mainly predaceous; others include only occasional species that have this habit. To single out accurately the most important family of predators, among the insects, is difficult with the present state of our knowledge. However, many authorities on the subject rank the lady beetles high in this respect in spite of the fact that there are a few species that are plantfeeders.

Beetles comprise approximately 40% of all insect species on the earth, and as a group include many species that feed on other insects. Two outstanding families are the lady beetles (Coccinellidae) and the ground beetles (Carabidae). Lady beetles (Figs. 32, 49, 58, and 59) are a large group, widely distributed, active in both larval and adult stages, with a considerable range of prey which includes some of our most destructive pests, notably the mites, thrips, aphids, scales, and mealybugs. Ground beetles are for the most part relatively large, active predators, and some species become quite numerous. The fact that they often feed at night tends to detract from an appreciation of their activity and value. Caterpillars are probably preferred hosts, but many life stages of insects are destroyed by both adults and larvae. Other beetles predatory in habit are the soldier beetles (Cantharidae), fireflies (Lampyridae), rove beetles (Staphylinidae), and the larvae of several blister beetles (Meloidae), which feed on the eggs of grasshoppers and play an important role in grasshopper control. Tiger beetles (Cicindelidae) in both the larval and adult stages, likewise are active as predators on insects.

The true bugs include quite a number of first-class predatory species. Many species of stink bugs attack various harmful insect life stages. Assassin bugs (Reduviidae) are, for the most part, predatory, but tend to attack species other than crop pests. A few members, however, are important enemies of potato beetle and asparagus beetle larvae, and caterpillars on crop plants. The minute

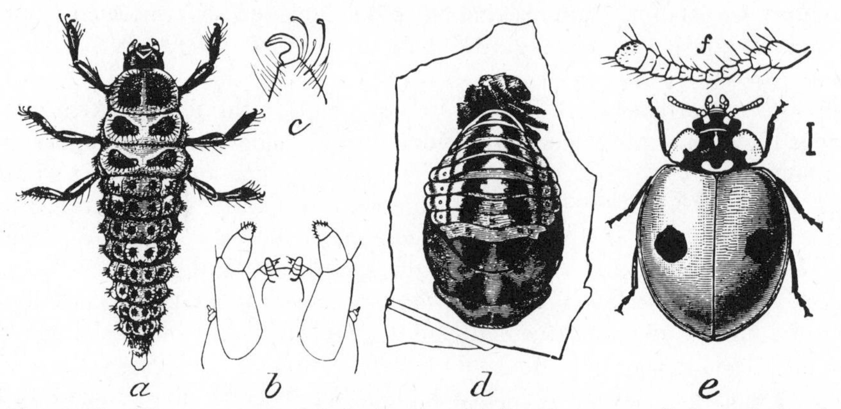

Figure 49. The two-spotted lady beetle, *Adalia bipunctata* (L.): *a,* larva; *b,* mouthparts; *c,* claw; *d,* pupa; *e,* adult; *f,* antenna; all enlarged. (Marlatt, USDA)

pirate bugs (Anthocoridae) are important predators of insect eggs, a common species being *Orius tristicolor* (White) (Fig. 24). The damsel bugs (Nabidae) are predatory and the two common species in North America are *Nabis americoferus* Carayon and *N. alternatus* Parshley (Fig. 25). The ambush bugs (Phymatidae) are represented by a few species that attack a number of insects, beneficial as well as destructive ones. A plant bug, *Cyrtorhinus mundulus* (Bredd.) (Miridae), is known to be an important predator of sugarcane leafhopper eggs.

Many predatory flies are known. The most important are the syrphid fly larvae (Syrphidae), which are active enemies of aphids (Fig. 50) and occasionally attack other insects. The adults of robber flies (Asilidae) are considered to be of some importance, because they destroy a number of insect species. Dipterous predators in other families are predaceous either as larvae or adults but are less important to agriculture so far as we now know (Fig. 51).

Lepidopterous predators are almost nonexistent. The larvae of some butterflies (Lycaenidae) are recorded as feeding on scale insects, leafhoppers, aphids, and treehoppers. A moth larvae (Pyralidae) is predatory on unarmored scale insects.

Many ants (Formicidae) are predaceous in habit in spite of the fact that they are thought of mainly as pests. They are omnivorous and readily become scavengers when insect food becomes scarce. Ants are more universally distributed probably than any other predators and doubtless far exceed in total numbers any insects approaching them in size. Besides the ants, there are other hymenopterous insects which are considered valuable predators. These are the wasps in the families Sphecidae and Vespidae.

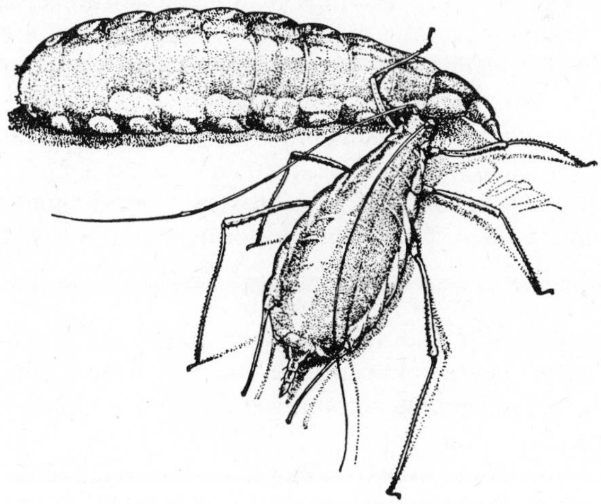

Figure 50. Larva of the syrphid fly, *Allograpta obliqua* (Say) attacking a pea aphid; 9×. (Walton, USDA)

Figure 51. (a) *Promachus vertebratus* (Say), an asilid predator of white grubs and other insects, natural size; (b) *Aphidoletes meridionalis* Felt, a predaceous cecidomyid fly; adult ♀; *a,* antenna of ♂, and *b,* abdomen of ♂, enlarged. (Walton, USDA)

The praying-mantids are entirely predatory and, although they are rather large and clumsy, they destroy numerous insects during their life cycle. They are the most important predators in the order Orthoptera. Both roaches and crickets have been known to feed upon other insects which they have captured and killed, but the habit is not common or extensive enough to reflect any particular credit on these usually pestiferous groups.

Among the members of the minor orders are a few groups that are effective and valuable predators. The lacewings (Chrysopidae) are perhaps the most important of these; their larvae are predators of aphids and are called aphid-lions. Many thrips are plant pests, but the order Thysanoptera includes predatory species of considerable value, viz., *Leptothrips mali* Fitch, *Haplothrips faurei* Hood, and *Scolothrips sexmaculatus* (Pergande), important predators of phytophagus mites. Young dragonflies, in their aquatic habitat, destroy other insects including quantities of mosquito larvae. Adult dragonflies capture insects in flight and eat them. Other aquatic forms may have slight value in the same way. Although many other insects are predatory, none has proved to have any measurable effect on pest populations.

The aggregate effect of predator populations cannot yet be measured or even estimated with a great degree of accuracy, but there is no doubt that they play a very important role in keeping certain pest populations in check to the point they are no longer serious problems.

A *parasite* is defined as an organism that lives in or on another usually larger living organism (host), and requires only one or a partial individual to complete its life cycle. For example, in some parasites of aphids only one individual develops in each aphid, whereas many braconid parasites may develop in a single caterpillar such as the tomato hornworm. In some kinds of parasitism, the host does not die, for example, a bed bug on man or a tapeworn in man, but usually death is the end result of insects parasitic on insect species, although it is not immediate. The insect parasites of insects are often called *parasitoid* species. Because of the confusion created in publications the term parasitoid will not be used further in this book.*

There are various kinds of parasites. When they attack the host externally they are known as *ectoparasites,* when internally they are known as *endoparasites. Permanent* parasite is descriptive of a species that spends all life stages on the host whereas an *intermittent* parasite is on the host only at feeding time. A *primary* parasite is any species attacking a particular host, and it is considered beneficial to man if the host is a destructive insect. A parasite that attacks another parasite is called a *hyperparasite.* When hyperparasites are abundant they may completely counteract the beneficial effects of a primary parasite. If the host is attacked by many individuals of the same parasite species, *superparasitism* results; when attacked by more than one species of a primary parasite the term applied is *multiple parasitism.*

The parasites of insects are conveniently grouped into the following categories: insects, nematodes, protozoa, bacteria, fungi, viruses, and rickettsiae. The last five groups are usually thought of as pathogens that produce diseases of insects.

On the whole, insects are generally considered the most important group of parasites that reduce insect populations. They are to be found in several orders, two of which are outstanding: Hymenoptera and Diptera.

There are many hymenopterous families which include species for the most part parasitic in habit. These have been discussed in Chapter 3, but further mention here seems appropriate.

The superfamily Chalcidoidea is composed of some 19 families, all more or less closely related. The vast majority of the species are parasitic in their larval stages on the eggs and larvae of other insects. The genus *Aphelinus* contains species that parasitize aphids and scales; the genera *Aphycus, Aphytis, Physcus, Aspidiotiphagus,* and *Prospatella* contain cosmopolitan species parasitic on scale insects, and the genus *Pteromalus* contains species that parasitize the chrysalids of cabbage butterflies and related forms (Fig. 52), *Trichogramma evanescens* Westwood, *T. minutum* Riley, and *Anagrus epos* Girault are egg parasites of considerable importance. They are minute wasps just large enough to be seen by the unaided eye. Their importance is based on the fact that, unlike many parasites, they have a large number of possible hosts, are widely distributed, and present most of the time. *Encarsia formosa* Gahan is a cosmopolitan species parasitic on whiteflies; *Brachymeria ovata ovata* (Say) and *Spilochalcis albifrons* (Walsh) are parasitic on many kinds of caterpillars (Fig. 53).

* See *Bul. Ent. Soc. Amer.* 22:135, 1976.

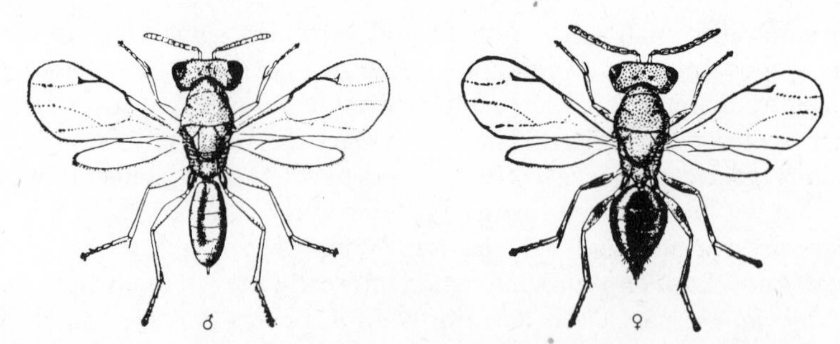

Figure 52. *Pteromalus puparum* (L.), a chalcid that parasitizes the cabbage worm and many other injurious insects; male and female; 12×. (USDA)

In the superfamily Ichneumonoidea are many important parasitic species. The rather large ichneumon wasps of the genera *Horogenes* (Fig. 53), *Ophion, Itoplectis* (Fig. 54), and *Pimpla* are among the more common parasites of caterpillars. Other genera such as *Apanteles, Microbracon, Ascogaster, Chelonus,* and *Macrocentrus* are all likely to attack caterpillars. The last-named includes parasites of the European corn borer, codling moth, and oriental fruit moth. *Apanteles glomeratus* (Linn.) is a well-known parasite of the larvae of the cabbage butterfly. The almost universal aphid parasite is *Aphidius testaceipes* (Cresson). It reproduces throughout the summer and is an effective check on many species.

Parasitic flies are often more important enemies of insects than are the Hymenoptera. *Blaesoxipha kellyi* (Aldrich) (Sarcophagidae) and several related species attack grasshoppers, very effectively at times. *Winthemia quadripustulata* (Fabr.) has a large number of caterpillar hosts and usually succeeds

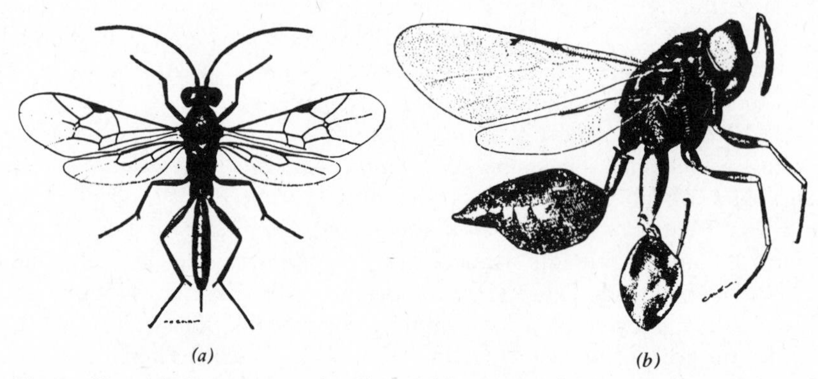

Figure 53. (a) *Horogenes plutellae* (Viereck), an ichneumonid parasite of the diamondback moth; (b) *Spilochalcis flavopicta* (Cresson), a chalcid parasite of boll weevil, sorghum webworm, and other insects; both greatly enlarged. (USDA)

Figure 54. Itoplectis conquisitor (Say), an ichneumon parasite of many caterpillars; enlarged. (Pettit, Mich. Agr. Exp. Sta.)

in keeping several of them under complete subjugation. It is a tachina fly. Another species in the family Tachinidae is *Trichopoda pennipes* (Fabr.). This is an important species and of special interest because of its somewhat unusual host range. True bugs, such as the squash bug, and beetles are both on the list.

Other members of this family, *Exorista larvarum* (L.), and *Compsilura concinnata* (Meigen) (Fig. 55), attack a wide variety of hairy caterpillars. Fly parasites of less importance are found in the families Pipunculidae (Fig. 56), Phoridae (Fig. 57), Bombyliidae, and Pyrgotidae.

Although insect parasites are known in other orders, none of them affects any great destruction among the major pests.

The nematodes have been studied and have proved on occasion to be controlling factors for some injurious insects. Some species attack boll weevils, face flies, grasshoppers, cockroaches, bumble bees, leaf beetles, mosquito larvae, May beetles, various lepidopterous caterpillars, and bark beetles. California entomologists have shown that *Heterorhabditis bacteriophora* Poinar can utilize many different hosts including those of major agricultural and medical importance.*

Many kinds of protozoa are associated with insects, but the greater part of those known to be pathogenic belong to the Microsporida. A good example is *Nosema bombycis,* which is the cause of the disease of silkworms known as pébrine. *N. apis* causes a disease of adult honey bees, *N. locustae* of grasshoppers, *N. algerae* of mosquitoes, and *N. trichoplusiae* of cabbage loopers. There are other species that are parasitic and cause disease in members of the orders Hymenoptera, Diptera, and Lepidoptera.

* *Calif. Agr.* 32(3):12, 1978.

Figure 55. (*a*) *Exorista larvarum* (L.), a parasite of the gypsy moth, and (*b*) *Compsilura concinnata* (Meigen), also an important parasite of both gypsy and brown-tail moths; enlarged. (USDA)

The bacteria rate high as parasites causing disease in insects. Some examples are: *Coccobacillus acridiorum* d'Herelle, which attacks grasshoppers; *Bacillus pluton* White, or European foulbrood, and *B. larvae* White, or American foulbrood, both of which attack the brood of honey bees; *B. thuringiensis* Berliner, which attacks many lepidopterous caterpillars; *B. popilliae* Dutky and *B. lentimorbus* Dutky, the organisms that produce milky disease of Japanese beetle larvae and other Scarabeidae; *B. sphaericus* Neide, that attacks mosquito larvae. These diseases are often sporadic and at such times greatly reduce insect populations. Environmental conditions undoubtedly control very largely the normal development of these diseases.

Many fungus diseases of insects are described, and some of them play an important role in controlling a number of our major pests. One of the best known is the white muscardine fungus, *Beauveria bassiana* (Bals.), which has been recorded as occurring on more than 70 insect species in North America. This is the fungus that controls the chinch bug when the weather is warm and humid. These same conditions are necessary for the proper development of most fungi.

Other fungi of considerable importance in reducing insect pest populations are: *Entomophthora sphaerosperma*, *E. fumosa*, *E. coronata*, *E. aphidis*, *Metarrhizium anisopliae* (green muscardine), *Cephalosporium aphidicola*, and species of *Cordyceps* and *Aspergillus*.

The viruses comprise a group of important disease-producing organisms, and research in this area has been quite active during the past decade. Most authorities divide them into two groups, polyhedral and nonpolyhedral diseases. Of the 250 virus infections recognized in nearly 175 species of insects and arachnids, about 170 of these are nuclear polyhedroses, 30 are cytoplasmic poly-

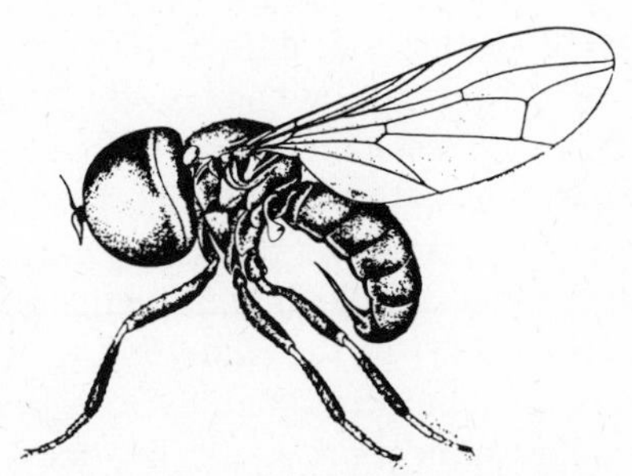

Figure 56. Tomosvaryella subvirescens (Loew), parasitic on adults and nymphs of beet leafhopper; enlarged, (Pipunculidae). (Knowlton, Utah Agr. Exp. Sta.)

hedroses, 35 are granuloses, and 8 do not appear to be associated with inclusion bodies of any kind. Some examples of polyhedroses are jaundice or grasserie of the silkworm and the so-called wilt diseases of various species of Lepidoptera and Hymenoptera larvae, such as cutworms, tent caterpillars, tussock moths, gypsy moth, cabbage looper, tobacco budworm, bollworm, codling moth, European spruce sawfly, European pine sawfly, Swaine jack pine sawfly, hemlock looper, armyworm, and alfalfa caterpillar. Sacbrood of honey bees is the only well-known nonpolyhedral disease that has been studied to any extent. This disease is troublesome to the beekeeper.

Species of rickettsiae have been described as attacking various saturniid moths, white grubs, and midges. Their importance in checking pest populations is yet to be determined.

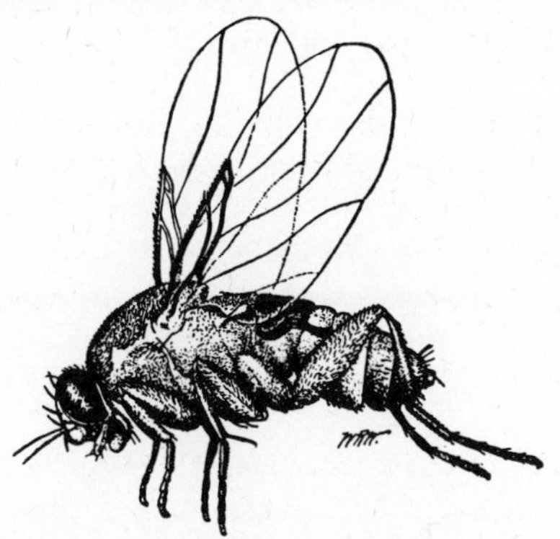

Figure 57. Megaselia perdita (Mallock), a phorid parasite of the alfalfa caterpillar; enlarged. (Walton, USDA)

Resistance when applied to plants or animals, may be defined as any factor or group of factors existing from natural causes which deter insects or prevent them from attacking. Types of resistance have been categorized into nonpreference, tolerance, and antibiosis. The factors involved are not too well understood but are thought to be physical and physiological, and most often genetic in character. Any naturally occurring animal or plant species, or their varieties, which are not attacked by insects to the same degree may be said to possess natural resistance. Many present-day plant hybrids resistant to insects had their beginnings in selections of some naturally occurring species that had exhibited resistance.

The last factor of a biotic nature to be considered is competition between the same or different species for a given food supply. When the food supply is exhausted mortality results especially for immature insects which, lacking a means of rapid locomotion, are less able to find additional or substitute food.

REFERENCES

Chapman, R. N., *Animal Ecology,* McGraw-Hill Book Co., New York, 1941.

Clausen, C. P., *Entomophagous Insects,* McGraw-Hill Book Co., New York, 1940.

DeBach, Paul, and E. I. Schlinger, *Biological Control of Insect Pests and Weeds,* Chapman and Hall, London, 1964.

DeBach, Paul, *Biological Control by Natural Enemies,* Cambridge Univ. Press, London, 1974.

Graham, S. A., *Principles of Forest Entomology,* 3rd ed., McGraw-Hill Book Co., New York, 1952.

Huffaker, C. B., *et al., Biological Control,* Plenum Press, New York, 1971.

Madelin, M. F., "Fungal Parasites of Insects," *Ann. Rev. Ent.* 11:423–448, 1966.

Painter, R. H., *Insect Resistance in Crop Plants,* The Macmillan Co., New York, 1951.

Poinar, G. O. Jr., "Nematodes as Facultative Parasites of Insects," *Ann. Rev. Ent.* 17:103–122, 1972.

Price, Peter W., *Insect Ecology,* John Wiley and Sons, New York, 1975.

Steinhaus, E. A., *Insect Microbiology,* Comstock Publishing Assoc., Ithaca, New York, 1946.

Steinhaus, E. A., *Principles of Insect Pathology,* McGraw-Hill Book Co., New York, 1949.

Steinhaus, E. A., *Insect Pathology,* Vols. I and II, Academic Press, New York, 1963.

Sweetman, H. L., *Principles of Biological Control,* W. C. Brown Co., Dubuque, Iowa, 1958.

USDA Yearbook of Agriculture, pp. 373–394, 422–436, 1952.

van den Bosch, Robert, and Kenneth S. Hagen, "Predaceous and Parasitic Arthropods in California Cotton Fields," Cal. Agr. Exp. Sta. Bul. 820, 1966.

Weiser, J., "Recent Advances in Insect Pathology," *Ann. Rev. Ent.* 15:245–256, 1970.

5

APPLIED CONTROL: MECHANICAL, CULTURAL, BIOLOGICAL, LEGISLATIVE

An applied control measure may be defined as any method employed by man to bring about reduction of insect damage. For convenience in our discussion the methods are divided into the following groups: mechanical, cultural, biological, legislative, and chemical. Some or all of these may be used when an economic pest threshold is reached that will justify the expense of the operations. The choice of method depends largely on its cost and effectiveness and the valuation of the crop or property to be protected. The ultimate goal of any control or management program is to shift, by the best means available, a balance of nature adverse to man to one that favors him, while minimizing any effects on the ecological system.

MECHANICAL CONTROL

A very important category for applied control is that of mechanical devices. Mechanical control involves the employment of special equipment or operations for the specific purpose of reducing insect populations or preventing attacks by them. To be effective these control measures must be initiated promptly, and the results are therefore immediate.

Barriers are devices to keep insects away from plants or animals. Window screens, screen doors, and mosquito netting are barriers to all kinds of insects; metal shields on building foundations prevent subterranean termites from entering; in the field, dusty furrows check the advance of chinch bugs, armyworms, and similar migratory pests; sticky bands on trees and other objects trap and prevent migration of insects; shrouding young trees with cheesecloth prevents cicada injury; paper or plastic collars around single plants are barriers to cutworms; framed screening over small garden plantings prevents adults of root maggots and leafminers from ovipositing on the crops; wrapping trunks of transplanted trees with paper prevents infestation of flatheaded borers; tree paints on pruning wounds are barriers to wood borers and carpenter ants; low metal, wood, or paper fences prevent migration of Mormon crickets and chinch bugs. Often chemicals are used in combination with many types of barriers.

Collection of insects by mechanical means may sometimes be the most practical method of control. Hand-picking egg masses, larval nests, or large insects often is the cheapest method of destroying them. Other devices for collecting or trapping insects are the following: hopperdozers for catching grasshoppers; inverting umbrellas for collecting curculios jarred from trees; corrugated paper bands to trap codling moth larvae; house fly traps; Japanese beetle traps and other types utilizing the positive response of insects to sound, light,* or odor, luring them into an inescapable chamber or electrocuting device. Male mosquitoes are attracted by sound waves, and a greater number of Japanese beetles are caught in yellow traps baited with geraniol, anethol, and eugenol (see attractants). Trap crops have been utilized in the control of insects. Attracting insects to plantings of their favorite food and then destroying them may result in the main crop remaining relatively free of injury.

Direct mechanical destruction may be accomplished by ensilage cutters, husker-shredders, heavy rollers, plows, and soil pulverizers. In areas where irrigation is possible, flooding reduces insect populations. Drainage ditches in marshes and swamps aid greatly in controlling such insects as mosquitoes and horse flies.

Artificially raising or lowering of the temperature may be employed as a means of control. Many insects are killed at temperatures of 110–115 F (43–46 C), and all are killed at exposures to 140 F (60 C) or higher. For example, some mills and grain elevators are equipped with heating systems enabling the temperatures inside to be raised to approximately 140–150 F (60–65 C) for several hours during periods of hot summer weather. This method is effective and often less costly than the application of fumigants. Infested clothing, baled cotton, cereals, and other foods may be effectively treated in this way if facilities are available for either large- or small-scale operations. It is well to keep in mind that tightly piled or compact products require longer exposure to allow for effective heat penetration. There is also shrinkage because of moisture loss, and the germination of seeds will be impaired by prolonged exposure at high temperatures. Sterilizing of soils with steam, dipping insect- and mite-infested bulbs in water at temperatures of 110–112 F (43–45 C), sponging the legs of horses with water at 115 F (46 C) to cause hatching of common bot fly eggs and resultant death of the larvae, flaming for control of hibernating alfalfa weevils, or rotating infested logs toward the direct rays of the sun for killing bark beetles are other examples of control by heat.

Most insects become inactive at temperatures below 55 F (13 C) and may withstand hibernating temperatures of −25 F (−32 C) or lower, but practically

* *Bul. Ent. Soc. Amer.* 20:279–281, 1974; USDA Tech. Bul. 1498, 1974.

all activity ceases at 40 F (22 C) or lower. The knowledge of this fact has resulted in the construction of cold-storage vaults for protecting woolen clothing and furs during the summer. Usually low temperatures are not as effective as high temperatures for insect control, but, since there is no activity at near-freezing, no damage results. Sudden changes in temperature usually give higher mortality.

Exposing puparia of the screwworm fly to the proper amount of gamma radiation results in sterile adults with normal mating behavior. Adult screwworm females mate only once and when sterile individuals are released in quantity over an infested area they react normally with the natural population, resulting in the production of many sterile eggs. Almost complete eradication of the screwworm in the United States has been achieved by this control method (autocidal). Other insect species eliminated in this manner were the melon fly *Dacus cucurbitae* Coq. from the island of Rota in 1962, and the oriental fruit fly *Dacus dorsalis* Hendel from Guam in 1964.

The use of microwaves and other radiofrequency energy to control insects has been studied and reviewed.* From the economic viewpoint their use is questionable.

CULTURAL CONTROL

Cultural control measures are another means by which insect populations are reduced, but the end results are usually not immediate. These involve the best cropping practices known for a given crop that may incidentally check possible insect populations from developing. Many cultural operations destroy insects by mechanical means even though they may not be performed specifically for that purpose.

Fall plowing has long been considered beneficial in insect control and, where it fits in with the farm operations, is still to be commended. In this operation the control results partly from mechanical injury to the insects and partly from their exposure to unsuitable environmental conditions.

Deep and thorough plowing is a sanitation measure that often results in burial of insects, making it impossible for them to escape; this has been done to control such insects as corn borer, chinch bug, hessian fly, and other pests of field and garden crops. Another sanitation measure is burning weeds and crop remnants which harbor insects. However, this practice is generally condemned because the value of the crop remnants in soil conservation greatly offsets the possible gain in reducing an insect population. Burning destroys both beneficial and destructive insect species, as well as the nesting sites and hibernating quarters of birds and other vertebrate predators. It also contributes to greater soil erosion.

* *Bul. Ent. Soc. Amer.* 19:157–163, 1973.

Frequent cultivation will kill insect life stages such as larvae and pupae of wireworms, and prevents pests such as corn root aphids and attendant ants from becoming established. Proper timing of cultivation may also be important in the destruction of less resistant life stages. Allowing land to remain idle or fallow is another means of control; specific examples are cutworms in Canada and sand wireworms in the southeastern states.

Some problems arise from the agronomic practices of no tillage or reduced tillage. These practices are popular because they lessen soil erosion and they save time and labor costs, especially since weeds can be controlled by chemicals. However, in many instances they have contributed to greater pest problems than where conventional tillage is practiced. In the Corn Belt, wheat, grain sorghum, corn, and soybeans are especially vulnerable crops.* University of Nebraska scientists solved some of the problems by what they term "*eco-fallow*." It differs from most reduced tillage practices in that a crop is planted into the residue of a different crop. This seems to be the key to avoiding some of the pitfalls of monoculture under reduced tillage.†

Regulation of stand affects humidity and temperature in microclimates which may be favorable or unfavorable to the pest or its natural enemies.

Insect damage may frequently be lessened by adjusting the dates of planting and harvesting. By delaying the planting of corn it is possible to avoid heavy infestations of corn earworm and corn borer. Sowing wheat on the "fly-free" dates to avoid hessian fly infestation is another example. Harvesting crops, threatened with injury, at a date somewhat earlier than usual may reduce insect damage, an example being pests of legume crops grown for hay. Leafhopper damage to alfalfa is often reduced by proper adjustment of the dates of cutting. The same is true for alfalfa weevil.

Crop rotation is the most widely employed farm-management practice for the control of pests. For those pests that do not readily move from place to place it may be the only control measure needed. The concentration of one crop tends to permit a build-up of the pest population peculiar to that crop. This is due to the favorable food supply for the pest and often the lack of a constant food supply for the predators and parasites that keep it checked. The natural enemies generally require alternate hosts to increase in numbers. Diversification in farming may be justified merely because it brings about conditions unfavorable to the development of destructive insect populations. The program is more effective if practiced throughout a given area.

Other cultural practices that may be unfavorable to the development of insect pests are the choice of good seed, proved varieties, proper seed-bed preparation, proper fertilizing and soil-conservation practices.

REFERENCE: *USDA Yearbook of Agriculture*, pp. 437–440, 1952.

* *Ent. Soc. Amer.* 22:289–304, 1976.

† Phytopathology 65:1021–22, 1975; Agri-fieldman 32(2):9, 1976.

BIOLOGICAL CONTROL

Biological control is considered by many to be the most important phase of applied control. Although it does not effect immediate reduction in insect populations as do some of our latest pesticides, over a longer period of time it is more effective and more economical than chemical control. Biological control may be defined as the artificial manipulation of natural biological phenomena for the purpose of reducing or checking destructive populations of insects, other animals, or plants. It includes the use of resistant strains or varieties of plants or animals that have been developed through extensive research programs taking advantage of our knowledge of genetics. It also implies the introduction, mass rearing, and liberation of large numbers of predaceous or parasitic animals or disease-producing organisms. In other words it might be termed "biological warfare" against the enemies of man.

Frequently resistant varieties of plants have proved to be the best and most economical way of reducing or avoiding insect damage. Although not too well understood, our present knowledge indicates that resistance is due to either chemical, physical, or physiological factors, or combinations of these. By selection and hybridization experiments a number of good quality hybrid varieties of field corn resistant to European corn borer have been developed, as well as sweet corn hybrids showing a high degree of resistance to the attack of corn earworm. Resistance of Monon, Redcoat, and Reed varieties of wheat to hessian fly attack has been demonstrated. Other examples of insect control by employment of resistant varieties of plants are given in the references at the end of this chapter.

Successful control of destructive insect populations by the liberation of large numbers of their natural enemies has been demonstrated. This type of control consists of the distribution of many of the predators or parasites discussed under natural control from areas where they are abundant to areas where they are scarce. This is normally done by rearing them artificially under laboratory or insectary conditions, or collecting them in natural habitats where they are abundant. For example, hibernating lady beetles (Fig. 58) are collected in great numbers from the foothills of the Rocky Mountains, held in storage during the winter months, and then liberated in the spring and summer in regions where destructive populations of aphids and scales exist. Likewise, both native and introduced species of parasites of such insects as the gypsy and brown-tail moths, Japanese beetle, European corn borer, codling moth, and oriental fruit moth have been reared artifically by governmental or private agencies and distributed free or sometimes sold.

The introduction of the Australian or vedalia lady beetle, *Rodolia cardinalis* (Muls.), into California and Florida for controlling cottony-cushion scale in citrus groves is an outstanding example of biological control by a predator (Fig. 59). Establishment of the woolly apple aphid parasite, *Aphelinus mali* (Halde-

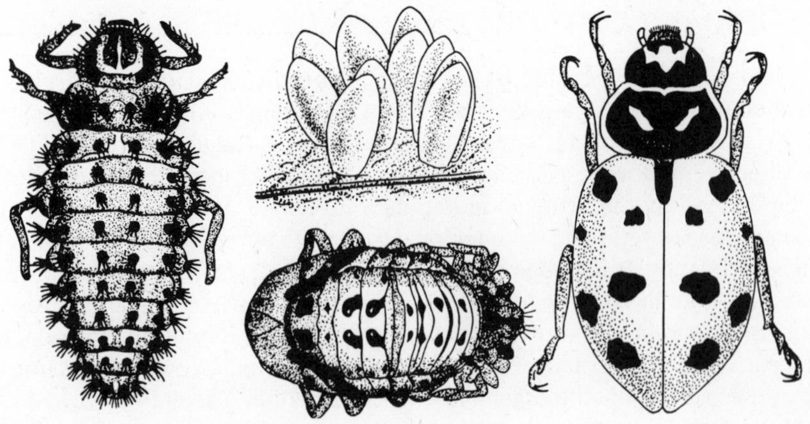

Figure 58. The convergent lady beetle, *Hippodamia convergens* Guérin; eggs, larva, pupa, and adult, family Coccinellidae; 4×. (Prince, Kansas Agr. Exp. Sta.)

man) (Fig. 280), in a number of countries is an excellent example of biological control by a parasite. Other examples of biological control by parasitic insects are the establishment of the fly, *Lydella thompsoni* Herting, and the wasp, *Macrocentrus grandii* Goidanich, for controlling the European corn borer; *Macrocentrus ancylivorus* Roh. and *M. delicatus* Cress. in peach orchards for controlling oriental fruit moth; *Aphytis holoxanthus* DeBach for control of Florida red scale; *Aphytis maculicornis* (Masi) and *Coccophagoides utilis* Doutt for olive scale, *Parlatoria oleae* (Colvée); *Aphytis lingnanensis* Compere, *A. melinus* DeBach, and *Prospatella perniciosi* Tower for California red scale; *Encarsia formosa* Gahan for whitefly; *Neodusmetia sangwani* (Rao) for rhodes grass mealybug; *Aphycus helvolus* Compere for black scale on citrus in California; *Coccophagus gurneyi* Compere and *Tetracnemus pretiosus* Timberlake for citrophilus mealybug; *Amitus spiniferus* (Brethes) and *Cales noachi* DeSantis for woolly whitefly, *Aleurothrixus floccosus* (Maskell) in southern coastal regions of California; and *Trichogramma minutum* Riley (Fig. 59), the almost universal egg parasite.

The introduction of *Cactoblastis cactorum* Berg into Australia, and other countries, to destroy the prickly pear; *Chrysolina quadrigemina* (Suffrian) into California and the Northwest for Klamath weed; *Coleophora parathenica* Meyrick into western North America for Russian thistle; *Dactylopius opuntiae* Lichtenstein into California for prickly pear are examples of biological control of weeds.

Many of our most serious pests are foreign invaders which were introduced without the parasites and predators that normally held them in check. Their change in habitat has resulted in a disturbance of the long-established balance of nature that had been reached in their original homes. To restore this balance we import many promising natural enemies after carefully investigating the

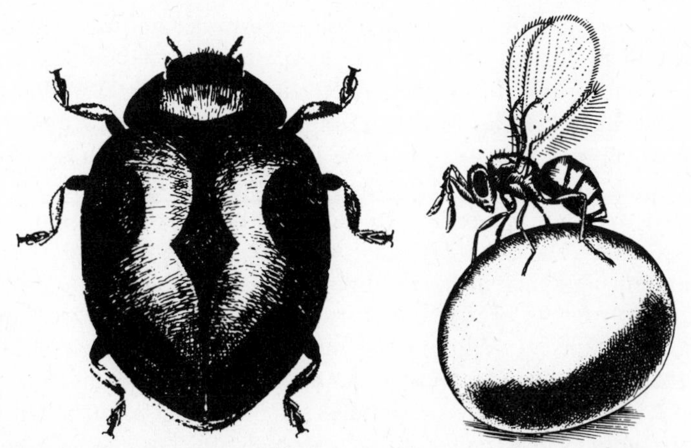

Figure 59. The Australian or vedalia lady beetle, *Rodolia cardinalis* (Muls.) (after Cook), and the most valuable egg parasite, *Trichogramma minutum* Riley; greatly enlarged. (USDA)

possibility that they may become a plant pest or may be predaceous or parasitic on beneficial insects already present. Sometimes the introduced parasites or predators have survived and become as effective in checking insect populations as in their former habitats; at other times the insects fail to become established or do not increase in numbers to the point where any benefit results from their presence. Although there are good examples of plant-feeding insects being held in check by only one parasitic or predatory species, it is usually a large and varied aggregation of such enemies, each operating when conditions change to favor it, that reduces the population most effectively.

A little-considered phase of parasite establishment is the deliberate cultivation of plants that may harbor insects serving as alternate hosts. Studies along this line are few but they may result in ways of increasing beneficial parasite populations. For a one-host parasite this method would be of no value.

A way of encouraging predators and parasites would be to refrain from practices that interfere with them. Many of our modern pesticides destroy great numbers of our beneficial insects. This has resulted in the increase of pests formerly held in check by their natural enemies. A good example was the remarkable population increase in plantfeeding spider mites and other pests following the introduction of DDT and other chlorinated compounds into various crop pest control operations. It was soon discovered these pesticides were killing the predaceous mites, and other predators and parasites that normally kept the plantfeeding species checked. This, and other known examples, definitely proves that, at times, *pesticide usage in some situations often does more harm than good*. More studies are needed concerning the habits and life histories of parasites and predators in order to integrate more effectively the joint action of both biological and chemical control measures.

Many disease-producing organisms discussed under natural control (pp. 69–71) are used in biological control. The estimated number of microorganisms known to be pathogenic to insects is 250 viruses, 80 bacteria, 460 fungi, 250 protozoa, and 20 rickettsia. Mostly Lepidoptera are affected by them. Some have not proved to be practical, but this is probably due to lack of knowledge of the exact conditions required for their effective utilization. These organisms are restricted by environmental conditions in the same way as plants or animals. For example, chinch bug fungus, *Beauveria bassiana* (Bals.), is naturally well distributed, and, when the conditions are favorable for its development, effective control results. The same is true for bacteria, protozoa, and viruses causing disease in insects. Failure to control a pest where these organisms are disseminated during periods unfavorable for their development should not be a reason for condemning them. Examples of biological control with disease-producing organisms (also called microbial pesticides) are the following: inoculating infested soil with spores of *Bacillus popilliae* Dutky and *B. lentimorbus* Dutky, the causal agents of milky disease of Japanese beetle and twelve other coleopterous grubs; spraying plants with water suspensions of the spores of the brown, red, and yellow fungi for whitefly and scale control on citrus in Florida; dissemination of the bacterium *Bacillus thuringiensis* Berliner, as control agent for 112 species of Lepidoptera, including the alfalfa caterpillar in California; successful control of the European pine sawfly in Canada and the United States by application of a virus either by hand equipment or by airplane.

Soil-conservation and farm-management practices that encourage the presence or increase the populations of insect-feeding birds and other predatory animals should be considered of value in biological control. Top feeding minnows have successfully controlled mosquito larvae when introduced into certain swampy areas, and the giant toad of Central America has controlled white grubs attacking sugarcane since it was introduced into the West Indies, the Philippines, and the Hawaiian Islands.

REFERENCES: H. D. Burges and N. W. Hussey, *Microbial Control of Insects and Mites,* Academic Press, N.Y., 1971; R. L. Ridgway, S. B. Vinson, *et al., Biological Control by Augmentation of Natural Enemies,* Plenum Press, N.Y., 1977; C. P. Clausen, *Entomophagous Insects,* McGraw-Hill Book Co., N.Y., 1940; Paul De Bach and E. I. Schlinger, *Biological Control of Insects and Weeds,* Chapman and Hall, London, 1964; R. H. Painter, *Insect Resistance in Crop Plants,* Macmillan Co., N.Y., 1951; E. A. Steinhaus, *Insect Microbiology,* Comstock Pub. Assoc., Ithaca, N.Y., 1946; E. A. Steinhaus, *Insect Pathology,* Vols. I and II, Academic Press, N.Y., 1963; H. L. Sweetman, *The Principles of Biological Control,* W. C. Brown, Dubuque, Iowa, 1958; Kans. Agr. Exp. Sta. Bul. 520, 1968; Paul DeBach, *Biological Control by Natural Enemies,* Cambridge Univ. Press, 1974; C. B. Huffaker *et al., Biological Control,* Plenum Press, N.Y., 1971; USDA Agr. Handbook 480, 1978.

LEGISLATIVE CONTROL

Legislation in connection with insect control authorizes quarantines and provides funds for their support; regulates pesticides; establishes tolerances for poisonous residues on foods; authorizes and supports extermination campaigns; and provides facilities and funds for investigational work needed to establish proper control practices.

Quarantines are designed to restrict the spread or introduction of insects from an infested region to one not yet infested, from a part of the world where a known pest exists to other parts of the world to which it is not native and where it has not become established. The efficacy of quarantines varies with the circumstances. In the absence of natural barriers an insect capable of flight may spread a certain distance each year. Such spread can seldom be limited by quarantines. Any insect may be transported, willfully or accidentally, through the activities of man, much greater distances than it would naturally migrate. Quarantines can never entirely prevent such spread, but they can do much to delay the establishment of a new pest in territory not infested. The work of quarantine officers is partly a police function; often it is largely educational in character. Officials have found that, where people are fully informed as to the purpose and need of a quarantine, the police powers need seldom be invoked and that a large measure of cooperation may be secured.

Quarantines designed to keep out of this country insects from foreign countries, other than Canada and Mexico, have a much better chance of success than those intended to restrict the movement of insects within the nation. Even though they cannot always succeed in preventing accidental introduction, the chances are reduced to a minimum through the operation of quarantines. Quarantines of this character forbid the importation of materials that may possibly carry insects, or they require that such shipments be inspected and effectively treated before being released. Much material of this type is subjected to vacuum fumigation, a most effective safeguard.

Interstate quarantines cannot operate as effectively as the international quarantines; when they are supplemented by natural barriers to the spread of insects the chances of success of interstate quarantines are greatly increased. Quarantines established by California and other far western states may, for this reason, be justified whereas similar quarantines between eastern regions would be useless impositions on commerce. Where the probable value is less than the cost of maintenance, quarantines should be discontinued. Most of the pests of Canada are also present in the United States, so that the question of international quarantines at this border has not been important. Certain insects present in Mexico constitute a menace to our citrus industry, and quarantines against them will probably have to be continued indefinitely. These quarantines seem to be effective, and the Mexican government has cooperated in the execution of the necessary measures to render them so.

Some of the insects against which domestic and foreign quarantines have been established are gypsy, brown-tail, and satin moths; boll weevil; citrus blackfly; imported fire ant; melon fly; navel orange worm and orange fruitworm; khapra beetle; pink bollworm; European chafer; alfalfa weevil; cereal leaf beetle; white-fringed beetle; vetch bruchid; Mexican fruit fly; Japanese beetle, and Mediterranean fruit fly. A more exhaustive discussion of quarantines may be found in publications on the subject.*

PESTICIDE REGULATIONS

The Federal Insecticide, Fungicide, and Rodenticide Act of 1947 (FIFRA) gave administrative authority to the USDA for registering and controlling the use of pesticides in the United States. The Federal Environmental Pesticide Control Act of 1972 (FEPCA), an amended version of FIFRA, expanded the scope of the original law and transferred authority for pesticide regulation to the Environmental Protection Agency (EPA). In 1975, significant amendments were added to FEPCA, which is the current law. The law requires that all pesticides shipped or sold in intra- or interstate commerce in the United States be registered with EPA. Before registration is granted the manufacturer must provide to EPA scientific evidence that the product (1) is effective against the pests listed on the label, (2) will not injure humans, crops, livestock, and/or cause unreasonable adverse effects on the environment when used as directed, and (3) when used according to instructions, will not leave illegal residues on food or feed.

Other important features of FEPCA-amended are: to approve all labels and graphic material on pesticide containers; to classify pesticides for general or restricted use; to certify the competence of pesticide applicators, permitting only certified applicators to use restricted pesticides; to suspend or cancel a registered pesticide if its proper use poses unreasonable adverse effects on people or the environment; to suspend immediately a pesticide that is an "imminent hazard"; to automatically review each pesticide product every five years for reregistration; to consult the Secretary of Agriculture and a Scientific Advisory Council when new EPA regulations are proposed; to solicit comments from a Scientific Advisory Council during cancellation and suspension proceedings.

Pesticide labels must show the following: brand name; type of formulation; common name; ingredient statement, amount of active and inert; net contents; name and address of manufacturer; registration and establishment numbers; signal words: danger, warning, or caution; the statement "Keep out of reach of children"; symbol, skull and crossbones; precautionary statement; environ-

* *Calif. Bul.* 553, 1933; *USDA Bur. Ent. Cir.* E–455, 1936; *Misc. Pub.* 80, 1946; *Cir.* 172, 1949; *USDA Yearbook of Agriculture,* 360–364, 1952; ARS 22–91, 1965.

mental hazards; statement of practical treatment; use classification, general or restricted; directions for use; misuse statement; re-entry statement; category of applicator; storage and disposal directions.

In addition to the federal laws each state also has laws governing pesticide use; both are applicable to users within a given state. The Cooperative Extension Service is the agency that trains users for certification in each state.

The Pesticide Amendment (Miller Bill) to the Federal, Food, Drug, and Cosmetic Act protects the public by setting tolerances for pesticide residues in their food. Tolerances are established for residues that might remain in or on harvested food or feed crops as a result of the application of a chemical for pest control. The employment of tolerances provides a method for establishing the amount of pesticide residue safely permissible on foods sold in the United States; they are expressed as parts per million (ppm), or the parts of pesticide per million parts of food product. The acceptable residue for each pesticide has been determined by careful laboratory animal feeding tests over a 2 year period. The final tolerance value has a hundredfold factor of safety below a dosage that shows no evidence of animal tissue damage. Pesticides giving any indication of being carcinogenic always have a zero tolerance. Establishment of tolerances is the responsibility of EPA; enforcing them is the responsibility of the Secretary of Health, Education, and Welfare through the Food and Drug Administration.

EPA has a wide range of responsibilities in regulating pesticides and anyone involved in the manufacture, labeling, distribution, sale, application, use, or disposal of them should contact the proper authority before proceeding in order to avoid penalties from possible violations of the law.

The Toxic Substances Control Act of 1976 includes pesticides and will be administered by EPA.

Extermination campaigns are initiated when a new pest becomes established or when a new method of eradication involving well established pests is likely to succeed. Perhaps the best example of a new method of eradicating an old pest was that of the screwworm from Florida and many other areas of the United States, except the lower tip of Texas. More examples of extermination by this method are the melon fly from Rota and the oriental fruit fly from Guam. This method is successful mainly because the normal mating habits are met for the target species in the infested areas by the great quantities of released sterile individuals, resulting in eggs that do not hatch. The chance of success is much greater when treating an island versus a country, which explains the constant reinfestation of screwworm in areas of Texas adjacent to Mexico even though releases have been made over Mexico.

Other examples of suppression and/or extermination campaigns were against the cattle tick in southern United States; Mediterranean fruit fly in Florida in 1929, again in 1956–57, and a few more times since then; date palm scale in Arizona and California; sheep scab mite and cattle grubs over large areas of the world; citrus blackfly from Key West, Florida in 1934 and southern Texas in 1955–56; and pink bollworm, Mexican bean beetle, khapra beetle, gypsy and browntail moths in several isolated locations in the United States where they have, from time to time, become established.

Extermination campaigns are costly and many entomologists feel existing programs should be critically reviewed to determine whether they should be continued or eliminated, and all new programs be thoroughly evaluated on a cost-benefit and probability for success basis before being initiated.

Research funds for insect control programs are also provided through legislation by the state and federal governments. Most of this work is done at the state agricultural experiment stations and the laboratory and field stations of the Entomology Research Branch, Agriculture Research Service, United States Department of Agriculture, and Departments of Agriculture in other countries.

6

APPLIED CONTROL: CHEMICAL

CHEMICAL CONTROL

One of the important methods of controlling insects is by means of chemicals. Application of chemicals for killing insects implies the use of insecticides. Since the word "insecticide" means insect-killer, it would seem to include any control measure that results in mortality of insects, but the usual interpretation is application of chemicals. The chemicals employed for killing mites are called *miticides* or *acaricides*. *Pesticide* is another term now widely adopted to describe any chemical used for controlling various kinds of pests, be they fungi, weeds, mites, or insects. It will be the term used in this book.

Chemicals for insect control are classified in various ways. One method of classification is based on their chemistry, that is, inorganic and organic compounds. A common method is to group them according to their type of action, for example, stomach poisons, contact poisons, fumigants, repellents, attractants, and chemosterilants.

STOMACH POISONS

A stomach poison is a material taken internally through the mouth, producing its effect by absorption from the digestive tract. Insects with chewing-type mouthparts are more often killed by these poisons. However, some flies having sucking mouthparts are killed by them. Also some chemicals when applied to plants or animals are absorbed into their systems and produce poisonous sap or blood, which will kill pests with either chewing or piercing-sucking mouthparts by stomach poison action. This group of chemicals is called *systemic poisons*. They must decompose fast enough so that unsafe residues are not present in any portion of plants or animals consumed as food or feed. Although some compounds listed as stomach poisons may have other types of action, the majority of them are primarily effective as stomach poison pesticides.

Arsenicals

The compounds of arsenic are considered to be the most important chemicals applied as stomach poisons. Although the use of these compounds has been decreasing, they still hold a place in the list of chemicals for insect control.

The arsenicals are manufactured primarily from the two oxides: arsenious oxide (As_2O_3) and arsenic oxide (As_2O_5). These materials in water form arsenious and arsenic acids, respectively. When reacted with basic compounds they form the corresponding salts. Those formed from arsenious acid are called "arsenites" and those from arsenic acid "arsenates." The arsenites are quite soluble in water, very toxic to both plants and animals, and are generally used in poisoned baits. The arsenates are the more stable product and the safer form to apply on plants.

Acid lead arsenate is the most widely used of the arsenical compounds and is one of the standard pesticides for apple orchard and shade tree pests. Other arsenicals, having limited usage, are basic lead arsenate and calcium arsenate.

Fluorine Compounds

These compounds are made from calcium fluoride and cryolite and were developed as substitutes for the arsenicals. At times they had a rather wide usage, especially during World War II, when many other pesticides were in short supply. Today they are still utilized to a limited extent. These compounds are cryolite or sodium fluoaluminate, sodium fluosilicate, and sodium fluoride.

Boron Compounds

The principal compounds in this group are *borax* and *boric acid.* Both have been used to kill house fly larvae in manure, and boric acid solution has served as a poison for cockroaches.

Antimony Compounds

The major compound is *potassium antimonyl tartrate,* or tartar emetic. It has been used in sweetened baits for control of some flies and moths and for controlling thrips on citrus.

Thallium Compounds

These are highly poisonous materials and are combined with a sweet or fatty substance for ant baits. The common forms are *thallous sulfate* and *thallous acetate.*

CONTACT POISONS

Substances that kill insects by contact are derived from many sources and act in different ways, not all of which are thoroughly understood. Generally it can be said that they enter either through the body wall or the respiratory system and thereby affect the nerve and respiratory centers and the blood stream. Contact poisons may be applied for controlling insects having any type of mouthpart. Many materials listed under this heading would also result in stomach poison action if taken internally through the mouth, but most of them affect the normal functioning of the insects before ingestion takes place; some have fumigant effects too.

Botanicals and Derivatives

Some of our most common pesticides come from plants. The seeds, flowers, leaves, stems, or roots may contain the toxic ingredients, and these are formulated for use in various ways. Although many plant poisons are quite toxic to certain insects, most of them do not leave very long-lasting residues toxic to humans and other animals. This is a desirable feature for a pesticide, especially when applied to edible crops that are about to be harvested.

Nicotine is one of the oldest materials to be used as a pesticide. It is the important toxicant of a number of alkaloids found principally in the leaves and stems of the commercial species of tobacco, *Nicotiniana tabacum* and *N. rusticum*. Nicotine is sold on the market largely as 40% *nicotine sulfate*.

Rotenone, $C_{23}H_{22}O_6$, is the principal toxic ingredient found in the roots of the leguminous plants known as derris, cubé (pronounced koo-bay), and cracca. There are other closely related compounds (rotenoids) that also possess pesticidal properties. These should be taken into consideration when formulating. The amount of these other toxicants is usually expressed as a percentage of the total extractives, using ether, chloroform, acetone, or some other solvent as the extracting agent.

Derris plants are grown in the East Indies and nearby countries, the principal species being *Derris elliptica,* and *D. malaccensis*. Cubé plants are grown primarily in South American countries, the common species being *Lonchocarpus utilis* and *L. urucu*. Cracca is found in the United States and is known as devil's shoestring, the principal species being *Tephrosia virginiana*. These plants are known as fish poisons because infusions of them in water paralyze fish. Derris was first applied in this way by the natives in the East Indies to obtain fish for food, the earliest reference to this being in 1665. The first record of its use as a pesticide was in 1848, when it was applied to control insects attacking nutmeg trees in Singapore.

Rotenone and products containing it are primarily contact poisons in their action although they have been shown to possess stomach poison properties. The residues remain toxic to insects for 3 to 5 days, but are considered nontoxic to higher animals.

Pyrethrum pesticides come from the flowers of species of plants belonging to the genus *Chrysanthemum,* the principal commercial species being *C. cinerariaefolium*. Pyrethrum has been known hundreds of years for its pesticidial properties; the finely ground flowers were commonly called "insect powder."

Gnadinger* states that it was first used as a pesticide in Persia. The flowers have been grown in various countries including Japan, United States, Brazil, Dalmatia, Ecuador, Zaire, Rwanda, Tanzania, and Kenya, the last three being the main sources of supply today. From the dried powdered flowers are extracted 6 esters, called "pyrethrins." These are: pyrethrins I and II, cinerins I and II, and jasmolins I and II. They are quite toxic to insects, causing rapid paralysis owing to their effect on the nervous system. Because of this paralytic action they are often combined with other toxicants to produce rapid knockdown. When dissolved in volatile oils they make excellent space sprays for fly and mosquito control. They are also valuable for controlling garden pests, especially when destructive populations are present near harvest, because the residue rapidly becomes nontoxic in the presence of air, moisture, and light. Pyrethrum pesticides are quite safe to use on plants or animals.

A derivative of cinerin I, named *allethrin,* has been synthesized. Although less toxic than the naturally occurring toxicants in pyrethrum, it is a more stable compound. Its preparation results in a uniform product which can be more easily standardized for commercial use than the naturally occurring pyrethrum toxicants. Other synthetic derivatives of pyrethrum are *cyclethrin, furethrin, barthrin, dimethrin, permethrin, resmethrin,* and *tetramethrin.* As a group they are called pyrethroids.

Sabadilla. These toxicants are derived from the seeds of a number of species of plants in the lily family, a common one being *Schoenocaulon officinale,* grown commercially in Venezuela. The active materials are a complex group of alkaloids related to veratrine. Sabadilla has served primarily in controlling plant-feeding bugs and as a louse powder. Exposed residues on plants quickly become nontoxic. The major objection to sabadilla pesticides is the property of being highly irritating to the mucous membranes.

Ryania pesticides are made from the ground stems and roots of a South American plant, *Ryania speciosa.* The toxic principles are complex alkaloids, the major one being *ryanodine.* These alkaloids are rather stable compounds and give longer residual toxic action than either pyrethrum or rotenone. The residue is less toxic to mammals than rotenone. Major uses of ryania have been for control of European corn borer and sugarcane borer.

Hellbore is made of ground dried roots of plants belonging to the genus *Veratrum,* the principal species being *Veratrum album,* grown in southern Europe and Siberia. The toxicants are a group of complex alkaloids. Because of high cost and variability in toxic action, this pesticide is now little used.

* C. B. Gnadinger, *Pyrethrum Flowers,* 2nd ed., McLaughlin Gormley King Co., Minneapolis, 1936; Supplement to 2nd ed., 1945; J. E. Casida, *Pyrethrum: the natural insecticide,* Academic Press, N.Y., 1973.

Oils

The word "oil" covers a wide range of liquid chemical compounds composed primarily of carbon and hydrogen. They are lighter than water, usually greasy, and soluble in such substances as toluene, carbon disulfide, carbon tetrachloride, chloroform, and ether. Oils may be classified into three major groups: *animal* and *vegetable* oils, *aromatic* or *essential* oils, and *petroleum* oils. Some examples of oils of animal origin are whale, fish, and neat's-foot oils; those of vegetable origin are soybean, castor, cottonseed, linseed, peanut, palm, corn, safflower, and olive oils. They all saponify easily with alkalies, and are little used as pesticides except in the form of soaps. Wintergreen, peppermint, citronella, camphor, eugenol, geraniol, anethol, and menthol are examples of aromatic oils. They are volatile, do not saponify in the presence of alkalies, and are not greasy. Citronella has been used as a mosquito repellent, and geraniol, anethol, and eugenol are Japanese beetle attractants. The petroleum oils are of mineral origin and are derived from sedimentary rocks. They comprise the most important group from the standpoint of insect control and are especially effective as contact pesticides for all insect life stages. Petroleum oils also serve as carriers for toxicants and often increase their effectiveness. Some products separated by fractional distillation and purification of crude petroleum, in order of volatility, are: naphtha, gasoline, benzene, kerosene, fuel oil, lubricating oils, petrolatum, paraffin wax, asphalt, tar, and pitch.

In order to understand the use of petroleum oils as insecticides certain specifications as to their properties should be considered. These specifications are based on viscosity, volatility, and degree of refinement.

Viscosity is defined as resistance to flowing and is measured by the Saybolt test, which is the number of seconds required for 60 cubic centimeters to flow through a standard opening at 100 F (37.7 C). Dormant spray oils have a Saybolt reading of 90–150 seconds and a summer spray oil of 45–85 seconds. Oils of low viscosity are safer on plant foliage.

Volatility is measured by the boiling points of the fractions as distilled. A light oil has a boiling point range of 160–300 F (71–149 C), a medium oil of 300–575 F (149–302 C), and a heavy oil from 575 F (302 C) upward.

Degree of refinement means the purity of the oil. Certain unsaturated hydrocarbons are present in oils and are removed by treatment with sulfuric acid. If such treatment does not remove any portion of the oil it would be 100% unsulfonatable (or have 100% purity). Dormant oils range from 65 to 85% in unsulfonatable residues for the "regular" type, and from 90 to 92% for the "superior" type. Most summer spray oils range from 90 to 98% in unsulfonatable residues.

Naphtha, gasoline, and benzene are not used as pesticides or solvents for pesticides because of their high volatility and flammability. Varous grades of kerosene serve as solvents for toxicants in livestock fly sprays, household pesticides, and mosquito larvicides. At one time kerosene-type oils were employed extensively for field spraying, but more recently oils having greater viscosity and lower volatility have been substituted.

Oils are injurious to plants when applied undiluted. Since they are insoluble in water they must first be made into "emulsions." This is done by mixing an emulsifier with the oil and water so that the droplets of oil remain uniformly suspended in the water until applied to the plant. There is a wide variety of substances suitable as emulsifiers, several of which will be mentioned in a later discussion. Miscible oils are solutions of oil and the emulsifier with certain other materials which make rapid emulsification possible when water is added. They contain very little if any water and therefore are not affected by freezing temperatures. They are recommended primarily for dormant use at strengths of 2 to 5% oil for controlling scale insects and the eggs of aphids and spider mites. Available are various types of commercial stock oil emulsions containing 80–90% oil, the remaining portion being water and emulsifier. They emulsify readily when mixed with water for spraying. Some have the consistency and appearance of mayonnaise, others are light-colored, highly refined oils called "white" or "summer" oil emulsions. These standard products are preferable to amateur preparations, and it is wise to follow the directions of the manufacturer since the oil content of each may vary. To avoid plant injury oil sprays must be thoroughly emulsified and should not be applied in freezing weather or on oil-sensitive plants.

Dinitrophenyl Compounds

These products are used in dormant oil sprays as ovicides for mite and aphid eggs, as blossom-thinning sprays on fruits, and as weed killers. Some of the products available are: binapacryl, dinitrocresol, dinitrocyclohexylphenol, dinocap, dinoseb, and dinobuton.

Sulfur Compounds

These have been used for controlling spider mites, chiggers, many true bugs, some leafhoppers, and livestock parasites, as well as for fungicidal purposes. One of the most valuable qualities seems to be a residual effect on insects, continuing for several days after the application of the material. *Dusting sulfur, wettable sulfur, colloidal sulfur, flotation sulfur, lime-sulfur,* and *dry lime-sulfur* are formulations generally available.

Phenothiazine

This sulfur-containing heterocyclic compound is used primarily as an anthelmintic for domestic animals. Fed in salt or mineral supplement it also controls horn and face fly larvae.

Chlorinated Compounds

Practically all the chemicals to be discussed from this point on in the book were made and developed as pesticides during and after World War II. Many of them give phenomenal control compared to those in use before this period; because of this, some pests are no longer considered difficult problems; others for which we had no control have now been successfully checked. Most of these compounds have persistent, toxic residues. This property has been described as *residual action*. Although primarily contact poisons, they also are toxic if taken into the alimentary canal, and many possess poisonous vapors.

However, continued use of most of these chemicals have resulted in pest populations tolerant or resistant to the chemical; in some instances they polluted the environment in adverse ways, and in others they killed off the natural enemies of the target pests, resulting in greater pest problems than before.

Those desiring more detailed information on toxicity, chemical, structural, and empirical formulas are referred to Kenaga and Morgan.* For pesticide names other than the approved ones used in this book, the reader is referred to the *Pesticide Handbook,* or *Farm Chemicals Handbook.*†

Some common closely related pesticides in this category include DDT, TDE, methoxychlor, dicofol, chlorobenzilate, chloropropylate, and another related group includes benzene hexachloride, lindane, chlordane, heptachlor, aldrin, dieldrin, endrin, endosulfan, toxaphene, and Strobane.

The use of many of the chlorinated pesticides has now been greatly limited or banned in the United States and other countries.

Organophosphorus Compounds

Some of this group were discovered in chemical laboratories in Germany near the end of World War II, and have since been developed as pesticides. Many possess an extremely high order of toxicity to insects and mites as well as to other forms of animal life; some have low mammalian toxicity. A person's respiratory tract, eyes, and skin must be protected during their application, as they are toxic if swallowed, inhaled, or absorbed through the skin. They are cholinesterase inhibitors, and therefore the symptoms of poisoning are those of marked parasympathetic stimulation. Nausea, vomiting, diarrhea, cramps, blurred vision, feeling of pressure under the breast bone, sweating, excessive salivation, headache, and muscular tremors are indications. If any of these symptoms develop during application of the compounds, operations should cease and a physician be consulted.

Some common chemicals in this category are: tepp, trichlorfon, naled, dichlorvos, fonofos, malathion, ethion, parathion, leptophos, fensulfothion, temephos, carbophenothion, crotoxyphos, dioxathion, phosmet, chlorpyrifos,

* *Commercial and Experimental Organic Insecticides,* Ent. Soc. Amer., College Park, Md., 1978.

† *Pesticide Handbook-Entoma,* 1978, Ent. Soc. Amer., College Park, Md. 20740; *Farm Chemicals Handbook,* 1979, Meister Publishing Co., Willoughby, Ohio 44094.

diazinon, and azinphosmethyl. Those systemic in animals are coumaphos, famphur, fenthion, ronnel, crufomate, and dimethoate; those systemic in plants are: mevinphos, phosphamidon, demeton, phorate, disulfoton, oxydemeton-methyl, dimethoate, and phosalone. Great care should be exercised in using any of these chemicals.

Sulfonate, Sulfone, Sulfide, and Sulfite Compounds

Most of the chemicals in this group are effective only as miticides. The miticides are: ovex, fenson, tetradifon, tetrasul, chlorbenside, and Omite.

Carbamate Compounds

This group of pesticides became popular with the relatively low mammalian toxicity and broad-spectrum characteristics of carbaryl or Sevin. Carbamates are inhibitors of cholinesterase enzymes, and atropine in an antidote. In addition to carbaryl the group includes aminocarb, bendiocarb, bufencarb, dimetilan, formetanate, methiocarb, mexacarbate, promecarb, propoxur, aldicarb, carbofuran, dimetan, and methomyl, the last four being systemic in plants. Some of the aforementioned chemicals have high mammalian toxicity.

Other Compounds

Oxythioquinox is useful for controlling insects, mites, and fungi in orchards. Cyhexatin is primarily a miticide. These compounds do not fit into any of the aforementioned categories.

Wetting, Spreading, Emulsifying, Adhesive, and Synergistic Agents

Combined with other toxicants, these adjuvants insure better coverage of the sprayed surface by reducing surface tension, assist in holding the toxicant to the surface, serve as formulating agents (emulsifiers), and increase effectiveness of pesticides generally. Many of them, if used alone and in concentrated form, will kill insects primarily by contact action. Wetting and spreading agents are usually not good adhesive materials and often reduce the deposit of toxicant. Most oily substances increase spray deposit.

Some examples of wetting, spreading, and emulsifying agents are soaps, saponins, and detergents or surfactants. Perhaps the detergents are now the most important group of chemicals that reduce surface tension in hard or soft water and have excellent emulsifying properties as well. They may be milled with pesticidal dusts to make wettable powders, that readily disperse into water suspensions. Commercial products such as Dreft, Tide, Tergitol, and Dupanol are examples of the sodium alkyl sulfates, and Igepon, Vatsol, Ultrawet, and Nekal are examples of sodium alkyl sulfonates. The sodium alkyl aryl sulfonates are represented by Santomerse, Nacconol, Breeze, and Fab. Closely related products are the Tritons, said to be sulfonated ethers and alcohols. The esters, formed by reacting fatty acids with the hexahydric alcohols, sorbitol and mannitol, are excellent emulsifying and wetting agents and possess considerable pesticidal activity as well. Commercial products called "Spans" and "Tweens" are examples.

More common adhesive agents include gelatin, glue, rosin residue, gum karaya, and gum arabic.

Synergism, where pesticides are involved, is usually defined as joint action of two materials resulting in a total toxic effect greater than the sum of their toxic effects when applied separately. Because this joint action usually is not known, the term "activator" has been suggested as a better term. Most materials having this property were developed in trying to prolong the toxicity of pyrethrum pesticides. Some of the promising synergists or activators are: piperonyl butoxide, piprotal, sesamin, sesamex, sesamolin, sulfoxide, propyl isome, and MGK 264. Any material that possesses the properties indicated could be placed in this category.

Dust Diluents

Large amounts of relatively inert materials serve as diluting agents or carriers in the preparation of pesticides to be applied as dusts. Because some of these materials possess pesticidal properties,* they can be grouped with contact poisons. In addition to being toxic many diluents are also abrasive in action; by abrading the waxy layers of the exoskeleton water loss results. Some also have affinity for water and this combined action may cause death by desiccation. Important properties to be considered in the selection of a diluent are: compatibility, pH, cost, stability, particle size, pesticidal activity, availability, absorptivity, abrasive action, and adhesiveness. Alkaline materials, like hydrated lime, are used in nicotine dusts to liberate the nicotine but should not be combined with certain chemicals because their active principles are unstable in the presence of alkalies. Pyrethrum marc, tobacco dust, and sulfur are diluents having considerable pesticidal activity to some insects and mites. Watkins and Norton† have classified pesticide diluents into two groups, botanicals and minerals. Examples of botanical diluents are the following: wheat and soybean flours, tobacco dust, pryethrum marc, and walnut shell flour. The mineral group includes such compounds as calcium carbonate, magnesium carbonate, gypsum, kaolin, sulfur, bentonite, pyrophyllite, diatomaceous earth, hydrated lime, talc, attapulgite, and silica aerogel.

FUMIGANTS

Fumigants are chemicals employed in gaseous form for killing insects or related pests, and are adapted primarily for use in enclosed spaces. Buildings to be

* *J. Econ. Ent.*, 40:215–219, 1947.
† *J. Econ. Ent.*, 40:211–214, 1947.

fumigated should be thoroughly sealed. Rapid and effective penetration of a fumigant results when liberated in a tight chamber under partial vacuum. This is called *vacuum fumigation* and is employed by quarantine officers to insure imported products being insect-free. Nursery stock, foods, and many products of commerce are also treated in this manner. Fumigants may be applied in the open for controlling insects on vegetation, but in such situations more effective results are secured if the gases are confined, even for a very short period of time. This is done by employing gastight tents, covers, or hoods. Citrus trees are fumigated while enclosed in tents; tarpaulins or covers confine soil fumigants.

The choice of a fumigant depends on several factors: cost of material and application, toxicity to insects, safety to the operator, possibility of fire hazard, plant and animal toxicity, possible bleaching effects to household items, tarnishing effects to metals, ease of application, penetration power, persistence and effects on materials to be treated, such as the germination of seeds, milling or baking properties of flour, tainting or preserving qualities of foods. Regardless of the fumigant chosen, its characteristics should be thoroughly understood, and all necessary precautionary measures taken to safeguard human life.

Some common fumigants are: dichloromethane, carbon tetrachloride, ethylene dichloride, tetrachloroethylene, dichloropropenes mixture, chloropicrin, methyl bromide, ethylene dibromide, ethylene oxide, propylene oxide, hydrocyanic acid, acrylonitrile, o-dichlorobenzene, p-dichlorobenzene, naphthalene, carbon disulfide, sulfur dioxide, sulfuryl fluoride, sodium methyldithiocarbamate, and carbon dioxide. Combinations of these chemicals are sold under various trade names.

REPELLENTS

Materials that prevent migration, oviposition, or feeding of insects are described as being repellent or having repellent properties. This effect is made manifest by the negative response of insects to the physical and chemical properties of these various substances when applied to plants, animals, buildings, soils, or anything to be protected from insect attack. Many of the better repellents are chemicals producing lethal effects, if the concentration is high enough and if the insects are unable to move away rapidly. Repellency to insects must be determined by experimentation since many substances obnoxious to man are not necessarily disagreeable to insects.

Bordeaux mixture is very repellent to potato flea beetles, leafhoppers and psyllids; *sulfur* applied to the body is a chigger repellent; *cattle fly sprays* protect livestock from annoyance and loss of blood; *coal tar creosote* repels chinch bugs, termites, powder post beetles, and wood borers; *paradichlorobenzene, oil of cedar,* and *moth balls* are repellent to carpet beetles, clothes moths, and other fabric pests; *pine tar oil* prevents screwworm flies from ovipositing in or near wounds of animals; *indalone, dimethyl carbate, ethyl hexanediol, deet, butoxy polypropylene glycol, dibutyl phthalate, dimethyl phthalate, benzyl benzoate,* and MGK-326 are repellents for mosquitoes, flies, ticks, and chiggers. During World War II, army uniforms were impregnated with dibutyl phthalate, benzyl benzoate, or dimethyl phthalate for protection against chiggers and related mites. Many of the aforementioned stomach, contact, or fumigant materials also possess some repellent properties.

ATTRACTANTS

An agent or substance that has the power of eliciting a positive response by insects is called an attractant. Most of the better-known stimuli in this category are olfactory, and therefore chemical, but some insects are also attracted by certain wavelengths of sound and light. Male mosquitoes are attracted to sound waves in the range 430–540 vibrations per second. Adult Japanese beetles are collected in greater numbers from yellow traps.

The major types of chemical attraction are sex, oviposition, and food. Sex attractant secretions in insects are called *ectohormones* or *pheromones*. It is well-known that odors from unfertilized female insects will attract the males of the same species. For some this attraction is from great distances. These odors, infinitesimal in amount, bring immediate reaction from the males. A concentration of less than 10^{-14} micrograms of female sex pheromone wafted into a cage of American cockroaches will start them running around madly, vibrating their wings and trying to mate with each other. Similar behavior has been observed in over 200 other species studied.

Some of the better known male sex attractants are *anisylacetone* and *cue-lure* for melon fly; *cue-lure* for Queensland fruit fly, *gyplure* and *disparlure* for gypsy moth, *grandisol* and *grandlure* for boll weevil, *gossyplure, hexalure* and *propylure* for pink bollworm, *riblure* for red-banded leafroller, *muscalure* for house fly, *trimedlure* for Mediterranean fruit fly, and *methyl eugenol* for oriental fruit fly. Because the flies devour methyl eugenol when attracted to it, some believe it to be a food lure too.

Food attraction odors and oviposition stimuli also have been clearly demonstrated. Protein hydrolysates are attractants to fruit flies and have been widely used in poisoned bait sprays. Other specific examples are *anethol, geraniol,* and *eugenol* in Japanese beetle traps; fermenting sugars and syrups attract many moths and butterflies; *metaldehyde* lures slugs and snails to calcium arsenate baits; *formalin* in low concentrations is both attractive and toxic to house flies; sheep blow flies are attracted to *keratin;* codling moth to *anethol;* oriental fruit moth to *terpinyl acetate;* fruit flies to *ammonium carbonate;* tobacco hornworm moths to *isoamyl salicylate;* and European chafer to *butyl sorbate*. There are many substances possessing attractant properties to insects. This is an area where further research should contribute much valuable information on new approaches to insect control. At present, attractants play a part in control primarily by luring the insects into traps or to poisoned food. The male annihilation technique depends on luring all males of a species to a sex attractant and poisoning them, leaving the females to die unfertilized. Attractants are also useful tools for survey work, and helpful in determining thresholds of pest species.

CHEMOSTERILANTS AND GROWTH REGULATORS

Chemicals capable of causing sterility in insects or other organisms are called chemosterilants. They may act in one of three ways: (1) inhibit the production of egg or sperm, (2) cause the death of the egg or sperm after having been produced, (3) bring about multiple dominant lethal mutations or severely damage the chromatin or genetic material in the egg or sperm. In the third type of action the egg and sperm remain alive and motile, but the zygotes, if formed, do not develop into normal offspring. This is a desirable type of action in insect population control because the affected males compete with normal males for the available females; thus the mating requirements of the females are met as if mating with normal males.

Some of the more widely tested chemosterilants are aphamide, apholate, metepa, methiotepa, morzid, tepa, thiotepa, tretamine, and methotrexate. A wide variety of insects and mites have been employed in tests with these compounds with some favorable results. However, activity of chemosterilants is not restricted to insects; for instance, some of them have been used in palliative treatment of cancer and many are analogs of carcinostatic compounds. These compounds must be carefully evaluated as to their effects on many other organisms.

At present no chemosterilants are recommended for insect or mite control. If any effective compounds are found safe for control purposes they will probably be used in combination with various attractants and poisoned baits. Potentially there should be greater population reduction with chemosterilants than with pesticides having similar intensity of action.

Growth regulators in insects and other arthropods are of three types: juvenile hormones (JH), which affect metamorphosis and prevent adulthood, ecdysone hormones, which regulate molting, and anti-chitin compounds, which prevent chitin formation at molting.

Juvenile hormones are not persistent in the environment nor toxic to mammals. They are biodegradable but chemical instability seems to be a problem associated with their field use. One (methoprene or Altosid) has been registered by EPA for control of mosquitoes. Another (kinoprene or Enstar) provides effective control of aphids and whiteflies in glasshouses. The ecdysones are complex steroid molecules. They are active in insects at less than 1 ppm of body weight. Very little information is available concerning their effects on higher animals. Anti-chitin compounds are specific for arthropods and considered non-hazardous to use around mammals. Dimilin or diflubenzuron has been registered for gypsy moth control and temporary permits granted for boll weevil, mosquito larvae, and Clear Lake gnat. This entire area of study requires additional research.

RESISTANCE TO PESTICIDES

Continuous use of the same pesticide on a given population of insects or mites often results in the development of a tolerance or resistance to that chemical. This is not a new discovery because it was known that San Jose scale had developed resistance to lime-sulfur sprays in the state of Washington by 1908. From 1908 to 1945 only 13 species of insects and ticks were recorded as being resistant to insecticides. Since 1945, with the increased use of the new synthetic organic pesticides, the number of resistant species has risen to over 200 in 1965, and is still rising.

Resistance within a pest population may be innate, behavioral, biochemical, and/or physiological but it becomes more pronounced with pesticide usage because the weak or susceptible individuals are eliminated from the population, leaving the tolerant individuals to reproduce. In general, pests with short life cycles develop resistance faster because the selection process proceeds more rapidly. Detoxication mechanisms within their bodies enable the animals to survive, and these survival factors can be passed on to subsequent generations through genetic material. The phenomenon of cross resistance also magnifies the problem. There is evidence too that a relationship exists between the diet of the animal and the degree of resistance. This phenomenon is not clearly understood.

Even though recommended dosages of pesticides are used, uneven distribution and dissipation of the toxicant between successive applications occur, enabling some of the tolerant individuals to survive. To cope with the resistance problem it is suggested that pesticide applications be as thorough as possible. When evidence of resistance is indicated a change should be made to another pesticide in a chemical group entirely different from the one being applied. This alternation of different pesticides in a spray program has been termed "rotational spraying." It is not the solution to the resistance problem but is of some help while study continues. Perhaps the best solution to the problem is to keep pesticide applications to a minimal number and substitute cultural, mechanical, and biological control methods in the management programs.

REFERENCES

Brown, A. W. A., *Insect Control by Chemicals,* John Wiley and Sons, New York, 1951.

Brown, A. W. A., "The Challenge of Insecticide Resistance," *Bull. Ent. Soc. Amer.* 7:6–19, 1961.

Brown, A. W. A., "Mechanisms of Resistance Against Insecticides," *Ann. Rev. Ent.* 5:301–326, 1960.

Bushland, R. C. *et al.,* "Development of Systemic Insecticides for Pests of Animals in the United States," *Ann. Rev. Ent.* 8:215–238, 1963.

Casida, J. E., "Mode of Action of the Carbamates," *Ann. Rev. Ent.* 8:39–58, 1963.

Dethier, V. G., *Chemical Insect Attractants and Repellents,* Blakiston, Philadelphia, 1947.

Gilbert, L. I., *et al., Juvenile Hormones,* Plenum Pub. Co., New York, 1976.

Gordon, H. T., "Nutritional Factors in Insect Resistance to Chemicals," *Ann. Rev. Ent.* 6:27–54, 1961.

Hall, Stanley, *et al., New Approaches to Pest Control and Eradication,* Advances in Chemistry ser. 41, American Chemical Society, Washington, D.C., 1963.

Jacobson, Martin, *Insect Sex Attractants,* John Wiley and Sons, New York, 1965.

LaBrecque, G. C., *et al.* Principles of Insect Chemosterilization, Appleton-Century-Crofts, New York, 1968.

Matsumura, Fumio, Toxicology of Insecticides, Plenum Pub. Corp., New York, 1975.

Metcalf, R. L., *Organic Insecticides: Their Chemistry and Mode of Action,* Interscience, New York, 1955.

Perry, A. S., and M. Agosin, *The Physiology of Insecticide Resistance by Insects* (*Physiology of Insecta* 6:3–121, M. Rockstein, ed.), Academic Press, New York, 1974.

Saunders, D. S., *Insect Clocks,* Pergamon Press, New York, 1976.

Shorey, H. H., "Pheromones and Behavior," *Ann. Rev. Ent.* 18:349–380, 1973.

Shorey, H. H., *Animal Communication by Pheromones,* Academic Press, New York, 1976.

Smith, C. N., G. C. LaBrecque, and A. J. Borkovec, "Insect Chemosterilants," *Ann. Rev. Ent.* 9:269–284, 1964.

7

PESTICIDE TOXICITY, FORMULATIONS, COMPATIBILITY, APPLICATORS, AND SAFETY

PESTICIDE TOXICITY

The extent or degree to which a chemical may be poisonous to humans is determined experimentally with test animals, usually rats, mice, or guinea pigs. Obviously, humans cannot be test animals. These animals are fed the chemical to determine (1) *oral toxicity;* their skin is exposed to the chemical to determine (2) *dermal toxicity;* or they are placed in a chamber where they breathe the chemical vapors to determine (3) *inhalation toxicity*. In addition, the effect of the chemical as an irritant to the eyes is determined.

Toxicity is expressed as LD_{50} (lethal dose) or LC_{50} (lethal concentration), which means the amount or concentration of a toxicant required to kill 50% of the test animal population under a standard set of conditions. Toxicity values of pesticides, based on a single dose, are expressed in milligrams of pesticide per kilogram of body weight of the test animal (mg/kg) or in parts per million (ppm). LD_{50} and LC_{50} values are useful in comparing the toxicity of the active ingredients of different pesticides. The lower the LD_{50} value of a pesticide the greater its toxicity.

Pesticides are categorized on the basis of their LD or LC values. Those considered *highly toxic* on the basis of oral, dermal, or inhalation toxicity data must have the signal words **DANGER** and **POISON** (in red letters) and a skull and crossbones displayed on the package label. Pesticides considered to be *moderately toxic* must have the signal word **WARNING** displayed on the product label; those classified as either *slightly toxic* or *relatively nontoxic* are required to have the signal word **CAUTION** on the label (see pesticide regulations p. 82).

Toxicity values are divided into four categories; the lower the LD_{50} value the more dangerous the pesticide. The homeowner may use pesticides in categories 3 and 4 with or without protective equipment according to the label recommendations.

Categories	Required Label Word	Acute Toxicity-LD_{50} mg/kg Oral	Acute Toxicity-LD_{50} mg/kg Dermal	Probable Oral Lethal Dose for 150 lb. Man
1. Highly toxic	Danger-Poison	0.1–50	0.1–200	A few drops
2. Moderately toxic	Warning	50–500	200–2000	A teaspoonful to 2 tablespoonfuls
3. Slightly toxic	Caution	500–5000	2000–20,000	Over 1 oz. to 1 pint
4. Relatively nontoxic	Caution	over 5000	over 20,000	Over 1 pint to 1 lb.

Source: Modified from New York State Insecticide, Fungicide, and Herbicide Recommendations, Cornell University, Ithaca, N.Y., 1979.

Toxicity of chemicals is also expressed as acute or chronic. The effect produced after administering or exposing a test animal to a single dose of a toxicant is called acute toxicity, whereas chronic toxicity is the effect produced after the test animal is administered or exposed to repeated small doses of a toxicant over a long period of time. Oral, dermal, and inhalation toxicities may also be described as acute or chronic.

PESTICIDE FORMULATIONS

Chemicals for insect control have to be properly formulated and diluted before they can be employed efficiently and without injury to animals or plants. Pure or undiluted, usually technical grade, chemicals are highly toxic to both animals and plants, and often their physical properties are such that they cannot be diluted easily. Without proper dilution excessive quantities are generally used, resulting in higher insect control costs. In addition, applications heavier than necessary leave longer-lasting residues that are hazardous in many situations, especially on edible crops.

The common types of pesticide formulations are: emulsions, low volume concentrates (LVC), ultra low volume concentrates (ULVC), solutions, wettable powders (WP), flowables, aerosols, dusts, granules, microencapsulation, and baits.

Emulsions are made by dissolving the toxicant and an emulsifying agent in an organic solvent thus making emulsifiable concentrates. These are diluted with water and applied as sprays. At recommended dosages they can be used with safety on most plants; however, higher dosages may result in phytotoxicity. There is little evidence of residue following application.

LVC are concentrated emulsifiables mixed with a small quantity of water and applied in reduced amounts. They conserve water and costly pesticide. In conventional spraying (dilute) large quantities of spray material (350–450 gals./A) are applied to the drip point (run-off). This falls to the ground and is wasted. With LVC only small quantities (40–150 gals./A) are applied. For success the application equipment must be very carefully calibrated. Phytotoxicity may also be a hazard. These same precautions should be followed in using the *ULVC* formulations which are undiluted emulsifiables applied at one-half gal. or less per acre.

Solutions consist of molecular mixtures of the toxicant with a solvent such as water, or with petroleum distillates such as kerosene. They may be applied by spraying, by dipping or impregnating, or by brushing. Oil solutions are phytotoxic and often constitute a fire hazard. The so-called soluble powders are in reality solutions.

Wettable powders are toxicants absorbed or adsorbed on powders that can be readily mixed with water because a wetting or conditioning agent has been added. These form suspension-type sprays and must be constantly agitated to give uniform coverage. At recommended dosages they are quite safe on most plants and are generally less costly. Abrasive action occurs on spray equipment much more rapidly when wettable powders are employed. *Flowables* are simply finely ground thick suspensions of a toxicant in water ready for further dilution.

Aerosols are air suspensions of solid or liquid particles of ultramicroscopic size which remain suspended for long periods. They may be the liquefied gas type utilizing the propelling agents Freon-12 or methyl chloride, or smokes, or mechanically generated oil clouds or fogs. The liquefied gas type is especially adaptable for the household and glasshouse and has given favorable results in outdoor situations where there is very little air movement. Mechanical fogging of alleys, wooded areas, swamps, and other breeding areas for fly and mosquito control has been practiced with some degree of success. Because of atmospheric contamination aerosols containing fluorocarbon propellants have been banned by EPA.

Dusts are usually made by diluting the toxicant with finely divided, ground plant materials, such as marc, tobacco dust, and walnut shell flours, or with minerals, such as talc, clays, and sulfur. Coarse or *granular* formulations of dusts are made for broadcast treatments by aircraft and also for ground applicators. Dusts drift; granules do not.

Microencapsulation is placing pesticides in pinhead-sized capsules having an inert coating that disintegrates gradually, releasing the chemical slowly over a long period of time. This new development makes hazardous chemicals safer to handle, prolongs the activity of both residual and nonresidual pesticides, and lowers environmental contamination problems. The coating must break down fast enough to provide protection from pests but slow enough to remain effective during the entire growing season. Much is yet to be learned about this formulation.

Baits consist of some attractive substance in combination with a poisonous chemical. Both liquid and dry baits have been formulated; which are employed depends on the pest and the circumstances under which it is to be used.

Miscellaneous formulations are liquefied gases, large pills or boluses, pesticidal shampoos, and pesticide-fertilizer mixtures.

COMPATIBILITY

Often the spraying or dusting operation is designed to control, by one application, several insect pests and some plant diseases as well; thus a mixture of materials can be used advantageously. In making such mixtures some knowledge of the effect of the different materials on each other is essential. If no deleterious effects are produced when different insecticides, fungicides, and sometimes fertilizers are combined and applied simultaneously, they are said to be compatible. If the combination causes any kind of injury to the host, or lowered effectiveness of the components, or if it results in any other adverse effects, it is said to be incompatible. Compatibility charts are obtainable from pesticide companies and from some agricultural magazines. The Meister Publishing Co., Willoughby, Ohio, 44094, prints a new chart in color each year, at a nominal cost.

PESTICIDE APPLICATORS

Numerous types of equipment are available for the application of pesticides. They range from small, hand-operated outfits to the large-capacity power machines capable of treating great acreages in a short time. From the selection available a careful choice should be made either of equipment designed for a special area of use or a versatile piece of equipment which may fill many needs. Multiplicity of function is an important factor to consider.

Many types of hand-operated equipment are illustrated (Fig. 60) for home and garden vegetation purposes. Small bellows-type dusters and hand atomizers or aerosols may be useful for the application of pesticides to control household pests or insects on house plants. Simple plunger-type dusters and sprayers of slightly greater capacity may suffice for small gardens. For the average suburban garden, dusters of 1- or 2-quart capacity, or the compressed-air or knapsack-type sprayers, may fulfill the needs. The duster may be preferred because it is more quickly made ready for treatments, and these are likely to be more frequent and timely, although dusts are generally more expensive. If more than

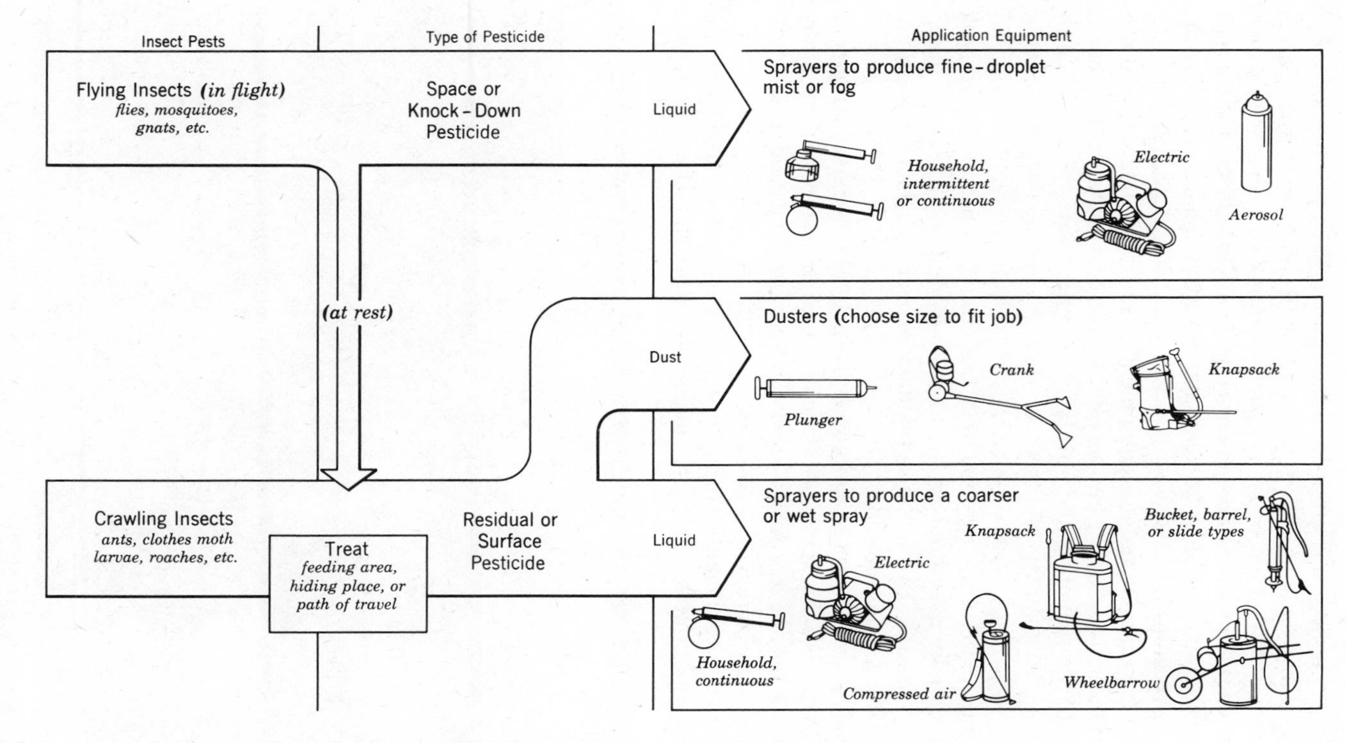

Figure 60. Guide to selection of hand equipment for application of pesticides. (USDA)

one pesticide is applied, it is often advantageous to have two or more of these small dusters. Farm gardens of large size and small commercial truck gardeners need equipment with added capacity, such as crank or traction-type dusters, and gasoline engine-powered equipment (Fig. 61). Regardless of the appliance selected, to be effective it must be adaptable for thorough coverage in order to give satisfactory control.

Larger farm operations require equipment of much greater size and power. Traction outfits may be used, but gasoline-powered equipment is more often employed with the mounting and nozzle arrangement varying according to the crop (Fig. 62). Attachments for treating several rows at once are commonly provided. Low-pressure, low-volume, or concentrate sprayers are popular for field crops and are often tractor-mounted (Fig. 63). High-pressure, high-volume sprayers are equally useful but necessitate the hauling of larger quantities of a diluted spray mixture. This is a disadvantage especially when applications are to be made after a rainy period.

Commercial orchardists must employ equipment of large size and capacity if the crop is to receive the maximum benefit. The optimum period for the pesticide application may be less than a week, and part of the period may be unsuitable for treatment. Sprayers capable of covering the entire acreage within a period of 2 or 3 days are highly desirable. This means that the large grower must have several small outfits, requiring many operators, or a few high-capacity machines.

Modern hydraulic orchard sprayers have a tank capacity of 300 to 500 gallons

Figure 61. A small power sprayer designed for home and garden use. (Courtesy of FMC Corporation)

Figure 62. A tractor-mounted high-clearance sprayer applying pesticide to corn. (USDA)

or more and are capable of delivering 30 to 50 gallons of spray per minute at a pressure of 400 to 800 pounds per square inch. Multiple cluster nozzles, large-capacity, single-nozzle spray guns, or other nozzle groupings which permit the effective use of such quantities of sprays are available.

Fixed booms or spray masts are also made by some manufacturers. They are preferable where the trees are not large, as in peach and cherry orchards. This

Figure 63. A tractor-mounted low-gallonage concentrate sprayer treating 8 rows of cotton simultaneously. (USDA)

Figure 64. A speed or air-blast sprayer in an orange grove. (USDA)

type of equipment is rapidly being replaced by the air blast- or mist blower-type machines which require less manpower. One of these can replace up to three large-capacity hydraulic sprayers at a great saving in labor costs. The air blast sprayer consists of a powerful motor-driven fan which discharges a large volume of rapidly moving air behind a series of nozzles supplied by a low-pressure, low-volume pump (Fig. 64). The air stream helps break up the liquid into small particles and carries them to the surface to be treated.

Truck- or trailer-mounted mist blowers are available for pesticide application to very tall trees or for treating a large number of row crops from the side or mid-field (Fig. 65). Attachments are provided on some machines to apply both dusts and sprays simultaneously. These are called spray dusters or vapor dusters. They offer the added advantage of increased adherence of dust particles to foliage, and a reduction in the amount of inert carrier

Airplanes and helicopters (Fig. 66) have been effective in the application of sprays and dusts in many situations. Their great advantage is speed and the fact that ground conditions do not hamper them. Cranberry bogs, rice fields, and forested areas, all inaccessible to ground sprayers, can now be treated with ease from the air. Aircraft is widely employed for pesticide applications on cotton, corn, legumes, and some vegetable crops. Similar treatment of orchards generally has not given as good results as when ground equipment is used. This is often true of other crops as well, but, in spite of these limitations, aircraft has a definite place in agricultural pest control. The first aerial application of pesti-

Figure 65. An air row crop sprayer applying chemicals from mid-field. A line of sprayers with swaths from 30 to 90 feet is available. (Courtesy of FMC Corporation)

cides in the United States was lead arsenate to control catalpa sphinx caterpillars.

EQUIPMENT CARE

Application equipment needs to be cleaned and oiled systematically. The corrosive and abrasive properties of spray and dust materials make it essential for tanks and dust hoppers to be cleaned after each period of use. Pumps and accessories should be washed thoroughly with water to reduce corrosion and nozzle clogging troubles. Tanks, pumps, and engines must be drained during freezing weather. Lubrication of all wearing parts, according to the instructions provided by the manufacturer, will contribute to smooth operation and longer useful life. Replacement of worn parts as needed is good economy.

PESTICIDE SAFETY

Since pesticides vary greatly in their relative toxicities, it is wise to become acquainted with the product being applied. *Read carefully all labels on the package before opening and follow the directions.* The following general precautions should be taken in handling and applying any pesticide. Avoid breathing dusts, spray mists, or vapors. For protection against the more toxic pesticides it is necessary to wear specially designed respirators or masks. Avoid skin contact with all pesticides especially those formulated in oil. If this should

Figure 66. Aerial spraying of forested areas by means of a helicopter. (USDA)

accidentally occur, wash immediately and thoroughly with soap and water. Wear protective clothing during each application and wash these garments thoroughly before wearing again. Do not smoke, eat, or drink until the face and hands have been washed thoroughly after handling pesticides. Exercise care in transporting pesticides and decontaminate all accidental spills. Store all pesticides where they cannot be reached by children. Check re-entry time to treated areas on the pesticide label. Dispose of empty pesticide containers in the manner described on the package (see pesticide labels p. 82).

THE METRIC SYSTEM

The following conversion factors will be helpful in changing from the English to the metric system and vice versa. A fully charged calculator is also an aid.

English/Metric		Metric/English
	Length	
Inches × 25.4 = millimeters		Millimeters × .04 = inches
Feet × 30.5 = centimeters		Centimeters × .033 = feet
Miles × 1.6 = kilometers		Kilometers × .6 = miles
	Weight	
Ounces × 28.35 = grams		Grams × .0353 = ounces
Pounds × .4535 = kilograms		Kilograms × 2.2 = pounds
Short tons × .9 = metric tons		Metric tons × 1.1 = short tons
	Liquid	
Ounces × 29.5 = milliliters		Milliliters × .035 = ounces
Pints × .47 = liters		Liters × 2.1 = pints
Quarts × .95 = liters		Liters × 1.06 = quarts
Gallons × 3.8 = liters		Liters × .26 = gallons
	Area	
Acres × .4 = hectares		Hectares × 2.5 = acres
	Temperature	
Fahrenheit − 32 ÷ 1.8 = Celsius		Celsius × 1.8 + 32 = Fahrenheit

REFERENCES

Anon, *Pesticide Index,* 5th ed., Ent. Soc. Amer., College Park, Md., 1976.

Anon, *Farm Chemicals Handbook,* Meister Publishing Co., Willoughby, Ohio 44077, 1979.

Colberg, W. J., and R. B. Widdifield, "Guide for Mixing Insect and Weed Control Chemicals," *N.D. Agr. Ext. Leaflet,* 1954.

Garman, P., "A Study of Spray Machines in Connecticut Orchards," *Conn. Agr. Exp. Sta. Bul.* 567, 1953.

Fulton, R. A., F. F. Smith, and R. L. Busbey, "Respiratory Devices for Protection Against Certain Pesticides," *USDA, ARS*-33-76-2, 1966.

Nelson, R. H., "Conversion Tables and Equivalents for Use in Work Relating to Insect Control," *USDA* E-517, revised 1952.

Neal, Jr., John W., *A Manual for Determining Small Dosage Calculations of Pesticides and Conversion Tables,* Ent. Soc. Amer., College Park, Md., 1974.

USDA, "How to Spray the Aircraft Way," *Farmers' Bul.* 2062, 1960.

USDA, "Power Sprayers and Dusters," *Farmers' Bul.* 2223, 1966; "Safe Use of Agricultural and Household Pesticides," *Agr. Handbook* 321, 1967; "Aerial Application of Agricultural Chemicals," *Agr. Handbook* 287, 1976. Available from Supt. of Documents, Wash., D.C.

8

ENVIRONMENTAL MANAGEMENT

The importance of a management approach to crop production, including protection from pests, is emphasized as world populations expand and demands for quality food and fiber increase.

Insect management is not an entirely new concept or an unknown operation of man. One of the oldest insect management operations in the world is the art of beekeeping. Other insect management operations involve not only the rearing of predators and parasites for liberation in biological control, but also the mass rearing of many insect species for classroom demonstration and teaching purposes, for use as test animals in evaluating new chemicals as pesticides, for cleaning vertebrate skeletons in museums, for pet food and bait stores, and for sterilization by chemicals or radiation and liberation in autocidal suppression and extermination campaigns.

Pest management has become popular terminology in recent years but when scrutinized carefully and thoroughly it has all the components of much of what was known, taught, and practiced years ago as "Good Farm Management." Over the years good judgment and sensible decisions about pest control have been shunted to the use of chemicals alone. This has caused many of the problems society is now trying hard to correct by proper management. Actually most of the operations involve manipulating the environment rather than the pest. Planting resistant varieties of corn, wheat, or alfalfa for limiting populations of European corn borer, Hessian fly, and spotted or blue alfalfa aphids; rotating crops to control northern and western corn rootworms; flooding land to control wireworms or draining wet lands to control mosquitoes; delaying the planting of wheat and corn to avoid infestations of Hessian fly and European corn borer; early cutting of alfalfa for hay to prevent large populations of potato leafhopper from developing; withholding a pesticide application in order that parasites and predators might develop unhampered; or applying a pesticide to poison the food plants of a pest are not management of the pests but management of the crops and environment of the pests. More descriptive terminology of what actually happens in such programs would be environmental management.

Integrated environmental management involves programs having a multidimensional approach to control. Entomologists should integrate their cultural, mechanical, biological, chemical, and legislative operations with the plant breeder, plant pathologist, agronomist, soil scientist, economist, and farm manager in employment of reasonable, compatible, and economical techniques to control a wide variety of pests. No one control measure will suffice because of the adaptive powers of insects, mites, nematodes, weeds, and plant pathogens and because of the many variables related to season, location, and cropping patterns. To manage several pests at the same time will require complex programs. The ultimate goal is to hold pest populations below damaging economic levels without serious disruption of the agroecosystem.

Environmental management is a system of encouraging not only the use of predators, parasites, and resistant varieties, but also crop rotation, and adjusted planting and harvesting dates, and at the same time abandoning the use of pesticides alone as the sole controlling agent. Complete reliance in the past on chemicals for pest control on such crops as cotton, citrus, deciduous fruits and vegetables resulted in insect and/or mite pests becoming difficult to control because of the development of resistance. Also, pesticides destroyed many beneficial predators and parasites often leading to a rapid resurgence of target pests and buildup of populations of previously unimportant species to economically harmful levels. Good environmental management involves not the complete elimination of pesticides but the wise use of them, thus reducing environmental hazards.

Effective environmental management programs are composed of at least three important elements:

1. Diagnosis of the pest problem by field checking, scouting, and trapping.
2. Determination of the necessity for pest suppression based primarily on economic damage thresholds.
3. Deployment of pest suppression by the most appropriate techniques available.

To determine when pest suppression is necessary, it is important to understand current terminology. *Economic injury level* is the lowest population density that will cause economic damage. The level will vary depending on the crop, season, area, and a human scale of values. *Economic threshold* is the density at which control measures should be initiated to prevent an increasing pest population from reaching the economic injury level. *Population equilibrium position* is the average density of a population over a long period of time in the absence of permanent environmental change (see Fig. 67).

As shown in Fig. 67, as long as the population of pests and their natural en-

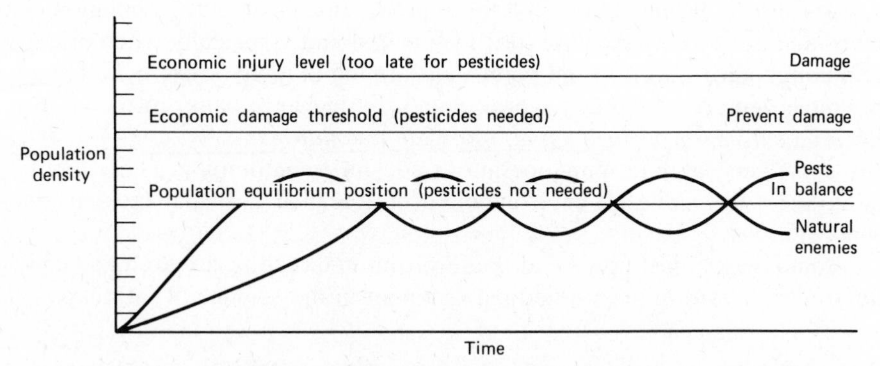

Figure 67. Schematic diagram showing need for pest control based on population dynamics. (Modified after Stern 1959)

emies are in the population equilibrium position but below the economic threshold, it is not necessary to apply chemicals. Should the pest population increase to the economic threshold, a chemical is usually needed to prevent economic loss. It is often too late to apply a chemical to avert loss when the economic injury level is reached. Special effort should be made to keep populations of pests and their natural enemies in the equilibrium position or in balance with each other below the economic threshold.

Successful environmental management programs must cover a wide area and enlist the cooperation of everyone concerned. Information is the key ingredient to proper development of these programs. One of the problems facing them is the extremely difficult task of establishing valid economic injury levels and economic thresholds for a wide range of crops and pests. At present relatively few of these levels have been established in the United States and Canada or for world agriculture and the main reason is lack of information. The programs are so new that suitable experimental techniques necessary for research on economic injury levels and economic thresholds for pest complexes have not been determined. Until substantial research effort is initiated the necessary knowledge for proper implementation will be lacking.

Nevertheless, with the knowledge at hand progress is being made, on a limited scale, for certain pests of cotton, corn, wheat, citrus, alfalfa, soybeans, tobacco, rice, deciduous fruits, and vegetables. Simple scouting counts as well as sophisticated computer plant pest models have now been implemented in the management programs of several states. As new information becomes available programs are modified accordingly.

The major effort in training scouts, checking fields, trapping, reporting, and assembling the information for a given area and crop is being coordinated by the Cooperative Agricultural Extension Service in the United States. This effort is being financed by federal grants and by participating growers. Some feel the growers should ultimately bear all these production costs but, as inflation continues and other costs escalate, this idea is in doubt, especially when prices of farm commodities drop to the point that farming in many areas is no longer a profitable venture. A hungry world cannot be fed by a bankrupt agriculture. *When incentive for profit is gone, motivation collapses.*

In some very large farming operations consulting entomologists are hired by the growers to handle all environmental or so-called pest management decisions.

Entomologists must now be at the forefront in directing research and education towards restoration of good management in the science of farming.

REFERENCES

Apple, J. L., and Ray F. Smith, *Integrated Pest Management,* Plenum Publishing Corp., New York, 1976.

Asquith, Dean, and Richard Colburn, ''Integrated Pest Management in Pennsylvania Apple Orchards,'' *Bul. Ent. Soc. Amer.* 17:89–91, 1971.

Bay, Ovid, ''Extension Has a Key Role in Pest Management,'' *Extension Service Review,* USDA Agr. Ext. Service, Wash., D.C., 1972.

Benham, Jr., G. S., ''Pest Management: A Student Commentary on Contemporary Problems,'' *Bul. Ent. Soc. Amer.* 20:319–326, 1974.

Blair, B. D., ''Corn Management Programs,'' *Proc. N.C. Branch Ent. Soc. Amer.* 29:36–38, 1974.

Campbell, F. L., and F. R. Moulton, ''Laboratory Procedures in Studies of the Chemical Control of Insects,'' Amer. Assoc. for Advancement of Science Pub. 20, Wash., D.C., 1943.

Campbell, R. W., and M. W. McFadden, ''Design of a Pest Management Research and Development Program,'' *Bul. Ent. Soc. Amer.* 23:216–220, 1977.

Geier, P. W., ''Management of Insect Pests,'' *Ann. Rev. Ent.* 11:471–490, 1966.

Glass, E. H., *et al.,* ''Integrated Pest Management: Rationale, Potential, Needs, Implementation,'' *Ent. Soc. Amer. Pub.* 75-2, College Park, Md., 1975.

Hall, D. C., ''The Profitability of Integrated Pest Management: Case Studies for Cotton and Citrus in the San Joaquin Valley,'' *Bul. Ent. Soc. Amer.* 23:267–274, 1977.

Holdsworth Jr., R. P., ''Aphids and Aphid Enemies: Effect of Integrated Control in an Apple Orchard,'' *J. Econ. Ent.* 63:520–535, 1970.

Holdsworth Jr., R. P., ''Codling Moth Control as Part of an Integrated Program in Ohio,'' *J. Econ. Ent.* 63:894–897, 1970.

Jeppson, L. R., ''Pest Management in Citrus Orchards,'' *Bul. Ent. Soc. Amer.* 20:221–222, 1974.

Knipling, E. F., ''Use of Population Models to Appraise the Role of Larval Parasites in Suppressing *Heliothis* Populations,'' USDA *Tech. Bul.* 1434, 1971.

Ledbetter, R. J., ''Principles and Concepts of Pest Management, Environmental Entomology Slide Set,'' USDA Extension Service, Wash., D.C., 1972.

Lincoln, Charles, and C. R. Parencia Jr., "Insect Pest Management in Perspective," *Bul. Ent. Soc. Amer.*, 23:9–14, 1977.

Metcalf, R. L., and W. H. Luckmann, *Introduction to Insect Pest Management,* John Wiley and Sons, New York, 1975.

Musick, G. J., *et al.,* "Crop Protection With Reduced Tillage Systems," *Bul. Ent. Soc. Amer.* 22:289–304, 1976.

National Academy of Sciences, *Pest Control: An Assessment of Present and Alternative Techniques,* 5 vols., Wash., D.C., 1976.

Needham, J. G., *et al., Culture Methods for Invertebrate Animals,* Comstock Publishing Co., Ithaca, N.Y., 1937.

Newton, C. M., and W. A. Leaschner, "Recognition of Risk and Utility in Pest Management Decisions," *Bul. Ent. Soc. Amer.* 21:169–172, 1975.

Pate, T. L., J. J. Hefner, and C. W. Need, "A Management Program to Reduce Cost of Cotton Insect Control in the Pecos Area," Tex. Agr. Exp. Sta. Misc. Pub. 1023, 1972.

Rabb, R. L., and F. E. Guthrie, "Concepts of Pest Management," Proc. Conf. N.C. State Univ., March 25–27, Raleigh, 1970.

Sanchez, F. F., "The Current Status of Rice Pest Management in the Philippines," *Bul. Ent. Soc. Amer.* 23:29–31, 1977.

Shepard, H. H., *Methods of Testing Chemicals on Insects,* Vol. I, 1958; Vol. II, 1960, Burgess Publishing Co., Minneapolis.

Singh, Pritham, *Artificial Diets for Insects, Mites, and Spiders,* Plenum Publishing Corp., New York, 1977.

Smith, E. H., and David Pimentel, *et al., Pest Control Strategies,* Academic Press, New York, 1978.

Stern, V. M., R. F. Smith, Robert van den Bosch, and K. S. Hagen, "The Integrated Control Concept," *Hilgardia* 29:81–101, 1959.

Stern, V. M., "Economic Thresholds," *Ann. Rev. Ent.* 18:259–280, 1973.

Van Gundy, S. D., "Nematode Pest Management," Proc. Nat. Ext. Pest Management Workshop, March 13-15, Baton Rouge, La., 1973.

Watson, T. F., L. Moore, and G. W. Ware, *Practical Insect Pest Management,* Freeman Co., San Francisco, 1975.

Watson, D. L., and A. W. A. Brown, *Pesticide Management and Insecticide Resistance,* Academic Press, New York, 1977.

9

PESTS OF VARIOUS CROPS AND TURF

Many of our major insect pests attack a variety of hosts. To avoid repetition this chapter covers some of them in detail; the reader will be referred to this section in later chapters. A few additional pests of turf are appropriately discussed in the next chapter.

GRASSHOPPERS OR LOCUSTS

ORDER ORTHOPTERA, FAMILY ACRIDIDAE OR LOCUSTIDAE

Grasshoppers or locusts are one of the most serious insect pests of crops in many areas of the world. Although there are many species responsible for much local destruction, only a few are important, causing about 90% of the grasshopper damage to cultivated crops in North America. These are the differential grasshopper, *Melanoplus differentialis* (Thos.); the red-legged grasshopper, *M. femur-rubrum* (DeG.); the two-striped grasshopper, *M. bivittatus* (Say); the Rocky Mountain grasshopper, *M. spretus* (Walsh); the migratory grasshopper, *M. sanguinipes* (F.); the devastating grasshopper, *M. devastator* Scudder; and the clear-winged grasshopper, *Camnula pellucida* (Scudder).

Other species of importance in some areas are the following: *Melanoplus dawsoni* (Scudder); *M. packardii* (Scudder); the Nevada sage grasshopper, *M. ruggelsi* Gurney; the High Plains grasshopper, *Dissosteira longipennis* (Thomas); the Carolina grasshopper, *D. carolina* (L.); the big-headed grasshopper, *Aulocara elliotti* (Thomas); the lubber grasshopper, *Brachystola magna* (Gir.); and the American grasshopper, *Schistocerca americana* (Drury). The desert locust, *S. gregaria* Forsk. is perhaps the most destructive and widespread. Size varies according to species, ranging from about 25 mm to 7 cm.

The Rocky Mountain grasshopper is considered the important migratory species in the United States and Canada. Other species develop what appears to be at least a partial migratory habit in seasons of great abundance and short food supply. Not all the conditions responsible for grasshopper outbreaks are thoroughly understood. They are, however, known to be climatic and biological in character. Prediction of probable infestations is done by making careful counts of eggs present in the areas concerned a few months in advance of hatching. This information helps in planning an effective control campaign.

The injurious species of grasshoppers agree in the main features of their cycle of development. Eggs of all are deposited in late summer in elongated masses or pods inserted in the soil (Fig. 68) of grain, grass, or other crops, but sod clumps receive the greater number of eggs of the differential, two-striped, and clear-winged species. These pass through the winter, and, on hatching in the spring, the young seek food in the immediate vicinity. As they increase in size and food becomes scarce, migration to other sources takes place. After molting 5 or 6 times, during a period of 40 to 60 days, the adults appear and continue feeding until cold weather kills them. If their food supply becomes exhausted, adult migration takes place by flight. Great hordes of migrating grasshoppers have been observed in other countries, but this habit is infrequent in North America. Oviposition begins shortly after the adults appear and, because of irregularities in hatching time and variation in rate of development, continues for nearly 3 months. The egg pods of the clear-winged and migratory species usually contain 15 to 20 eggs, and those of the two-striped and differential species have 50 to 75. A single female lays 200 to 400 eggs over a period of several weeks. The number of pods laid by each female varies according to the species, the food supply, and weather conditions. Some species have more than 1 generation per year in the South.

Grasses are said to be the normal food of grasshoppers, but they feed on nearly any kind of vegetation and when numerous may destroy every green plant within their line of march. After regular crops and all grasses are destroyed they frequently feed on the bark and leaves of deciduous trees.

Fluctuation in population of the hoppers indicates variation in the operation of the natural controlling influences. A succession of warm dry seasons is favorable for the development of serious grasshopper outbreaks. Cool wet weather with long periods of high humidity is unfavorable for grasshopper development. A season with the following characteristics would favor heavy mortality: weather warm enough in early spring to cause considerable premature hatching, followed by temperatures low enough to prevent normal growth; a short period of hot weather late in the spring to insure complete hatching of the remaining eggs, with long periods of cloudy, wet weather favorable to grasshopper diseases; a cool summer and early fall to delay maturity and to shorten the time for egg-laying.

Several important natural enemies reduce grasshopper populations. Blister beetles, ground beetles, anthomyiid flies, and bee flies lay their eggs in the soil

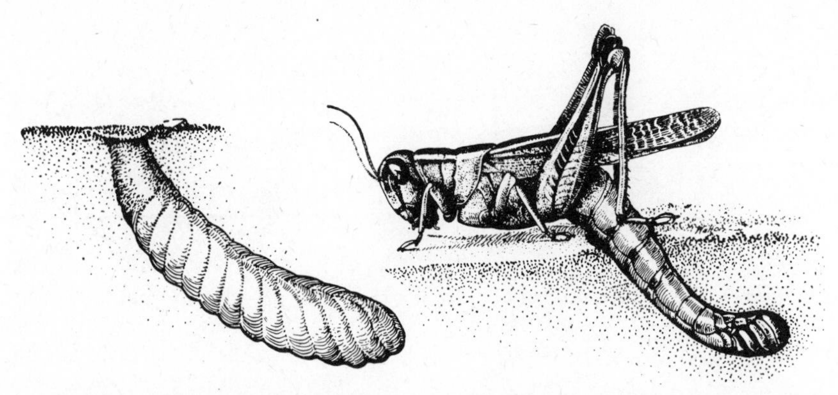

Figure 68. Drawing of egg pod in soil (left) and adult *Melanoplus bivittatus* (Say) in act of ovipositing (right). (Walton, USDA)

near or in the egg pods. Their larvae have been known to destroy from 40 to 60% of all grasshopper eggs laid over large areas. Many flesh flies, including *Blaesoxipha kellyi* (Aldrich), deposit active larvae upon grasshoppers, often while in flight, resulting in high mortality.

Threadworms are frequently important as internal parasites. Many birds and mammals, as well as predatory insects, feed on grasshoppers. When the climatic factors are favorable for the development of the common and widely distributed fungi, *Beauveria bassiana* (Bals.) and *Empusa grylii* Fresenius, grasshoppers become diseased and die.

Applied control is practiced in many ways. Collection of adults by hopperdozers, or plowing and disking to destroy overwintering eggs are of some value in control but are not widely practiced at the present time. Planting resistant crops is of value where feasible. Sorghums have been known to be resistant to grasshoppers since 1877. The more recently developed chemicals have revolutionized grasshopper control. Application of recommended materials to the hatching areas gives effective kill. An infestation is considered economically significant if 8 grasshoppers per square yard are found.

REFERENCES: *USDA Tech. Bul.* 774, 1941; 1165, 1957; 1167, 1958; *Farmers' Bul.* 2193, 1975; *Agr. Inf. Bul.* 287, 1976; *Canada Dept. Agr. Pub.* 1036, 1958; *Ann. Ent. Soc. Amer.* 64:574–580, 1971.

MORMON CRICKET

Anabrus simplex Haldeman, Family Tettigoniidae

The Mormon cricket is not a true cricket but is more closely related to katydids and meadow grasshoppers. Its common name comes from the fact that an outbreak of the insect did vast damage during the early days of the Mormon colony in Utah. According to legend this outbreak was suppressed by sea gulls, and a

Figure 69. Mormon crickets; *a*, female; *b*, male; natural size. (USDA)

monument was erected and unveiled in Temple Square, Salt Lake City, October 1, 1913 in commemoration of this event. It is an insect of long standing as a pest, but one which has usually been of minor importance. Presumably the conditions that favor great grasshopper outbreaks also are responsible for a marked increase and abundance of this pest.

Mormon crickets are found in the intermountain states of Colorado, California, Kansas, Wyoming, Montana, Idaho, Utah, Nevada, Oregon, Washington, North Dakota, Minnesota, New Mexico, Nebraska, and bordering Canadian Provinces.

The flightless adults resemble crickets (Fig. 69) and reach a length of 25 mm. They deposit eggs singly or in small groups in the soil during late summer, and hatching takes place the following spring. Feeding and 7 successive molts follow, the adult stage appearing again in early summer 60 to 90 days after hatching. All plants in the infested area may be attacked, but grasses are preferred, and, as the food supply becomes exhausted, migration to other areas takes place. The injury to plants is due to removal of vegetative and reproductive parts, breaking of the stalks by the weight of the feeding insects, or girdling of twigs.

Natural control results from a variety of predators such as wasps, birds, and small mammals. Applied control is accomplished by chemical treatment of infested areas.

REFERENCES: *USDA Tech. Bul.* 161, 1929; *Tech. Bul.* 1202, 1959; *Cir.* 575, 1940; *Tech. Bul.* 866, 1943; *Farmers' Bul.* 2081, 1962; *Ida. Ext. Bul.* 100, 1936.

FIELD CRICKET

Gryllus spp. Family Gryllidae

Although this cricket is often regarded as a household pest of minor importance, it can cause serious damage to garden and field crops. Widely distributed in both North and South America, it has been found damaging cotton seedlings,

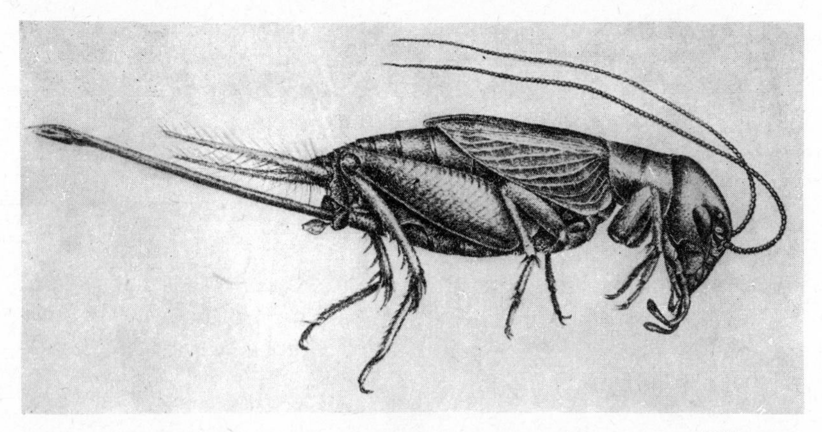

Figure 70. The field cricket, *Gryllus* spp., a general feeder; 3× (Gilbertson, S. D. Agr. Exp. Sta.)

seeds of alfalfa and grain crops, strawberries and garden vegetables, especially tomatoes. In houses they may eat holes in clothing and furnishings, or feed on stored tubers or fruits.

In the latitude of South Dakota crickets produce one generation per year, overwintering as eggs in the soil. There may be as many as 2 or 3 generations per year in the Imperial Valley in California and in some of the Gulf states, where they are usually active year round. In the northern states, egg-hatching begins in late May or during June, and after 8 to 12 nymphal instars the dark brown-to-black adults (3 cm long) (Fig. 70) appear by the latter part of July. Mating, feeding, and ovipositing occur thereafter and continue until cold weather kills the adults.

Control measures in the field are the same as for grasshoppers. In houses, crickets can be effectively eliminated by applying an approved pesticide.

REFERENCES: *USDA Tech. Bul.* 642, 1939; *Agr. Inf. Bul.* 237, 1962; *S.D. Bul.* 295, 1935.

WIREWORMS

ORDER COLEOPTERA, FAMILY ELATERIDAE

Wireworms are the familiar shiny, slender, cylindrical, hard-bodied, wirelike, yellow-to-brown larvae, found at all times of the year and in almost any kind of soil. When fully developed they vary in length from 12 to 35 mm, depending on the species. Adults of these larvae are known as click beetles, so named because of the habit of snapping and flipping their bodies into the air when turned upside down. The tan-to-black beetles vary from 7 to 25 mm in length, but the most common pest species are nearly 13 mm (Figs. 71 and 72).

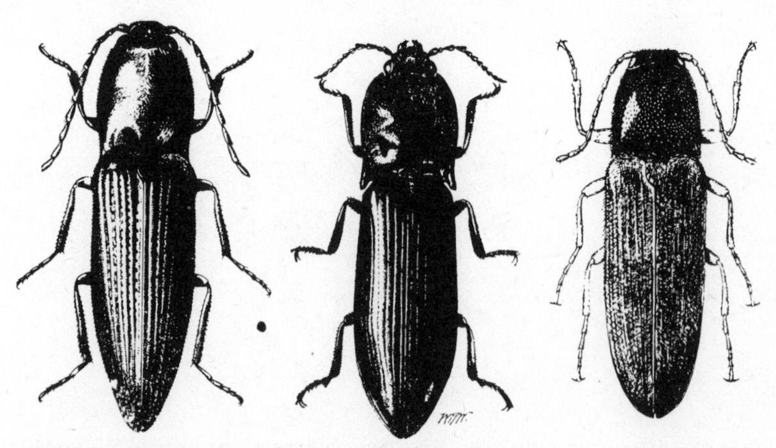

Figure 71. Wireworm adults, from left to right: the sand wireworm, *Horistonotus uhlerii* Horn; the Great Basin wireworm, *Ctenicera pruinina* (Horn); and the eastern field wireworm, *Limonius agonus* (Say); enlarged. (Walton, USDA)

The food plants of wireworms are many and the estimated loss to farmers because of their feeding runs into several million dollars annually. They injure crops by devouring seeds in the soil, by cutting off small underground stems and roots, and by boring in the larger stems, roots, and tubers. No crop is known to be entirely immune to their attack and such crops as grass pastures, potatoes, tobacco, carrots, onions, corn, sweet potatoes, cotton, lettuce, melons, peas, beans, cowpeas, and sugar beets are particularly susceptible.

Some of the important species are the wheat wireworm, *Agriotes mancus* (Say); sugar-beet wireworm, *Limonius californicus* (Mann.); Pacific Coast wireworm, *L. canus* LeC.; eastern field wireworm, *L. agonus* (Say); western field wireworm, *L. infuscatus* Mots.; Columbia Basin wireworm, *L. subauratus* (LeC.); sand wireworm, *Horistonotus uhlerii* Horn; prairie grain wireworm, *Ctenicera aeripennis destructor* (Brown); Puget Sound wireworm, *C. aeripennis aeripennis* (Kirby); Great Basin wireworm, *C. pruinina* (Horn); Gulf wireworm, *Conoderus amplicollis* (Gyll.); tobacco wireworm, *C. vespertinus* (Fab.); and southern potato wireworm, *C. falli* Lane.

Wireworms have varied life histories but many of them agree in some points. The sand wireworm and Gulf wireworm overwinter as larvae and complete their life cycles in 1 year; other species complete their life cycles in 2 to 5 years and overwinter as eggs, larvae, or adults in the soil. Usually the eggs are laid singly in soil 1 to 6 inches deep during the spring or early summer by either newly emerged overwintering adults or, as with the sand wireworm, by adults which have developed from the overwintering larvae. Hatching takes place in 2 to 4 weeks, and the young larvae begin working their way through the soil in search of food. The heaviest natural mortality occurs during this early period of

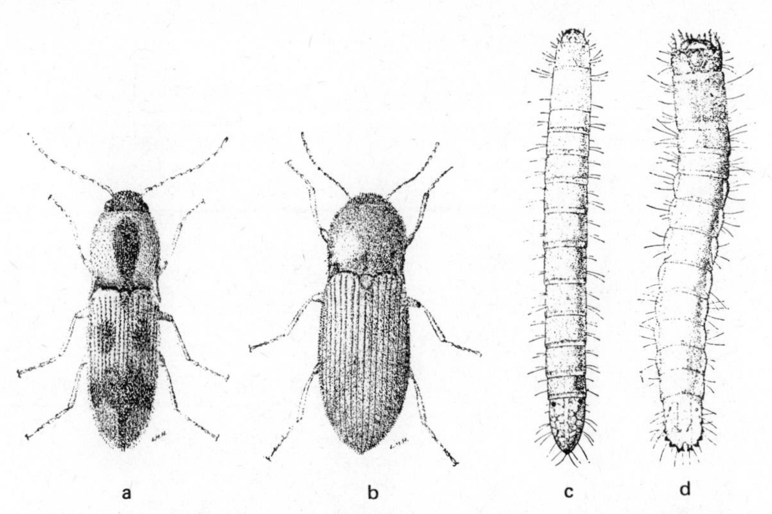

Figure 72. Adults and larvae of wireworms: *b* and *c*, the wheat wireworm, *Agriotes mancus* (Say); *a* and *d*, a common species known as *Drasterius dorsalis* (Say); enlarged. (Forbes.)

life because of unsatisfactory environmental conditions and lack of proper food. Little damage is caused to crops the first season from the 2 to 5-year species, but those that survive attain sufficient size to do considerable damage the succeeding years of larval life. There is always an overlapping of generations, wireworms of all sizes and ages being present in the soil throughout the year.

A parasitic nematode, a fungus, and birds are the major natural enemies. Some wireworm larvae are cannibalistic, which reduces populations; sudden cold periods in winter also kill larvae.

Cultural practices to reduce wireworm damage are an important phase of control. Plowing or cultivating infested soil in late summer or in autumn kills or exposes to predators the larval, pupal, and adult stages. In the irrigated areas of the Pacific Slope states regulation of the irrigation water can control wireworms. Either flooding or drying of the soil during the summer season will kill many wireworms, especially the current season's brood, thus reducing their numbers and subsequent damage. Withholding water in irrigated fields of such crops as alfalfa, which can withstand this treatment, is another way of reducing wireworm populations. Rotation of crops is also a recommended practice since continuous growing of vegetable or field crops, especially potatoes or wheat, on the same land has a tendency to increase wireworm numbers. A cropping plan, as shown in Fig. 73, is suggested for the sand wireworm. In other areas, alfalfa in the rotation is unfavorable to wireworms, and red clover and sweet clover sometimes favor wireworms. The most effective rotation system depends on the area concerned and has to be determined accordingly.

Chemicals are effective for wireworm control and the increased yields after

Two-Year Cropping Plan

First-year planting			Second-year planting	
November to late February or early March	March to July 10	July 10 to October	November to late February or early March	March to September
Winter Cover	Spring	Summer	Winter Cover	Main Crop
Oats or rye	Grain followed by stubble	Cover crop, hay, or sweetpotatoes	Austrian winter peas or vetch	Cotton or corn

Three-Year Cropping Plan

First-year planting			Second-year planting		Third-year planting	
November to late February or early March	March to July 10	July 10 to October	November to late February or early March	March to September	October to February or March	March to October
Winter Cover	Spring	Summer	Winter Cover	Main Crop	Winter Cover	Spring
Oats or rye	Grain followed by stubble	Cover crop, hay, or sweetpotatoes	Austrian winter peas or vetch	Cotton or corn	Austrian winter peas or vetch	Velvetbeans, tomatoes, or early watermelons

Figure 73. Cropping plans for reducing the damage done to crops by the sand wireworm. The 2-year plan is for land having a heavy wireworm infestation, and the 3-year plan is for land having a wireworm infestation known to be light. (USDA)

their application often justify their cost. They may be mixed with the top several inches of the seedbed, used as seed treatments or as soil fumigants.

REFERENCES: *USDA Farmers' Bul.* 2220, 1977; *Leaflet* 212, 1941; 534, 1965; E-765, 1948; E-786, 1949; EC-6, 1948; EC-19, 1951; *Farmers' Bul.* 2040, 1952; *Tech. Bul.* 1172, 1958; 1443; 1971; *J. Econ. Ent.*, 29:288–296, 1936; 38:643–645, 1945; 45:548–549, 1952; 46:1075–1083, 1953; 51:690–692, 1958; *Canada Dept. Agr. Pub.* 942, 1955; 979, 981, 1956.

FALSE WIREWORMS

ORDER COLEOPTERA, FAMILY TENEBRIONIDAE

False wireworms range from Canada to Texas but are more commonly found in the drier wheat-growing areas of the region between the Mississippi River and the Pacific Coast. Wheat is a favored host plant, but other grasses, cotton, legumes, sugar beets, and garden crops may be attacked. The sprouting seeds and young seedlings are eaten by the larvae, and the adult beetles eat various plants above ground but to a lesser degree. The greatest damage occurs in autumn.

Species recorded as causing damage are *Eleodes opacus* (Say), *E. suturalis* (Say) (Fig. 74); *E. tricostatus* (Say); *E. hispilabris* (Say); *Blapstinus sub-*

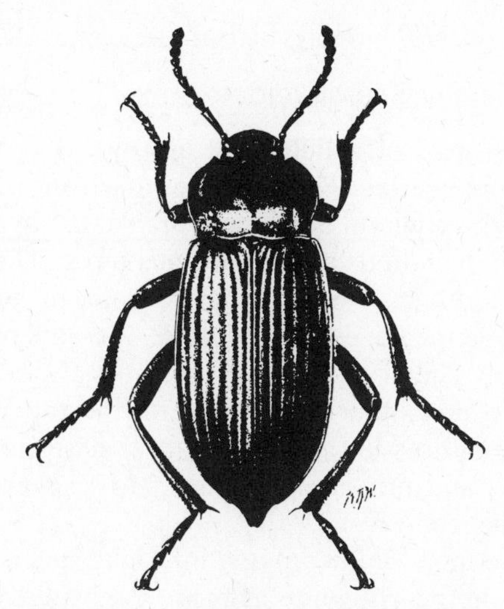

Figure 74. A false wireworm, *Eleodes suturalis* (Say), family Tenebrionidae; enlarged. (Walton, USDA)

striatus Champion; *B. discolor* Horn; *B. pimalis* Casey; *B. histricus* Casey; and *Embaphion muricatum* Say. Populations fluctuate considerably; these insects are usually of minor importance.

The dark or black adults, called darkling beetles, resemble ground beetles in general aspect and measure scarcely 5 to 35 mm in length, depending on the species. The larvae resemble wireworms in shape and coloration but can be distinguished from them by the prominent, thickened antennae and larger front legs. False wireworms are also more active than wireworms.

The life cycles of the different species vary to some extent. *Eleodes opacus* has a one-year cycle; *E. hispilabris,* in the Pacific Northwest, may require as long as 2 years. The eggs are laid in spring and/or late summer and fall, and the most active feeding period of the larvae is immediately after hatching. The partly grown larvae overwinter, pupate in the spring, with adult emergence continuing over a period of several weeks. The overwintering stage may also be the adult since they live for as long as 3 years.

The control measure commonly practiced is to follow a rotation so that corn or other cultivated crops occur for 2 years between plantings of wheat. Clean culture and accumulated soil moisture associated with summer fallowing tend to reduce injury. Pesticides are seldom needed but seed treatment as recommended for wireworms has given satisfactory control.

REFERENCES: *Ark. Bul.* 185, 1923; *Ida. Res. Bul.* 6, 1926; *Mont. Bul.* 269, 1932; *Kans. Agr. Exp. Sta. Bul.* 367, 1954; *J. Econ. Ent.* 12:212, 1919; 30:670–675, 1937.

WHITE GRUBS OR MAY BEETLES

Phyllophaga spp., Family Scarabaeidae

White grubs or grubworms of which there are over 100 different species, are larvae of May and June beetles. These beetles are from 12 to 25 mm in length, vary from light to dark brown in color, and are robust in form (Fig. 75). They eat the leaves of both deciduous and coniferous trees. The larvae or grubs are white, curve-bodied, with brown heads and 3 pairs of legs. The hind part of the abdomen usually appears darker because the soil particles inside show through the body wall. The grubs feed on the roots of bluegrass (Fig. 76), timothy, corn, soybeans, and other crops, and on the tubers of potatoes. They often ruin bluegrass pastures in the north central states and may become serious as pests of lawns and nursery plantings. Most severe damage occurs on crops that follow grass sod.

The life cycle of the more abundant and injurious species extends over three years. In late spring the pearly white eggs are deposited 1 to 8 inches deep in the soil. Since the beetles are attracted to trees, they tend to concentrate their egg-laying in the higher portions of timothy and bluegrass sod near wooded tracts. About 3 weeks later the eggs hatch into young larvae that feed on living roots and decaying vegetation throughout the summer. In the autumn they migrate downward and remain inactive until the following spring when they return to feed on the roots of plants near the soil surface. The greatest damage occurs during this period. The next autumn they again go deep into the soil, returning near the surface in the spring of the third year when they feed, as before, until about June. Then oval earthen cells are made and pupation follows. The adult beetles emerging from these pupae a few weeks later remain in the pupal cells through the winter and emerge from the soil the following year in May and June when feeding, mating, and egg-laying take place. The new adults are attracted in great numbers to lights at night. In the northern range of these

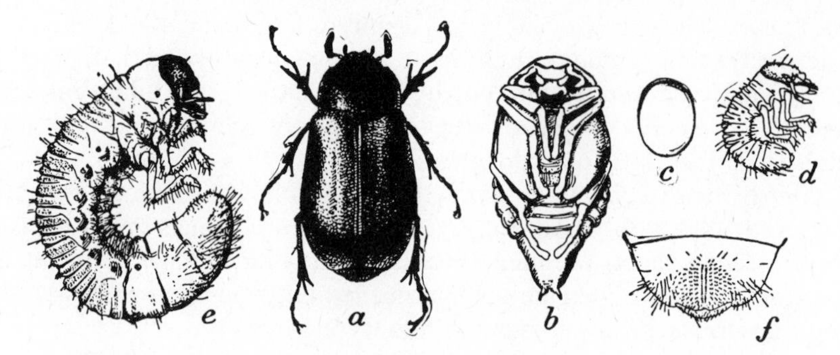

Figure 75. A May beetle, *Phyllophaga fervida* (Fabr.), *a*, adult; *b*, pupa; *c*, egg; *d*, newly hatched larva; *e*, fully grown larva; *f*, anal segment of larva from below. All enlarged. (Chittenden, USDA)

Figure 76. White grubs underneath sod which has been loosened by their feeding. (Courtesy Ohio Agr. Res. and Dev. Center)

insects, 4 years are sometimes required for complete development, whereas in the latitude of Texas the period from egg to adult seems to be 2 years for most species. The annual white grub or northern masked chafer, *Cyclocephala borealis* Arrow, completes its life cycle in 1 year and is commonly a pest of lawns (see p. 132).

Since the common species have a 3-year cycle it follows that there must be 3 broods if the adults are seen every year. The broods are not equally extensive; the heaviest infestation is from brood *A,* adults of which will appear in 1980, 1983, 1986, and every 3 years thereafter; next in abundance is brood *C,* adults of which appeared in 1979; and brood *B,* which is of little importance at present, appearing only infrequently. The greatest damage to crops occurs the year after the appearance of the adults.

Populations of white grubs can be reduced by planting deep-rooted legumes, such as alfalfa, sweet clover, or other clovers, in rotation with more susceptible crops, such as timothy and small grains. Legumes are most effective when planted during the year the flight of the dominant brood occurs. Corn or potatoes may follow clovers in the rotation but should not follow grasses, especially in the year after a heavy beetle flight. A rotation of oats, barley, or wheat with clover and corn has proved satisfactory in some regions.

Late summer or early fall plowing destroys many larvae, pupae, and adults in the soil and also exposes these stages to predators. To be effective, the plowing must be done before the grubs go below plow depth.

Some of the natural enemies of white grubs are the vertebrate predators such as hogs, poultry, birds, and skunks. The parasitic wasps in the genera *Tiphia* and *Myzinum,* and *Pelecinus polyturator* (Drury), and the fly, *Pyrgota undata*

Wied., all destroy some grubs. The fungus parasite in the genus *Cordyceps* infects white grubs and is a factor in natural control.

Inoculating soils with spores of *Bacillus popilliae* Dutky and *B. lentimorbus* Dutky reduces grub populations. No-tillage or reduced-tillage crop management enhances grub populations.

Where crop management has failed and a valuable crop is likely to be seriously damaged, chemical control measures may be necessary.

REFERENCES: *J. Econ. Ent.*, 31:340–344, 1938; 44:359–362, 1951; *USDA Farmers' Bul.* 1798, 1961; *Tech. Bul.* 1060, 1953; *Home and Garden Bul.* 53, 1971; 159, 1977; *N.A. Scarab. larvae*, Ore. Univ. Press, 1966; *Bul. Ent. Soc. Amer.* 19:92–94, 1973; 22:302–310, 1976.

JAPANESE BEETLE

Popillia japonica Newman, Family Scarabaeidae

The Japanese beetle was first observed in this country about the year 1916 in Riverton, New Jersey. Since that time its continuous range has been extended, and it has become established in many isolated colonies from the Carolinas to Missouri northward into the Canadian provinces of Ontario and Nova Scotia; it has also been found in California. The larvae feed in the soil, devouring the roots of a large number of plants. They have been especially injurious to turf in lawns, parks, golf courses, and pastures, and are sometimes troublesome in nurseries and gardens. Adults feed on foliage, flowers (Fig. 77), and fruits dur-

Figure 77. Japanese beetle, and injury to a rose. (Ohio Agr. Exp. Sta.)

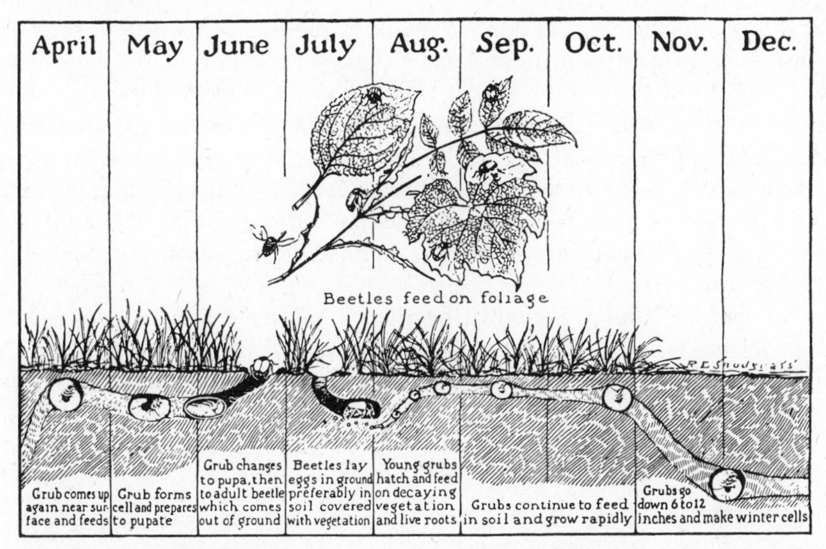

Figure 78. Diagrammatic representation of the life cycle of the Japanese beetle in New Jersey. (Safro, N. J. Dept. of Agr.)

ing their period of activity, which is only on warm sunny days. More than 275 kinds of plants, including fruit and shade trees, ornamental shrubs, flowers, small fruits, and garden crops are often damaged; about 100 species are moderately to severely damaged. Although the larvae have a broad tolerance range to soil pH, female beetles lay more eggs in soils of high acidity, the optimum being pH 4.5 downward. Heavily infested areas are in soils that have pH values of 5.3 or less.

The beetles are somewhat less than 13 mm in length, shiny metallic green with coppery brown elytra. There are six tufts of white hairs on each side of the abdomen near the distal margins of the wing covers. Larvae resemble white grubs, and the layman cannot be certain, usually, whether they are Japanese beetle larvae or young white grubs (see Fig. 80).

The winter is spent in the soil in the partly grown larval stage. In the spring they migrate near the soil surface and feed. By late May or early June pupation begins, and adults make their appearance in late June and continue activity until September. White eggs are laid in the soil at a depth of an inch or more and hatching takes place in about 2 weeks, the young larvae feeding until cold weather. There is 1 generation per year. The complete life cycle is illustrated (Fig. 78).

Quarantines are maintained to restrict the movement of this insect, but they can be expected only to delay the spread rather than to prevent it. Trapping the adult beetles in yellow traps, baited with 9 parts of anethol and 1 part of eugenol, has resulted in the capture of many beetles, but it does not prevent damage to plants. Biological control involves the use of parasites, predators, and disease organisms and is of some value. Two imported parasitic wasps, *Tiphia vernalis* Rohwer and *Tiphia popilliavora* Rohwer, (Tiphiidae) which attack the larval stages, and a parasitic fly, *Hyperecteina aldrichi* Mesnil (Tachinidae), which attacks the adults, have been introduced into the heavily infested areas. A bacterial disease, *Bacillus popilliae* Dutky, commonly known as milky disease of the larvae, has caused marked reductions of the insects in some of the older infested areas. Moles, skunks, and birds also consume a large number of grubs.

For immediate effective control pesticides are necessary. Because adults feed on so many hosts they are difficult to suppress. Since the grubs are primarily root-feeders of grass plants, treating such areas with soil poisons is more often the recommended control measure.

REFERENCES: *Conn. Agr. Exp. Sta. Cir.* 184, 1953; *USDA Farmers' Bul.* 2151, 1960; *Leaflet* 500, 1968; *Agr. Handbook* 236, 1970; *Home and Garden Bul.* 53, 1971; 159, 1969; *USDA* PA-375, 1972; *Tech. Bul.* 1449, 1972; 1545, 1976; *Bul. Ent. Soc. Amer.* 22:305–310, 1976.

EUROPEAN CHAFER

Rhizotrogus majalis (Raz.), Family Scarabaeidae

The European chafer was discovered in Wayne County, New York, in 1940. By 1958 it had become established in 9 counties in western New York State, and in 1959 it was found infesting a large part of Brooklyn and the New York harbor area. Isolated infestations occur in Connecticut, New Jersey, West Virginia, Ohio, Pennsylvania, Massachusetts, and Rhode Island.

The root-feeding grubs damage and sometimes destroy meadows, pastures, lawns, winter grains, and legumes.

Adults, resembling May beetles, are about 13 mm long and light brown. They are most abundant from mid-June to mid-July. Mating flights occur about sunset when thousands of these insects swarm around trees, shrubs, and light poles. In about half an hour they settle on objects and mate. Then the females burrow into the soil 2 to 6 inches deep, lay their white eggs, and die soon afterwards. The beetles are harmless to plants or man.

Hatching occurs in 2 to 3 weeks, and the white curved larvae (see Fig. 80) with brown heads attain maximum growth in approximately 3½ months. They burrow below the frost line in the fall and return near the surface of the soil in the spring. After feeding a short period most larvae change to pupae and emerge as adults. Some spend a second summer in the soil and pupate the following spring. The life cycle usually is completed in one year, but occasionally requires 2 years.

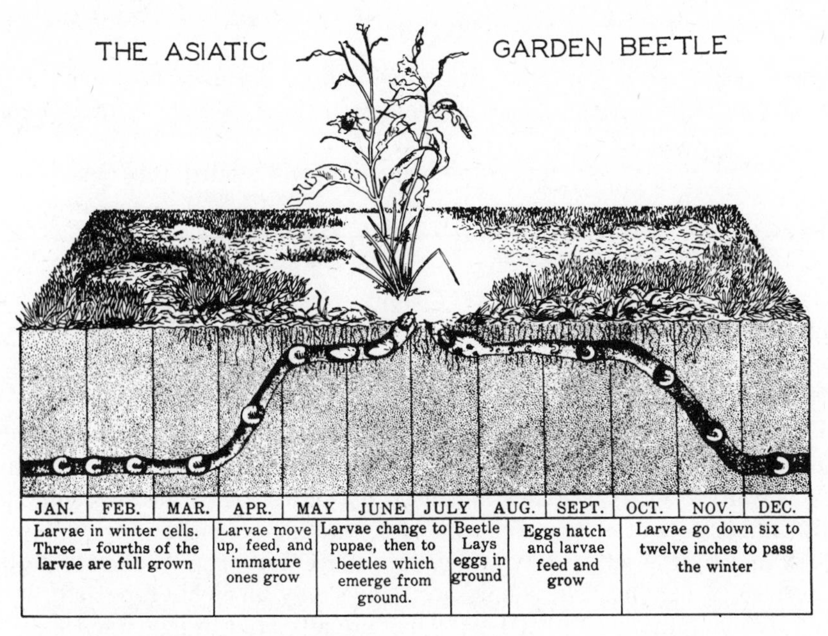

Figure 79. Diagrammatic representation of the life cycle of the Asiatic garden beetle. (USDA)

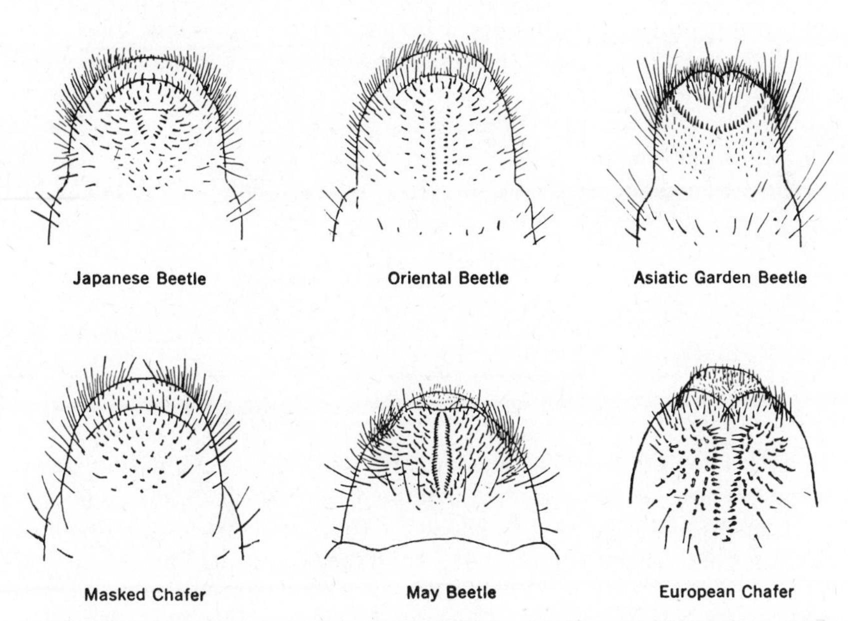

Figure 80. Ventral view of last abdominal segment (raster) of common grubs showing rastrel pattern, useful in species determination. (Courtesy of USDA and Cornell Univ.)

Some infested areas are under State quarantines designed to prevent interstate spread of the insect. Natural enemies include poultry, birds, skunks, moles, shrews, and some predaceous beetles.

Application of pesticides to grub-infested soil will control this pest.

REFERENCES: *USDA* PA-455, 1961; *N.Y. Agr. Exp. Sta. Farm Res.* 33:8–9, (2)(3), 1967; *J. Econ. Ent.* 36:345–346, 1943; 38:169–173, 1946.

ASIATIC GARDEN BEETLE

Maladera castanea (Arrow), Family Scarabaeidae

First reported from New Jersey in 1921, this beetle has now spread to other localities of the Atlantic coastal states, from Massachusetts to South Carolina, and westward to Ohio. Damage is caused by the larvae or grubs feeding primarily on the roots of grasses and, to a lesser extent, on garden vegetables and flowers, and by the adults devouring the foliage and flowers of both woody and herbaceous plants. The adults are active only at night, especially when the air temperatures are rather high. They are attracted to lights and may become a nuisance when populations are high.

Overwintering larvae deep in the soil become active in the spring, migrate near the surface and feed. Pupation begins in late May, and velvety brown adults nearly 13 mm in length appear in late June or early July. Eggs are laid principally in grassy areas, and the white curved larvae feed until cold weather, then migrate downward in the soil where the winter is passed. Only 1 generation develops each year. The complete life cycle is shown in Fig. 79. For larval identification, see Fig. 80.

Chemical control measures are directed toward killing the adults and larvae in the same manner as for Japanese beetles.

REFERENCES: *USDA cir.* 246; *Conn. Agr. Exp. Sta. Cir.* 184, 1953.

ORIENTAL BEETLE

Anomala orientalis Waterhouse, Family Scarabaeidae

So far as is known this insect was imported into Connecticut from Japan in 1920, and has now spread into several of the other eastern coastal states. Major injury is caused by the larvae feeding on the roots of plants, especially grasses; serious damage to lawns often results. Adult feeding on the foliage and flowers of plants is considered unimportant. The oriental beetle has also been a serious pest of sugarcane in Hawaii.

Partly grown white curved larvae with brown heads hibernate deep in the soil, coming near the surface in the spring and feeding until fully grown. Pupa-

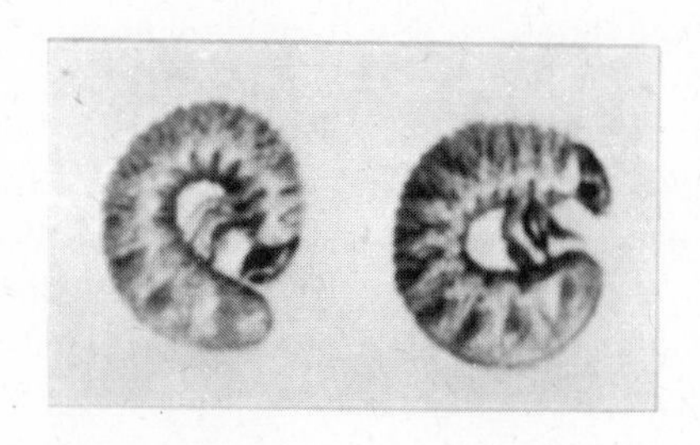

Figure 81. The oriental beetle, adults (left); larvae (right). (Friend, Conn. Agr. Exp. Sta.)

tion takes place in early June, and new adults begin to appear in early July, continuing into August. They are 13 mm or more in length and marked as illustrated (Fig. 81). Eggs are laid in the soil, and the newly hatched larvae feed and develop until cold weather, then burrow down to a depth of almost 1 foot where they pass the winter. Only 1 generation occurs each year (see Fig. 80 for larval identification).

Populations of this insect often occur in the same areas with the Japanese beetle and the Asiatic garden beetle. Experimental results show that the same pesticides for controlling these insects will control oriental beetle.

REFERENCES: *Conn. Agr. Exp. Sta. Bul.* 304, 1929; *J. Econ. Ent.*, 41:905–912, 1948; 42:366–371, 1949; *USDA Home and Garden Bul.* 53, 1971.

NORTHERN MASKED CHAFER

Cyclocephala borealis Arrow, Family Scarabaeidae

This annual white grub is found over most of the United States as far south as Alabama. It is especially injurious to the turf of golf courses and lawns, the larvae burrowing through the soil and eating off the roots. Injury becomes more conspicuous after an invasion of moles, skunks, and birds, which tear up the loosened turf to feed on the grubs.

The adult is nearly 13 mm long, chestnut brown, and covered with fine hairs. Larvae are curved, white with brown heads, resembling small white grubs (see Fig. 80).

Winter is passed as larvae in the soil at a depth of about 16 inches. In the spring they migrate upward and feed until about June 1, then move down to an average depth of 6 inches, transform to pupae, and begin emerging as adults nearly 2 weeks later. In Ohio, most adults are present from June 25 to July 25.

They are nocturnal, usually emerging about dusk, and are active during early evening hours. No food is taken since their mouthparts are nonfunctional. Oviposition begins a few days after emergence, and the white eggs are placed 4 to 6 inches below the soil surface. Hatching occurs in 3 weeks; the larvae feed until cold weather, then migrate downward in the soil. One generation occurs each year.

A related species, with similar habits and life cycle, is the southern masked chafer, *C. immaculata* (Oliv.).

Experiments indicate that these insects can be readily controlled by soil pesticides.

REFERENCES: *J. Agr. Research,* 62:79–86, 1941; *J. Econ. Ent.,* 31:340–344, 1938; 42:626–628, 1949; 45:347–348, 1952; *USDA Home and Garden Bul.* 53, 1971.

ARMYWORM

Pseudaletia unipuncta (Haw.), Family Noctuidae

The armyworm is a caterpillar which, in seasons of unusual abundance, crawls in large numbers from field to field devouring grasses and grain crops. Wheat, corn, oats, barley, and rye are among its favored food plants. Outbreaks are more common following cold, wet, spring weather, and damage may occur from late April to late June. It occurs throughout most of the United States east of the Rocky Mountains, and it has also been found in New Mexico, Arizona, California, and Canada.

The buff or sand-colored moth has a wing expanse of about 4 cm with a small white dot in the center of each fore wing and somewhat darker margins on the hind wings (Fig. 82). The dot is a convenient recognition mark and the basis for the specific name. Adults feed on the nectar of flowers. The full-grown larva is a nearly hairless, smooth, striped caterpillar, about 5 cm long. Its general

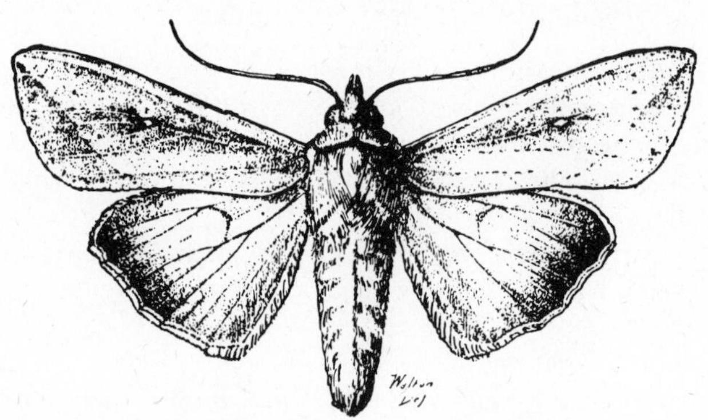

Figure 82. The armyworm moth, *Pseudaletia unipuncta* (Haworth); 2×. (USDA)

color is green to brown, and the stripes, one along each side and a broad one down the back, are dark and often nearly black. The stripe along the back usually has a fine, light-colored, broken line running down its center. The head is pale brown with a green tinge and mottled with dark brown (Fig. 83).

Partly grown caterpillars hibernate in the soil or debris at the surface and complete their growth in the spring. In the latitude of central Ohio pupation takes place the latter part of April, and about 2 or 3 weeks later adults emerge and lay eggs. The moths are attracted to lights at night, and their relative numbers may serve as an indication of probable larval abundance. The eggs are laid in large masses at night, usually in the folded blades or under the leaf sheaths of grains and grasses. They resemble small white beads, each smaller than the head of a common pin. Many moths seem to congregate and lay their eggs in the same locality. In 8 to 10 days tiny greenish caterpillars hatch from the eggs and begin feeding. This is the first generation of larvae and it usually causes the most damage. After molting several times they become fully grown in about 3 or 4 weeks, then pupate and emerge as adults. There are usually 3 generations of caterpillars each year in the northern states but seldom 2 successive outbreaks in any given locality. Some evidence indicates that all stages may be present during the winter in the extreme South.

One of the important natural enemies of the armyworm is *Winthemia quadripustulata* (Fabr.), commonly called the red-tailed tachina fly (Fig. 84). It oviposits on the caterpillars, and its larvae bore into the body and devour the inside portions. The flies multiply rapidly and often become numerous enough to completely control the armyworm in some localities.

The egg parasite, *Telenomus minimus* Ashmead, (Sceleonidae) and the braconid wasps, *Apanteles laeviceps* Ashmead, *A. marginiventris* (Cresson), and *A. militaris* (Walsh), also play an important role in natural control. Other natural enemies are the ground beetles, sphex wasps, birds, toads, skunks, and domestic fowls.

Pesticides are helpful when natural enemies do not keep it in check, especially in no-tillage corn following sod or small grains. Armyworm problems in

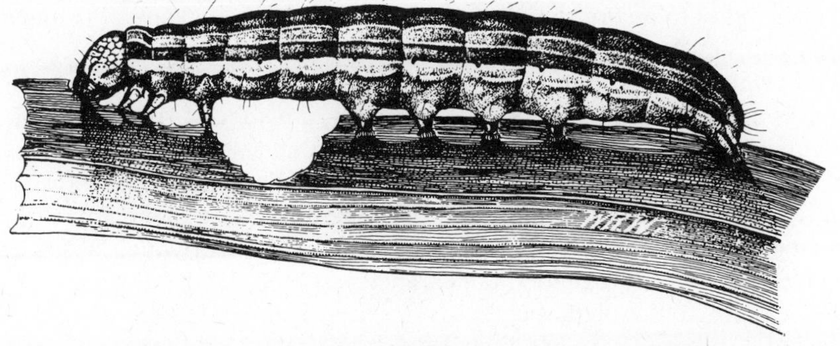

Figure 83. Armyworm feeding on corn; 2×. (USDA)

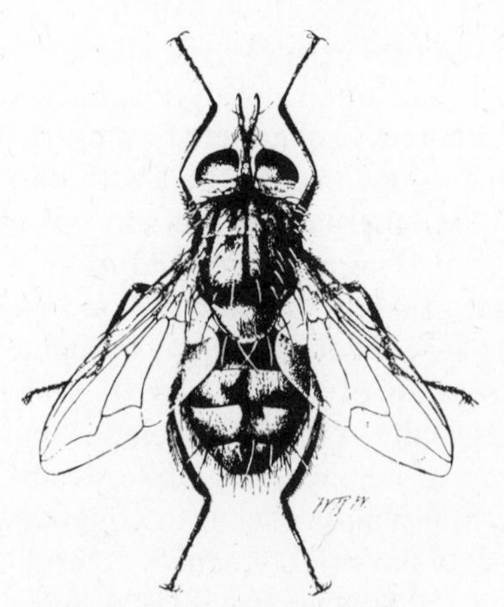

Figure 84. Winthemia quadripustulata (Fabr.), one of the most important armyworm parasites; 4×. (Walton, (USDA)

no-tillage corn planted in sod appear to result from eggs deposited on grass before it is killed with a herbicide, leaving young corn plants as the only food; corn near adjacent sod areas may be severely damaged under such crop management.

REFERENCES: *USDA Farmers' Bul.* 1850, 1947; *leaflet* 494, 1961; Cir. 849, 1950; *Pesticide News* 26:38–44, 1973; *Bul. Ent. Soc. Amer.* 22:302–304, 1976.

FALL ARMYWORM

Spodoptera frugiperda (Smith), Family Noctuidae

The fall armyworm is essentially a southern insect which extends its attacks during some seasons to a large portion of the United States and Canada. It migrates northward each year and sometimes becomes abundant in the northern states but only in late summer and fall. Other areas in which it is found are Central and South America.

Food plants are much more varied than those of the true armyworm. Known mainly as a pest of corn, especially sweet corn, it also feeds on cotton, alfalfa, clover, peanuts, grasses, tobacco, and many garden crops. The damage results when the larvae devour plant parts above ground. The shanks and ears of corn are often extensively damaged by the larvae boring inside them. It is probably of greater importance as a pest than the true armyworm which it somewhat resembles in habits and general appearance.

The eggs are laid at night on grasses or other plants in masses of as many as a hundred or more. These hatch into larvae in 2 to 10 days, molt six times before becoming fully grown in about 20 days, and enter the soil and transform to pupae. The inactive pupal stage lasts about 10 days, after which the adult moths emerge and often migrate many miles before the females lay their eggs. In its northern range there are one or two generations per year, whereas in the South there may be several. None of the life stages survive the winters in the North; in the South all stages may be present but are more or less inactive in areas where no freezing temperatures occur. When abundant, and the food supply becomes exhausted, the larvae migrate as do true armyworms. The larvae have a prominent, white inverted Y-shaped suture on the front of the head and longer hairs arising from black tubercles on the body, which are more conspicuous than those on the true armyworm. The adults resemble cutworm moths, having dark gray fore wings mottled with light and dark spots and grayish white hind wings (Fig. 85).

Besides the effects of climatic factors on the abundance of fall armyworms, the parasitic enemies are numerous and greatly reduce populations. Some of these parasites are the ichneumon wasp, *Ophion bilineatus* Say; the braconid wasps, *Chelonus texanus* Cress. (Fig. 86), *Meterous laphygmae* Vier., and species of *Apanteles;* the egg parasite, *Trichogramma minutum* Riley; a chalcid wasp in the genus *Euplectrus;* and the tachina flies, *Winthemia quadripustulata* (Fabr.) and *W. rufopicta* (Bigot). Predators include ground beetles, birds, and vertebrate enemies of caterpillars of all types.

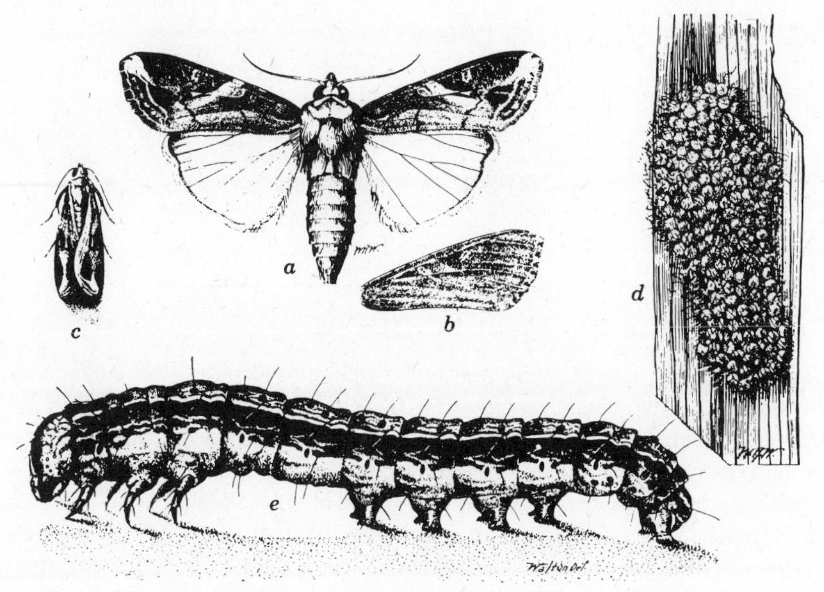

Figure 85. The fall armyworm: *a,* ♂ moth; *b,* right front wing of ♂ moth; *c,* ♀ moth in resting position; *d,* egg mass; *e,* larva. Natural size except *d* and *e,* which are about 2×. (Walton, USDA)

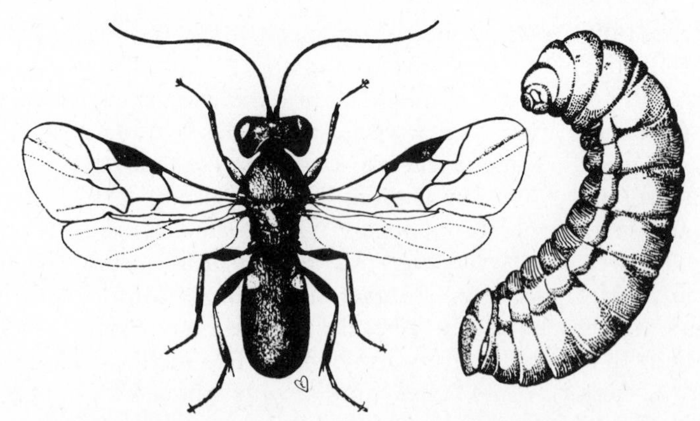

Figure 86. Adult and larva of *Chelonus texanus* Cresson, a braconid parasite of the fall armyworm, enlarged. (Walton and Luginbill, USDA)

Other species in the genus *Spodoptera* with somewhat similar habits and life cycles are: southern armyworm, *S. eridania* (Cramer); nutgrass armyworm, *S. exempta* (Walker); beet armyworm, *S. exigua* (Hübner) (Fig. 203); lawn armyworm, *S. mauritia* (Boisduval); yellowstriped armyworm, *S. ornithogalli* (Guenée); and western yellowstriped armyworm, *S. praefica* (Grote). The host range is rather broad and usually these species are more serious pests in southern, southwestern, and western United States.

Chemical control is often necessary to suppress them. A reported economic threshold is 15 larvae (12 mm long) per sweep of an insect net.

REFERENCES: *USDA Tech. Bul.* 34, 1928; *Tech. Bul.* 138, 1929; *Leaflet* 494, 1961; *J. Econ. Ent.*, 42:502–506, 1949; *Ohio Res.* and *Dev. Ctr. Res. Cir.* 227, 231, 1977.

CUTWORMS

ORDER LEPIDOPTERA, FAMILY NOCTUIDAE

Larvae that cut off low-growing plants at, or slightly below, the surface of the soil, are called cutworms. Species that feed above ground on any part of the plant are called climbing cutworms. Fruit trees and many other plants are sometimes injured by these species. When a high population develops and the food supply becomes exhausted, migration to other plants takes place as with armyworms. These species are sometimes called army cutworms. Some species are more specific than others as to the hosts on which they feed. Our common cutworms are most injurious to garden vegetables, corn, cotton, tobacco, and similar crops grown in rows or hills, whereas small grains and forage crops are damaged to a lesser extent.

Some better-known cutworms are the following: black cutworm, *Agrotis ipsilon* (Hufnagel); claybacked cutworm, *A. gladiaria* (Morrison); palesided cutworm, *A. malefida* Guenée; pale western cutworm, *A. orthogonia* Morrison; glassy cutworm, *Crymodes devastator* (Brace); spotted cutworm, *Amathes c-nigrum* (L.); army cutworm, *Euxoa auxiliaris* (Grote); darksided cutworm, *E. messoria* (Harris); redbacked cutworm, *E. ochrogaster* (Guenée); white cutworm, *E. scandens* (Riley); striped cutworm, *E. tessellata* (Harris); variegated cutworm, *Peridroma saucia* (Hübner) (Fig. 87); bristly cutworm, *Lacinipolia reingera* (Stephens); bronzed cutworm, *Nephelodes minians* Guenée; black army cutworm, *Actebia fennica* (Tauscher); dingy cutworm, *Feltia ducens* Walker; granulate cutworm, *F. subterranea* (Fabricius); and western bean cutworm, *Loxagrotis albicosta* (Smith).

The adults of all cutworms are moths with dark gray fore wings, often variously marked with darker or lighter spots and narrow bands, and lighter colored hind wings. When at rest the wings are folded over the back. They feed at dusk by sucking nectar from flowers, and are often attracted to lights at night. The moths lay many hundreds of eggs, and most of these are on plants in grassy sod or weedy fields. On hatching, the larvae molt several times and when fully grown are nearly 5 cm long. They then tunnel into the soil where they form cells in which pupation occurs. The moths emerge from the pupae and crawl out of the ground through the tunnels made by the larvae. The time required to complete the life cycle varies with the different species. The glassy, army, dingy, bronzed, and black army cutworms overwinter as partly grown larvae and have 1 generation per year. The pale western cutworm overwinters

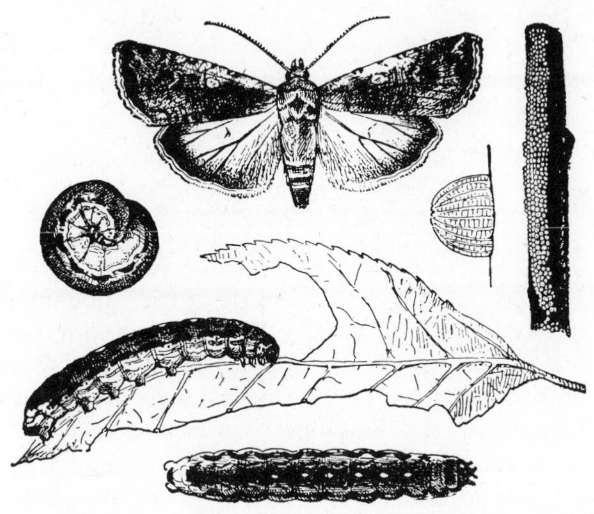

Figure 87. Life stages of the variegated cutworm, *Peridroma saucia* (Hübner); enlarged. (Howard, USDA)

as an egg with 1 generation per year. The black cutworm overwinters as a larva or pupa and has 2 generations in Canada and several in southern United States. The spotted and variegated cutworms have 2 to 4 generations per year and overwinter as larvae or pupae. The black, claybacked, and glassy cutworms are below ground feeders.

Greatest injury from 1-generation cutworms results from the spring feeding of those species wintering as larvae. If grassy sod land or weedy fields are plowed in the spring, the food supply of the cutworms present is greatly reduced and, if this soil is then prepared and set to such plants as tomato, tobacco, or cabbage, serious damage usually follows. Damage at other times in the summer occurs with species having 2 or more generations per year or with overwintering stages other than larvae.

Natural and applied control measures for cutworms are very similar to those for armyworms (p. 135). Where they fit into the general farming practices, cultural control methods are of value. For instance, land kept free of weeds and crops during late summer is rarely infested; various life stages may be destroyed by fall plowing; cultivation of infested fields in the spring after vegetation has appeared in them and has grown to a height of 2 inches followed by seeding, deferred at least ten days, results in starvation of many larvae; a rotation in which hill or row crops do not follow a grassy sod is also a means of avoiding damage.

Parasitism of soil-infesting cutworms has reached 24% where the braconid, *Microplitis feltiae* Muesebeck is present. Where no-tillage or reduced-tillage without crop rotation has become popular, pesticides are usually necessary to suppress damage.

REFERENCES: *Mont. Bul.* 225, 1930; *Can. Bul.* 59, 1938; *Tenn. Exp. Sta. Bul.* 159, 1936; *USDA Tech. Bul.* 88, 1929; *Leaflet* 417, 1957; *Cir.* 849, 1950; *Tech. Bul.* 1135, 1956; *Ann. Ent. Soc. Amer.* 63:645–648, 1970; *J. Econ. Ent.* 61:961–973, 1968; 65:734–737, 1972; *Bul. Ent. Soc. Amer.* 21:229–234, 1975; 22:409–415, 1976; *OARDC Res. Cir.* 227, 1977.

STALK BORER

Papaipema nebris (Guen.), Family Noctuidae

The stalk borer is widely distributed east of the Rocky Mountains. Its larvae feed on a great variety of plants, especially the stalks of corn, cotton, potato, tomato, dahlia, hollyhocks, tiger lilies, tobacco, and giant ragweed. The damage is done primarily by boring and tunnelling in the stalks. Cultivated crops adjacent to weedy areas of the favorite host, giant ragweed, are most often attacked.

The moth is fawn gray or mouse-colored, with the outer third of the fore wings paler and bordered within by a white cross line. Two clusters of white spots, which are variable in distinctiveness, appear on the darker two thirds of the fore wings. The spots in the fore wings are obscured in the variety *nitela*. The hind wings are gray-brown to smoky for both *nebris* and *nitela*. Wingspread is about 5 cm. The appearance of the larva in its earlier instars is purplish to light brown with 5 white stripes, 1 along the middle of the back and 2 on each side. These side stripes are absent on the first 4 segments of the abdomen, giving the larva an appearance of having been injured. The last instar is light in color and only faintly striped.

The overwintering eggs are laid on the leaves of the host plants during late summer and early fall. They hatch the following May or June and the larvae pass through 7 to 16 instars, most of them completing their development in seven to nine. Pupation takes place in the soil or the larval burrows in plants; the moths emerge in late August and September and deposit their eggs, completing the life cycle. Only 1 generation occurs each year.

Climatic conditions, birds, mammals, insect predators and parasites, and diseases are all responsible for reducing the numbers of this insect.

Applied control measures consist of destruction of the weeds which harbor the larvae or eggs by deep plowing. Mowing weeds along fence rows in late August renders the locality undesirable for oviposition and in turn reduces the population in such areas the following season. No-tillage crop management enhances this pest.

REFERENCES: *Iowa Res. Bul.* 143, 1931; *Bul. Ent. Soc. Amer.* 22:302–304, 1976.

SOD WEBWORMS OR GRASS MOTHS

Crambus spp., Family Pyralidae

Sod webworms are widely distributed throughout the United States and Canada. They become important pests of grassland, lawns, golf courses, and occasionally cultivated crops, such as corn or tobacco, planted in soil which was previously a grassy sod. The varied yellow-to-white larvae spin silken threads as they crawl about and feed, webbing the soil particles and leaves together near the surface, often forming silken tubes within which they live. Most of the damage results from their feeding at the soil surface or slightly below. The pale brown moths vary from 13 to 20 cm in length, hold their wings tightly over the body when at rest, and the labial palpi project forward from the head, forming a snout, which is the basis for another suggested common mame, the snout moths.

Some of the more common species are: the corn root webworm, *Crambus caliginosellus* Clemens (Fig. 88); bluegrass webworm, *Crambus teterrellus* (Zincken); vagabond crambus, *Crambus vulgivagellus* Clemens; larger sod webworm, *Crambus trisectus* Walker; striped sod webworm, *Crambus muta-*

Figure 88. The adult of the corn root webworm, *Crambus caliginosellus* Clemens; enlarged. (USDA)

bilis Clemens; silverstriped webworm, *Crambus praefectellus* Zincken, and *Crambus topiarius* Zeller (Fig. 89).

The usual overwintering stage is larvae in cases of silk covered with soil particles near the ground level or slightly below. They feed in the spring, becoming fully grown after passing through 7 to 20 instars. This is followed by pupation close to the feeding tunnels inside the cases which are lined with soft gray silk and covered with soil and grass particles interwoven with silk. Adult moths appear in the summer months, the females at dusk dropping their eggs over grassy areas while either in flight or at rest. These eggs hatch in about 7 days and the larvae begin the cycle anew. The corn root webworm has only 1 generation per year; the bluegrass webworm has 2 or 3, and the larger sod webworm normally has 2, but a partial third generation may occur at its southern limits of distribution.

Natural enemies include many vertebrate and insect predators, especially birds and ants, a number of dipterous and hymenopterous parasites, as well as some pathogenic diseases, viz., *Beauvaria bassiana* and *Bacillus thuringiensis*.

Rotation of crops so as to avoid planting corn on soil that had been in grass sod the previous year, ample fertilization, early fall plowing to prevent egg deposition in the field, or later fall plowing followed by cultivation to kill overwintering larvae are cultural and mechanical control measures. Resistant cultivars show promise.

Chemical control is now practical, especially on lawns and golf courses.

REFERENCES: *USDA Tech. Bul.* 30, 1927; 117, 1930; *Home and Garden Bul.* 53, 1971; *Ann. Ent. Soc. Amer.* 60; 1014–1018, 1967; 61:1481–1486, 1968; 64:116–119, 1971; *Fla. Ent.* 6:49–55, 1923; 11:12–14, 1927; *Hilgardia* 17, 267–307, 1947; 22:535–565, 1954; *J. Econ. Ent.* 33:886–890, 1940; 45:114–118, 1952; 52:966–969, 1959; 53:670–672, 1960; 62:703–708, 1969; *J. Agr. Res.* 24:399–425, 1923; *Bul. Ent. Soc. Amer.* 19:94–95, 1973; 20:11–23, 1974.

Figure 89. Sod webworms. Above, left to right, *Crambus vulgivagellus* Clemens, the vagabond crambus, and *C. mutabilis* Clemens, the striped sod webworm; below, the bluegrass webworm, *C. teterrellus* (Zincken), and the silver-striped webworm, *C. praefectellus* Zincken; enlarged. (Ainslie, USDA)

RHODES GRASS MEALYBUG

Antonina graminis (Maskell), Family Pseudococcidae

This pest was first identified in the United States in 1942 by Harold Morrison from specimens submitted by an agronomist at the King Ranch, Kingsville, Texas. The mealybug occurs in tropical and subtropical regions of the world with distribution confined to the area between the 30° north and south latitudes. In the United States it is recorded from Alabama, Arizona, California, Florida, Louisiana, Mississippi, New Mexico, and Texas, with possible infestations in adjacent areas. Since 1942, 94 hosts have been recorded in North America, all members of the family Gramineae, and over 122 hosts from 26 foreign countries. Rhodes grass, Johnson grass, bermuda grass, and St. Augustine grass are preferred hosts of economic importance. Most other hosts are only lightly infested.

Ranges, lawns, and golf courses are damaged by the mealybug, but it is difficult to ascribe damage data caused by the insect alone because of the interrelating factors of drought, mowing, and overgrazing all operating at the same time. Heavily infested grasses turn brown and die because of removal of plant sap by the mealybugs.

Adult mealybugs are parthenogenetic and reproduce ovoviviparously, the reproductive period of an individual averaging 50 days. Males have never been observed. There are 3 nymphal instars but only the first is active, it having well-developed legs and antennae. The second and third instars and the adult are sessile on the host and are enclosed in a felted waxy sac. The adult is broadly ovate, purplish brown, and average 3 by 1.5 mm in size. The life cycle ranges from 60 to 70 days; there are 5 generations annually, the winter generation being longer. In Texas the largest populations occur about July 1 and November 1.

Dispersion is accomplished by first instars being blown by wind, carried by grazing animals, or transported in infested sod or grass cuttings.

Temperatures near 100 F (38 C) retard mealybug development, and 108 F (42 C) for 24 hours is fatal. A temperature of 32 F (0 C) also retards development, and 28 F (−2 C) for 24 hours is fatal. Optimum development temperatures are 85–90 F (29–32 C).

The use of resistant or highly tolerant grasses is recommended. Distributing the introduced encyrtid parasite, *Neodustmetia sangwani* (Rao), by aircraft has been highly successful in controlling this pest.

REFERENCES: *USDA Tech. Bul.* 1221, 1960; C. B. Huffaker *et al., Biological Control,* Chapt. 10, Plenum Press, New York, 1971.

BLACK TURFGRASS BEETLE

Ataenius spretulus (Haldeman), Family Scarabeidae

Described in 1848 from specimens collected in the Middle Atlantic States of North America, this insect has not been recorded as a pest except in recent years where damage was to turf, especially in golf courses. Fairways of annual bluegrass and bentgrass were most heavily damaged, but Kentucky bluegrass was also injured. Only scattered reports of damage by the grub stage were reported prior to 1973, but since then it has been reported from New Hampshire to Minnesota, Colorado to Virginia, and in Ontario, Canada. One can only speculate as to the reasons for its recent abundance. Evidence of the presence of this insect are discolored, wilted, or dead patches of grass.

The black beetles, 4–5 mm long, pass the winter 1–2 inches deep in well-drained soil near wooded areas of golf courses. In late March and April they begin emerging and flying about. In May and June egg clusters, each containing 8–12 eggs, are laid in the lower thatch and first half inch of soil. Larvae from these eggs are tiny white curved grubs which become most numerous within the next month. Pupation then occurs 1 to 2 inches deep in the soil. New adults, appearing by mid-July, are red-brown at first but darken to black on aging. They begin laying eggs that produce a second generation which reaches maturity during September and October. These adults mate and seek overwintering sites.

Larval populations as high as 629 per square foot have been recorded in Ohio. Population thresholds for larvae have not been determined. It is known that the deeper rooted grasses tolerate higher grub populations.

A promising control measure consists of treating the grasses of golf courses with diazinon in April or May to kill adults before they lay their eggs. Only one application is necessary but proper timing is important.

REFERENCES: *J. Econ. Ent.* 69:345–348, 1976; *W. Va. Agr. Exp. Sta. Rept.* 62, 1978.

10

PESTS OF GRASSES AND CEREAL GRAINS

This chapter contains a description of the major insect enemies of corn, wheat, oats, rye, barley, rice, sugarcane, and sorghums. Many of these crops are essential the world over as a source of food for man and other animals. The pests in the previous chapter may also damage these crops.

CORN EARWORM

Heliothis zea (Boddie), Family Noctuidae

Corn earworm is cosmopolitan and considered to be one of the half-dozen most destructive insects attacking corn, especially sweet corn. Its range in North America is almost coincident with that of the corn plant, although in the northern part of the corn-growing area the insect may be very rare. Because of high winter mortality north of 39° N latitude it is not a permanent resident throughout its range but migrates from the South and reinfests northern areas each year.

Besides corn, it is an important pest of cotton and is then called the bollworm. On tomatoes it is known as the tomato fruitworm. Some other hosts are all kinds of beans, alfalfa, clover, vetch, tobacco, pepper, lettuce, sorghum, and peanuts.

Damage to corn results when the larvae feed in the tips of the ears (Fig. 90), devouring the kernels and fouling them with excrement, sometimes destroying the silks before pollination is completed. Rarely do the larvae bore through the cob or husks like the European corn borer, entrance being made through the silk channel. Secondary loss, from molds that follow the feeding of the worms, is often extensive. Injury to tomatoes, cotton bolls, and bean pods is caused by the larvae boring through these plant parts. Foliage of these and other hosts is often eaten.

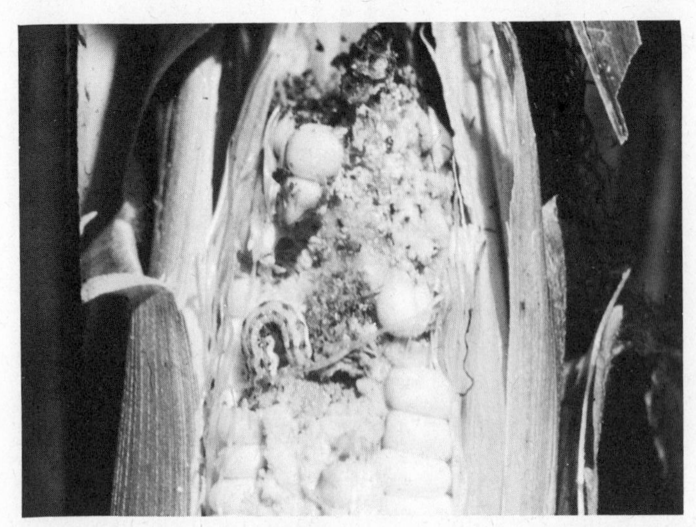

Figure 90. Earworm damaging sweet corn. About natural size. (Courtesy of L. D. Anderson, Univ. of Calif.)

In the areas where this insect survives the winter brown pupae hibernate in the soil. Moths having a wing expanse of about 4 cm emerge in February for the Gulf Coast states and in June for 39° N latitude. The moths are light buff with irregular darker lines and spots near the outer margins of the fore and hind wings (Fig. 91). They fly at dusk and deposit their eggs on the corn silks and other parts of the plant. Each female moth may deposit 1000 or more eggs during her life span. Hatching occurs a few days later, and the larvae feed downward through the silks into the ear tip, become fully grown in about three or four weeks, then leave the ear and enter the soil forming a cell in which pupation takes place. Fully grown larvae are about 4 cm long, variable in color ranging from light green through brown to almost black, with light and dark stripes running lengthwise of the body which is lighter on the underside. Development from egg to adult requires about 30 days in midsummer and a longer period in early spring or late autumn. Usually 2 generations develop in the North, with as many as 5 or 6 in the extreme South.

Natural control results from cannibalism of the larvae; the egg parasite, *Trichogramma minutum* Riley; the larval parasite, *Winthemia quadripustulata* (Fabr.); the predatory bug, *Orius tristicolor* (White); the braconid, *Microplitis croceipes* (Cresson); the chalcid, *Brachymeria ovata ovata* (Say); a virus; a number of birds; and other predaceous insects. Over a period of years in central Virginia *Orius tristicolor* destroyed 38% of the eggs of the adult corn earworm that were deposited on corn.

Injury to both field and sweet corn can be reduced by growing strains with long tight husks which extend beyond the tips of the ears. Earworm-resistant

Figure 91. Corn earworm moths and larva. Moths about normal size; larva 2×. (USDA)

sweet corn varieties also show promise. Planting at a time that results in the corn silking between the two generations in the northern range averts damage. In the areas where overwintering occurs, early- or late-planted sweet corn usually suffers the greatest damage. Fall or spring plowing and disking of the soil reduces the number of moths that emerge in the spring from overwintering pupae. This operation has some control value if done on a community-wide basis. The infestation is reduced by clipping the silks, including about an inch of the husks, after pollination is completed. The clippings must be removed from the field.

Usually no chemicals are applied to control corn earworm in field corn, but they are widely used on sweet corn. Direct the pesticide to the silks about 1 day after 7 to 10% of the ears are silking, with repeat applications about 2 or 3 days apart.

A compressed-air hand sprayer is satisfactory for treating garden plots of sweet corn; for large commercial plantings high-clearance power sprayers, adjusted to give coverage of the ears and silks, are required. Dusts can be applied with a stencil brush or hand duster for garden plantings.

REFERENCES: *USDA Farmers' Bul.* 1651, 1953; *Leaflet* 411, 1961; *Tech. Bul.* 561, 1937; *Tech. Bul.* 1160, 1957; *Tech. Bul.* 838, 1942; E–780, 1951; *J. Econ. Ent.,* 41:928–935, 1948; 45:105–108, 137–138, 931–933, 1952; 52:1111–1114, 1959; 53:22–24, 1960; *Ann. Ent. Soc. Amer.* 63:67–70, 1261–1265, 1970.

EUROPEAN CORN BORER

Ostrinia nubilalis (Hbn.), Family Pyralidae

It is thought that the corn borer was introduced into the United States about the year 1909 in shipments of broom corn from Europe, where it had long been known as a pest of corn and other crops. However, the first record of its presence was in Massachusetts in 1917. It has since spread to practically all the major corn-producing areas of the United States and Canada, and yearly losses are estimated to be high. Early plantings of sweet corn are often a complete loss in years of corn borer abundance. One borer per cornstalk reduces the yield 2 to 3 bu./acre.

The major host is corn, with some indication that sweet corn is the favorite kind. Other hosts are chrysanthemum, dahlia, gladiolus, eggplant, pepper, beet, bean, potato, tomato, oat, soybean, and many kinds of weeds.

Damage to corn is caused by the early larval instars chewing the leaves, resulting in destruction of the leaf surface and midrib breakage. Later instars tunnel all parts of the stalks and ears, resulting in broken stalks and tassels, poor ear development, and dropped ears. Other hosts are damaged primarily by the tunnelling of the stalks or stems by the larvae. The first indication of the presence of the borers is leaf damage, followed by larval borings protruding from holes in the stalks.

The borers pass the winter as fully grown larvae concealed in parts of the plants on which they have been feeding. They are flesh-colored, about 3 cm long, and marked with brown spots (Fig. 92). Pupation takes place in late spring, with the adult moths appearing in May and June. The moths are buff-colored, with dark brown wavy bands across the wings and a wing expanse of about 25 mm. The male is smaller in size and darker in color. After maturing the females (Fig. 93) begin laying white eggs, in masses of 15 to 35, on the undersides of the lower leaves of the host plants.

The tallest earliest maturing varieties of corn receive the greatest number of eggs. Each female may lay 500 or more eggs during her life span. These overlapping scalelike eggs (Figs. 93 and 94) hatch in about 7 days, and the black heads of the young larvae can be seen inside before hatching (Fig. 94). The larvae feed at first on the leaves, and then the leaf sheath and whorl, often causing damage to the unfolding tassel. They are fully grown in approximately 35

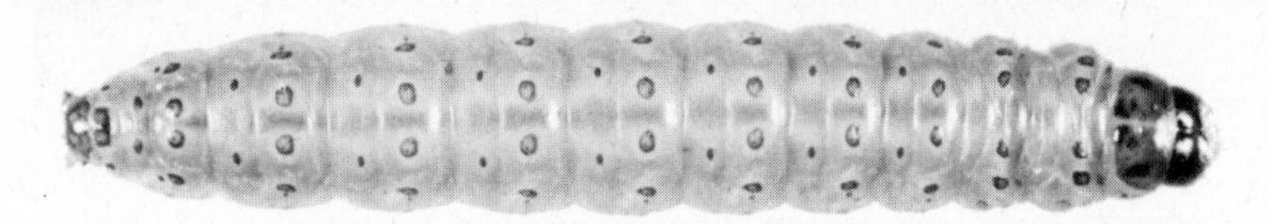

Figure 92. Dorsal view of the European corn borer larva; 3×. (Courtesy of W. D. Guthrie, USDA)

Figure 93. Adult female European corn borer and egg mass; 2×. (Courtesy of W. D. Guthrie, USDA)

days. The percentage of first-generation larvae that pupate, emerge as adults and produce succeeding generations is variable in different latitudes of the corn belt and increases from North to South. A third generation occurs in some areas and is often a pest of fall potatoes. Dr. R. N. Hofmaster has observed a partial fourth generation in Virginia.

The cause of this variation or diapausing condition of some larvae is correlated with temperature and photoperiod (sunrise to sunset). Mid-summer pupation of corn borer larvae under field conditions is closely associated with length of photoperiod. Borers reaching maturity after photoperiods shortened to less than 14.75 hours tend to enter a state of diapause. This length of photoperiod occurs at Madison, Wisconsin, on July 25. Pupation occurs if borers attain the fifth instar before this date. Prolonged periods of high temperature before this date increases pupation, whereas periods of low temperature reduces pupation and increases the number of diapausing larvae.

Cool, rainy weather during the month of June greatly reduces the infestation of borers because it inhibits oviposition and washes the tiny hatching larvae from the plants. Very dry summers and extremely cold winters also are unfavorable to borer development and survival.

Some degree of biological control has been achieved with the tachina fly, *Lydella thompsoni* Herting. Other insect parasites of some value are the braconids *Macrocentrus grandii* Goidanich and *Chelonus annulipes* Wesm.; the ichneumons *Eriborus terebrans* (Gravenhorst), and *Phaeogenes nigridens* Wesmael; and the eulophid, *Sympiesis viridula* (Thomson). A protozoan parasite, *Perezia pyraustae* Pail., is widely distributed in hibernating larvae. It also

Figure 94. European corn borer egg mass about to hatch. (USDA)

infects other life stages, retarding female oviposition and larval development. The fungus, *Beauvaria bassiana* (Bals.), also kills corn borers.

Planting borer-resistant or tolerant hybrid varieties of corn adapted for particular areas is a recommended cultural practice. Consult the state extension service for its recommendations. Some hybrids resist stalk breakage and ear dropping under light-to-medium borer infestation and are therefore harvested more easily. Avoiding very early or very late planting keeps the borer infestation at a low level. The ideal planting date for yield in the latitude of northern Ohio is May 7; for more southern areas of that state, 2 weeks earlier. Corn planted at the end of May may be severely damaged by second-generation larvae. Mechanical destruction of stalks and stubble to kill overwintering larvae is of little value unless carried out on a community-wide basis, and even then it is questionable. Plowing is the suggested method for eliminating crop remnants that harbor borers. The larvae in infested corn cut for silage are destroyed by the silage cutter. All these control measures are of value, but using resistant hybrids and planting them at the proper time are especially important, and in most years are all that is needed to keep damage at a low level in field corn. No-tillage crop management enhances the borer.

The need for chemical treatment is determined by the value of the crop and the intensity of infestation. Early sweet corn or hybrid seed corn plantings usually compensate for the increased costs of chemicals and their application. One or 2 treatments are recommended for early-planted field corn of a susceptible hybrid when examination in the field reveals that 75% of the plants have new larval feeding marks in the whorls. In the 2-treatment schedule, the first is made 1 week after egg-hatching begins and the second a week or 10 days later. The single application is recommended 10 to 12 days after first hatching occurs.

If there are 100 egg masses per 100 plants, treatment should be made when hatching begins. Sweet corn should be treated when 25% of the plants show larval feeding marks or if 20 or more egg masses per 100 plants are found on corn 1 week before tasselling, or 50 or more egg masses per 100 plants for corn 14 to 20 days before tasseling. Make 3 or 4 applications about 5 days apart, beginning when the first eggs are hatching. Avoid using chemicals that kill the important natural or introduced enemies of corn borer.

REFERENCES: *USDA Farmers' Bul.* 2190, 1967; *Dept. Bul.* 1476, 1927; *Conn. Agr. Exp. Sta. Bul.* 462, 1942; *Iowa Agr. Exp. Sta. Pamphlet* 164, 1950; 176, 1952; *Ohio Agr. Exp. Sta. Bul.* 429, 1928; 916, 1962; *J. Econ. Ent.*, 54:550–558, 1961; 56:804–808, 1963; *Bul. Ent. Soc. Amer.* 22:302–304, 1976.

SOUTHWESTERN CORN BORER

Diatraea grandiosella (Dyar), Family Pyralidae

A native of Mexico, the southwestern corn borer was found in Arizona, New Mexico, and Texas about 1913, and has now spread to Oklahoma, Colorado, Kansas, Nebraska, Arkansas, Louisiana, Mississippi, Alabama, Tennessee, Missouri, and other Corn Belt states.

Figure 95. Typical stalk girdling by the southwestern corn borer. (Kansas Agr. Exp. Sta.)

Damage to corn results from the larvae feeding on the leaves, severing the terminal bud causing "deadheart," tunnelling in the stalks and ears, and girdling the stalks near the soil surface, causing stalk breakage (Fig. 95). Other hosts of lesser importance are forage- and grain-type sorghums.

Mature larvae about 4 cm long and white with faint spots, pass the winter in the bases of corn stubble. Brown pupae appear in the larval tunnels in early June and adults emerge about a week later. The adults are approximately 2 cm long and of a soiled white-to-pale yellow color, the male being slightly smaller and somewhat darker. The wings are folded over the body when at rest (Fig. 96).

The eggs are laid during the evening, either singly or overlapping one another in chains or masses, on both upper and lower surfaces of leaves. Hatching larvae at first feed on the leaves and later enter the stalks. All larvae are dull white with a regular pattern of conspicuous dark brown to black spots, except the overwintering forms. Pupation follows, with adult emergence by early August. These moths deposit eggs, and the larvae hatching therefrom becomes fully

Figure 96. The southwestern corn borer: (above) eggs; (below) spotted borers are the summer form, white borer is the overwintering form which girdles the stalk; (right) adult moth; all enlarged. (Kansas Agr. Exp. Sta.)

grown and form the overwintering stage. There are 2 or more generations per year.

Research shows the following will reduce borer damage: substitution of sorghum for corn, early planting, deep plowing of stubble, low cutting of stalks which removes hibernating larvae from the field, fall treatment to dislodge the stubble and expose hibernating larvae to atmospheric changes, resistant varieties, and natural enemies such as birds,* insect predators, and parasites. *Trichogramma minutum* Riley and *Apanteles diatraeae* Muesebeck are perhaps the most important parasite species. With no-tillage crop management it is usually necessary to use pesticides.

REFERENCES: *USDA Tech. Bul.* 388, 1933; *Kansas Agr. Exp. Sta. Bul.* 339, 1950; *Ark. Agr. Exp. Sta. Bul.* 553, 1955; *J. Econ. Ent.*, 50:103, 1957; 54:16–21, 1961; *Ann. Ent. Soc. Amer.* 63:701–706, 1970; *Bul. Ent. Soc. Amer.* 22:302–304, 1976; 23:185–190, 1977.

SOUTHERN CORNSTALK BORER

Diatraea crambidoides (Grote), Family Pyralidae

This destructive corn pest is found primarily in the region south of a line from the states of Virginia to Kansas. Other hosts are the sorghums and several related grasses.

Damage to corn closely resembles that done by the European corn borer. The tiny larvae feed on the leaves and midrib, and, as they increase in size, boring and tunnelling inside the stalks take place. Much stalk breakage results from the feeding of the larger larvae. Because the larvae resemble closely those of the sugarcane borer, both are illustrated together (Fig. 99) for comparative purposes.

The 25 mm long, almost white, pale spotted larva passes the winter in the roots of corn stubble. Pupation takes place in the larval galleries in early spring, followed by moth emergence seven or more days later. The moths are straw-colored, and hold their wings tightly over the body when at rest, but when extended have an expanse of 3 cm. White oval eggs are deposited either singly or several overlapping one another, on both upper and lower leaf surfaces. The hatching larvae feed first on the leaves and later on the stalks; on becoming fully grown they pupate and then emerge as adults several days later. There may be as many as 4 generations per year in the warmer regions where this insect occurs.

Natural control results from several parasites and predators. The egg parasite *Trichogramma* has been reported as being important. Recommendations for

* The yellow-shafted flicker has removed as many as 80% of the overwintering larvae in some areas.

applied control include destruction of the stalks immediately after harvest, fall and spring plowing of the stubble, late planting, and crop rotation.

REFERENCES: *S.C. Bul.* 294, 1934; *N.C. Bul.* 274, 1920; *J. Econ. Ent.*, 46:176, 1953; *USDA Leaflet* 363, 1954.

LESSER CORNSTALK BORER

Elasmopalpus lignosellus (Zeller), Family Pyralidae

This insect has been sporadically numerous in the South for many years; its range extends from Maine to southern California, and in Mexico, Central and South America. Corn and cowpeas are favorite hosts, but it also attacks sorghums, wheat, beans, peas, peanuts, turnips, and a variety of grasses.

The moth is brownish gray with a wing expanse of less than 2.5 cm. The fore wings of the female are darker than those of the male (Fig. 97). The larvae are slender caterpillars, about 2 cm in length. The prevailing color is a light green, but there are faint stripes and more prominent transverse bands of brown.

The winter is passed in the soil as larvae or pupae. Adult moths emerge in the spring and lay their eggs on the host plants. The hatching larvae feed first on the leaves and then bore into the stalks, resulting in damage similar to that of the southern cornstalk borer. After feeding for about 3 weeks they leave their

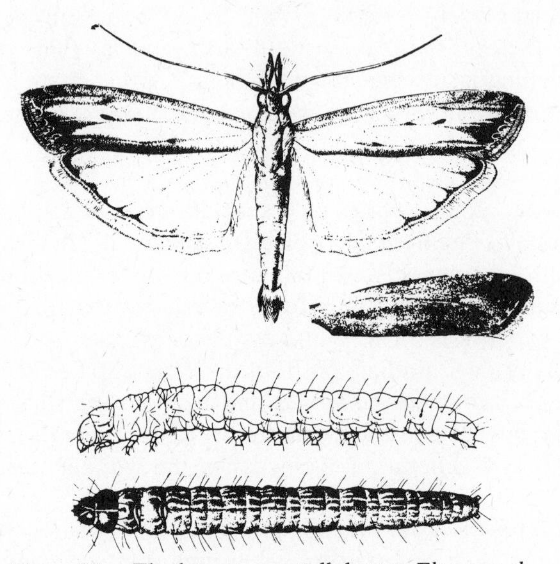

Figure 97. The lesser cornstalk borer, *Elasmopalpus lignosellus* (Zeller); (above) male moth and fore wing of the female; (below) side and top view of a larva; 3.6×. (Luginbill and Ainslie, USDA)

burrows and pupate in silken cocoons under debris at the soil surface. Adults emerge and begin laying eggs that become the next generation. There are 2 or more generations per year over most of the range they inhabit.

Control by sanitation measures, late fall plowing, and crop rotation are recommended and are usually adequate.

REFERENCE: *USDA Bul.* 539, 1917.

SORGHUM WEBWORM

Celama sorghiella (Riley), Family Noctuidae

This insect is frequently responsible for severe damage to grain sorghum crops. In 1882 it was first recorded in Alabama and has since been reported from most South Atlantic and Gulf Coast states. It is known to range northward to Nebraska, Illinois, and Indiana. Injury is caused by the chewing larvae that attack and destroy the seed. Other hosts besides sorghums are Sudan grass, Johnson grass, broom corn, and rye. Outbreaks are periodic and most likely to occur during seasons characterized by prolonged periods of rainy weather.

The adult is a moth with wings expanding a little more than 15 mm. It is nearly white in color with irregular dark markings along the anterior margins of the fore wings. Larvae are 13 mm long, green with distinct darker stripes, and on each segment several bristle-bearing tubercles.

The winter is passed in the mature larval stage, on or in the food plant. Pupation follows in March, and normally the moths start emerging about April 1. Oviposition begins in a few days and continues throughout the growing season. The eggs are laid singly and cemented to the host plants. Hatching occurs in a few days, the larvae becoming fully grown in 2 weeks, after molting 4 to 7 times, but normally 5. Pupation takes place within a cocoon spun by the larvae on the host, and in about a week adult moths of the next generation begin emerging. Six or more complete generations may develop each season.

Control measures practiced are primarily cultural or mechanical. Plowing under all crop residues in late fall to destroy the overwintering larvae, eliminating Johnson grass in the vicinity of sorghum fields during the winter, and planting early to mature the crop before injurious infestations develop are perhaps the most important. Although the insect is attacked by several parasites, these are not effective in preventing the development of injurious populations. The most effective natural control is dry weather accompanied by high temperatures.

REFERENCES: *Texas Agr. Exp. Sta. Bul.* 559, 1938; *J. Econ. Ent.*, 56:483–484, 1963.

SORGHUM MIDGE

Contarinia sorghicola (Coquillet), Family Cecidomyiidae

This midge is considered the most important insect attacking grain sorghums. It is believed to have been introduced into the United States from southern Asia. Its hosts include 48 varieties of grain and sweet sorghums, corn, Johnson grass, Sudan grass, and allied plants. Injury results from the feeding by the larva on the developing seeds. The midge is apparently well-established throughout most of the sorghum-growing regions of North America and the world.

Larvae hibernate within cocoons in the spikelets of their host plants. In the spring these change to pupae and emerge as adults, which are tiny fragile flies (Fig. 98), similar to the adults of hessian fly and wheat midge. If conditions are not favorable some larvae remain dormant until the following year. Emergence continues for several weeks with the result that subsequent generations are produced about every 15 days. These overlap to such an extent that the effect is a continuous production of all life stages throughout the summer. In the fall, a large number of larvae form cocoons in which they pass the winter.

The eulophid parasite, *Aprostocetus diplosidis* Crawford (Fig. 45), is of some value in biological control. Yield losses can be minimized by the following management practices: plow under crop remnants containing hibernating larvae; plant early using seed of uniformly blooming cultivars; eliminate Johnson grass where possible; remove from the field "out-of-season" blooming heads; clean up fields after harvest; avoid planting near earlier sorghums, Johnson grass, or other hosts; prepare a good seed bed and cultivate the field to produce as uniform a crop as possible; avoid cutting hay while sorghum is blooming to prevent migration of egg-laying adults to the sorghum. Significant yield losses did

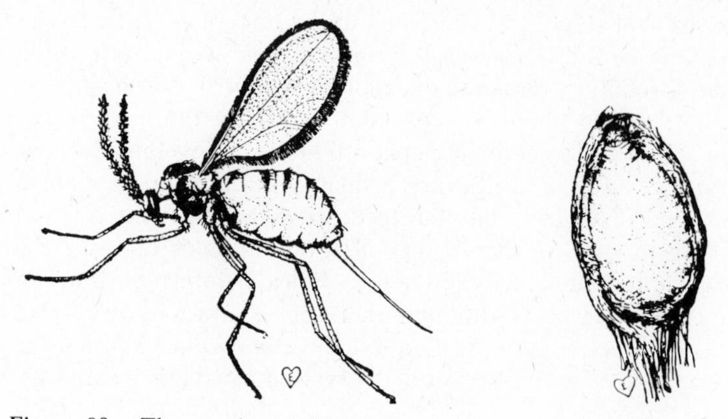

Figure 98. The sorghum midge, *Contarinia sorghicola* (Coq.). Adult female and larva in its cocoon; 30×. (Gable, Baker, and Woodruff, USDA)

not occur until the midge infestation exceeded 40 adults emerging per head from plantings on June 29 in California.

REFERENCES: *USDA Tech. Bul.* 778, 1941; *Farmers' Bul.* 1566, 1959; *J. Econ. Ent.*, 65:796–799, 851–854, 1972; *Proc. N.C. Branch Ent. Soc. Amer.*, 18:86, 1963; *Calif. Agr.* 29(9):4–5, 1975.

SUGARCANE BORER

Diatraea saccharalis (F.), Family Pyralidae

This insect is said to be native to the West Indies, Central and South America, but it has been known in North America since before 1856 and is now established in parts of Florida, Mississippi, Louisiana, and Texas. It is the principal pest of sugarcane and in the infested area causes loss amounting to a high percentage of the crop. Other hosts often seriously damaged are corn, sorghums, rice, and some wild grasses. This borer is closely related to the southern cornstalk borer and the southwestern corn borer.

Damage to all hosts is due to the larvae tunnelling inside the stalk, thus decreasing growth and weakening the plant until parts of it may die or break over, especially in high winds. Sugarcane damage is characterized by deadhearts of young plants, dead tops in older plants, broken stalks, loss in weight of sucrose, and injury to seed cane.

The sugarcane borers overwinter as larvae in tunnels of their host plant. As the temperature rises in the spring they become active and extend their tunnels toward the plant surface until the covering of tissue remaining is very thin. Pupation follows and the straw-colored moths (Fig. 99), with a wing expanse of 25 mm, emerge from the stalks about a week later. After mating, the females deposit their eggs on the leaves of the plants. These overlapping eggs are in rows or clusters averaging about 25 or more per mass. Hatching takes place in 4 to 9 days, and the tiny larvae at first feed on the leaves or in the whorl and then bore into the stalk. At the end of 20 to 30 days larval development is complete, pupation follows, and in about 7 days the next generation of adults appears. There are 4 or 5 generations annually. Fully grown larvae are about 25 mm long, pale yellow to white with brown spots, although in the winter these spots are almost absent and the color is a deeper yellow (Fig. 99).

Fewer borers pass the winter successfully if low temperatures and heavy rainfall prevail. Late freezes, heavy rains during the hatching period, and prolonged periods of dry weather also adversely affect the borers.

The egg parasite, *Trichogramma minutum* Riley, is of value in reducing borer damage. The braconid, *Agathis stigmaterus* (Cresson), and the tachinid fly, *Lixophaga diatraeae* (Townsend), are other parasites of some value. Many species of birds are listed as predators of sugarcane borers and their near relatives. Ants, spiders, and ground beetles are helpful predators.

Applied control measures consist of more thorough clean-up of pieces of millable cane to eliminate overwintering quarters; burning or plowing under trash after harvest to destroy hibernating larvae; cutting low to decrease the number of borers overwintering in stubble; selecting seed cane as free from borers as possible; soaking seed cane in cold water for 72 hours kills a high percentage of the larvae; where agronomically practical plant resistant varieties of cane.

If natural and cultural controls fail to provide adequate check of this pest, a pesticide will be needed. Treatment is suggested to control only the second- and third-generation borers. Make the first application after the joints have started to form, and when at least 2% of the plants are infested with young larvae that have not bored into the stalks.

REFERENCES: *USDA Tech. Bul.* 41, 1928; *Farmers' Bul.* 1884, 1941; *Cir.* 878, 1951; *Leaflet* 479, 1960; *J. Econ. Ent.*, 41:914–918, 1948; 52:821–824, 1959; 56:407–409, 1963; 57:350–353, 1964; 62:620–622, 1969.

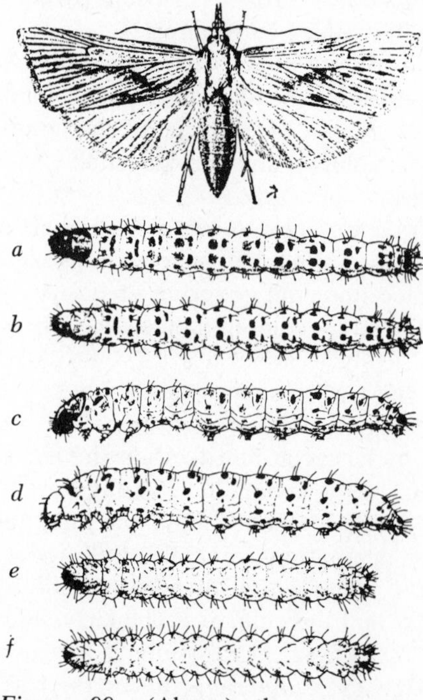

Figure 99. (Above) the sugarcane borer adult; (below) dorsal and side views of larvae; *a* and *c*, the sugarcane borer, summer larvae; *e*, same, winter form; *b* and *d*, the southern cornstalk borer, summer form, and *f*, same winter form; 2×. (USDA)

SUGARCANE BEETLE

Euetheola rugiceps (LeConte), Family Scarabaeidae

The sugarcane beetle (also called rough-headed cornstalk beetle) is found in most of the southern states. A serious pest of sugarcane wherever grown, except in southern Florida, it causes great loss of rice and corn in some localities and also attacks cotton, strawberries, roses, and wild grasses. The adult beetles injure the host plants by chewing on the young shoots and leaves at ground level or slightly below. In rice fields this damage occurs before the fields are submerged.

Adult sugarcane beetles are black and 12 mm in length. They overwinter in the soil of well-drained sod land and become active during the first warm days of February and March, with feeding and ovipositing continuing through June. Most eggs are deposited in sod land near cultivated fields. The larvae (Fig. 100) are white, fleshy grubs with curved bodies, about 32 mm long when fully grown. They feed on the roots of grasses. Under laboratory conditions the average incubation period for eggs is 14 days, for larval development 65 days, and for the pupal period 9 days. New adults appear in August and September and feed for a short time before hibernation. There is only one generation per year.

Many natural enemies such as birds, toads, frogs, skunks, and a few insect predators and parasites reduce the population slightly but cannot be depended upon for control. Eliminating sod land breeding areas, especially near rice or sugarcane plantings; submerging rice fields; and planting sugarcane early in August, which gives a better stand and consequently less noticeable beetle damage, are all recommended control measures.

REFERENCES: *USDA Cir.* 632, 1942; 878, 1951; *Leaflet* 520, 1963; *J. Econ. Ent.*, 51:631–633, 1958.

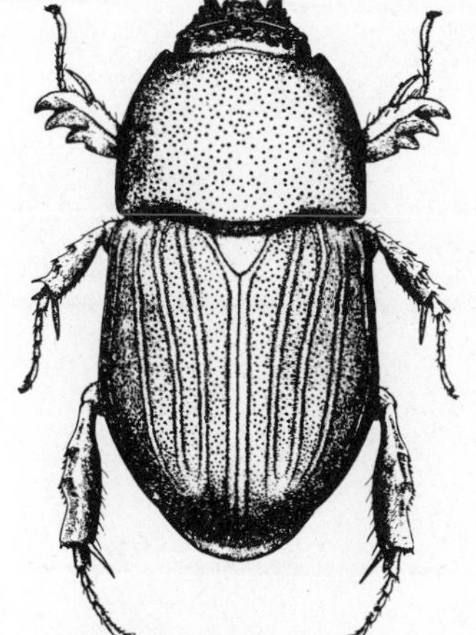

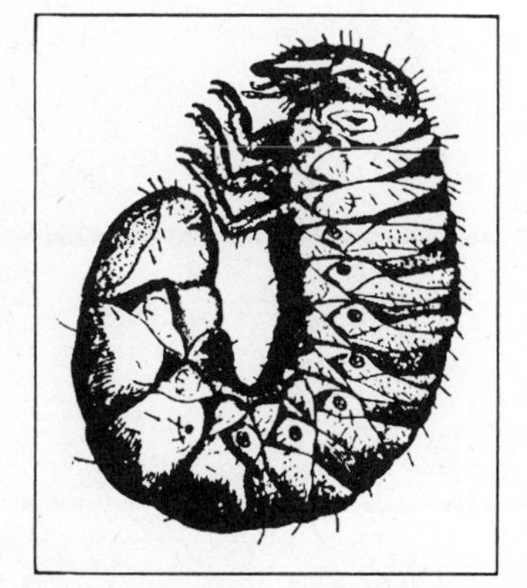

Figure 100. The sugarcane beetle, *Euetheola rugiceps* (LeC.): (left) adult; (right) larva; 3×. (USDA)

SUGARCANE MEALYBUGS

ORDER HOMOPTERA, FAMILY PSEUDOCOCCIDAE

These insects are troublesome throughout southern Florida and Louisiana and other areas of the world where sugarcane is grown. Damage is caused by the nymphs and adults removing sap from the cane plant, most often between the sheath and the stalk. Their presence also makes syrup manufacture more troublesome and generally lowers the quality of the product.

The common species are the pink sugarcane mealybug, *Saccharicoccus sacchari* (Cockerell), and the gray sugarcane mealybug, *Dysmicoccus boninsis* (Kuwana). Both are somewhat similar in habit and life cycle so only the gray species is described.

The female mealybug, which is more commonly seen than the male, is a soft-bodied, grayish insect about 3 mm long when fully grown, and of a flattened oval shape, with small protruding processes near the margin of the body (Fig. 101). Its eggs are deposited within a white, cottony substance. Hatching nymphs migrate over the plant, feeding on the sap, with continuous generations occurring throughout the year. The eyes of fall-planted seed cane may be killed by the mealybugs, thus reducing the stand.

Suggested control measures are: planting uninfested seed cane in new fields some distance from those known to be infested; destroying all scraps of cane left on the fields and around the mills after grinding; when harvesting, cutting close to the ground so that less cane remains on which winter feeding may take place; avoiding contamination of new areas by thoroughly cleaning out trucks and wagons employed in hauling infested cane to the mills. Seed cane may be freed from infestation by soaking in hot water at 122–126 F (50–52 C) for 30

Figure 101. The gray sugarcane mealybug; 10×. (USDA)

minutes, or soaking in water at ordinary temperature for seven days. Ants are attracted to the honeydew secretions of mealybugs, and perhaps eliminating the ants would contribute much toward reducing pest populations.

REFERENCES: *USDA Cir.* 878, 1951; *Bul. Soc. Entomol. Egypt* 53:21–39, 41–62, 499–516, 1970.

CHINCH BUGS

Blissus leucopterus (Say), Family Lygaeidae

The chinch bug is widely distributed throughout the United States, southern Canada, Mexico, and Central America but causes the greatest damage in the regions drained by the Ohio, Missouri, and Mississippi Rivers. It is mainly a pest of corn and sorghums but also injures small grains and other grass crops. Lawns and golf courses may be severely damaged in years of chinch bug abundance. Native to North America, this insect apparently fed on the prairie grasses before large acreages of cultivated grass crops were planted by man.

The damage results when the piercing-sucking nymphs and adults become numerous and remove plant sap, thus causing retardation of growth and sometimes death of the plant.

The black and white adults are scarcely 5 mm long, typically true bug in shape, with whitish wings which are marked by a dark triangle on their outer margins. Both long- and short-winged forms are found, but the long-winged form prevails throughout the central states. The legs and base of the antennae are red. Young bugs are bright red but become darker as they near the adult stage (Fig. 102).

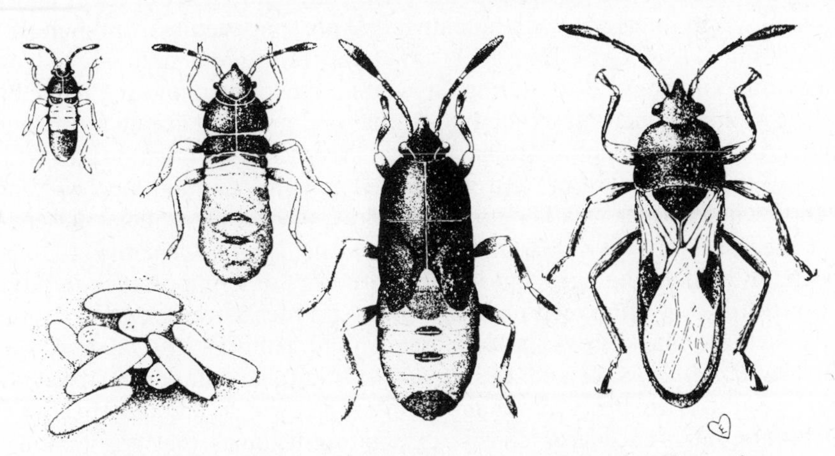

Figure 102. Eggs, nymphs, and adult of the chinch bug, *Blissus leucopterus* (Say); 9×. (USDA)

Adult bugs hibernate in any shelter available but are most often found in greater abundance in heavy grass sod along roadsides, fence rows, and pasture lands. They emerge from hibernation quarters with the coming of warm spring days and oviposit on small grains and grasses. Each female deposits several hundred yellow-white eggs, at the rate of 15 to 20 per day, behind the lower leaf sheaths or in the soil around the plants. These hatch in 1 to 3 weeks into nymphs which feed and pass through 5 instars before reaching the adult stage. As barley, rye, and wheat ripen the plants become less succulent, and the chinch bug nymphs crawl to adjacent corn or sorghum fields where serious damage may result. It is during this migration period that barrier-trap control is effective. The new adults reach maturity about the middle to latter part of June and fly throughout corn and sorghum fields, laying their eggs, which hatch to form the next generation. By late summer or early fall second-generation adults appear which become the overwintering stage. In the extreme South, where activity begins earlier in the spring and continues later in the fall, a third generation usually develops. A partial third generation may occur as far north as Iowa.

The hairy chinch bug, *Blissus leucopterus hirtus* Montandon, is sometimes considered only a variety of the other species. Primarily a turf pest, it is predominantly short-winged and is, at times, abundant in lawns in northeastern and midwestern United States. Its life cycle is similar to that of *B. leucopterus*. The southern chinch bug, *B. insularis* Barber, is the most troublesome form attacking grasses in lawns and golf courses in the South. In south Florida there are 6 or 7 generations per year.

Weather is the chief factor governing the abundance of chinch bugs. They are most susceptible to mortality during the period of hatching. Frequent heavy driving rains during this period beat the young bugs into the soil, cover the eggs with mud which interferes with hatching, and prevents adults from ovipositing their full number of eggs. Extremely low temperature and sudden changes in temperature kill many adults during the winter and spring months. Chinch bugs thrive in prolonged periods of hot dry weather but are killed during warm humid weather by the green muscardine fungus, *Beauvaria bassiana* (Bals.). The spores of this fungus are present wherever bugs are common, and will cause infection when proper weather conditions prevail, making artificial dissemination unnecessary. A tiny wasp, *Eumicrosoma beneficum* Gahan (Fig. 103), is parasitic in chinch bug eggs and is considered of some importance in natural control. Birds that often feed on chinch bugs are meadowlarks, brown thrashers, red-winged blackbirds, starlings, quails, and catbirds. The big-eyed bugs, *Geocorus bullatus* (Say) and *G. uliginosus* (Say), are important predators. Other predators are *Orius tristicolor* (White), lacewings, and lady beetles.

Cultural control measures consist of planting legumes that are immune to attack, or varieties of corn and sorghums that are decidedly resistant to attack. Growing soybeans or cowpeas in the field of corn, or clover in small grains,

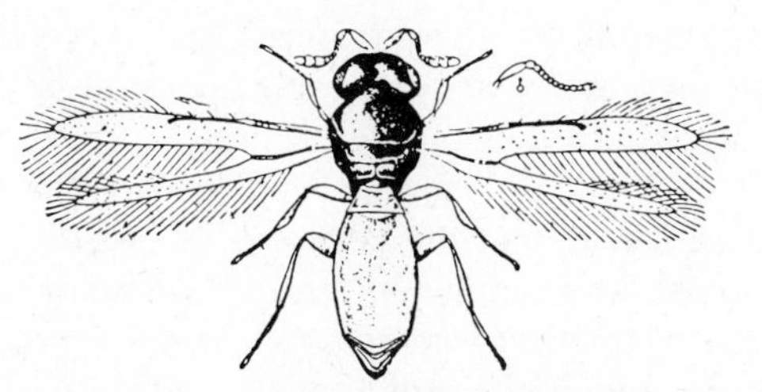

Figure 103. Adult female of *Eumicrosoma beneficum* Gahan, a chinch bug egg parasite. Insert of male antenna. Greatly enlarged. (Gahan, USDA)

produces a dense shade at the base of the plants which is unfavorable to the bugs. Damage may be reduced by avoiding adjacent plantings of small grains and corn.

Plowing under ruined fields of small grains or corn may be of some value in control if the soil is immediately disked and prepared into a dust mulch, thus preventing escape of the bugs. Burning over areas where the bugs are likely to hibernate may do more harm than good and is not recommended. Heavy watering of lawns reduces populations.

A control recommendation is the employment of a pesticide barrier about 50 to 60 feet wide where fields of corn are adjacent to small grains. Half this treated area may be in the corn field and half in the small grain. The chemical should be applied beyond the end of the field for about 150 feet to prevent nymphs from migrating around the barrier. Treat the soil just before the nymphs begin crawling toward the corn; they are killed when they come in contact with the chemical. If bugs are found feeding on a few of the outer rows of corn, spraying them with chemical may be necessary. Several chemicals are now used for treating lawns.

REFERENCES: *USDA Farmers' Bul.* 1780, 1937; *Leaflet* 364, 1954; *Tech. Bul.* 585, 1937; *Cir.* 508, 1938; *Conn. Agr. Exp. Sta. Bul.* 677, 1966; *Cir.* 233, 1970; *J. Econ. Ent.*, 61:523–525, 1968; *Ohio Agr. Exp. Sta. Bul.* 122, 1963; *Fla. Entomol.* 55:231–235, 1972; *Bul. Ent. Soc. Amer.* 19:91–92, 1973.

SAP BEETLES

ORDER COLEOPTERA, FAMILY NITIDULIDAE

The common name of this family of beetles is based on the habits of the species that are attracted to the wounds of trees where they feed on sap. At times some species become very serious pests of sweet corn grown for processing or for the roasting ear market. They enter the tip of the ear and feed on the developing kernels, particularly in ears already infested with corn earworm or damaged by birds. Adults are attracted to the more mature ears in a field, but under heavy infestation they will oviposit in any available, regardless of maturity. Some species also feed in ripened fruits and vegetables.

The primary pest species attacking sweet corn are the dusky sap beetle, *Carpophilus lugubris* Murray and the corn sap beetle, *C. dimidiatus* (Fab.). Others are *C. antiquus* Melsheimer, *C. brachypterus* (Say), *C. corticinus* Erichson, *C. freemani* Dobson, *C. hemipterus* (L.), *Cryptarcha ampla* Erichson, and *Glischrochilus quadrisignatus* (Say).

The dusky sap beetle is dull black and about 4 mm long with short elytra and capitate antennae (Fig. 104). Winter is passed as adults in protected places above ground, and as adults and pupae in the soil. Activity begins with warm spring weather, and early feeding is on sap from wounds on trees or on decomposing plant material. Adults are noticed in corn fields about the time tassels appear. Mating and egg-laying begin; the tiny white eggs are deposited in earworm frass, wet accumulations of pollen in leaf axils, corn silks, and in smut galls. Hatching occurs in 2½ days at 75 F (24 C), and the white larvae pass through 3 instars in a period of 14 days, then drop to the soil where pupal cells are formed a few inches below the surface. After a period of approximately 14 more days new adults emerge and the cycle is repeated. The total developmental period is about 30 days and 3 to 4 generations occur annually in the latitude of Illinois. Some adult females lay over 300 eggs and live for 147 days. Except for *G. quadrisignatus* the rest are similar in appearance and life cycle.

Delayed planting averts damage in some regions. Damage is less where varieties with long tight husks are grown and where earworm is controlled by pesticides.

Glischrochilus quadrisignatus is known by several common names but picnic beetle seems appropriate when it becomes a nuisance in picnic areas of homes

Figure 104. The dusky sap beetle, *Carpophilus lugubris* Murray; 15×. (Drawing by C. A. Triplehorn, courtesy of W. A. Connell, Univ. of Delaware.)

and parks. Abundance is attributed to breeding in melon rinds, tomatoes, sweet corn, and other fruits and vegetables discarded in picnic areas and at roadside fruit and vegetable markets.

Adults are nearly 6 mm in length, black with 4 orange red spots on the elytra. Overwintering beetles become active by May and begin to oviposit. New adults reach abundance in July and September; they feed on damaged fruits, vegetables, and field crops until cold weather. Two generations develop each year.

Disposing of breeding materials contributes to lower populations of the picnic beetle.

REFERENCES: *J. Econ. Ent.*, 49:539–542, 1956; 52:640–642, 1959; 53:174–175, 1960; 55:671–674, 922–925, 1962; *Del. Agr. Exp. Sta. Bul.* 318, 1956; *Proc. N. C. Branch Ent. Soc. Amer.* 18:38–43, 1963.

CEREAL LEAF BEETLE

Oulema melanopus (L.), Family Chrysomelidae

This pest, distributed widely in Europe, also occurs in Iran, Turkey, Morocco, and Tunisia. It was first found in the United States in July 1962, in Berrien County, Michigan. Subsequent surveys have shown it to be present in many eastern and central states of the United States and in southern Canada.

Both adults and larvae damage the young tender leaves of oats, wheat, barley, rye, corn, and grasses, particularly timothy and quack grass. The preferred host seems to be oats. The adults feed primarily on grain shoots and adjacent grasses, whereas the larvae eat out long narrow strips of tissue between the leaf veins. In some infested areas this damage has resulted in almost complete loss of the oat crop. It is also a vector of maize chlorotic mottle virus (MCMV) causing the disease, corn lethal necrosis.

The adult beetle is slender and about 4 mm in length; the elytra and head are metallic blue-black and the legs and prothorax are red; males are slightly smaller than the females. Overwintering beetles appear in the spring; following mating the females lay tiny cylindrical eggs that are rounded at the ends and yellow at first, darkening to almost black before hatching. Eggs are placed on the upper surfaces of the host plant leaves. The larva has a pale yellow body with brown-black head and legs, but it is usually covered by a globule of fecal matter that obscures its coloration except for the head and legs. Pupation occurs in earthen cells in the top 2 inches of soil, and adults begin emerging 20 to 25 days later. The entire life cycle may be completed in 46 days. New beetles feed on grasses for a short period, then pass into summer diapause till autumn; then they hibernate under crop remnants until the following spring. One generation occurs per year.

Quarantines have been imposed in the infested areas to prevent spread of the insect in the movement of hay, straw, fodder, grains, sod and harvesting machinery. Fumigants have been useful in eliminating infestations in grains, forage, and straw.

Chalcid and ichneumonid parasites of the larval stage have reduced populations of this pest in its native country. An introduced eulophid parasite, *Tetrastichus julis* (Walker), and the mymarid *Anaphes flavipes* (Förster) are now established in the infested area and parasitization is high. Lady beetles, especially *Coleomegilla maculata* and *Hippodamia convergens,* are primary factors in destroying eggs and larvae of this pest. The fungus, *Beauvaria bassiana,* attacks adults when conditions are favorable. Rainstorms are of major importance in causing larval mortality. Resistant varieties are helpful in reducing crop loss. On oats, if 1 larva per stem completes its development, yields will be reduced at least 3 bu/acre. However, infestations below 2 larvae/stem do not justify the cost of chemical control measures.

REFERENCES: *USDA* PA-550, 1964; *J. Econ. Ent.* 62:699–702, 1969; *Ann. Ent. Soc. Amer.* 63:52–59, 1970.

CORN ROOTWORMS

Diabrotica spp., Family Chrysomelidae

The larvae of several species of beetles that feed on the roots of corn are called corn rootworms. Weakened root systems result and the plants are easily blown down by strong winds giving a "sled-runner" effect. The adults are similar in size and form and eat the foliage and silks of corn. Some species attack other plants including cotton, legume, and vegetable crops, which they may damage more often than they do corn. Additional damage results to corn and cucumbers should the beetles carry and transmit the organisms of bacterial wilt to these plants.

The southern, northern, and western species are also vectors of maize chlorotic mottle virus (MCMV) that causes the disease, corn lethal necrosis. The disease results when a plant is infected with MCMV in combination with either maize dwarf mosaic virus or wheat streak virus.* A brief discussion of some common rootworms follows.

Southern Corn Rootworm, *Diabrotica undecimpunctata howardi* Barber, is a widely distributed pest throughout southern Canada and the United States east of the Rocky Mountains, and it has a large host range. The adult is better known as the spotted cucumber beetle, since it often attacks cucumbers; it is greenish yellow with 11 black spots (Fig. 105). Adults overwinter under crop remnants, become active in the spring, and lay their eggs in the soil next to the host plant. The tiny white larvae burrow in the roots, crowns, and stems and may even feed externally on the roots. Injury is especially serious to very

* Information from Dr. L. R. Nault, Virus Vector Lab, OARDC, Wooster, Ohio.

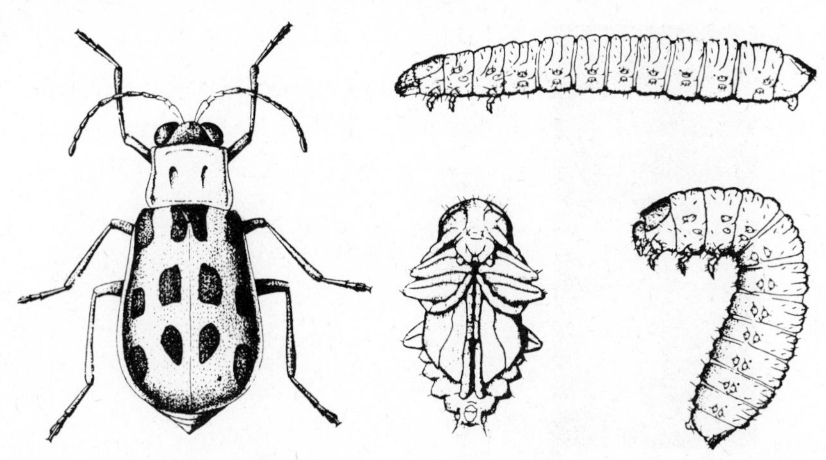

Figure 105. The southern corn rootworm, *Diabrotica undecimpunctata howardi* Barber. Adult, larvae, and pupa; 5×. (Isley, Ark. Agr. Exp. Sta.)

young plants, and replanting is often necessary. There are 2 generations over most of its range, although in the northern limits there may be only 1. This insect is a vector of cucumber mosaic virus and bacterial wilts of corn and cucumbers. It also causes serious damage to peanuts in the southern states.

Northern Corn Rootworm, *Diabrotica longicornis* (Say), is found in much of the range of the spotted species but does most damage in the Corn Belt. Adults are scarcely 6 mm in length (Fig. 106) and are uniformly pale green or yellowish green. In late summer and fall each female may lay 300 eggs in the soil of corn fields, and these remain unhatched throughout the winter. By late spring and early summer newly hatched whitish larvae appear and migrate through the soil and feed in or on corn roots should they be present. So far as is known they attack no other important food plants, and if corn roots are not available they die. Pupation takes place in the soil in the summer, and adults appear in late July and are present until frost. Adults often congregate and feed on fresh corn silks, resulting in ears with few kernels (Fig. 107). There is one generation per

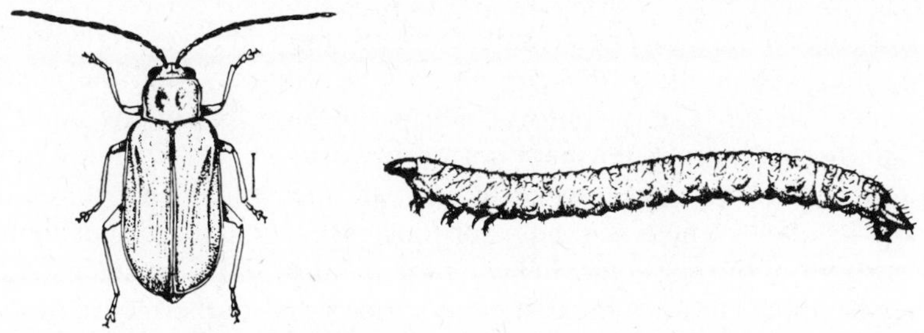

Figure 106. Adult and larva of the northern corn rootworm, *Diabrotica longicornis* (Say); 4×. (USDA)

Figure 107. Injury to sweet corn by adult corn rootworms. (USDA)

year. Since this insect attacks only corn and the eggs are laid in corn fields, rotation of crops is an effective means of control.

Western Corn Rootworm, *Diabrotica virgifera* LeC. (Fig. 108), is an important species in Nebraska, Iowa, South Dakota, Minnesota, Colorado, and Kansas, but is spreading rapidly eastward and southward and now infests most of the Corn Belt. The adults are strong flyers and voracious feeders of corn leaves and silks. Overwintering occurs as eggs in the soil, and 1 generation per year is typical. Life cycle is similar to *D. longicornis*.

Banded Cucumber Beetle, *Diabrotica balteata* LeC., has been observed feeding on corn and sweet potatoes but it is usually considered a general feeder with a southern range of distribution. Two to three generations occur per year in Gulf States.

Western Spotted Cucumber Beetle, *Diabrotica undecimpunctata undecimpunctata* Mannerheim feeds, in the larval stage, on the roots of corn, rice, grasses, and other plants. It is occasionally numerous enough in the western states to cause serious damage. The life cycle is similar to that of the southern corn rootworm. Reports indicate that it does not overwinter in areas where the mean January temperature drops below 32 F (0 C).

Natural control of rootworms results from ground and rove beetle predators; also from the tachina fly, *Celatoria diabroticae* (Shimer), and the nematode, *Howardula benigna* (Cobb), both of which are parasites of the adult stage. Protozoan diseases attack northern corn rootworms.

Cultural control measures suggested for all species consist of rotation of crops, but its effectiveness is greater for those species that are specific to corn and overwinter in the egg stage; delaying the planting date is a means of avoiding attack from the larvae of the first generation of the southern corn rootworm; plowing in early spring or in the fall and frequent cultivation preceding planting reduce populations of most species. Reduced or no-tillage crop management

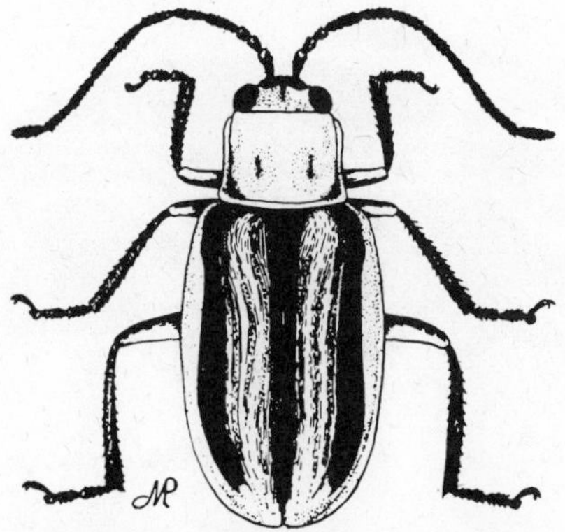

Figure 108. Western corn rootworm, *Diabrotica virgifera* LeC.; 6×. (Colo. Agr. Exp. Sta.)

enhances rootworm problems, especially for species overwintering as eggs. Using soil pesticides is necessary with such management practices.

REFERENCES: *Ark. Bul.* 232, 1929; *W. Va. Agr. Exp. Sta. Cir.* 102, 1957; *J. Econ. Ent.* 55:904–906, 1962; 62:541–543, 1969; 65:1697–1700, 1714–1718, 1972; *Bul. Ent. Soc. Amer.* 22:302–304, 1976.

CORN FLEA BEETLES

ORDER COLEOPTERA, FAMILY CHRYSOMELIDAE

Widely distributed, flea beetles may cause severe injury to corn, sorghums, broom corn, and small grains. The larvae destroy the roots; adults eat the green tissue between leaf veins, usually on the upper surface, resulting in silvery streaks. Some species eat small holes in the leaves as adults; others serve as disseminators of bacteria causing wilt of corn or Stewart's disease (*Bacterium stewarti* Smith) and the virus causing corn lethal necrosis.

Important species are the corn flea beetle, *Chaetocnema pulicaria* Melsheimer; toothed flea beetle, *C. denticulata* (Illiger); sweetpotato flea beetle, *C. confinis* Crotch (Fig. 211); desert corn flea beetle, *C. ectypa* Horn; pale-striped flea beetle, *Systena blanda* Melsheimer (Fig. 227); red-headed flea beetle, *S. frontalis* (Fabr.); and western black flea beetle, *Phyllotreta pusilla* Horn (Fig. 200). The corn flea beetle, *C. pulicaria* (Fig. 109) is the major species responsible for harboring corn wilt bacteria over winter and the secondary spreading of these organisms during the corn-growing season. Experiments have shown that *C. denticulata* ranks a close second in spreading bacteria causing Stewart's disease. In addition, *C. pulicaria* and *S. frontalis* are vectors of maize chlorotic mottle virus (MCMV) which in combination with either maize

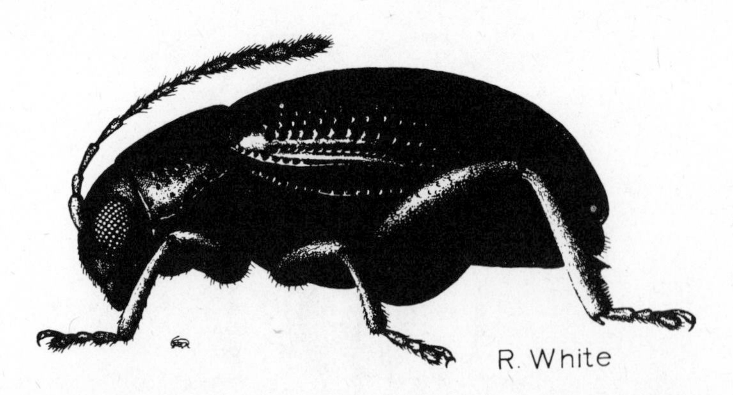

Figure 109. The corn flea beetle, *Chaetocnema pulicaria*; actual size indicated by small figure. (Courtesy of R. E. White, USDA)

dwarf mosaic virus (carried by aphids or leafhoppers) or wheat streak virus (carried by mites) will cause the disease known as corn lethal necrosis.

Most flea beetles overwinter as adults, emerge from hibernation in the spring, mate and lay their eggs on the leaves or on the ground near the base of the host plants. The tiny gray-white larvae with brown heads feed on the roots. Some species have a single generation per season, others having 2 or more. Adult beetles are 1.3–3.5 mm long, shiny black to dark gray in color, except the pale-striped flea beetle which is larger and has a broad, white stripe on each wing cover (Fig. 329). Flea beetle adults jump readily when disturbed.

Since the flea beetles also develop on many weeds, keeping fields and field borders as free of them as possible is of value in control. Frequent cultivation destroys or disrupts the activity of the larvae. Delaying the planting date and cultivation of wilt-resistant hybrid varieties of corn decrease the loss from Stewart's disease. Low winter temperatures will reduce overwintering populations. There is some correlation between the winter temperature index and incidence of corn wilt. This index is the sum of the average temperatures for December, January, and February in the area under consideration. An index of 90 or more Fahrenheit, (32 Celsius) is usually indicative of corn wilt outbreaks. The temperature data can be obtained from the local weather office.

Chemical control of adults with certain pesticides has been considered satisfactory if applied at the time of seedling emergence. Several applications about five days apart may be necessary. Treatment is not necessary every year but may be justified in years of flea beetle abundance after a mild winter.

REFERENCES: *USDA Tech. Bul.* 362, 1937; *J. Agr. Research,* 52:585–608, 1936; *Yearbook of Agriculture,* p. 587, 1952; *J. Econ. Ent.,* 55:1008–1009, 1962, *CEIR 19* (32):634, 1969.

SEEDCORN BEETLES

Order Coleoptera, Family Carabidae

Most species of ground beetles are beneficial but there are at least two that sometimes become pests of newly planted corn seeds. They are *Stenolophus lecontei;* (Chaudoir) and *Clivina impressifrons* LeConte, the latter known as the slender seedcorn beetle. Soft kernels with high moisture content, at or near the soil surface, are most often attacked. The beetles are widely distributed in North America and become pests mainly when low vitality seed has been used and when very cool weather prevails, delaying germination.

The life cycles of the beetles have not been determined, but both species evidently overwinter as pupae or adults because adults appear early in the spring and are attracted to light. *S. lecontei* is broad, dark-brown, and striped, *C. impressifrons* quite slender and brown; both species are about 8 mm in length.

Damage is best avoided by planting seed of high vitality late enough to insure quick germination. Treating seed corn with diazinon is sometimes recommended. Reduced tillage cropping systems enhances these insects.

REFERENCES: *USDA Bul.* 85, Part II, 1909; *S. C. Agr. Exp. Sta. Bul.* 478, 1960; *Ohio Coop. Ex. Bul.* 545, 1979.

BILLBUGS

Sphenophorus spp., Family Curculionidae

Billbugs, or snout beetles, are listed as pests of corn, small grains, other grasses, rushes, sedges, and, in the South, peanuts, rice, and sugarcane.

Injury is caused by the larvae eating the roots and crowns of plants and by the adults feeding on stems and foliage. The characteristic feeding punctures in corn and coarse grasses show up as a series of transverse holes of the same size and shape in the leaf. They are the result of a single puncture through the leaf in the bud stage before it has unfolded. Punctures in the stems are likely to cause more damage even though they are less noticeable.

Billbugs are more likely to be numerous in lowland areas, which are wet or subject to overflow from adjacent streams. Distribution of the various species is throughout the cultivated and grassland regions of the United States and Canada. The most common forms include the maize billbug, *Sphenophorus maidis* Chittenden (Fig. 110); the clay-colored billbug, *S. aequalis* Gyll. (Fig. 111); the bluegrass billbug, *S. parvulus* Gyll.; the southern corn billbug, *S. callosus* (Oliv.); the corn or timothy billbug, *S. zeae* Walsh; the hunting billbug, *S. venatus vestitus* Chittenden; and the cattail billbug, *S. pertinax* Oliv.

The maize billbug is one of the largest forms and sometimes exceeds 15 mm in length; other species range from 6-14 mm. The adults vary in color from light

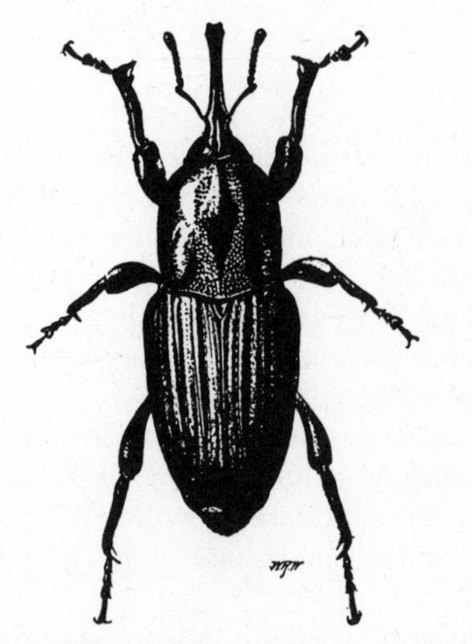

Figure 110. Maize billbug, *Sphenophorus maidis* Chittenden; 3×. (USDA)

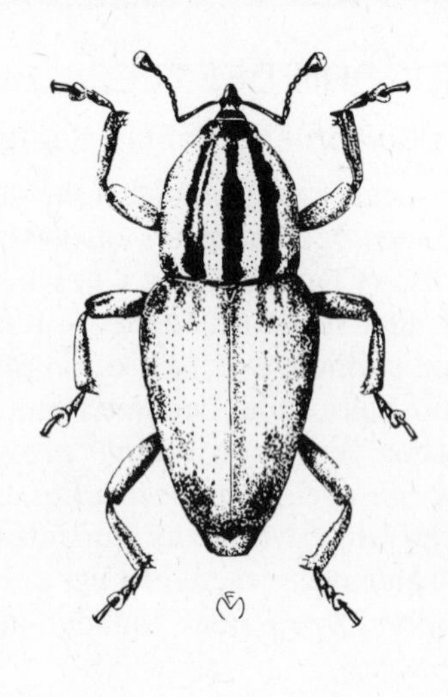

Figure 111. The clay-colored billbug, *Sphenophorus aequalis* Gyllenhal; 3×. (USDA)

olive-yellow to brown and black. The prolongated head is quite characteristic of the adults and is the basis for the name "snout beetle." Larvae are white, short, thick-bodied, curved, legless grubs.

Billbugs have one generation per year. Overwintering adults are produced in late summer or fall and may be active and feed for a period, or may remain in the pupal cells until spring. They feed for a considerable period in the spring, and may still be found when larvae are well grown or even after the next generation adults begin to appear. Eggs are laid in feeding punctures in the food plants, and the larvae feed in crowns and larger roots. In midsummer pupation takes place in the soil or in feeding cavities near the base of plants. The adult stage is reached soon afterward, thus completing the life cycle.

Natural enemies are of little value in control. Applied control measures have been very largely cultural or mechanical in scope. A crop rotation in which corn does not follow grassy sod is recommended where the clay-colored, bluegrass, or corn billbugs are a problem. Corn should not follow corn where the maize or southern corn billbugs are causing damage. Good drainage, fall plowing, cultivation, destruction of rushes and sedges, all are recommended practices. Lawns should be treated with a pesticide when one adult per minute is seen crawling on adjacent paved surfaces.

REFERENCES: *USDA Farmers' Bul.* 1003, 1932; *N. C. Tech Bul.* 13, 1917; *S. C. Bul.* 257, 1929; 452, 1957; *J. Econ. Ent.* 9:120–130, 1916; 50:707–709, 1957; *Bul. Ent. Soc. Amer.* 19:92–94, 1973.

RICE STINK BUG

Oebalus pugnax (F), Family Pentatomidae

Practically all rice fields in Louisiana, Texas, and Arkansas suffer loss from the attack of rice stink bugs. This bug is also found in all rice-growing states east of the Mississippi River but is not found in California. Besides rice, bullgrass *(Paspalum urvillei)* is the favorite wild host, but other grasses in or near rice fields are attacked.

Damage is caused by the piercing-sucking nymphs and adults feeding on the rice kernels in the milk and dough stages. Complete removal of the grain content in the milk stage results in an empty seed coat, and partial removal results in shrivelled kernels with spots varying from light yellow to black. Such injury is known to the rice trade as "pecky rice." This rice is graded down, lowering the market value, with the result that the average annual loss to the rice growers is approximately $500,000.

The adult is a straw-colored, shield-shaped bug (Fig. 112), 12 mm long, which passes the winter in heavy grass near the ground surface. The bugs emerge from the winter quarters in April and early May, and begin depositing eggs on the blades of rice or grass. These light green, short, cylinder-shaped eggs are arranged in two rows on the upper surface, the number in a single cluster varying from 10 to 47. Their color develops a red tinge before hatching, which usu-

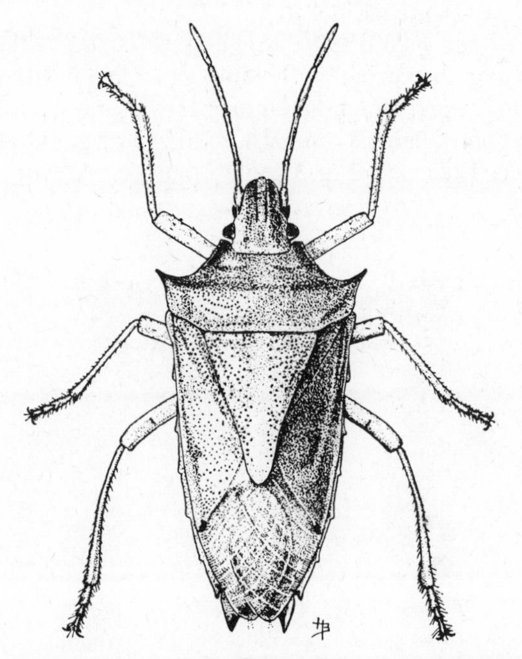

Figure 112. The rice stink bug; 2½×. (USDA)

ally requires about 5 days. The nymphs molt 5 times in a period of 15 to 28 days and then become adults. There may be as many as 4 or 5 generations annually on grass, and 2 or 3 on rice. Other stink bugs also attack rice (see p. 233).

Natural control results from their inability to survive the cold of winter or periods of intense heat of summer; also included in natural control are the bird and spider predators, and the egg parasites, *Telenomus podisi* Ashmead and *Ooencyrtus anasae* Ashmead. Applied control consists of plowing or burning heavy grasses in which the bugs hibernate.

REFERENCES: *USDA Cir.* 632, 1957; *J. Econ. Ent.*, 55:648–651, 877–879, 1962; 56:197–200, 1963.

RICE STALK BORER

Chilo plejadellus Zincken, Family Pyralidae

In habits this insect closely resembles the sugarcane borer, but the injury it causes is never as severe. Occasional injury from its feeding is noted in rice fields. It has also been found in corn.

The adult moth is similar to the sugarcane borer moth in size (25 mm wingspread) and shape (Fig. 113), and is not unlike it in coloration, being straw-colored with a golden tinge and marked with a sprinkling of minute black dots. Mature larvae are 25 mm in length, white and without stripes, the latter being characteristic of the sugarcane borer.

The winter is passed in the fully grown larval stage in the stubble. These transform to pupae in the spring, with adults emerging in June. At night the female moths lay their eggs, which hatch in 7 days or more, with larval development complete in approximately 1 month, followed by pupation and adult emergence 7 days later. There are 2 or 3 generations per year in rice fields.

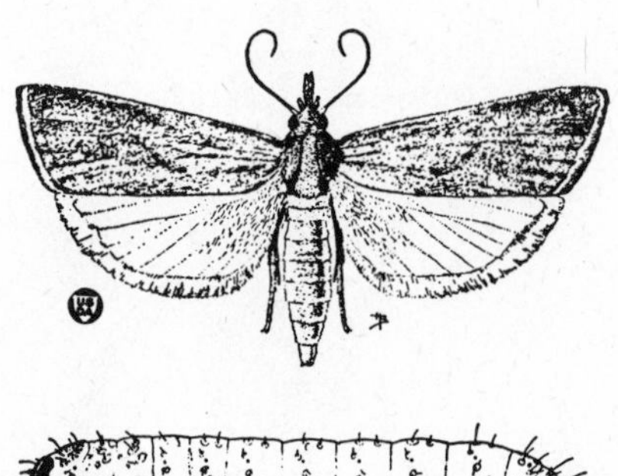

Figure 113. The rice stalk borer, *Chilo plejadellus* Zincken, 2×. (Webb, USDA)

Pasturing, plowing, or burning rice stubble in the winter greatly decreases the number of overwintering borers. A combination of dragging and submerging stubble fields in winter has resulted in 100% mortality of the hibernating larvae. Growing resistant cultivars is recommended.

REFERENCES: *USDA Cir.* 632, 1957; *J. Econ. Ent.* 65:711–713, 1972.

RICE WATER WEEVIL

Lissorhoptrus oryzophilus Kuschel, Family Curculionidae

This native insect, which feeds normally on semiaquatic grasses from New England to the Gulf states, is a pest of the rice crop in most places where it is grown. Damage results from the larvae chewing off the roots of the young plants and the adults causing slitlike feeding scars on the leaves that give the appearance of shredding. As the season advances the injury becomes less important.

The chewing adults are dark-colored, 3 mm long, and shaped as illustrated (Fig. 114). They hibernate under grass or crop remnants during the winter and become active in the spring, the females depositing their white eggs in the roots soon after the rice fields are flooded. The slender, white, legless, aquatic larvae are nearly 1 cm in length when fully grown. Larval development requires about 5 weeks, followed by pupation inside an oval cell of mud attached to the roots, with adult emergence occurring a week or more later. The total length of the life cycle is from 35 to 50 days, with 2 generations developing each year.

Experiments have shown that drainage undertaken for the purpose of killing the larvae is not worthwhile. However, draining may be necessary in order to correct other unusual ills said to be caused by the larvae of the weevil. It is also possible that the larvae are associated with heavy root rot injury.

REFERENCES: *USDA Cir.* 632, 1942; *J. Econ. Ent.*, 47:676–680, 1954; 54:710–712, 1961; 56:826–827, 893–894, 1963; 65:1380–1383, 1972; *Ark. Agr. Exp. Sta. Bul.* 624, 1960; *Calif. Agr.* 13(8): 10–11, 1959; *Cir.* 555, 1970.

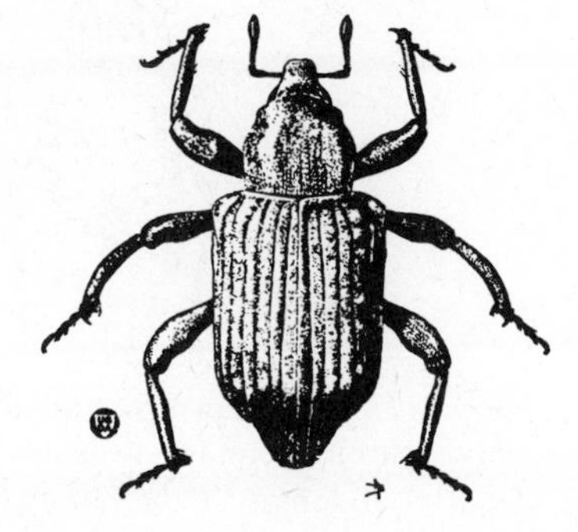

Figure 114. The rice water weevil; 12×. (Webb, USDA)

CORN ROOT APHID

Anuraphis maidiradicis (Forbes), Family Aphididae

When growth of corn is retarded, the leaves becoming yellow- or red-tinged coupled with a general lack of vigor, the grower may well suspect the presence of corn root aphids. These tiny blue-green wingless aphids, scarcely 2 mm long, that cluster on the roots of corn, cause injury by extracting the plant sap with their piercing-sucking mouthparts. The common host is corn, but other species of grasses, cotton plants, and many weeds, especially smartweed, are also attacked. This aphid can transmit maize dwarf mosaic virus.

The general appearance of the winged and wingless adults is shown in Fig. 115. The winged female has a black head, a dark brown thorax, and a pale green abdomen bearing 3 or 4 dark marginal spots and small dark specks over the surface. The antennae and legs are almost black.

Glossy black, oval eggs overwinter in the soil nests of the cornfield ant, *Lasius alienus* (Förster). These hatch in the spring into wingless nymphal females. which are carried by the ants to the roots of plants where they feed. The aphids become fully grown in 2 or 3 weeks and give birth to more of their kind. Since the female sex only is present, reproduction is by parthenogenesis. Winged females appear in the summer; they fly to other corn fields and start new infestations, if ants are present to care for the young. The period from birth to maturity is 8 to 16 days in the summer, with approximately 12 generations occurring during the season. In late September and in October wingless males and females are produced. These mate and the females lay eggs, which are carried by the ants to their nests in the soil. This cooperation on the part of the ants and

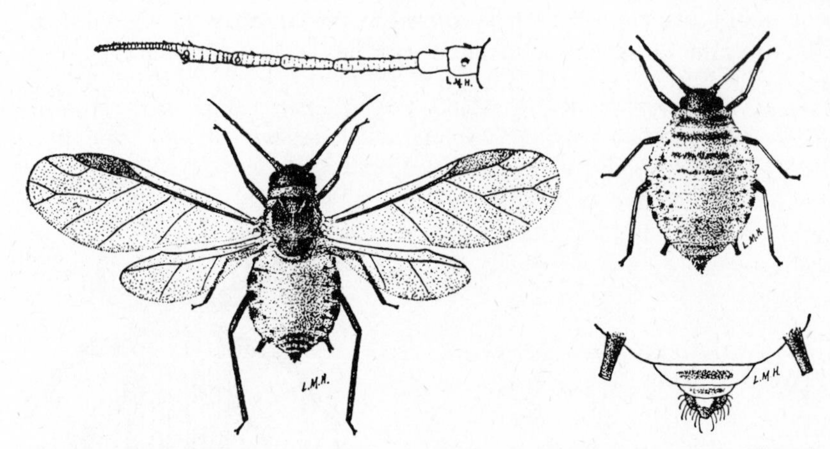

Figure 115. The corn root aphid, *Anuraphis maidiradicis* (Forbes). Winged ovoviviparous female, wingless ovoviviparous female; 12× and enlarged detail of abdomen of the latter. (Top left) antenna, greatly enlarged. (Hart.)

aphids seems to be necessary for their survival. The ants feed on the excrement from the aphids, called "honeydew." It contains sugars from the plant sap on which the aphids feed. In turn, the ants care for the tiny helpless aphids, place them on plant roots to feed, and protect the eggs during the winter period.

Control of the corn root aphid can be achieved by rotation of crops, fall plowing to destroy the ant nests, spring plowing and cultivation to destroy ant nests and weeds on which the hatching nymphs feed. Soil pesticides that kill ants usually solve this pest problem. No-tillage cropping systems favor high ant populations and, in turn, result in more root aphid damage.

REFERENCES: *USDA Farmers' Bul.* 891, 1917; *Ohio Agr. Res. and Dev. Ctr. Cir.* 193, 1965; *Bul. Ent. Soc. Amer.* 22:302–304, 1976.

GREENBUG

Schizaphis graminum (Rondani), Family Aphididae

A native of southern Russia, the greenbug has also been called the spring grain aphid; it is doubtless the most destructive of the aphids that attack small grains in North America. It has been reported from Canada to the Gulf states and from the Atlantic to the Pacific, but it is most likely to be destructive from Texas northward to Canada and eastward in the region north of the Ohio River. The insect has caused severe damage to barley, oats, and wheat in Texas, Oklahoma, Colorado, Kansas, and Nebraska. Its principal host plants are wheat and oats, but it can and does live on several kinds of grasses and other grain crops. In some years it is extremely damaging to grain sorghums from South Dakota to Texas and California. It is a vector of sugarcane mosaic.

Development of the greenbug in numbers great enough to make it the very serious pest that it occasionally becomes is the result of weather conditions which cannot be foreseen. It is capable of multiplying at temperatures as low as 40 F (4.5 C), whereas its insect parasites and predators, which usually keep it under control, can thrive only at temperatures above 65 F (18 C), so that during long periods of cool weather the greenbug increases to enormous numbers without much interference from its natural enemies. This relationship is apparently responsible for greenbug outbreaks in years when a cool spring follows a mild winter.

Epidemics of the greenbug often start in Texas and progress northward with the season. Their spread may continue until they reach the Dakotas and Canada, or they may be checked earlier.

In the South greenbugs remain active and produce living young throughout the year, resulting in more destruction in this area. Farther north adults may be found at all times, but they have what may be termed a hibernation period; still farther north shiny black eggs are laid on the food plant in the autumn, and the winter is passed in this condition. These hatch in the spring into female greenbugs. When fully grown, nearly 2 mm long, they are pale green with a dark green stripe down their backs; the eyes, antennae, and tips of the appendages are black. The cornicles are moderately long and converge toward the tapering tip of the abdomen rather than projecting outward as is usual with many aphids. The nymphs may develop into winged or wingless adults which give birth to young throughout the summer, resulting in numerous generations. In the North, with the approach of cold weather, winged males and females are produced, and after mating the females deposit the overwintering eggs (Figs. 116 and 117).

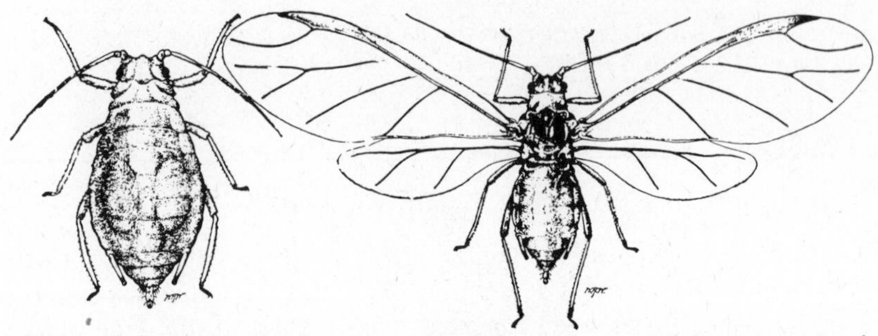

Figure 116. The greenbug, *Schizaphis graminum* (Rondani); wingless oviparous female and winged ovoviviparous female; 15×. (Walton, USDA)

Natural control is due to parasites of which the braconid wasp, *Aphidius testaceipes* (Cresson), is by far the most important (Fig. 118). The eulophid, *Aphelinus mali* (Haldeman) is also very important, at times. The usual predators, lady beetles, aphidlions, and syrphid fly larvae, contribute to keeping the aphids in check.

Cultural control consists of the destruction of volunteer grains, where feasible, by disking, plowing, or clean fallowing from harvest until seeding time. Chances of infestation are reduced sharply, so it is reported, when these measures are undertaken cooperatively by all the farmers in a given territory. Resistant varieties show promise for certain crops. Winter barley varieties, Dicktoo and Kearney, possess high resistance to greenbug. The economic threshold for barley is considered to be 25 to 30 aphids per tiller.

REFERENCES: *USDA Farmers' Bull.* 1217, 1921; *Leaflet* 309, 1951; *Tex. Agr. Exp. Sta. Bul.* 845, 1956; *J. Econ. Ent.*, 44:954–957, 1951; 53:278–299; 473–474, 798–802, 1960; 54:606–607, 1171–1173, 1961. 65:764–767, 1972; 68:161–164, 1975; *Bul. Ent. Soc. Amer.* 18:161–173, 1972.

OTHER GRAIN APHIDS

ORDER HOMOPTERA, FAMILY APHIDIDAE

Besides the greenbug, other species of aphids attack grain crops and cause damage by removal of the plant sap with their piercing-sucking mouthparts,

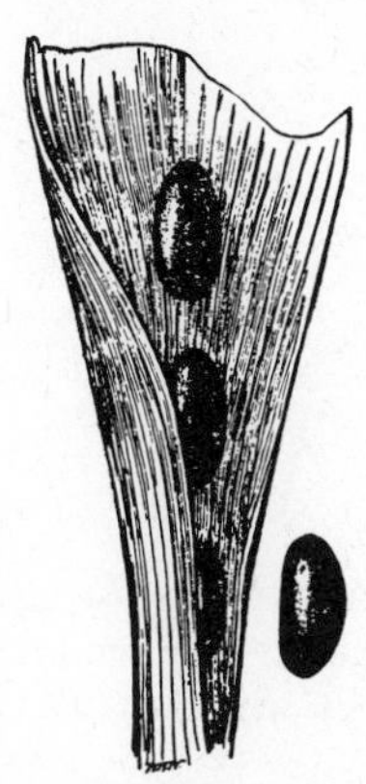

Figure 117. Greenbug eggs on leaf; greatly enlarged. (Walton, USDA)

and by aiding in the dissemination of various virus disease-producing organisms. Their honeydew secretion often attracts ants and serves as a medium for sooty fungus.

Apple Grain Aphid, *Rhopalosiphum fitchii* (Sanderson), second to the greenbug in importance as an aphid grain pest, is also known as a minor pest of apple trees and related hosts. This aphid is a vector of yellow dwarf virus of barley. Mature wingless females (stem mothers), yellowish green with a distinct dark stripe down the back and several dark stripes across the body, develop from the overwintering eggs. These give birth to young, all females and wingless. After 2 to 4 generations of these forms on apple trees, winged individuals develop which fly to grass and grain crops, and throughout the summer produce succeeding generations which are wingless and ovoviviparous. In the fall the winged individuals are again produced in abundance. They fly to apple and related plants, and produce wingless females which are sexual forms. After mating, these wingless forms deposit the overwintering, shiny, black eggs. In the spring the eggs of the apple grain aphid hatch before those of other apple aphids, which is a way of determining the species present. Since all summer generations are females, reproduction is parthenogenetic.

Corn Leaf Aphid, *Rhopalosiphum maidis* (Fitch), is an occasional pest of corn foliage in late summer just at the time of tassel emergence. Severely injured tassels may turn white and fail to produce pollen. Infested leaves and ear shoots often become a mottled yellow or reddish yellow. Heaviest populations have been observed on hybrid varieties of field corn susceptible to the European corn borer. Other hosts are sorghums, barley, sugarcane, broom corn, and cultivated as well as wild grasses. This aphid is said to be worldwide in distribution.

The life cycle of these greenish blue aphids (Fig. 118) does not differ materially from those of other species. It is uncertain whether they winter in the

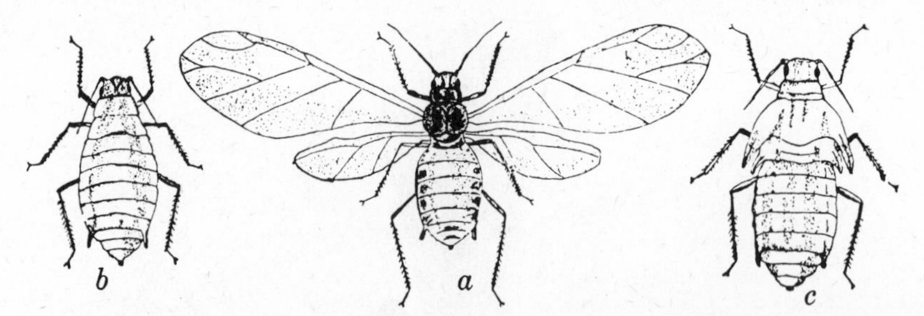

Figure 118. The corn leaf aphid, *Rhopalosiphum maidis* (Fitch). *a,* winged, ovoviviparous female; *b,* wingless, ovoviviparous female, and *c,* last stage of nymph of winged form; 7×. (Forbes.)

North as eggs or become established annually by migrations from the South. As many as 9 ovoviviparous generations may develop in Illinois, and up to 50 have been observed at Brownsville, Texas.

This aphid is a vector of the virus causing mosaic disease of sugarcane and its control where that crop is grown is therefore much more important than was formerly realized. It is also a vector of yellow dwarf virus of barley and maize dwarf mosaic virus of corn. The most dependable way to avoid corn leaf aphid injury is to plant resistant hybrid varieties of corn. Late-planted corn is usually damaged most.

English Grain Aphid, *Macrosiphum avenae* (Fabricius), is a native of Europe where it was first described. It is widely distributed throughout the United States and southern Canada, and attacks barley, oats, rye, wheat, corn, timothy, and other grasses.

Its feeding in early spring may kill young plants and, after grain crops are heading, shriveled kernels may result. This aphid is an important pest in this country. It is sometimes mistaken for the greenbug; however, the cornicles of this aphid are entirely black, whereas those of the greenbug are only tipped with black.

Another species of aphid widely distributed on grain crops and many wild grasses is *Rhopalosiphum padi* (L.). Its importance as a pest has not been accurately determined, but it is a vector of the virus diseases, maize dwarf mosaic, and yellow dwarf of barley. A common name, often used, is bird-cherry aphid.

Natural and applied control measures for all these species of aphids are the same as for the greenbug. Important parasites are the braconids *Aphidius testaceipes* (Cresson) (Fig. 119), *A. avenaphis* (Fitch) (Fig. 43), and *A. nigripes* Ashmead.

REFERENCES: *Ohio Sta. Bul.* 464, 1930; *USDA Tech. Bul.* 306, 1932; *J. Agr. Res.,* 7:463–480, 1916; *J. Econ. Ent.,* 53:197–200, 924–932, 1960; 57:22–23, 1964.

Figure 119. Aphidius parasite in act of depositing eggs in the body of a grain aphid; much enlarged. (Webster, USDA)

MINOR PESTS OF GRAINS AND GRASSES

Wheat Head Armyworm, *Faronata diffusa* (Walker), widely distributed in North America, is occasionally a pest of developing heads of wheat and other small grains, but more often causes greater damage to the heads of timothy and related grasses. The dark caterpillars have rather broad, yellow and brown stripes along the sides of the body. When fully grown they reach a length of slightly more than 25 mm. After the pupae have been in the soil from September to April the moths begin emerging, lay their eggs, and die. The larvae begin pupating in July, followed by adult emergence the latter part of the month and into August. The second-brood larval feeding period lasts from August until October, with fully developed larvae entering the soil and pupating during this time. Early harvesting of crops usually attacked, early fall plowing, and early fall pasturing are cultural control measures.

REFERENCES: *Iowa Agr. Exp. Sta. Bul.* 122, 1911; *USDA Cir.* 849. 1950.

Grass Thrips, *Anaphothrips obscurus* (Müll.), or "oat bugs," sometimes extremely abundant and obscurely injurious, feed on the foliage, flowers, and developing heads of many grains and grasses with their rasping-sucking mouthparts. Slight injury is indicated by the appearance of scattered, tiny light gray spots, but with high populations the entire plant may become pale green or gray. Thrips commonly hibernate in plant remnants during the winter and become active in late spring and summer, with overlapping generations occurring until autumn. In the summer they may migrate into buildings, their presence causing annoyance. Their numbers may be greatly reduced by plowing under crop residues during the hibernation period.

REFERENCES: *Me. Agr. Exp. Sta. Bul.* 83:97–128, 1902; *J. Econ. Ent.* 41:701–706, 1948; *Ga. Agr. Exp. Sta. Res. Bul.* 86, 1971.

Range Crane Fly, *Tipula simplex* Doane, like other crane flies, is similar in adult form to a giant mosquito. The larval stage is a dark, leathery-appearing, footless worm, often called "leatherjacket." It feeds on plant roots and other organic matter. This and related species are sometimes destructive in the range grasses of the Southwest; when abundant they may greatly reduce the pasturage over large areas. The European crane fly, *T. paludosa* Meigen, has become a serious pest of lawns and pastures in British Columbia and the State of Washington. Eggs of the flies are laid in late winter and remain unhatched in the soil throughout the intervening dry season. Fall rains stimulate hatching, as well as root growth and feeding by the larvae which overwinter in the soil. Pupation occurs in February and adults emerge in March. Two days after adult emergence the wingless females lay as many as 100 eggs. They die soon afterward.

REFERENCES: *Pflanzenschutz-Nachrichten-Bayer,* 17:1–24, 1964; *USDA Cir.* 172, 1929; *Can. Ent.* 89:288, 1957.

Brown Wheat Mite, *Petrobia latens* (Müller), is a tetranychid mite that sucks the sap from small grain crops, especially wheat. It occurs in most areas where wheat is grown in North America, but it does not cause serious damage every year. At times, chemical control measures may be necessary.

Winter Grain Mite, *Penthaleus major* (Dugès), is a eupodid mite that attacks small grains over a wide area of North America. In some years, and in some areas, damage becomes serious enough to require chemical control measures.

Wheat Curl Mite, *Eriophyes tulipae* Kiefer, is an eriophyid mite that sucks sap from grain crops, especially wheat. It is a vector of wheat streak mosaic virus and causes kernel red streak in corn.

Corn Planthopper, *Peregrinus maidis* (Ashmead), is a worldwide representative of the Delphacidae. Reports of its abundance in South Carolina, Texas, Florida, and nearby states have been made at various times. The nymphs and adults suck sap from the buds and leaves of late-planted and late-maturing corn. During the early part of the season damage is often unnoticed because of the insect's small size (3 mm). The appearance of the nymph and adult is indicated by Fig. 120.

Leafhoppers of several kinds suck the sap from grass plants and some are also vectors of viruses causing plant disease. Leafhopper vectors of corn stunt

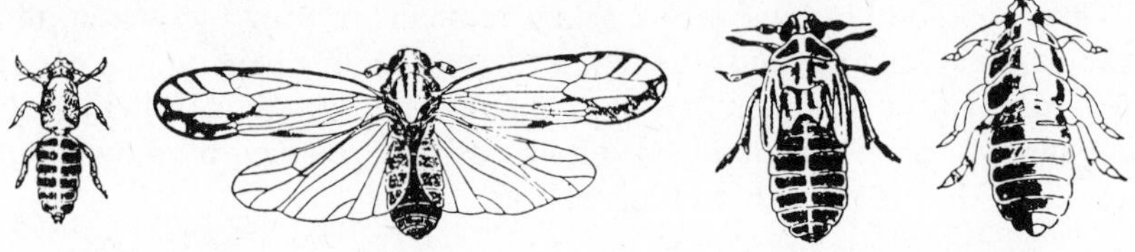

Figure 120. The corn planthopper, *Peregrinus maidis* (Ashmead); 7×. (Thomas, S. C. Agr. Exp. Sta.)

mycoplasma in the United States are *Dalbulus maidis* (DeL. and Wolcott) and *D. elimatus* (Ball). Predominant species are in the genus *Draeculacephala,* with *antica, producta, prasina, portola, minerva,* and *mollipes* (Fig. 121) rather common. *D. portola* is a vector of chlorotic streak virus of sugarcane. Other species, often abundant, are *Endria inimica* (Say), *Tylozygus bifidus* (Say), *Latalus sayi* (Fitch), *Psammotettix striatus* (L.), *Chlorotettix viridius* Van D., *Paraphlepsius irroratus* (Say), *Flexamia picta* (Osborn), and *Parabolocratus flavidus* Signoret. The aggregate damage caused by their feeding may be very

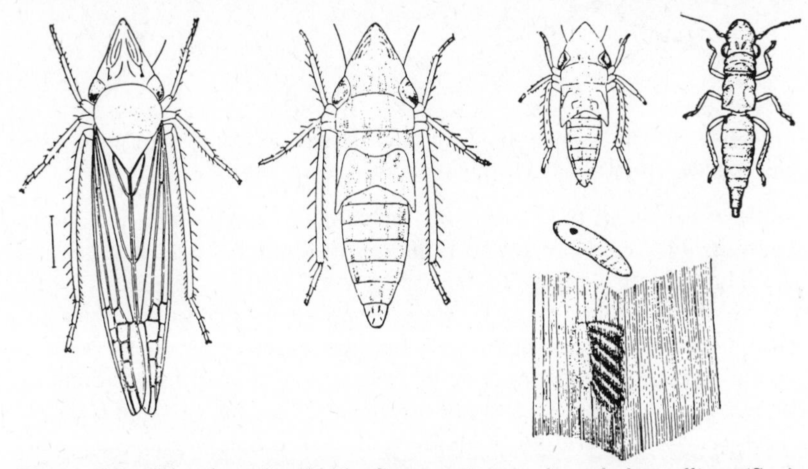

Figure 121. The sharp-headed leafhopper, *Draeculacephala mollipes* (Say); adult, nymphs, and eggs; 7×. (USDA)

great but this has been little studied. Life cycles of leafhoppers vary, some overwintering as eggs, others as adults. The number of generations also varies from one or more per year. Many leafhopper eggs are parasitized by *Anagrus epos* Girault.

REFERENCES: *USDA* 108, 1912; *Bul.* 254, 1915; *Cir.* 241, 1932; *Tech. Bul.* 1198, 1959; *Ann. Ent. Soc. Amer.* 63:789–792, 1790.

HESSIAN FLY

Mayetiola destructor (Say), Family Cecidomyiidae

The hessian fly is one of the most destructive of all insects that attack the wheat plant. It was introduced from Europe during the Revolutionary War in straw bedding used by the Hessian soldiers, and was first found in the vicinity of a camp that they occupied on Long Island about 1779. Widely distributed in the wheat-growing regions of the world, its principal range in the United States is in the winter wheat belt, but it also occurs in most all wheat-growing regions of North America, including the West Coast.

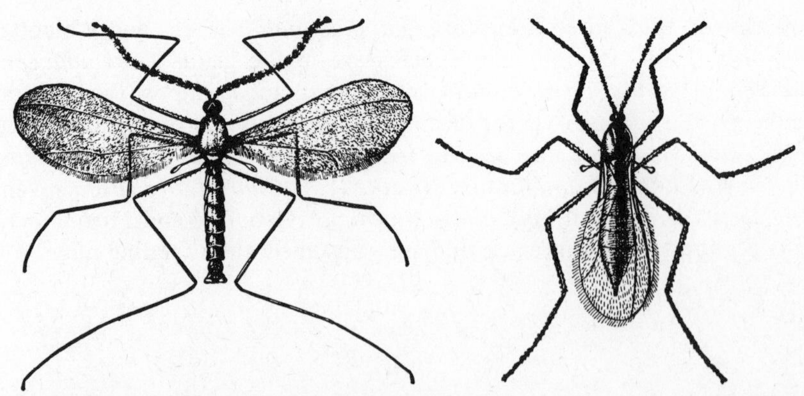

Figure 122. The hessian fly, *Mayetiola destructor* (Say); adult male, (left) and female, (right); 7×. (USDA)

In addition to wheat, the hessian fly occasionally damages barley and rye to a slight extent, and has been found breeding in emmer, spelt, and several wild grasses. The principal injury is caused by the larvae or maggots feeding between the leaf sheath and the stalk. By salivary secretion and intermittent sucking action they cause weakened stunted plants which often die in the winter. A central stalk frequently fails to develop. Damage in the spring is similar, except that the weakened stalks often break over just before harvest and the heads are small and poorly filled with low-quality grains. Losses of grain in bushels per acre, resulting from counts of stems infested by the hessian fly in the spring, show a range of 0.04 bushels with a 1% infestation to 15.7 bushels when 100% infested.

Figure 123. "Flaxseeds" or puparia of the hessian fly on young wheat; enlarged. (Pettit.)

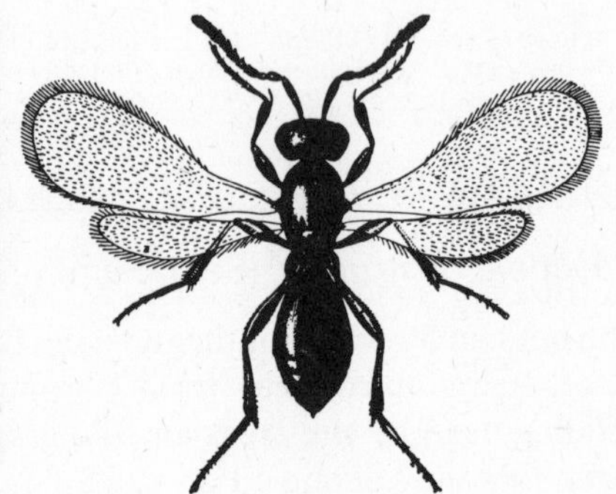

Figure 124 Platygaster hiemalis Forbes, a parasite of the hessian fly; greatly enlarged. (Walton, USDA)

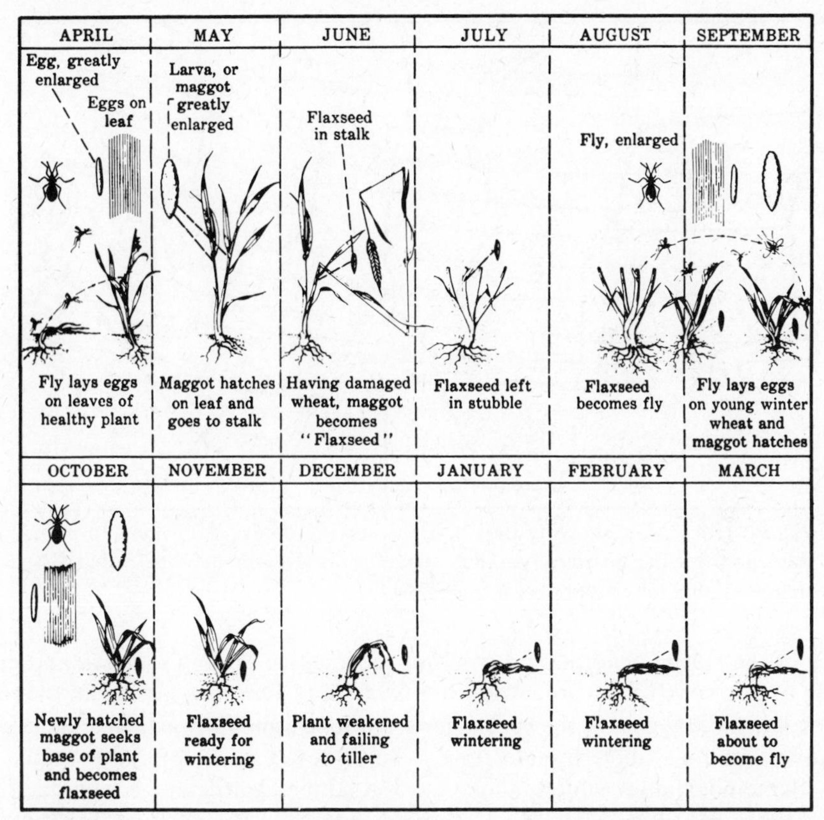

Figure 125. Diagram representing the life cycle and activities of the hessian fly throughout the year. (USDA)

There are normally 2 generations per year, but irregular generations may develop under the stimulus of unusual weather conditions, so that at times there may be 3 or even more. The usual designation is the spring or first generation and the fall or second generation.

The adult female is a minute fragile fly, about 4 mm in length, with the abdomen reddish-tinged; the male is slightly smaller, almost black in color with a pair of claspers at the tip of the abdomen (Fig. 122). Fall generation flies in late summer lay their tiny, elongated, yellowish red eggs end to end on the upper side of leaves of the host plants. These hatch in 3 to 12 days into legless maggots which are reddish at first, becoming white in the later instars. The tiny maggots migrate to the base of the leaf sheath where they feed between it and the stalk for a period of 4 to 6 weeks, after which they change to puparia. These are often referred to as the "flaxseed" stage because they resemble a true flaxseed in size, shape, and color. Overwintering puparia (Fig. 123) contain larvae until early spring when actual pupation occurs. This is followed by adult emer-

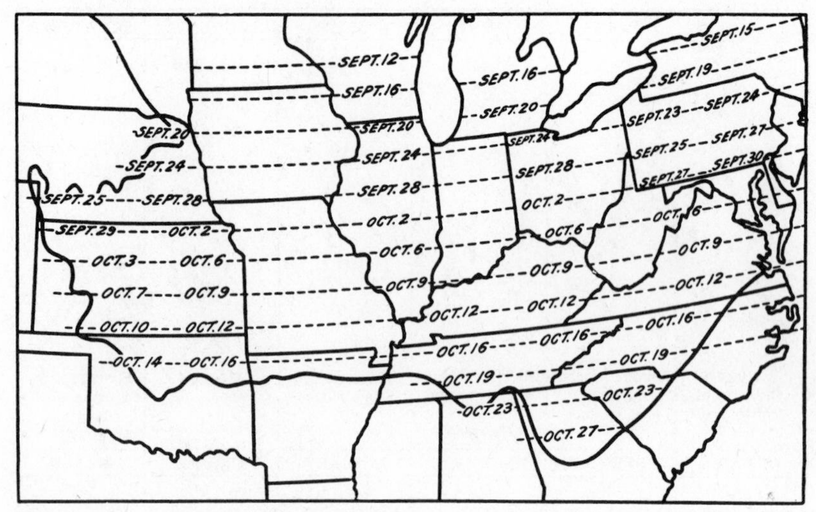

Figure 126. Map showing the "fly-free sowing dates" for wheat to avoid damage from the hessian fly. These dates represent the average rather than definite dates for any one year. (USDA)

gence of the spring generation a week or more later. Mating follows, and egg-laying begins soon afterward. Adult life is of short duration, lasting an average of 2 or 3 days. Maggots of the spring generation become fully grown and change to puparia in the stubble by late June, where they remain until late August or early September, after which fall generation adults again emerge (see diagram of the life cycle, Fig. 125). Most of the spring generation larvae feed in the region of the first and second nodes but may occur higher, depending upon the maturity of the plants at the time of infestation.

Many parasites play an important part in natural control of hessian fly. Those of special value are *Platygaster hiemalis* Forbes (Fig. 124), *P. zosine* Walker, *Eupelmus allynii* (French), and *Ditropinotus aureoviridis* Crawford. Other natural controls of value are climatic in nature, the most important being very hot, dry weather.

Applied control measures consist of sowing wheat on the "fly-free" dates (Fig. 126); destroying volunteer wheat on which the fall brood flies may oviposit, develop, and overwinter; plowing under infested wheat stubble soon after harvest, and disking the ground completely to prevent adult emergence; planting of resistant varieties of wheat, and selection of good seed, proper seed-bed preparation, and fertilizer. The safe sowing dates are based on the fact that the flies live only a few days, and the dates suggested have enough margin of safety so that all the adults will have emerged and died before the wheat comes through the ground. Community-wide cooperation is necessary for successful control. In California, adults emerge in March and April; therefore, early sowing is recommended, along with any cropping practice that favors production of vigorous growth. The mechanical destruction of volunteer wheat or infested stubble cannot be practiced where wheat is grown as a nurse crop for legumes or grasses. Varieties resistant to hessian fly attacks in the major wheat belt are the following: Pawnee, Ponca, Omaha, Warrier, Ottawa, Todd, Monon, Ace, Dual, Redcoat, Georgia 1123, Reed, Knox 62, Ben Hur, Logan, and Arthur. They are recommended especially for regions where it is impossible to get huge acreages of wheat planted at the proper time if sowing is delayed until the "fly-free" dates. In California, where the resistant varieties Big Club 43 and Poso 42 are grown, hessian fly is no longer a problem.

REFERENCES: *Kansas Agr. Exp. Sta. Tech. Bul.* 11, 1923; *USDA Farmers' Bul.* 1627, 1953; *Tech. Bul.* 81, 1928; *Tech. Bul.* 689, 1939; *Cir.* 663, 1943; *Leaflet* 533, 1965; *J. Econ. Ent.*, 49:182–184, 1956; 56:702–706, 1963.

WHEAT JOINTWORM

Harmolita tritica (Fitch), Family Eurytomidae

In the wheat-growing regions east of the Mississippi River, wheat jointworm at times ranks high as a pest. It has been found, but is of less importance, in Missouri, Utah, Iowa, Montana, Oklahoma, Kansas, Nebraska, North Dakota, South Dakota, Oregon, and California.

Heads of infested wheat plants have fewer and smaller kernels, and heavily infested fields show many "elbowed" straws. Occasional loss of heads occurs in harvesting.

Adult jointworms are about 3 mm long, black-bodied, 4-winged, and wasp-like (Fig. 127), with the joints of the legs yellow. They appear in wheat fields from April to July, according to the latitude, and lay their eggs in the stems near the joints. Several eggs are usually deposited in each place. These soon hatch into small legless grubs and their feeding causes the stem to develop gall-like thickenings just above the second or third joint from the ground. By harvest the galls or wartlike swellings become woody and brittle. Inside are found cells, each containing a single yellowish larva about 3 mm long. From 3 or 4 to 20 or more of these larvae may occur in each gall, where they remain until winter when pupation takes place. In the spring the new adults emerge through small circular holes which they make by gnawing through the wall of their cells. These mate and fly to new wheat fields in the vicinity to lay eggs, thus completing the cycle. There is only 1 generation per year.

Infested wheat stubble plowed under in late summer or early fall to prevent escape of adults the following spring gives effective control. Objections to control by plowing, because it interferes with the growing of red clover and other hay or pasture crops, may be met by temporary substitution of such crops as soybeans and sweet clover for forage and green manure. In Oregon it is recom-

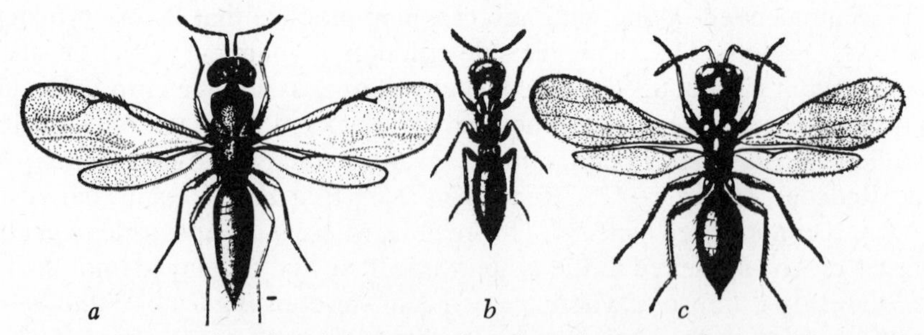

Figure 127. *a,* the wheat jointworm; *b* and *c,* wingless and winged forms of the wheat straw-worm; 10 ×. (Knowlton, Utah Agr. Exp. Sta.)

mended that winter barley or winter oats be substituted for wheat as a nurse crop for red clover. When infestation is present, planting wheat far from the wheat fields of the previous year will reduce damage. Infested straw for bedding of livestock usually is thoroughly trampled and well rotted, thus killing most of the jointworms. Hauling this manure in the winter to the fields that are to be planted in corn and plowing it under before April also helps to eliminate jointworms. Burning straw and stubble will destroy the insects but is not recommended. Several parasitic wasps contribute to jointworm control.

REFERENCES: *USDA Tech. Bul.* 518, 1936; 784, 1941; *Farmers' Bul.* 1328, 1938; *Leaflet* 380, 1954; *Utah Bul.* 243, 1933; *J. Agr. Res.,* 21:405–426, 1921.

WHEAT STRAW-WORM

Harmolita grandis (Riley), Family Eurytomidae

This insect is most numerous and causes the greatest damage in the wheat-growing areas west of the Mississippi River, but it is found throughout most of North America where wheat is grown, since this is the only host.

Winter is passed as larvae or pupae in the stubble or straw from the previous crop. In early spring the wingless, antlike, dark brown adults (Fig. 127) emerge and insert their eggs in the tissues of the host. On hatching, the tiny, yellow, legless larvae feed in the crown of the plant, causing stunting, and often destroy the central portion that later develops into the head. Pupation takes place within the stems in early May, and the winged adults of the second generation emerge a week or so later. These adults are larger and fly about depositing their eggs higher up on the plant, usually between the upper two nodes. Hatching takes place a few days later, and the larvae feed in the stems, weakening them and seriously interfering with head formation. By harvest time most of the larvae are fully grown; they remain in the stems and pupate there in the fall, passing the winter in this stage. There are two generations annually.

Natural enemies include some predators and parasites, which often attack jointworms as well. The most important parasite is the chalcid wasp, *Eupelmus allynii* (French). Another chalcid parasite, *Ditropinotus aureoviridis* Craw., has been recorded as important in Utah. The small elongate mite, *Pyemotes ventricosus* (Newport), feeds as a predator on the larvae in the stems and may be of considerable benefit in control. It is said that this mite is also predaceous on the parasites of the straw-worm.

If wheat is planted 60 to 80 yards from stubble or straw of the previous season, the wingless females of the first generation will not reach the new wheat. This forms the basis for the simplest of the control measures, which is crop rotation. Where wheat must follow wheat, growers are advised to plow under the stubble and straw, and plant a late crop such as soybeans or cowpeas, disking this crop under before wheat-seeding time. Where only spring wheat is grown the destruction of volunteer wheat in the early spring will prevent infestation, since wheat planted at the regular time will not be up soon enough to be attacked by the adults from the overwintering pupae.

REFERENCES: *USDA Farmers' Bul.* 1323. 1923; *Dept. Bul.* 808, 1920; *Utah Bul.* 243, 1933.

OTHER JOINTWORM OR STRAW-WORM SPECIES

ORDER HYMENOPTERA, FAMILY EURYTOMIDAE

Wheat Sheath Jointworm, *Harmolita vaginicola* (Doane), is similar in appearance to *H. tritici*. However, the larvae do not develop in the tissue of the main stem but confine their attack to the leaf sheath. Although widely distributed, they are seldom abundant enough to cause serious damage, and no special control measures are needed.

Rye Jointworm, *Harmolita secale* (Fitch), **Rye Straw-Worm,** *H. websteri* (Howard), and **Barley Jointworm,** *H. hordei* (Harris) are related species usually considered only of minor importance as pests of small grains. All are widely distributed, and have many parasites.

REFERENCES: *J. Agr. Res.*, 34:483–488, 1927; *J. Econ. Ent.*, 25:1171–1172, 1932; *Utah Bul.* 243, 1933; *USDA Bul.* 808, 1920.

WHEAT MIDGE

Sitodiplosis mosellana (Géhin), Family Cecidomyiidae

Sometimes in the delivery spout with the grain from the combine harvester may be found small maggots, oval and flattened in shape and of a deep orange-red color. These maggots, often called "red weevil" by the growers, are the larvae of the wheat midge, a near relative of the hessian fly. First reported from Brit-

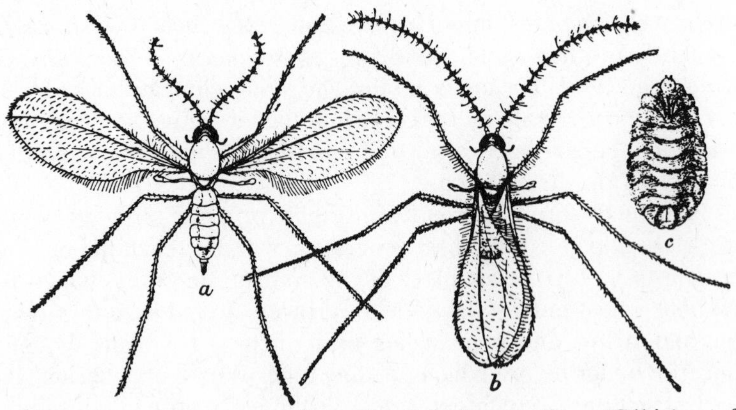

Figure 128. The wheat midge, *Sitodiplosis mosellana* (Géhin): *a*, female fly; *b*, male fly; and *c*, larva; 7 ×. (Marlatt, USDA)

ish Columbia in 1904, it has since spread to other areas of the Pacific Northwest.

The very small and fragile mosquitolike adults (Fig. 128) appear in May and June, and lay their eggs inside the chaff scales protecting the developing kernel while it is in the milk or early dough stage. The larvae feed in the kernels, preventing their proper development. A heavy infestation materially reduces the yield. Outbreaks are infrequent and local; rye, oats, and barley also serve as hosts. About the time of wheat ripening, mature larvae begin dropping to the ground and changing into puparia. They remain in the ground as puparia until the following spring.

Farm practices designed to control the hessian fly should aid in reducing populations of wheat midge, particularly plowing under the old stubble when this fits the cropping program. Early seeding of quick-maturing spring wheat varieties also reduces midge damage. Fall-seeded winter wheat is not often damaged.

REFERENCES: *Purdue Agr. Exp. Sta. Cir.* 82, 1918; *USDA Cir.* 732, 1945.

WHEAT STEM SAWFLY

Cephus cinctus Norton, Family Cephidae

This native insect of wheat is found most abundantly in the northern portion of the Mississippi Valley and in adjoining provinces of Canada. Other common hosts are spring rye, barley, timothy, and native grasses. The adult insect (Fig. 129) is a small, slender-bodied sawfly of black and yellow color; the larva is a slender, yellowish, almost legless caterpillarlike worm that tunnels up and down inside the stems, weakening them enough to reduce the yield of grain or cause loss by stalk breakage. Fully grown larvae attain a length of 10 mm. By

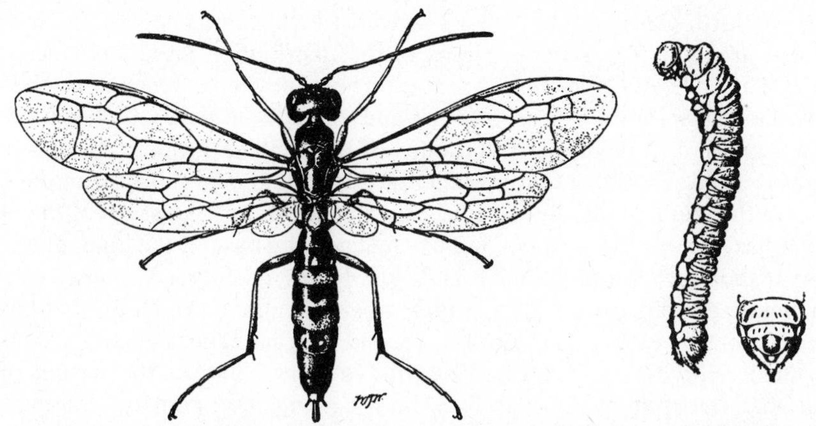

Figure 129. The wheat stem sawfly, *Cephus cinctus* Norton; adult female and larva with detail of the last abdominal segment; 4 × (Walton, USDA)

late July the larvae move to the base of the stems and gnaw a ring around each from the inside, weakening the straws which easily break off at ground level. Each infested stub is then plugged at the top with frass and lined with silklike material forming a chamber in which the larvae overwinter. Late in May of the following year pupation occurs, and adults begin emerging about June 10 and are present until about July 15. Egg-laying in the stems begins during this period with hatching taking place a few days after oviposition. This insect spends most of the year in the larval stage. Only 1 generation occurs annually.

Control may be secured by plowing under the stubble in the fall and working the soil to prevent escape of adults. Grasses that serve as alternate hosts should be destroyed where feasible. These include *Elymus condensatus, E. canadensis,* species of *Agropyron, Bromus inermis,* and timothy. Rotation of crops should also be of some value in checking the insect. Parasites that reduce the sawfly populations are *Bracon cephi* (Gahan), *B. lissogaster* Muesebeck, and *Eupelmus allynii* (French). Planting the resistant spring wheat varieties Rescue and Chinook has resulted in economical production in spite of this pest. Other resistant varieties are Golden Ball and Stewart.

REFERENCES: *USDA Tech. Bul.* 157, 1929; 1153, 1956; 1350, 1966; *Can. Dept. Agr. Pamphlet* 6, n.s., 1922; *Sci. Agr.,* 26:216–224, 231–247, 1946; *Can. Ent.,* 86:159–167, 1954; 89:272–275, 363–364, 1957; *Proc. N. C. Branch Ent. Soc. Amer.,* 15:100–101, 1960; *Can. J. Agr. Sci.* 36:196–202, 1956.

BLACK GRAIN STEM SAWFLY

Trachelus tabidus (F.), Family Cephidae

First collected in this country before 1899, this insect has spread from its original location in New Jersey as far west as central Ohio and as far south as the

Virginia-North Carolina line. Injury to wheat has been more serious in newly infested areas than in other sections of the distribution area. Infested wheat shows blasted heads, and stems break off where they have been cut by the sawfly larvae causing loss from shattering and difficulty in harvesting.

Eggs are laid in the upper internodes of the stem about the time the wheat begins to head. Hatching follows in a few days and the larva feeds on the inner lining of the stem, gradually working downward through the septa of the nodes until it has become fully grown and has reached the base of the plant at harvest time. It then forms a plug inside and, just below it, cuts a V-shaped incision completely around, with a thin outside fiber left intact, which holds the stem erect. Directly below this incision a second plug is made, beneath which is a silk-lined cell where the larva spends the winter. Wind, or the weight of the head, causes the stem to break off. The remaining stub contains the inactive, overwintering larva which pupates in the spring, with adult emergence occurring in May. There is only one generation per year.

The European wheat stem sawfly, *Cephus pygmaeus* (L.), first recorded in New York in 1887, has spread southward through the eastern half of Pennsylvania to Maryland and Delaware. The damage and life cycle of this species and *T. tabidus* are similar, except that the adults of *C. pygmaeus* appear about a week earlier. Both species attack wheat and to a lesser extent rye. The larvae of *C. pygmaeus* do not completely cut the stem, and as a result an irregular, ragged edge is characteristic at the point where the stem is broken, whereas *T. tabidus* cuts the stem more completely, leaving a finely serrated edge at the point of breakage. The cut made by *C. cinctus* is similar to that of *T. tabidus*.

The major parasite of *T. tabidus* is the eulophid, *Pleurotropis benefica* Gahan; the major parasite of *C. pygmaeus* is the braconid wasp, *Heterospilus cephi* Rohwer, and the eupelmid, *Eupelmus allynii* (French).

Recommended control measures are deep plowing of the stubble, with a clean turnover of the furrow slice to bury the overwintering larvae, where such procedure is feasible. To be successful this endeavor must be community-wide. Encouraging a strong stand by application of fertilizers and suitable cultural practices also aids in reducing loss. Where infestation is heavy much loss may be avoided by harvesting the wheat just before maturity.

REFERENCES: *USDA Cir.* 607, 1941; *Tech. Bul.* 1350, 1966; *Calif. Ins. Surv. Bul.* 11, 1969.

OTHER WHEAT MAGGOTS

Wheat Stem Maggot, *Meromyza americana* Fitch, is widely distributed in the United States, Mexico, and Canada. It is native to North America, but its origin was probably in the southern portion of the continent. Food plants include all the small grains, timothy, and other species of grasses. Injury resulting from the maggots feeding in the stem above the upper node causes the heads of wheat and other hosts to ripen before the kernels are produced. These white heads in contrast to the green unripened grain make it easy to determine the presence of this insect. Wheat scab fungus also produces the same symptom but can be distinguished by the presence of the pinkish mold. Loss from damaged heads by stem maggot rarely exceeds 2 or 3%. In the winter wheat belt, damage by the fall generation is similar to that of the hessian fly.

Adult flies, nearly 5 mm long with 3 dark stripes on the thorax and abdomen, appear about June and deposit 25 to 30 eggs, singly, over a period of 2 or 3 weeks. Under favorable conditions hatching takes place in 5 days, the tiny white maggots complete their development in 18 to 22 days, and the pupal stage lasts 12 to 16 days. There are 2 or 3 generations annually in South Dakota, and a greater number are produced farther south. Winter is passed as larvae in winter wheat, rye, barley, or grasses.

Many natural enemies are important in keeping this insect reduced in numbers. Several parasitic wasps have been found attacking the stem maggot larvae, parasitism sometimes being as high as 65%.

Rotation of crops, use of trap crops, destruction of volunteer grains and grasses, and late planting of fall grains should be considered in the applied control program.

REFERENCES: *S.D. Agr. Exp. Sta. Bul.* 217, 1925; *J. Agr. Res.* 55:215–238, 1937; *J. Econ. Ent.* 64:1129–1131, 1971.

Frit Fly, *Oscinella frit* (L.) (Fig. 130), is a widely distributed fly similar to the wheat stem maggot in appearance and in the injury that it causes. This species is of European origin. The American frit fly, *Oscinella soror* (Macquart), is very similar to the imported species. Methods of control suggested for the

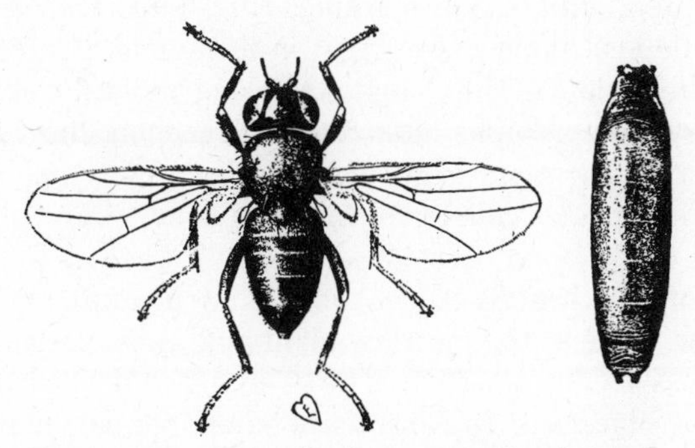

Figure 130. The frit fly, *Oscinella frit* (L.); enlarged, family Chloropidae. (Aldrich, USDA)

wheat stem maggot should reduce the damage caused by larvae of these flies. Silver topped panicles of bluegrass are caused by the frit flies, *O. neocoxendix* Sabrosky and *O. coxendix* (Fitch). Very few seeds are produced by such plants. At times frit fly larvae become damaging to turf in lawns and golf courses.

REFERENCES: *Conn. Agr. Exp. Sta. Cir.* 212, 1960; *J. Econ. Ent.*, 55:865–867, 1962.

PLANT BUGS

ORDER HEMIPTERA, FAMILY MIRIDAE

Several species of plant bugs that attack grass and forage crops also cause serious damage to vegetables, flowers, nursery plantings, and orchard fruits. The injury results from the removal of plant sap by the piercing-sucking mouthparts of the nymphs and adults. Legume and grass crops in bloom are often damaged to such an extent that very little seed is produced. Fruits such as apples, peaches, and strawberries are frequently misshapen.

Meadow Plant Bug, *Leptopterna dolabrata* (L.), feeds primarily on bluegrass, timothy, redtop, and orchard grass. The adults are scarcely 10 mm in length, green with fuscous markings, and 2 longitudinal black strips over the pronotum and scutellum. Two forms of females occur, one with short wings and the other with long wings (Fig. 131). Overwintering, yellow, curved eggs in the stems of grasses (Fig. 132) begin hatching as early as April 1, in the latitude of Lexington, Kentucky. The yellow-green nymphs with black markings transform to adults in May. Egg-laying takes place 10 to 14 days later, after which the adults soon die. There is 1 generation per year.

Tarnished Plant Bug, *Lygus lineolaris* (P. de B.), feeds on a wide variety of plants. In its northern range the overwintering adults are found under debris or in other protected places. These adults are scarcely 6 mm long, brown or yellow and sometimes green, with darker markings (Fig. 156). They become active in early spring and deposit their curved eggs in the stems, petioles, midribs, and blossoms of the host plants. Hatching takes place about a week or more later, and the green-to-yellow nymphs molt 5 times, reaching the adult stage in approximately 30 days. There are 3 to 5 generations or more annually, depending on the latitude. This species causes cat-facing of peaches and misshapen apples and strawberries.

Blue Grass Plant Bug, *Amblytylus nasutus* (Kirschbaum), is reported to be as numerous in some years as the meadow plant bug. It causes serious damage to Kentucky bluegrass seed production. The overwintering eggs are found only in stems of Kentucky bluegrass. In April these hatch into pale green nymphs, and 30 to 35 days later the green-yellow, winged adults appear, deposit their eggs, and soon die. In Kentucky, egg-laying begins about May 24, and all adults are

Figure 131. The meadow plant bug, *Leptopterna dolabrata* (L.); (left) short-winged female; (right) long-winged female; 4 × (Jewett and Townshend, Ky. Agr. Exp. Sta.)

dead by mid-June, there being only 1 generation per year. Twelve bugs per sweep of an insect net warrants chemical treatment.

REFERENCES: *Ky. Agr. Exp. Sta. Bul.* 508, 1947; *J. Econ. Ent.*, 56:532–533, 555–539, 1963; 57:181, 1964; *Bul. Ent. Soc. Amer.* 21:119–121, 1975; 23:277–287, 1977.

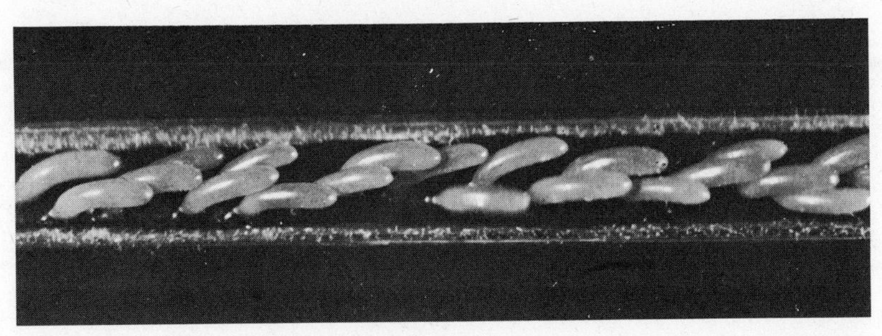

Figure 132. Eggs of *Leptopterna dolabrata* (L.) in a stem of Kentucky bluegrass; greatly enlarged. (Jewett and Townshend, Ky. Agr. Exp. Sta.)

11

PESTS OF COTTON

The growing of cotton is one of the major sources of income for the southern farmer. Because of this, it is essential to know the habits of the many insects that attack the crop, the damage they do, and their control. This chapter deals with major species of insects and mites attacking the cotton plant. However there are other insects, sometimes destructive to cotton, that are discussed in detail in other parts of the book. See Chapter 9.

The Agricultural Extension Service in each cotton-producing state issues an annual guide for controlling cotton pests. Obtain a copy of this guide for the management procedures in your area.

BOLL WEEVIL

Anthonomus grandis Boheman, Family Curculionidae

Originally occurring in Mexico, this insect invaded Texas before 1892 and has since gradually spread until it occupies almost the whole of the cotton belt except California, New Mexico, and Arizona. One of the most destructive insects ever known in America, the boll weevil has caused enormous losses annually in cotton fiber and cottonseed.

Boll weevils pass the winter as adults in debris, crop remnants, or other protected places near cotton fields. They are brown, or gray brown-to-nearly black, snout beetles, about 6 mm in length, although there is considerable variation in size. The front legs are disproportionately large and prominent with 2 spurs near the end of the femora (Fig. 133). Weevils emerging from hibernation feed on the foliage of cotton plants but do their greatest damage when the squares and bolls appear. They puncture the squares and bolls by boring into them with their chewing mouthparts, causing many to drop off. Each female may deposit over 250 eggs during her life span, usually placing a single egg in

Figure 133. The boll weevil on a cotton square showing a feeding puncture; 18 ×. (Courtesy of Frank J. Benci and Theodore B. Davich, Boll Weevil Res. Lab., USDA)

a feeding puncture of each square or boll. After 3 to 5 days the eggs hatch into white, curved-bodied, legless larvae with brown heads and chewing mouth-parts. These feed within the squares or bolls for 7 to 14 days and then transform into the pupal stage, which lasts about 5 days. Adults emerge and eat their way out; after feeding for about 4 days, egg-laying begins anew. The cycle from egg to adult requires about 3 weeks' time, and there may be 7 or more complete generations annually. Weevils are known to migrate 25 miles in 4 days.

Low winter temperatures and hot dry summers help control the boll weevils. Usually the weevils increase rapidly and severe damage results during rainy periods in the summer when the mean temperature is in the low 80's F (27 C).

Many birds, insect predators, and parasites assist in reducing the boll weevil population but, up to the present time, control by such biological means is not considered to be of great value. About 80% of all parasites reared from boll weevil larvae have been *Bracon mellitor* Say (Fig. 134). Another parasite of importance is *Eurytoma tylodermatis* Ashmead (Fig. 135).

Cultural practices that help set bolls quickly will aid in reducing damage. Those recommended are: planting early on good land, properly fertilized and well-prepared; choice of an early-maturing variety suited for the locality; plants spaced closely, cultivated frequently, picked early and clean; after picking, further fruiting should be prevented by chemical defoliants and desiccants, and by plowing, cutting, or grazing the cotton stalks early in the fall in order to

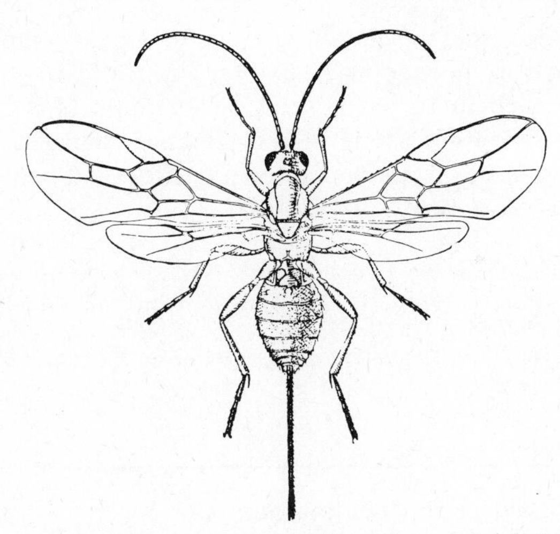

Figure 134. Bracon mellitor Say, one of the most important parasites of boll weevil larvae; enlarged. (Hunter and Hinds, USDA)

check the development of a high population of overwintering weevils. Any practice that tends to reduce shelter for hibernating weevils also contributes to the control program.

Variations in the toxicity of pesticides approved for boll weevil control have been observed in local areas of the cotton belt, and the proper choice must be determined by their effectiveness in a given area. Start applications when 10% of the squares are punctured and repeat every 4 or 5 days until the infestation is brought under control. Weekly inspections thereafter should be made to determine when additional treatment is needed. Select a pesticide recommended for your area that causes the least damage to the natural enemies and will control other possible pests of the cotton plant.

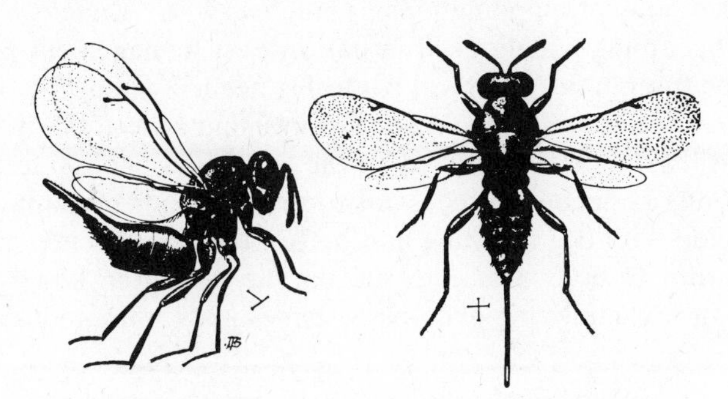

Figure 135. Eurytoma tylodermatis Ashmead: (left) male, (right) female. Parasitic on the boll weevil; actual size indicated on illustration. (USDA)

Some entomologists feel we do not have the technical competence to eradicate the boll weevil. Hence we may have to learn to live with it utilizing the best management techniques available. Consideration of a community-wide program of suppression consisting of reproduction diapause control, pheromone trapping, trap crops, resistant cotton varieties, and boll weevil sterilization, might be a wise venture.

REFERENCES: *USDA Farmers' Bul.* 1329, 1923; 1688, 1932; 2147, 1969; *Yearbook of Agriculture*, pp. 501–504, Plate I, 1952; *Ark. Bul.* 271, 1928; *Agr. Chemicals*, 8:34–35, 1953; *J. Econ. Ent.*, 54:622–624, 966–970, 1961; 55:688–692, 1962; 56:74–76, 350–356, 1963; 57:121–123, 1964; *Bul. Ent. Soc. Amer.* 21:6–11, 1975; *USDA-ARS-S-*71, 1976; *Miss. Sta. Univ. Ext. Pub.* 343, 1976.

BOLLWORM

Heliothis zea (Boddie), Family Noctuidae

Since this insect is the corn earworm, and its life history and description have been discussed under that heading (p. 147), it will be necessary only to call attention to the fact that, as a cotton pest, it always has several generations per year. These later generations damage cotton wherever it is grown with losses reaching serious proportions in some years. Besides the relatively unimportant eating of the foliage, each larva destroys a large number of bolls and squares. The bolls may be completely hollowed out, and this damage often occurs so late in the season that it is impossible for the plants to mature another crop. Heavy populations greatly reduce the yield.

Deep plowing and cultivating are helpful in destroying the overwintering pupal stages in the soil. Early planting, along with early maturing varieties, is a means of avoiding heavy damage. Extremely hot, dry, windy weather is unfavorable to bollworms. Lady beetles and other predators often destroy many eggs and tiny larvae. Mass releases of *Chrysopa carnea* eggs or larvae on cotton has reduced bollworm populations as much as 96%. A nuclear polyhedrosis virus is now becoming widely used in cotton pest management programs.

Early season (prebloom) chemical control is needed when 25% of the squares show damage. In mid- and late-season begin using a pesticide when eggs are present and at least 4 newly hatched larvae are found per 100 terminals. Continue applications as needed. Successful control depends on timing the applications to coincide with egg hatching and before the worms enter the bolls. Important predators of bollworm eggs and larvae are often killed by pesticide applications, for example, big-eyed bugs, chrysopids, minute pirate bugs, and spiders.

REFERENCES: *Ark. Bul.* 320, 1935; *USDA Leaflet* 462, 1960; *Tech. Bul.* 1454, 1972; *J. Econ. Ent.*, 55:143–144, 688–692, 1962; 56:442–444, 1963; 62:177–180, 1969; *Calif. Agr. Exp. Sta. Bul.* 660, 1942; *Cir.* 187, 1944; *Miss. Sta. Univ. Ext.* Pub. 343, 1976.

COTTON APHID

Aphis gossypii Glover, Family Aphididae

This species is widely distributed in both North and South America, and is considered the most important of several aphids that attack cotton and okra. It is also known as the melon aphid since it is a pest of melons, cucumbers, pumpkins, and squash. Damage, caused by the nymphs and adults removing plant sap, results in stunting of growth, curling of leaves, or death of tiny plants early in the season. Heavy feeding later in the summer causes the curled leaves to fall off before the bolls are mature, thus reducing the yield and grade of cotton. These aphids also excrete honeydew which drops on the open bolls and leaves, and serves as a medium for the development of sooty fungus. Open bolls covered with honeydew contain gummy lint, which is difficult to gin.

Other species of aphids sometimes found on cotton are the cowpea aphid, *Aphis craccivora* Koch, and the corn root aphid, *Aphis maidiradicis* Forbes (p. 178).

The cotton aphid varies from pale yellow to light or dark green, with dark leg joints, black cornicles and eyes (Fig. 191). In the South only parthenogenetic females that give birth to young are found, with over 30 generations developing throughout the year. In the northern states both sexes occur, and overwintering eggs hatch in the spring, all progeny being wingless parthenogenetic females that give birth to their young. Repeat generations of the same type of individual occur through the summer, with occasional winged stages developing and flying to other areas. With the coming of cold weather sexual forms are produced which mate, the females depositing the overwintering eggs.

Natural controls consist of aphidlions, syrphid fly larvae, lady beetles, and other predators; insect parasites, of which *Aphidius* spp. are the most important; diseases; and hot, dry, summer weather.

Chemical control measures are seldom necessary. However, populations often increase following the use of certain pesticides for other cotton pests, especially those chemicals that kill the natural enemies of aphids.

REFERENCES: *USDA Farmers' Bul.* 1499, 1926; *Leaflet* 467, 1960; 389, 1966; *Fla. Bul.* 252, 1932; *J. Econ. Ent.*, 56:326–333, 900–901, 1963; *Tex. Agr. Exp. Sta. Bul.* 257, 1919; *Miss. Sta. Univ. Ext. Pub.* 343, 1976.

SPIDER MITES

ORDER ACARI, FAMILY TETRANYCHIDAE

The increased use of organic pesticides for cotton insect control has resulted in higher populations of spider mites owing, at least in part, to the killing of their natural enemies. Damage to the plants results from the removal of sap by the nymphs and adults. This feeding is primarily on the lower surface of the leaves, appearing as tiny white spots. Under heavy infestation this causes the leaves to become entirely gray, curl, turn brown, and drop off. Loss of leaves causes shedding of small bolls and may prevent the lint from developing properly in large bolls. Over 200 host plants are recorded, including garden and field crops, ornamentals, and weeds.

Some common species attacking cotton are the carmine spider mite, *Tetranychus cinnabarinus* (Boisduval); strawberry spider mite, *T. turkestani* (Ugar. and Nikolski); two-spotted spider mite, *T. urticae* Koch; Pacific spider mite, *T. pacificus* McG.; desert spider mite, *T. desertorum* Banks; tumid spider mite, *T. tumidus* Banks; Schoene spider mite, *T. schoenei* McG.; four-spotted spider mite, *T. canadensis* (McG.); and the brown wheat mite, *Petrobia latens* (Müller).

Spider mites are less than 1 mm long, variable in color, usually of green, yellow, or red shades, with darker pigmented spots. The first instars are 6-legged and of a yellow color. Later instars have 8 legs, the mature females being the larger of the 2 sexes and usually showing more pigmentation (Fig. 136). The males are recognized by their smaller size and narrow, more pointed abdomen. The green spherical eggs are more often deposited on the undersides of the leaves where most feeding and spinning of delicate webs take place. Hot dry weather is favorable to rapid development, and there may be as many as 17 to 20 generations and more per year, these overlapping considerably. Many of the mites attacking cotton overwinter in the active stages, often as mature females, but eggs may be found in regions with very mild winters.

Spread of the mites takes place not only by crawling but by transportation with the aid of insects, birds, air currents, and the normal activities of man.

Natural control is from predaceous insects and mites, an important species of the latter being *Amblysieus fallacis* (Garman). This mite is easily killed by many of our new chlorinated hydrocarbon insecticides, resulting in greater numbers of plant-feeding species. Heavy rains have a checking effect on spider mite populations.

Chemical control measures are many and their effectiveness varies with the species of mite concerned. Check the field periodically for mites and when their presence is indicated begin application of the proper miticide. Two or more applications at 5- to 7-day intervals may be needed.

REFERENCES: *USDA Bul.* 416, 1917; *Leaflet* 502; 1962; *Hilgardia,* 22:203–234, 1953; *J. Econ. Ent.,* 46:224–233, 693–696, 1953; 51:710–712, 1958; 57:145–148, 1964; 62:732–733, 1969; *Miss. Sta. Univ. Ext. Pub.* 343, 1976.

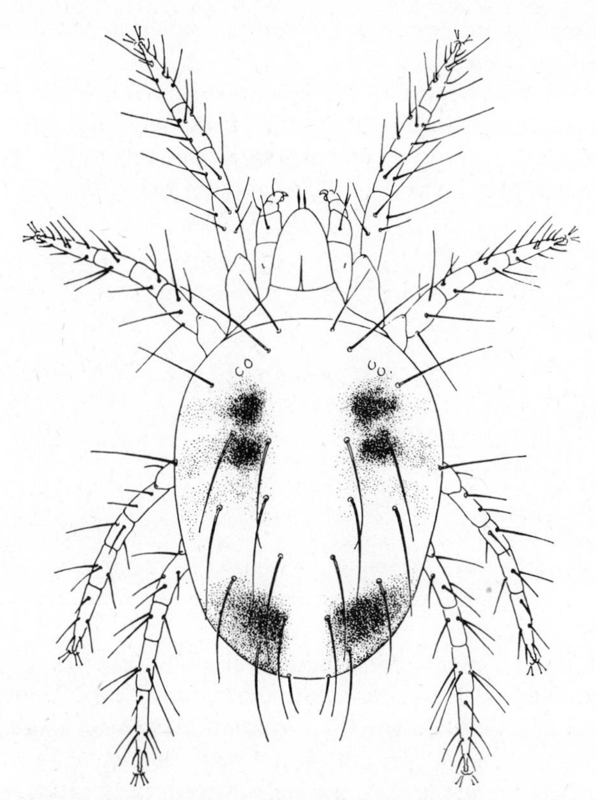

Figure 136. Pacific spider mite, *Tetranychus pacificus* McG.; a female showing markings when feeding; greatly enlarged. (Courtesy of A. E. Pritchard.)

COTTON FLEAHOPPER

Pseudatomoscelis seriatus (Reuter), Family Miridae

This bug is widely distributed throughout the cotton-producing regions of the United States. Besides cotton, horsemint, evening primrose, croton or goatweed, sage, and numerous weeds are attacked by the fleahoppers.

The piercing-sucking nymphs and adults interfere with normal vigorous growth, cause the death of very young squares or blossom buds, and prevent blooming. The feeding punctures stimulate the plant to produce numerous spindly branches or suckers, indicating that the salivary secretions injected while sap is being sucked are toxic. First feeding is primarily on weeds but as these become less succulent migration to cotton takes place. Rainfall is favorable to their breeding on cotton and when the squaring season is passed, the fleahoppers return to the weed hosts.

The fleahopper overwinters in the egg stage in the stems of the hosts. In early spring these hatch into nymphs that become adults after molting four times. The winged adult bug is 3 mm in length, pale green with tiny black spots (Fig. 137). A generation spans a period of about 3 weeks with 6 to 8 developing

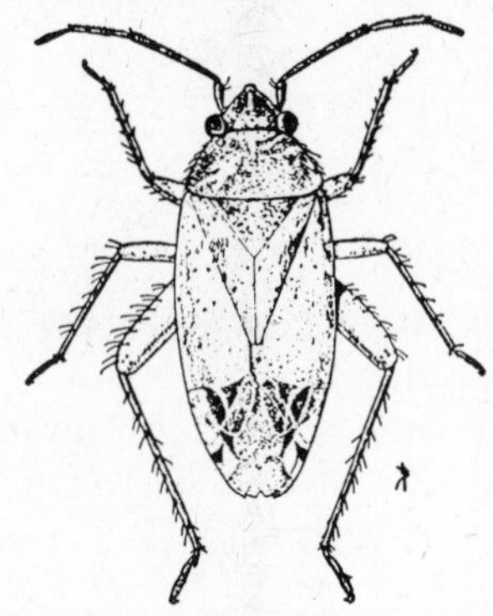

Figure 137. The cotton fleahopper; 10 ×. (USDA)

each season. Generations overlap so much they become indistinguishable as the season progresses.

Plowing under host plants containing overwintering eggs reduces potential fleahopper populations. Treat pre-squaring or young cotton plants not squaring normally with a pesticide, only if 15 or more of these insects are found per 100 terminals. After first bloom, on cotton not squaring normally, treat only if 30 or more insects are found per 100 terminals. Repeat the treatment only if necessary.

REFERENCES: *S.C. Bul.* 235, 1947; 251, 1928; *Texas Bul.* 339, 1926; 380, 1928; *USDA Yearbook of Agriculture,* Plate IV, 1952; *Leaflet* 475, 1960; *J. Econ. Ent.*, 54:966–970, 1961; *Miss. Sta. Univ. Ext. Pub.* 343, 1976.

PLANT BUGS

ORDER HEMIPTERA, FAMILY MIRIDAE

Plant bugs cause shedding of cotton squares and young bolls by puncturing and feeding with their piercing-sucking mouthparts. Older bolls are not shed but the lint is discolored and clings to the bur; also, the oil content and germination of the seeds are reduced. Damage to cotton is frequently serious when adjacent alfalfa fields are cut, causing migration of adults to cotton. The tarnished plant bug, *Lygus lineolaris* (P. de B.) (Fig. 156), is a cotton pest in the southeastern states. Related species are *L. elisus* Van Duzee and *L. hesperus* Knight, which cause damage in the Southwest (Fig. 138). The rapid plant bug, *Adelphocoris rapidus* (Say) (Fig. 157), and superb plant bug, *A. superbus* (Uhler) are widely

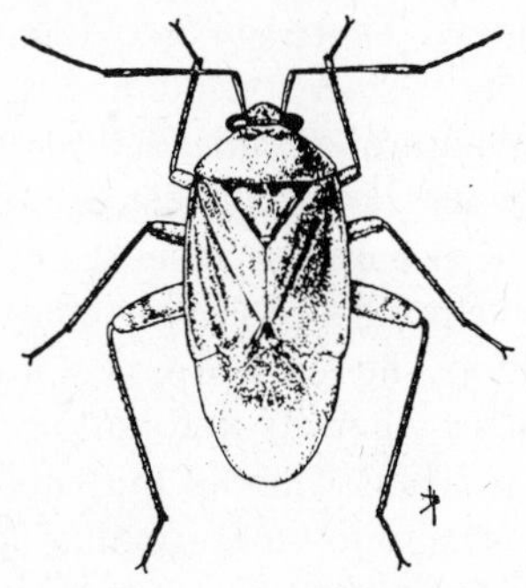

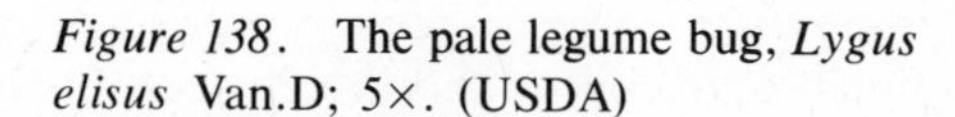

Figure 138. The pale legume bug, *Lygus elisus* Van.D; 5×. (USDA)

distributed in the United States and Canada and are pests of cotton adjacent to alfalfa, sweet clover, and weedy growths. In the latitude of Utah plant bugs may produce 3 or 4 generations per year, each requiring about 7 weeks; in southern Arizona and Texas a generation is passed in 30 days or less, and the insects breed most of the year. In cooler climates, the *Lygus* bugs overwinter as adults, *A. rapidus* in the egg stage.

Treat pre-squaring cotton if 15 or more bugs are found per 100 terminals; in mid-season (after first bloom) treat only if 30 or more bugs are found per 100 terminals. Some resistant cotton genotypes show promise.

REFERENCES: *USDA Leaflet* 503, 1969; *Miss. Sta. Univ. Ext. Pub.* 343, 1976; *Bul. Ent. Soc. Amer.* 23:19–22, 277–287, 1977.

COTTON LEAFWORM

Alabama argillacea (Hübner), Family Noctuidae

This is a tropical insect not known to survive the winters in the United States. Damage is done by the larvae that feed on the leaves of the cotton plant and when abundant also destroy the squares and bolls. Adult moths may feed on ripe grapes, peaches, or other fruits. The proboscis is fitted with spines on the tip enabling them to puncture the skin of the fruits.

The brown moths, marked with a few darker, wavy transverse bars on the fore wings, have a wingspread of about 38 mm. Each spring they fly in from the tropics and lay their eggs on cotton leaves. These hatch in a few days into pale dingy yellow caterpillars that molt five times. As growth progresses their color varies; some are yellow-green without prominent stripes, others have a broad black stripe with a narrow yellow stripe down their backs. All the caterpillars have four black dots that form a square on the dorsal side of each body segment. They become fully grown in two weeks or more and pupate on the plant, usually inside a leaf fold. Adults emerge several days later and begin the cycle anew. There are 3 to 7 generations per year. Adult migration northward occurs as the season advances, great numbers sometimes reaching Canada.

Many natural enemies help control the leafworm, for example, birds, wasps, ants, spiders, robber flies, predaceous bugs and beetles, mantids, aphidlions, and several parasitic insects and diseases.

When a program is followed for boll weevil control this insect never becomes a problem. The leafworm is sometimes considered beneficial, especially late in the season when its defoliation of the plants, after the crop is made, helps to starve the boll weevils that normally become the overwintering population. Orchardists following a regular spray program are usually not bothered by adult damage to fruits.

REFERENCES: *USDA Leaflet* 468, 1960; *J. Econ. Ent.*, 51:259, 1958; 55:688–692, 1962.

PINK BOLLWORM

Pectinophora gossypiella (Saunders), Family Gelechiidae

The pink bollworm, believed to be a native insect of India, is a serious pest of cotton in many parts of the world. It was first discovered in the United States in Texas in 1917, presumably having come from nearby Mexico where it was found before that date. Heaviest losses occur in Texas, but it also is present in Arizona, Arkansas, New Mexico, California, Oklahoma, and Louisiana. Okra is the only other widespread, commercially grown host plant.

Damage is done by the larvae which eat out the seeds in the green bolls and thus reduce the yield, weight, viability, and oil content of the seeds. They also reduce the quantity and quality of the lint. In heavily infested fields many of the bolls are so badly damaged that they are not picked.

The pink overwintering larvae may hibernate in a single seed, or two hollowed-out seeds united by silken threads, in bolls or trash in the field, or at the gin. Pupation takes place in the spring, with adult emergence occurring a week or more later. The adults are small gray-brown moths with narrow wings fringed with hairs (Fig. 139). Each moth lays 100 to 200 oval white eggs, usually in masses, beneath the calyx at the base of the boll. Hatching occurs in about 5 days, and the new white larvae with brown heads eat their way into the squares or bolls, where feeding occurs for 10 to 14 days, after which the fully grown pink larvae, now 12 mm in length, make exit holes and either pupate within the bolls or on the ground. Development from egg to adult requires 25 to 30 days in midsummer, and there may be as many as 6 generations per year in areas with long growing seasons.

Methods of control include planting early-maturing varieties on designated dates for your area; defoliating and desiccating the mature crop in late August and early September; harvesting early and destroying stalks immediately after

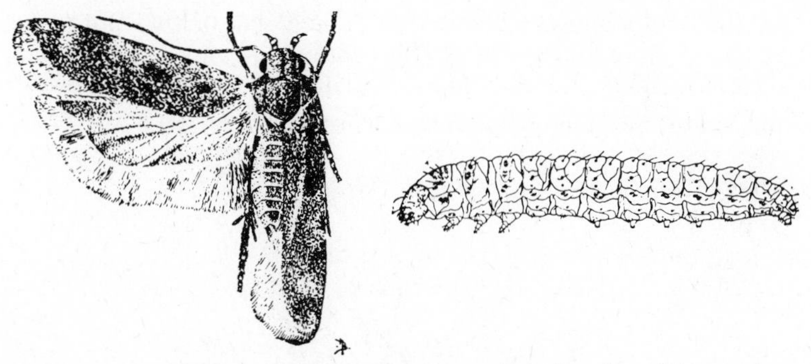

Figure 139. (Right) the pink bollworm: outline drawing of larva; 4 ×. (Left) the pink bollworm adult; 4 ×. (Busck.)

harvest by special shredders and plowing; heating cottonseed to a temperature of 145 F (63 C); burning gin waste; fumigating gins, warehouses, and seeds; and applying a pesticide only if needed. After square set, a 10 to 15% infestation is considered an economic threshold. Applications should be repeated until most of the bolls are open or the infestation is controlled. California entomologists found that fewer pesticide applications were needed when hexalure-baited traps were used to determine the degree of infestation. Widespread usage of the sex pheromone gossyplure in California, Arizona, and northern Mexico has decreased crop damage by confusing the male moths and reducing the chances of female fertilization.

Where cultural control measures have been carried out on a community-wide basis with the cooperation of all cotton growers, larval mortality has been more than 90%. In Texas, this combination of management practices has reduced the pink bollworm to minor pest status.

REFERENCES: *J. Agr. Res.* 9:343–370, 1917; *USDA Farmers' Bul.* 2207, 1965; *Tech. Bul.* 1304, 1964; 1454, 1972; *Prod. Res. Rept.* 34, 1960; 73, 1963; *J. Econ. Ent.* 57:148–150, 181–182, 1964; 62:682–685, 1969; 55:949–951, 1962; *Tex. Agr. Exp. Sta. Misc. Pub.* 444, 1960; *Calif. Agr.* 24(1):12–14, 1970; 30(6):12–13; (8)14–15, 1976.

SALTMARSH CATERPILLAR

Estigmene acrea (Drury), Family Arctiidae

The larva is a red-brown hairy caterpillar over 5 cm in length when fully grown, often called a woollybear. It feeds gregariously in the early instars on the foliage of cotton and many other garden plants. At times it can be a very serious pest. Overwintering occurs as pupae in brown hairy cocoons in debris on the ground. Adults appear in the spring. They are beautiful moths having a wingspread of 5 cm. Wings of the female are white above and yellow below; the males are similar except the hind wings are yellow both above and below. Both sexes have numerous black spots on the wings (Fig. 140). The spherical yellow eggs are laid in masses on plant foliage. There are usually 2 generations per year in central and southern United States and 1 in Canada.

REFERENCES: *USDA Bul.* 44, 1904; *Tech. Bul.* 1454, 1972.

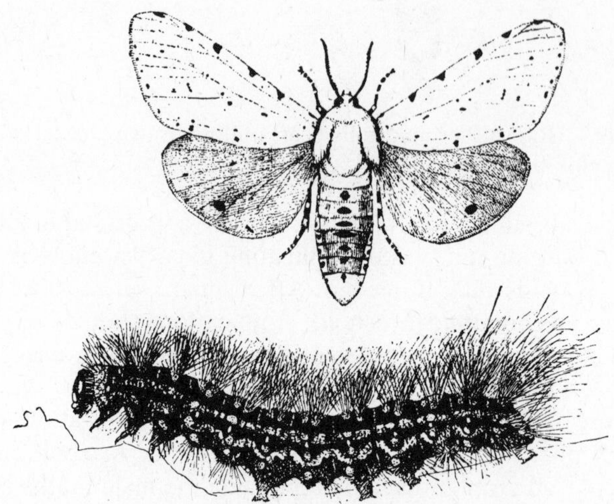

Figure 140. Saltmarsh caterpillar and male moth, slightly enlarged. (Chittenden, USDA)

MINOR PESTS OF COTTON

Cotton Square Borer, *Strymon melinus* (Hbn.), the larva of a small butterfly in the family Lycaenidae, is occasionally a pest of cotton in Texas and adjacent states. Damage is caused by the larvae boring into and hollowing out the squares. These larvae are sluglike, almost footless, light green in color, and densely pubescent. Both adults and larvae hibernate; egg-laying begins in March, and as many as 3 overlapping generations may develop, but natural enemies usually keep them checked. The pesticides normally used for boll weevil will prevent injury from the square borer.

Thrips of several species attack cotton. They may at times be more injurious than the fleahopper. Although thrips injury is usually observed on seedling cotton, damaging infestations will sometimes be found on older cotton, especially in areas where vegetables, legumes, and small grains are grown near-by. Common species found on cotton are tobacco thrips, *Frankliniella fusca* (Hinds) (Fig. 173), and *F. exigua* Hood; flower thrips, *F. gossypiana* Hood, *F. occidentalis* (Pergande), and *F. tritici* (Fitch); bean thrips *Caliothrips* *fasciatus* (Perg.); onion thrips, *Thrips tabaci* Lind.; and soybean thrips, *Sericothrips variabilis* (Beach.).

Both nymphs and adults cause injury with their rasping-sucking mouthparts. The life cycle is completed in about 2 weeks and overlapping generations occur throughout most of the year where cotton is grown. In some areas thrips are only one of several pests attacking cotton early in the season. When one or more thrips per plant are found, it is advisable to apply a pesticide. Many of the pesticides recommended for other cotton pests will also kill thrips.

REFERENCES: *S. C. Bul.* 271, 1931; 306, 1938; *USDA Leaflet* 516, 1962; *J. Econ. Ent.*, 55: 516–518, 1962; *Miss. Sta. Univ. Ext. Pub.* 343, 1976.

Leafhoppers of several species may damage cotton. Some common large species are *Homalodisca triquetra* (Fabr.) (Fig. 29), *Aulacizes irrorata* (Fabr.), *Oncometopia undata* (Fabr.), and *Cuerna costalis* (Fabr.) (Fig. 141). In addition many small green species in the genus *Empoasca* will attack cotton. When numerous, all species cause extensive injury by sucking plant sap.

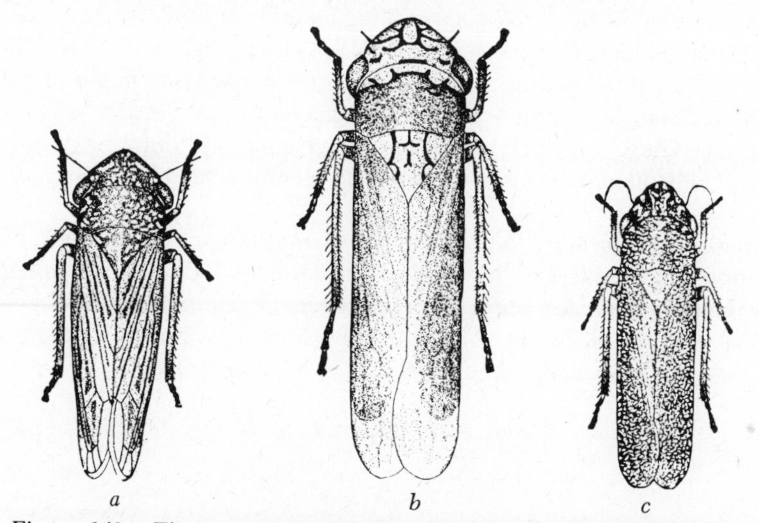

Figure 141. Three cotton leafhoppers commonly called sharpshooters: *a, Cuerna costalis* (Fabr.); *b, Oncometopia undata* (Fabr.); *c, Aulacizes irrorata* (Fabr.); all enlarged. (USDA)

Thurberia Weevil, *Anthonomus grandis thurberiae* Pierce, is a variety of the boll weevil, native to this country and found infesting a species of wild cotton in Arizona. It has spread from this host (*Thurberia* sp.) to cultivated cotton and seems to be increasing in numbers. Only 1 and a partial second generation occur on wild cotton, but this increases to 3 generations on cultivated cotton.

Control measures advocated for the boll weevil are recommended for this insect.

Stink Bugs attack a wide variety of plants, and several species seriously damage cotton in all areas where it is grown. These species are the conchuela, *Chlorochroa ligata* (Say), the green stink bug, *Acrosternum hilare* (Say), the western brown stink bug, *Euschistus impictiventris* (Stål), the southern green stink bug, *Nezara viridula* (L.), and the Say stink bug, *Chlorochroa sayi* Stål. Most damage is done when the cotton plants are fruiting. The piercing-sucking mouthparts are inserted into the bolls and the sap removed from the immature seeds. These punctures cause shedding of small bolls; those more mature are retained but the lint is stained, short, weak, and of little market value. Most species pass the winter as adults and produce 1 or more generations per year depending on the latitude. (See p. 233.)

REFERENCES: *Ariz. Agr. Exp. Sta. Bul.* 140, 1960; *J. Econ. Ent.*, 57:60–62, 1964; *Ann. Ent. Soc. Amer.* 62:1246–1247, 1969.

Cotton Stainer, *Dysdercus suturellus* (Herrich-Schäffer), is a bug that punctures cotton bolls with its piercing-sucking mouthparts and interferes with their proper development. It, as well as related species, causes yellow discoloration of the lint in the more mature bolls. As a rule the stainer is of minor importance. A species causing similar injury is the Arizona cotton stainer, *Dysdercus mimulus* Hussey. These bugs belong to the order Hemiptera, family Pyrrhocoridae. They are controlled by the same chemicals recommended for the major cotton pests.

Cotton Leaf Perforator, *Bucculatrix thurberiella* Busck, is the larva of a very small moth that feeds first by mining and later by eating small holes in the leaves. Hot dry climates seem favorable to its development; it has been very injurious in parts of Texas, Arizona, New Mexico, California, Mexico, and South America, especially where pesticides have been used.

REFERENCES: *USDA Agr. Handbook* 290, 1965; *J. Econ. Ent.* 54:67–70, 1961; 65:1765–1766, 1972; *Bul. Ent. Soc. Amer.* 23:195–198, 1977.

12

PESTS OF LEGUMINOUS CROPS

In addition to the pests covered in this chapter legumes may be damaged by whiteflies, spider mites, snails, slugs, nematodes and some of those discussed in Chapter 9.

MEADOW SPITTLEBUG

Philaenus spumarius (Linn.), Family Cercopidae

Spittlebugs are widely distributed in North America and have been known for a long time but only in the last two decades have they increased in numbers to a point where serious damage has been noted, especially to such perennial plants as strawberries, nursery, and legume forage crops. The injury occurs when the plant sap is extracted by the piercing-sucking nymphs and adults. Stunting of growth, shortening of internodes, dwarfing, rosetting, general loss of vitality, and low yields are some of the known effects.

The adults are about 1 cm in length and resemble robust leafhoppers but can be distinguished from them by stout spines on the hind tibiae and a circlet of spines at the base of each. Considerable variation in the color pattern exists, some are light tan, others mottled with dark brown, and still others almost black (Fig. 142). These have been called color varieties; field collections have shown that copulation between them does occur, and inheritance is undoubtedly involved in the variation. The eggs, white at first, then turning a light brown color, are laid in rows between the sheaths and stems (Fig. 143) or in cracks of the stems or stubble of plants near the soil surface. The yellow nymphs pass through five instars, becoming tinged with green as they approach the adult stage (Fig. 144). The normal excretion from the alimentary canal is mixed with air, forming the spittle mass inside which all nymphal instars must develop (Fig. 145).

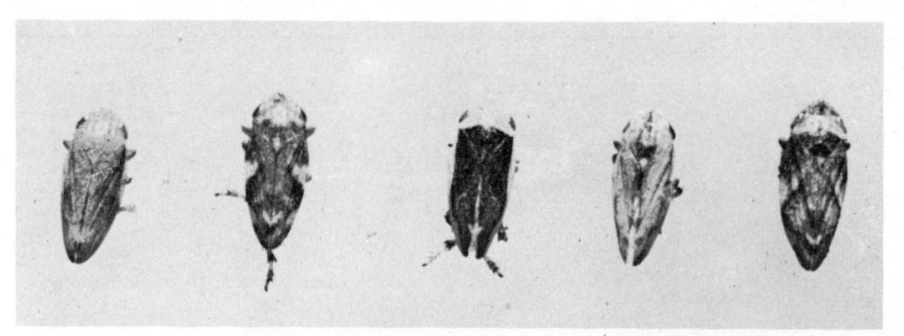

Figure 142. Meadow spittlebug adults, showing color varieties; 2 ×.

The overwintering eggs begin hatching about the middle of April in the latitude of Columbus, Ohio, reaching the adult stage in about 45 days. Adults are present throughout the summer feeding on a variety of plants, but their damage is not as evident as that of the nymphs in the spittle masses. Adults migrate by jumping or by flying short distances. Whenever a crop of hay is cut they migrate into adjacent areas causing a noticeable increase in numbers. Egg-laying begins about the first of September and continues until the females are killed by cold weather.

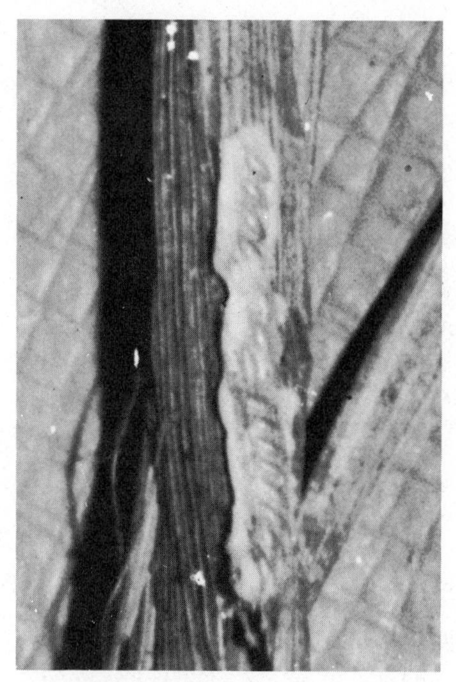

Figure 143. Egg mass of the meadow spittlebug; greatly enlarged. (Courtesy of D. D. Ahmed.)

Figure 144. Spittlebug nymph in spittle mass; 15 × (Courtesy of D. J. Borror.)

Figure 145. Spittle mass, greatly enlarged. (Courtesy of Roger M. Thomas.)

Plowing under any crop that harbors overwintering eggs will eliminate the possibility of nymphal damage to the new planting on that soil.

The need for chemical control in the spring can be predetermined by surveying the crop in autumn. An average of 1 adult spittlebug per sweep of a standard insect net in September will result in 1 nymph per stem in the spring. Control measures are usually justified where nymphal populations are at this level or higher.

REFERENCES: *J. Econ. Ent.*, 39:299–305, 1946; 43:905–908, 1950; 44:163–166, 289–293, 1951; 52:240–242, 904–907, 1959; 53:960–961, 1960; 55:184–188, 718–722, 828–830, 1962; *Ohio Res. Bul.* 741, 1954; *USDA Leaflet* 514, 1962.

ALFALFA WEEVIL

Hypera postica (Gyllenhal), Family Curculionidae

The alfalfa weevil is a European insect that appeared in Utah about 1904 and has since spread into all the western states; in 1951 it was found in Maryland and has now spread throughout the eastern and central states wherever alfalfa is grown.

Damage is caused by the adults and larvae feeding on the growing tips, leaves, and buds of alfalfa, lowering its value as a hay crop or preventing profitable production of seed. The insect is essentially a pest of first-growth alfalfa but the second growth is also attacked. Other clovers and vetch are sometimes slightly damaged. After damaged leaves dry, they appear gray as if frosted.

The adult is a snout beetle about 4 mm long, varying in color from brown to black with faint lighter markings (Fig. 146). Overwintering occurs primarily in alfalfa fields and chiefly as adults. With the coming of warm spring temperatures the females begin depositing tiny oval lemon-yellow eggs which darken as hatching approaches. At first the eggs are deposited in fragments of dead stems on the ground but after spring growth takes place egg-laying is gradually shifted to the alfalfa stems. They are placed in clusters of 25 or more with each female averaging about 400 eggs, most of which are laid in April and May. Hatching generally begins in April, but larvae do not become numerous until late April or May or about the time that first-growth alfalfa produces buds. Each larva molts 3 times and attains a length of 9 mm at maturity, when the head is black and the body is green with a white stripe along the middle of the back and a fainter parallel stripe on each side. By May 15 some larvae complete their growth and spin lacelike cocoons on the plant or on the ground in which transformation to pupae occurs. One or 2 weeks later new adults emerge, do some feeding, then estivate during the summer and go into hibernation with the coming of cold weather. Some weevils become sexually mature and lay eggs before cold weather arrives. These fall-laid eggs do not hatch until spring. Viability of overwintering eggs is low in areas having low winter temperatures. One generation occurs per year but in some areas a partial second generation sometimes develops.

Like all insects this weevil is limited by all the factors in natural control. Introduced species of ichneumonid wasps, *Bathyplectes curculionis* (Thomson) (Fig. 147) and *B. anurus* (Thomson), attack the larvae of the alfalfa weevil, and in some areas approximately 90% parasitism occurs when the first crop of al-

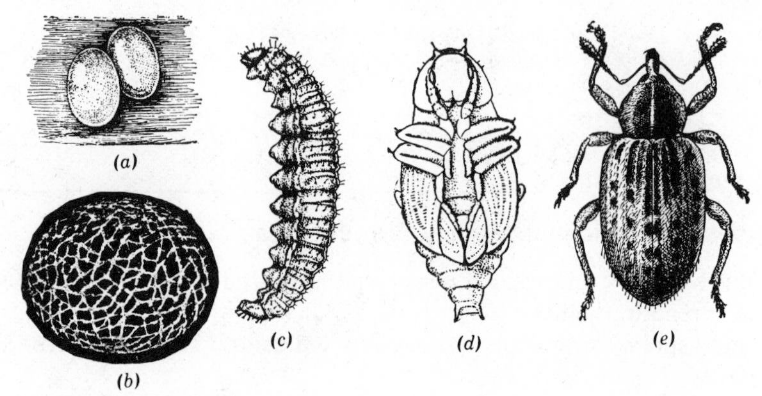

Figure 146. The alfalfa weevil, *Hypera postica* (Gyllenhal): *a*, eggs; *b*, cocoon; *c*, larva; *d*, pupa; *e*, adult—all much enlarged. (Webster, USDA)

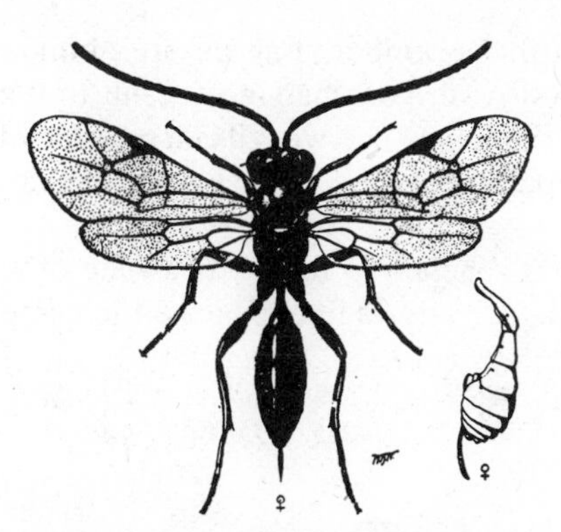

Figure 147. Bathyplectes curculionis (Thomson), a larval parasite of the alfalfa weevil. (Right) side view of abdomen, all enlarged. (Walton, USDA)

falfa reaches the flower-bud stage. Other parasites of importance are the braconid, *Microctonus aethiopoides* Loan, that attacks adults, and the eulophid, *Tetrastichus incertus* (Ratz.). An *Entomophthora* fungus kills many larvae.

By integrating biological, cultural, and mechanical control measures the use of chemicals, which kill natural enemies, can often be avoided. Some cultural and mechanical practices of merit are: planting tolerant or resistant varieties; maintaining a dense vigorous stand; cutting the first and second hay crops when most plants are in the bud stage, mowing the field clean and removing the hay as soon as cured. Chemical sprays are warranted when 25 larvae per net sweep or 2 larvae per stem can be seen without touching the plant and 50 to 75% of the leaves show feeding injury.

REFERENCES: *USDA Farmers' Bul.* 1930, 1943; *Leaflet* 368, 1965; *J. Econ. Ent.*, 46:178–179, 1953; 61:916–918, 1050–1054, 1968; 62:509–510, 1969; 63:554–557, 1970; *Bul. Ent. Soc. Amer.* 14:285–288, 1968; 21:251–255, 1975; *Va. Agr. Exp. Sta. Bul.* 502, 1959; *W. Va. Agr. Exp. Sta. Cir.* 104, 1958; *Conn. Agr. Exp. Sta. Bul.* 621, 1959; *Calif. Agr. Exp. Sta. Ext. Cir.* 85 and 86, 1964; *Ohio Sta. Univ. Ext. Bul.* 545, 1977.

EGYPTIAN ALFALFA WEEVIL

Hypera brunneipennis (Boheman), Family Curculionidae

This weevil was discovered in the United States near Yuma, Arizona, in 1939, and has spread over much of southern California and other adjacent areas where clover and alfalfa are grown.

The size, appearance, life history, and habits of this weevil are similar to those of *H. postica*. Adults estivate during the summer months and begin to appear in alfalfa fields in December. Larvae are abundant from February to April. During this period severe damage may result to the growing tips as well as skeletonizing of entire plants. The weevil is a problem during the first cutting although damaging populations may persist until the time of the second cutting.

When the larval count reaches 15 to 25 per sweep of an insect net chemical control is usually justified. Natural enemies are the same as for *H. postica*.

REFERENCES: *J. Econ. Ent.*, 48:297–300, 1955; *Calif. Agr. Exp. Sta. Leaflets* 85, 86, 1964; *Bul. Ent. Soc. Amer.* 18:102–108, 1972; 21:251–255, 1975; *Calif. Agr.* (9) 29:10–11, 1975.

SPOTTED ALFALFA APHID

Therioaphis maculata (Buckton), Family Aphididae

The spotted alfalfa aphid, native to the Middle East, was accidentally introduced into New Mexico in 1954 and subsequently has spread over most of the United States. It is especially destructive to alfalfa grown west of the Mississippi River, particularly in the states of Arizona, California, and Nevada. To a much lesser extent the aphid will feed on some clovers and other forage legumes.

Damage results from both nymphs and adults sucking sap from the leaves and stems of the host plants. The first evidence is seen as whitening of the leaf veins; curling, stunting, and yellowing follow, and under heavy infestation the plant will die. In addition, the aphids excrete copious quantities of honeydew, which serves as a medium for the development of sooty fungus; this condition complicates harvesting procedures and lowers the quality of hay.

The aphid is about 2 mm long and pale yellow, with 6 or more rows of spiny black spots on the dorsal side of the abdomen (Fig. 148). Winged forms have smoky areas along the wing veins.

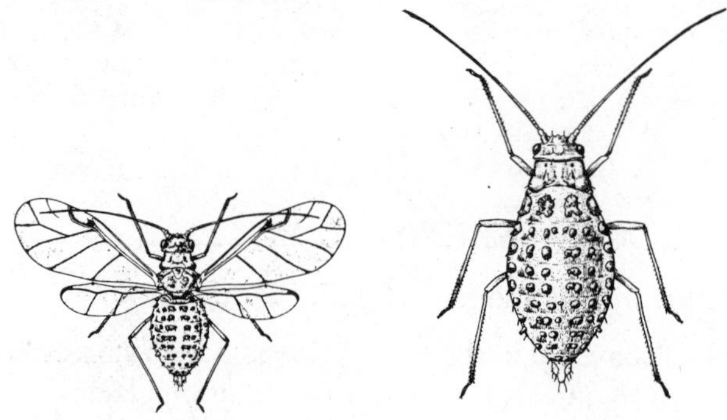

Figure 148. Winged and wingless adults of the spotted alfalfa aphid, enlarged. (Courtesy of Celeste Greene and Ray F. Smith, Calif. Agr. Exp. Sta.)

In warmer climates reproduction is parthenogenetic throughout the year and each female may give birth to over 140 nymphal aphids. The nymphs pass through 4 instars and mature in 1 to 2 weeks when temperatures are high and in 3 to 4 weeks when cool weather prevails. There may be 20 or more generations each year. In colder regions sexual forms may develop in autumn, and after mating the females deposit the overwintering eggs.

Natural control from lady beetles, syrphid fly larvae, lacewing larvae, big-eyed bugs (*Geocoris* sp.), minute pirate bugs (*Orius* sp.), damsel bugs, parasitic wasps, and fungus diseases often keeps populations of this aphid checked. Three introduced parasites, *Praon palitans* Muesebeck, *Aphelinus semiflavus* Howard, and *Trioxys utilis* Muesebeck are contributing to control in areas where they have been released. Growing aphid-resistant varieties such as Washoe, Lahontan, Moapa, or Sonora is a recommended procedure.

At times it will be feasible to apply pesticides to check a rapid increase in aphid populations. Treatment is necessary in seedling plantings where 1 or more aphids are found per plant and in older fields when a count in various areas average 10 to 30 aphids per stem.

REFERENCES: *USDA Leaflet* 422, 1965; *Hilgardia*, 24:93–117, 1955; 28:647–684, 1959; *J. Econ. Ent.*, 48:668–671, 1955; 50:352–356, 805–807, 817–821, 1957; 52:136–141, 714–719, 1959; 53:89–94, 234–238, 655–659, 1960; 54:1144–1147, 1961; 55:292–294, 900–904, 1962; 56:84–85, 1963; 57:71–76, 1964; *Calif. Agr.*, 17, June, 1963.

BLUE ALFALFA APHID*

Acyrthosiphon kondoi Shinji, Family Aphididae

This insect, native to Japan, was first identified in North America by California entomologists in 1975. The initial outbreak was discovered in the states of Arizona, California, and Nevada. Since then it has also been found in Utah, New Mexico, and as far east as Kansas. Further surveys may reveal a wider distribution.

Like most aphids the piercing-sucking nymphs and adults cause stunting and distortion of the plant, and finally chlorosis. Most severe feeding is on the tender terminal growth. Damaged portions of the plant never seem to recover, indicating a possible toxin is involved.

Among the important aphid pests of alfalfa in Arizona the spotted alfalfa aphid caused the severest damage followed by the blue alfalfa aphid and then the pea aphid.

The blue alfalfa aphid resembles the pea aphid (4 mm long) but is slightly smaller with a bluish-green body color and a uniformly brown third antennal

* Information on this pest was supplied through the courtesy of Dr. C. G. Summers, Univ. of Calif., Berkeley, and Dr. M. W. Neilson, USDA Forage Insect Res. Lab., Tucson, Arizona.

segment, whereas the pea aphid is bright green with the third antennal segment having a dark band at the tip. Winged forms of the blue aphid have a blackish-brown thoracic area. On the pea aphid this same area is light brown.

Fecundity was greatest at 60 F (15.5 C) but decreased rapidly at temperatures above that figure. Longevity was also affected by temperature. Maximum survival at 45 F (7.2 C) was 124 days, while at 85 F (29.5 C) it was only 13 days. The blue alfalfa aphid has several generations per year with peak numbers occurring from February through June, whereas pea aphid population peaks are in both spring and fall.

Natural enemies include parasites and predators, especially *Hippodamia convergens*. Development of resistant varieties of alfalfa appears to be a very promising control measure. The economic threshold appears to be lower than the level set for the pea aphid.

REFERENCES: *Proc. 5th Calif. Alfalfa Symp.*, pp. 24–28, 31–35, 1975; *6th Symp. Univ. Calif. Pub.* 3209, 66–67, 1976; *J. Econ. Ent.* 69:471–472, 1976; *Prog. Agr. Ariz.* pp. 18–19, spring, 1976; *Calif. Agr.* 31(8):4–5, 1977.

ALFALFA BLOTCH LEAFMINER

Agromyza frontella (Rondani), Family Agromyzidae

This pest was described in 1874 from specimens collected in the vicinity of Parma, Italy. It is widely distributed in Europe but has attracted little attention as a pest. Host plants are species of *Medicago* with preference shown for alfalfa.

Leafminer damage to alfalfa grown in North America has been noticed as early as 1963 in Virginia. In 1972 *A. frontella* was confirmed as the species involved. Primary distribution in the United States is the area from New England south to Virginia and west to Ohio. It also occurs in adjacent Canadian provinces. Intensity of the population varies considerably within the infested area and from year to year.

Damage is caused by the tiny yellow larvae that hatch from eggs inserted in the leaflets. Mines usually begin near the base of the leaf and broaden toward the tip as larval growth occurs. The end of the mine is expanded into a distinct blotch. Under heavy infestations the leaflets may absciss leaving only the stems. Pinhole feeding punctures in the leaves in June indicate presence of the fly. In more northern areas damage is heavier to the first cutting, whereas in southern areas the second and sometimes third cuttings are damaged most. The economic injury level is near 50 blotches per plant stem or a 20% infestation.

The adult is a tiny black-bodied fly with a dull sheen, about 2 mm long. The face, cheeks, and basal antennal segments are yellow to brown. The overwintering stage is a puparium in the soil. Three generations per year are indicated.

Since most agromyzids are heavily parasitized natural control plays an important role in keeping infestations checked. Perhaps some recent management operation with alfalfa is causing the increase in damaging populations in the United States.

REFERENCES: *USDA CEIR* 22:132–137, 1972; *Ont. Factsheet*, Dec. 1977.

CLOVER LEAF WEEVIL

Hypera punctata (Fabr.), Family Curculionidae

This weevil, a native of Europe, has long been present in North America and is now well established in most areas where clover and alfalfa are grown. Besides various clovers and alfalfa the weevils are also known to feed on soybeans, snap beans, and occasionally the leaves of timothy, wheat, and corn. Damage is caused by the larvae and adults devouring the foliage.

The adult (Fig. 149) is a large brown snout beetle with indistinct stripes formed by light-colored pubescence. It varies in size but averages about 6 mm long. The yellow oval eggs are deposited in various places about the host plant, usually in autumn. A few weevils overwinter and deposit eggs during mild periods in winter or early spring. Most eggs hatch in the fall, but some remain through the winter. The young green larvae from fall-hatched eggs feed throughout the winter on mild days, becoming fully grown in late April or May. At this time they are almost 12 mm long, light green with a pale stripe down the middle of the back. Pupation takes place within an oval, netlike cocoon just beneath the surface of the soil or in debris near the base of the plants. About 11 days later the new adults emerge, feed for a short period, and become inactive most of the summer, then oviposit in the fall. There is commonly only 1 generation each year, but under exceptional weather conditions a second generation may develop and produce weevils before cold weather arrives. These weevils overwinter and deposit their eggs in the spring.

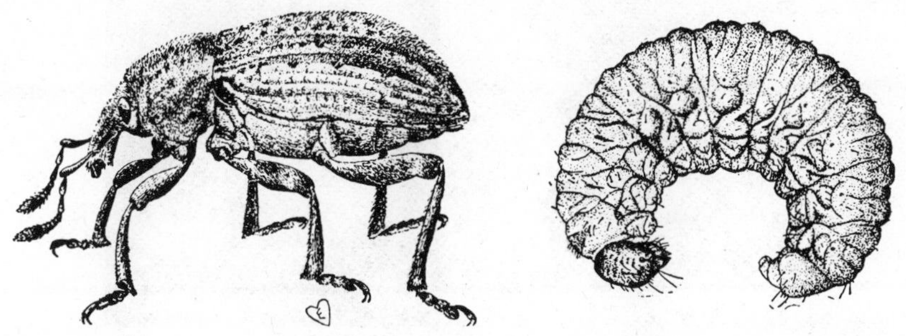

Figure 149. The clover leaf weevil, *Hypera punctata* (Fabr.); adult and larva, 8 ×. (USDA)

The larvae are subject to the attacks of a fungus, *Empusa sphaerosperma* Fres. Warm temperatures and high humidity favor the development of the fungus. In some areas parasitization of weevil larvae by the ichneumonid wasp, *Biolysia tristis* (Gravenhorst), has reached 90%. Both these natural control agents usually make chemical control measures unnecessary.

REFERENCES: *USDA Farmers' Bul.* 1484, 1949; *Cornell Agr. Exp. Sta. Bul.* 411, 1922; *J. Econ. Ent.*, 43:438–443, 1950; 47:927–928, 1954; 51:195–198, 1958; 55:831–833, 1962.

LESSER CLOVER LEAF WEEVIL

Hypera nigrirostris (Fabr.), Family Curculionidae

Supposedly indigenous to Europe, this weevil was introduced into North America about 1875. Since then it has spread throughout most of the regions where the favorite host, red clover, is grown. Other clovers and alfalfa are hosts of lesser importance. Damage is caused by both larvae and adults devouring the leaves, stems, and buds (Fig. 150).

The adult is scarcely more than 3 mm in length, dark greenish yellow with faint golden stripes on the wing covers, and of more slender form than the clover leaf weevil. Hibernating adults become active in the spring and begin depositing eggs, usually in or on stems and leaflets, from late March to May, depending on the season and latitude. Hatching occurs in approximately 2 weeks and the white-to-tan, curved-bodied larvae with brown heads feed almost a month, and pass through 4 instars, then spin cocoons on the ground or on the plant and transform to pupae. New adults emerge in 7 or more days, and, after a few weeks' activity, estivate until cold weather, then pass the winter as hibernating adults. One generation per year is typical but in some regions there

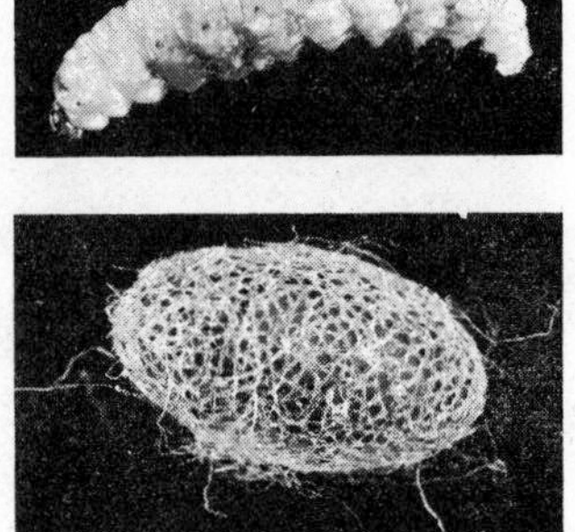

Figure 150. Adult, larva, and cocoon of the lesser clover weevil; 12 ×. (Ohio Agr. Exp. Sta.)

is a partial second generation; where this occurs, oviposition may begin as late as September.

Larvae and pupae are killed by *Empusa sphaerosperma* Fres. Several parasites have been observed attacking this pest with the braconid, *Bracon mellitor* Say, being the most numerous.

REFERENCES: *USDA Bul.* 85, 1911; *J. Econ. Ent.*, 44:785–791, 1951; 49:542–544, 1956; 50:224, 1957; *Ohio Agr. Exp. Sta. Res. Bul.* 956, 1963.

CLOVER HEAD WEEVIL

Hypera meles (Fabr.), Family Curculionidae

This species was first collected in North America at Rockaway Beach, N.Y., in 1907 and now is found widely distributed in North America. It infests the heads of various clovers, particularly crimson, and feeds among the florets, lowering seed production. It also has been reported as skeletonizing bean foliage. In appearance and size it is about the same as the lesser clover weevil.

CLOVER SEED WEEVIL

Tychius picirostris (F.), Family Curculionidae

The clover seed weevil, a pest of European origin, entered North America prior to 1920 and was first recorded from Puyallup, Washington, in 1934. It is now widespread in the United States, Canada, Finland, and other European countries. It feeds on alsike, Ladino, and white Dutch clover varieties; the major damage is by the larvae.

Adults average 2 mm in length, are black to gray-brown with white scales. This stage overwinters in debris in uncultivated areas and begins emerging in May and migrating to clover fields. Glossy white eggs are inserted into seed pods in June with hatching occurring in about 7 days. By mid-June white legless larvae can be found feeding on the seeds in the pods. When fully grown they leave the pods, drop to the ground and pupate in the soil; new adults emerge about 14 days later. The length of the life cycle is 40 days and 1 generation develops each year.

Cold winter weather without snow kills many of the hibernating adults. The pteromalid parasite, *Trimeromicrus maculatus* Gahan has been reared from infested pods.

Populations of 13 or more weevils per sweep of an insect net (12 inches in diameter) caused almost 100% infestation of clover heads; 90–100% of the pods, and one to all seeds (up to 7) in these infested pods were damaged. Monetary losses from the reduced seed yields approach $100 per acre.

Two weevils per sweep of an insect net is the threshold justifying chemical control operations.

REFERENCES: *J. Econ. Ent.*, 27:1103–1104, 1934; *Proc. N. C. Branch Ent. Soc. Amer.*, 10:53, 1955; *Wash. Agr. Exp. Sta. Bul.* 587, 1958; *Tech. Bul.* 53, 1967.

SWEET CLOVER WEEVIL

Sitona cylindricollis Fahraeus, Family Curculionidae

This weevil, first recorded in Canada in 1927 and in the United States in 1934, is now widely distributed in North America. Damage is caused by the adults eating stems of tiny seedlings and crescent-shaped areas from the leaves, and the larvae feeding within or on the nitrogen-fixing nodules of the rootlets and roots. When abundant, the adults may completely destroy seedling sweet clover in the spring. Other host plants, seldom attacked, are alsike clover, alfalfa, and black medic.

The adult is a dark gray snout beetle about 5 mm long, with pubescent elytra. This stage hibernates in the upper inch of soil and crop remnants. In the spring the weevils become active, and egg-laying may begin by the middle of March and continue into June. The eggs are dropped on the soil and a single female may deposit over 1700 during her life span. They are oval pearly white tinged with yellow at first, and in 24 hours become black in color. Hatching occurs in 1 week or more, depending on the temperature, and the tiny white legless larvae burrow into the soil and feed on the roots for 30 to 40 days. Following the fourth larval instar pupation occurs; after a period of 7 to 12 days, the new adults emerge, thus completing the cycle in 45 to 70 days. There is 1 generation per year.

Populations are limited in some areas by the fungus *Beauvaria bassiana*.

Cultural and mechanical control measures consist of shallow surface tilling in the summer after the hay crop is removed, rotating crops, making new seedlings some distance from old plantings, and plowing old plantings as a green manure crop. If the last operation is done after May 15 in Ohio, it effectively destroys as high as 85% of the new generation while in the egg or larval stages.

Chemical control measures are directed toward killing the overwintering adults that migrate from old plantings and destroy new seedling sweet clover.

REFERENCES: *Can Dept. Agr. Pub.* 72, 1948; 943, 1955; *Purdue Exp. Sta. Cir.* 369, 1951; *Minn. Agr. Ext. Folder* 180, 1954; *Ohio J. Sci.*, 53:105–112, 1953; *Ann. Ent. Soc. Amer.*, 56:831–835, 1963; *J. Econ. Ent.*, 45:316–319, 1952; 56:716–717, 1963; 61:391–394, 1968.

CLOVER ROOT CURCULIOS

Order Coleoptera, Family Curculionidae

The clover root curculios, *Sitona hispidulus* (Fabricius) (Fig. 151) and *S. flavescens* Marsham, are widely distributed in North America and attack various clovers, alfalfa, and other leguminous plants. Adults are leaf feeders and the larvae feed within or on the nodules and roots, frequently girdling the tap root. It is thought that root damage favors the development of bacterial wilt in alfalfa.

Both species resemble the sweet clover weevil in habits, size, and life cycle. Overwintering, however, occurs in both the adult and egg stages. Spring laid eggs hatch in one week, whereas fall laid eggs hatch the following spring. New adults emerge in June and July and live nearly 1 year.

REFERENCES: *USDA Bul.* 85, 1910; 649, 1915; *Ann. Ent. Soc. Amer.*, 56:831–835, 1963; *J. Econ. Ent.*, 23:334–342, 1930; 27:807–814, 1934; 48:184–187, 1955; 52:1155–1156, 1959; 55:906–908, 1962; 61:576–577, 1968.

CLOVER SEED CHALCID

Bruchophagus platypterus (Walker), Family Eurytomidae

Although many of the tiny wasps in the superfamily Chalcidoidea are beneficial parasites, some are pests. This one is a seed destroyer. It is one of the most important pests encountered in the production of clover seed. A related species, *B. roddi* (Gussakovsky), is the alfalfa seed chalcid, a major pest of the seeds of that plant. Both insects are cosmopolitan and of major importance in

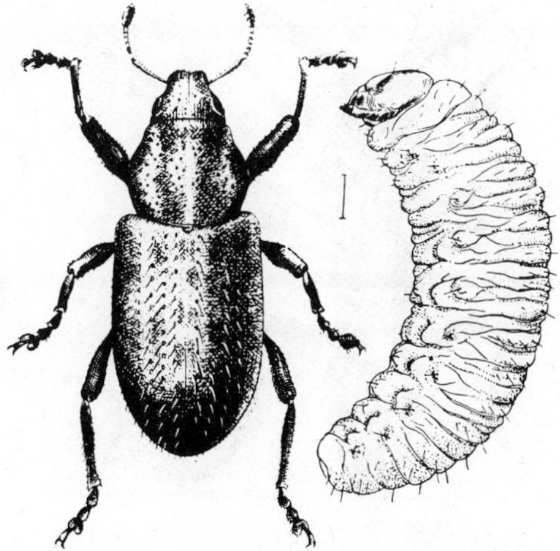

Figure 151. The clover root curculio, *Sitona hispiclulus* (F). adult and larva; 5 ×. (USDA)

western and southwestern United States where clover and alfalfa are grown for seed in North America. Where seed production is not the goal, the injury from these pests is negligible.

The adult is a tiny black-bodied wasp with 4 membranous wings (Fig. 152). It appears in the field in early summer and infests the early-blooming host plants by laying eggs in the developing seeds. The white legless larvae devour the inside portion of the seed, change to pupae, and emerge as new adults in approximately 30 to 40 days. This is repeated throughout the growing season with one to three generations developing per year, depending on the latitude. All development stages occur inside the infested seeds. Larvae overwinter within the harvested seeds or those that fall to the ground during harvest. Infested seeds from volunteer host plants along fence rows and similar habitats also serve as hibernation quarters.

Any practical sanitation measure that will help reduce the overwintering population will be of value in controlling this pest. Cooperative community effort is desirable in this type of control program. Plowing, harrowing, and disking in the winter are farm operations designed to bury infested seed or cause it to mold and decay. Fumigation or other treatment of infested seeds is also necessary. Badly infested seed crops should be utilized as hay rather than left in the field to add to the infestation. Pesticides kill the pollinators, and the seed chalcid parasites which are often quite important in reducing damage.

REFERENCES: *USDA Farmers' Bul.* 1642, 1931; *Wash. Agr. Exp. Sta. Bul.* 587, 1958; *J. Econ. Ent.*, 57:105–110, 1964; *Proc. Utah Acad. Sci.* 11:241–244, 1934; *Ann. Ent. Soc. Amer.* 63:744–749, 1970.

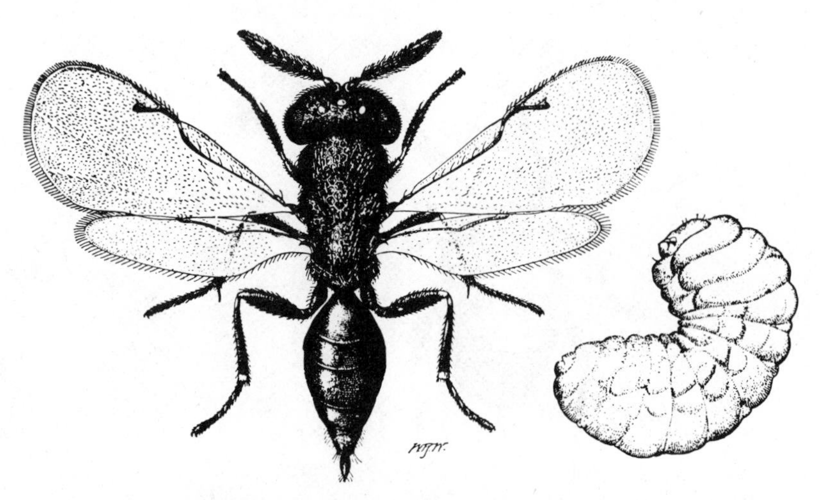

Figure 152. The clover seed chalcid, *Bruchophagus platypterus* (Walker); 17 ×. (Walton, USDA)

CLOVER ROOT BORER

Hylastinus obscurus (Marsham), Family Scolytidae

The clover root borer is an important pest of red and mammoth clovers. Introduced into the United States from Europe sometime before 1878, it was first reported attacking clover in western New York and since has spread to all parts of northern United States and southern Canada wherever red clover is an important crop. It has also been recorded feeding on alsike, crimson, and sweet clovers, alfalfa, vetch, garden peas, and field beans, but is of little importance on these hosts.

The roots of clover become infested during the spring of the crop year and damage is most apparent about the time of harvest of the first cutting for hay. The larvae tunnel within and throughout the roots, often killing the plant. Many infested plants are pulled out by the mower during cutting and others remain so weakened that they make little new growth. Such conditions result in a reduced second crop of hay or reduced seed yields. Where this pest is abundant root rot and virus diseases are also more prevalent. An average of 1½ borers per root reduces hay yields 5½%.

Adult clover root borers are roughly 3 mm in length (Fig. 153), varying from light red-brown to darker shades as they age. This stage overwinters in the roots of old clover and in the spring emerges and flies to new clover and deposits oval white eggs in the roots. These hatch into legless, curved, cream white larvae with straw-colored head capsules. Pupation occurs in the roots from July to September, with the pupal period lasting 10 or more days. There is but 1 generation per year.

REFERENCES: *USDA Dept. Bul.* 1426, 1926; *Cornell Univ. Memoir* 376, 1961; *Ohio Agr. Exp. Sta. Bul.* 827, 1959; *J. Econ. Ent.*, 47:327–331, 1954; 48:190–191, 1955, 50:255–256, 1957; 51:491–492, 1958; 53:449–450, 865–867, 1960; 54:631–635, 1058–1059, 1961.

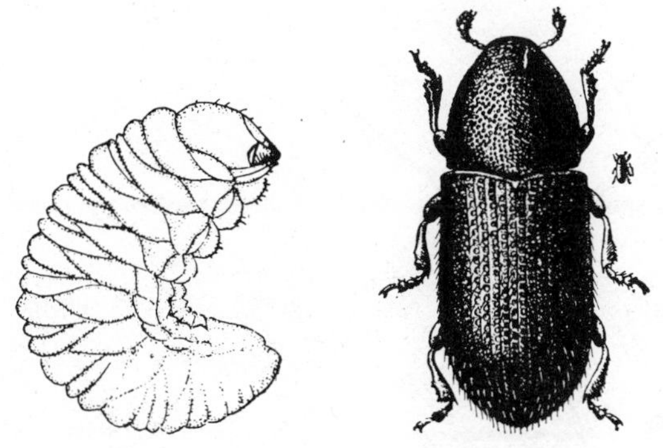

Figure 153. Larva and adult of the clover root borer. Small figure extreme right is adult natural size. (USDA)

CLOVER APHID

Nearctaphis bakeri (Cowen), Family Aphididae

In the Pacific Northwest extensive damage has been done at various times by this aphid. It is generally distributed north and west from Oklahoma but has been of the first magnitude only in the region first indicated. Red and alsike clovers are the common hosts attacked.

Serious injury results when great numbers of the piercing-sucking nymphs and adults remove plant sap; this causes stunting and irregular growth, particularly to the growing tips as well as reduced seed yields. These aphids also excrete great quantities of honeydew that causes the seeds to adhere in pellets, thus lowering their market value. Transmission of viruses causing alfalfa and bean mosaics is attributed to this aphid.

Overwintering eggs are found on apple and related woody plants. As the buds begin to open in the spring the eggs hatch into wingless, parthenogenetic, ovoviviparous females (Fig. 154) that produce succeeding generations of similar individuals, after which winged forms appear and fly to the clover plants. Here repeated generations of wingless and sometimes winged individuals develop throughout the summer. With the advent of cool autumn weather the winged fall migrants develop and fly back to apple and related hosts, where both males and females are produced and, after mating, the overwintering eggs are deposited.

Natural control from lady beetles, syrphid fly larvae, aphidlions, parasitic wasps, and fungus diseases usually keeps this aphid checked.

REFERENCES: *USDA Farmers' Bul.* 1128, 1926; *Mont. Bul.* 484, 1953; *J. Econ. Ent.*, 53:113–115, 1012–1015, 1960; 54:414–416, 1961; 55:460–462, 1962.

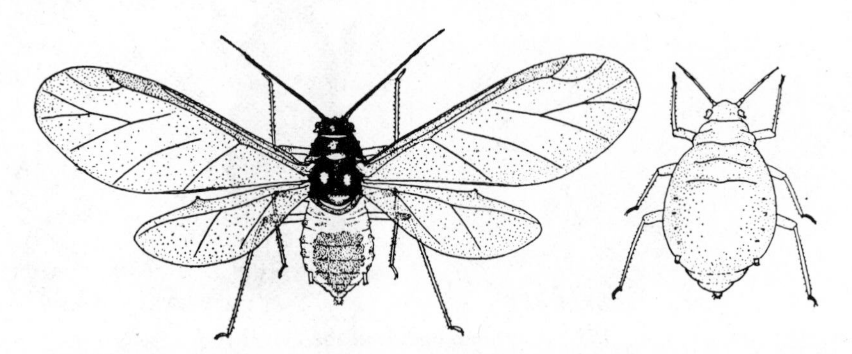

Figure 154. The clover aphid, *Nearctaphis bakeri* (Cowen): (left) spring migrant; (right) stem mother; 12 ×. (Idaho Agr. Exp. Sta.)

VETCH BRUCHID

Bruchus brachialis Fahraeus, Family Bruchidae

Widely distributed throughout the world, this insect is supposed to be native to Europe. It was first discovered in New Jersey in June, 1930 by L. J. Bottimer. Since then it has been found in most of the eastern coastal states, some of the Gulf Coast states, and in a limited area in Oregon, Washington, and Illinois.

The bruchid attacks only the developing seeds in the pod, often entirely destroying them. Hence it is only important from the standpoint of seed production, since vetch for forage is little damaged. Observations indicate that it attacks only plants in the genus *Vicia*. Breeding does not occur in dry seeds in storage.

The adult is about 3 mm in length, brown-black, with irregular spots of white pubescence. This stage emerges from hibernation in April, May, and June, depending on the prevailing temperatures, and begins to deposit pale yellow eggs on the pods of the host plant. As many as 42 eggs have been found on a single pod. Hatching occurs in approximately 1 week, and the white legless larvae enter the pods and then the seeds. Development continues for 2 weeks, the larvae passing through 4 instars, with pupation taking place within the seeds. Five days later the new adults begin emerging. Only 1 generation occurs per season.

Fumigation of all harvested seeds in sacks will kill both emerged and unemerged adults. Use the fumigants recommended for control of stored-product pests (p. 501). Limiting movements of infested hay to uninfested areas is a means of checking spread.

REFERENCES: *J. Agr. Res.* 46:739–751, 1933; *J. Econ. Ent.*, 30:621–632, 1937; 44:993–994, 1951; 52:955–957, 1959; 53:555–558, 1960.

LEAFHOPPERS

Order Homoptera, Family Cicadellidae

Leafhoppers comprise an important group of little noticed sap-sucking insects that cause great damage to a wide variety of crops. Their feeding by means of piercing-sucking mouthparts causes spotting, yellowing, leafcurling, stunting, and browning of the foliage, depending on the species. Sometimes leafhoppers are responsible for transmitting the organisms causing virus diseases in plants.

Potato Leafhopper, *Empoasca fabae* (Harris), is one of the most important hoppers attacking alfalfa in the eastern half of the United States. It has been found farther west with the possible exception of the Northwest. The feeding of this leafhopper causes yellowing and dwarfing of alfalfa foliage and, in heavy attacks, severe wilting, resulting in lower yield, quality, and stand. Besides alfalfa it has been found living on nearly 200 other kinds of plants, with other

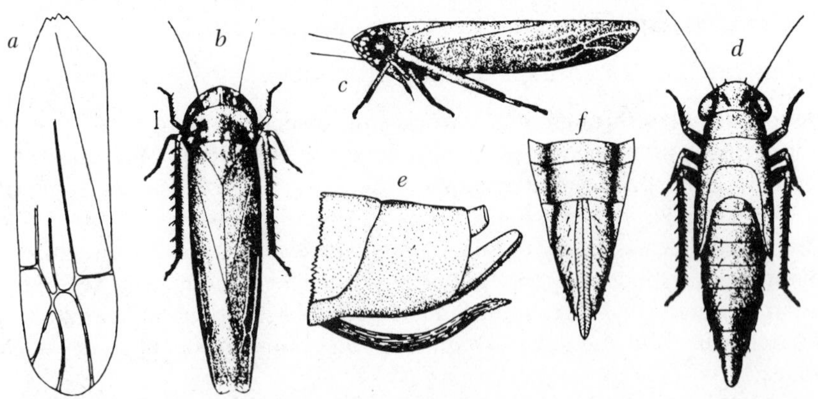

Figure 155. The potato leafhopper, *Empoasca fabae* (Harris). *a,* enlargement of wing; *b,* dorsal view of adult; *c,* side view of adult; *d,* dorsal view of nymph; *e,* side view of female abdominal tip; *f,* ventral view of female abdominal tip; 13 ×. (Utah Agr. Exp. Sta.)

forage legumes, potatoes, beans, cowpeas, and deciduous nursery stock the most seriously damaged.

The adult is a pale green, wedge-shaped insect about 3 mm long (Fig. 155). It is very active, jumping or flying when distrubed. Both nymphs and adults can run backwards or sideways as rapidly as they move forward. The females deposit slender white eggs within the stems and larger veins of the leaves. In summer hatching occurs in 6 to 9 days and the pale green nymphs molt 5 times before they become fully grown and transform to winged adults. Soon after the adults appear mating occurs, followed by egg laying. The period from egg to adult is about 3 weeks in warm summer temperatures. There are several generations per year and these greatly overlap. The potato leafhopper has not been found overwintering north of the Gulf states where it breeds throughout the year; it migrates northward with the warm spring winds annually. In the latitude of Washington, D.C., the first adults usually appear in late April or early May (see p. 259).

Other species of leafhoppers that attack legumes and vegetable crops in the western states are *Empoasca abrupta* DeL., *E. arida* DeL., *E. solana* DeL., and *E. filamenta* DeL. These species closely resemble *E. fabae* in general appearance and have a similar life cycle.

Clover Leafhopper, *Aceratagallia sanguinolenta* (Prov.) is generally distributed throughout North America. Although clovers are the common host it also attacks alfalfa, beans, cowpeas, other leguminous plants, and occasionally grasses.

The adult is about 3 mm in length, robust with dark mottling (see Fig. 10). In the Northern States overwintering occurs in the adult stage under plant remnants in the field; in the extreme South and Southwest the leafhopper is active throughout the year. The adult female places her eggs in leaves and stems, these hatching in 5 to 12 days during the summer months in the latitude of southern Illinois. The nymphs pass through 5 molts and become adults in 25 or more days depending on the latitude. There are 3 generations annually in southern Missouri, and more in subtropical areas.

Leafhopper Control

Many predaceous insects, spiders, mites, birds, and parasitic insects attack leafhoppers, but none of them is abundant enough to give satisfactory control where huge populations occur.

Cultural control measures for legumes are of value and can be accomplished by manipulating the date of harvest according to the region under consideration. This practice will remove many eggs with the hay. Tiny nymphs from hatched eggs also starve from lack of food. In northern latitudes early spring seedings allow the seedlings to become vigorous before potato leafhoppers arrive.

Using pesticides is the most effective method of preventing leafhopper damage to legumes. To determine the need for treatment make 20 sweeps in each of 5 different parts of the field with a standard insect net. Capturing an average of 1 or more potato leafhoppers per sweep is enough to justify pesticide application on alfalfa. Application should be made to other cuttings of alfalfa when the new growth is 6 to 12 inches tall and one adult per net sweep is collected.

REFERENCES: *J. Agr. Res.*, 43:267–285, 1931; *USDA Tech. Bul.* 231, 1933; 850, 1943; *Farmers' Bul.* 737, 1916; *Leaflet* 521, 1963; *J. Econ. Ent.*, 50:493–497, 1957; 55:828–830, 973–978, 1962; *Ohio Sta. Univ. Ext. Bul.* 545, 1977.

PLANT BUGS

ORDER HEMIPTERA, FAMILY MIRIDAE

Several species of plant bugs attack alfalfa, clover, cotton, sugar beets, soybeans, many garden vegetable crops, and weeds. Damage on legumes is primarily to the flower buds and the developing seeds and is caused by the piercing sucking nymphs and adults. Effects produced are distorted growing tips and buds, flower fall, and phytotoxic reactions attributed to the salivary juices injected into the plant. These insects also cause oviposition injury to the plant stems. The end result of all this damage is lowered seed and hay yields.

Alfalfa Plant Bug, *Adelphocoris lineolatus* (Goeze), was first reported in North America from Nova Scotia in 1917. The first record of its occurrence in the United States was from Iowa in 1929. Since then it has spread gradually through the midwest, north central and North Atlantic states. Its common hosts are alfalfa and sweet clover.

Overwintering eggs in the stems of the host plants begin hatching in May,

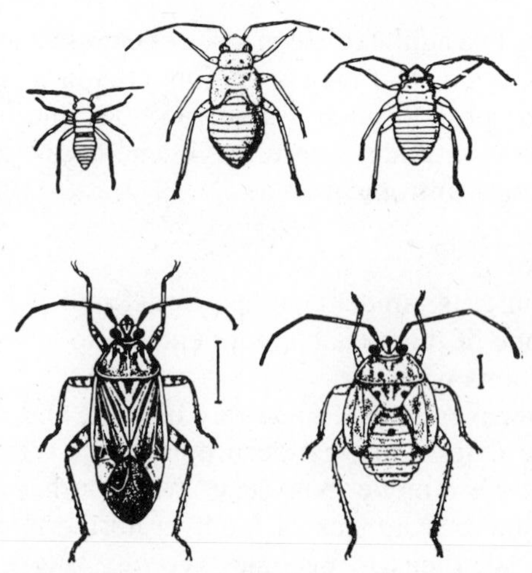

Figure 156. Nymphs and adult of the tarnished plant bug, *Lygus lineolaris* (P. de B.); 4 ×. (USDA)

and the tiny green nymphs with red eyes pass through 5 instars in approximately 30 days, then transform to adults. These adults are light green tinged with brown and about 9 mm long. In 2 weeks they begin egg-laying that results in a second generation by late August or early September. Only 2 generations occur in the latitude of St. Paul, Minnesota, and these overlap considerably.

Tarnished Plant Bug, *Lygus lineolaris* (P. deB.), is widely distributed in the United States and Canada, and feeds on many hosts, alfalfa, clover, cotton, sugar beets, and garden crops being the most common. Related species are *L. hesperus* Knight and *L. elisus* V.D., which are most abundant in the western and southwestern states where damage to alfalfa seed production is serious, especially from *L. hesperus*.

These bugs are scarcely 6 mm long, brown or tan and sometimes greenish with darker markings (Fig. 156). They overwinter as adults, becoming active in the spring and depositing eggs in the stems, petioles, and midribs of the hosts. Hatching occurs in about 7 days and the greenish nymphs molt 5 times, reaching the adult stage in approximately 30 days. There are 3 to 5 generations annually depending on the latitude.

Rapid Plant Bug, *Adelphocoris rapidus* (Say), occurs in the United States and Candad primarily east of the 110th meridian where it feeds on many plants including alfalfa and sweet clover. The size, shape, habits and life cycle are the same as those of the alfalfa plant bug. In coloration the adult rapid plant bug is dark brown with yellow costal margins and a yellowish pronotum having 2 black spots near the base. The nymphs are red-tinged (Fig. 157). The superb plant bug, *A. superbus* (Uhler), is another closely allied species with similar habits and life cycle.

In California strip cutting of alfalfa is a cultural control measure for lygus bugs. Natural enemies are minute pirate bugs, big-eyed bugs, and spiders.

Chemical control of plant bugs on legumes is usually recommended only for seed crops. The established thresholds for alfalfa are: prior to seed set, where 12 or more bugs are collected per sweep of a standard insect net (nymphs count as 2); after seed set, 20 or more per net sweep.

REFERENCES: *USDA Tech. Bul.* 741, 1940; *Minn. Agr. Exp. Sta. Tech. Bul.* 161, 1943; *Can. Dept. Agr. Pub.* 949, 1959; *Calif. Agr.*, 18 (4):4–6, 1964; 28:8–10, 1969; *Agr. Exp. Sta. Leaflet* 86, 1964; *J. Econ. Ent.*, 56:532–533; 823–825, 1963; 57:225–230, 1964; *Bul. Ent. Soc. Amer.* 23:19–22, 277–278, 1977.

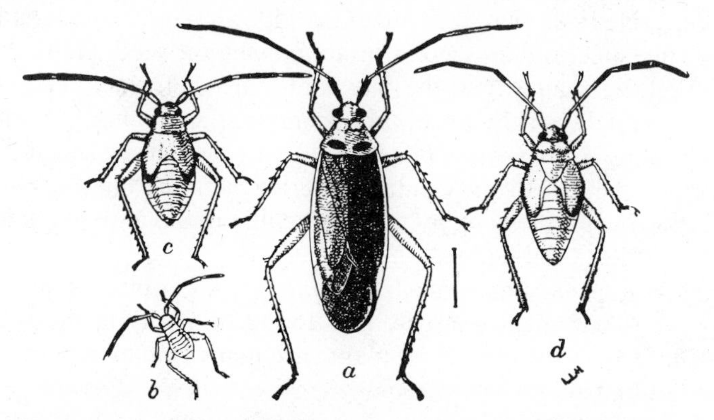

Figure 157. The rapid plant bug, *Adelphocoris rapidus* (Say); *a*, adult; *b, c, d,* stages in growth of nymph, 4 ×. (USDA)

STINK BUGS

ORDER HEMIPTERA, FAMILY PENTATOMIDAE

Stink bugs are important pests of alfalfa grown for seed in south-western United States. Besides alfalfa they also attack barley, wheat, oats, rice, soybeans, peaches, seed beets, okra, cotton, grain sorghum, squash, beans, peas, corn, cowpeas, tomatoes, vetch, and many weeds. Damage is caused by the nymphs and adults sucking sap, primarily from the pods, buds, blossoms, fruits, and seeds. Removing the liquid contents of the developing seeds causes them to become flattened and shrivelled; fruits are deformed and dimpled.

Most stink bugs overwinter as adults in sheltered places such as fence rows, roadsides, ditch banks or other places where plant remnants are abundant. In early spring, when temperatures reach 70 F (21 C) or above, they become active and begin egg-laying. A female may deposit 300 to 500 eggs in clusters, each containing an average of 30 over a period of a month or more. Hatching occurs in approximately a week and after 5 nymphal instars the adult stage is reached in almost 6 weeks, depending on the species and the prevailing temperatures. Repeat generations occur at 5 to 6-week intervals during the summer. Adults usually live 40 to 60 days in spring and summer, but overwintering individuals may live 7 to 9 months.

Say Stink Bug, *Chlorochroa sayi* Stål, is a widely distributed species in the western states, from Mexico to Canada. It is the most destructive species on seed alfalfa in Arizona. Adults vary in color from dark green in the spring to light green in midsummer, and to olive or red-brown in the fall and winter. The most characteristic markings are 3 large pale spots on the anterior margin of the scutellum (Fig. 158).

Western Brown Stink Bug, *Euschistus impictiventris* Stål, occurs mainly in Mexico, Texas, New Mexico, Arizona, Colorado, Utah, Nevada, and southern California. This species was second in abundance on seed alfalfa in Arizona. Adults are slightly smaller than the Say stink bug, uniformly yellowish brown with the lateral angles of the pronotum prominent and sharply pointed.

Green Stink Bug, *Acrosternum hilare* (Say), is a large green species (Fig. 159), widely distributed in North and Central America but more injurious in the South than in the North. It attacks many wild and cultivated plants, including alfalfa, lima beans, peaches, and cotton.

Southern Green Stink Bug, *Nezara viridula* (L.), is quite similar in appearance to *A. hilare* but slightly larger. It is found primarily in the southeastern states and Gulf Coast regions. Host plants are many, including legumes, vegetable crops, and citrus fruits.

Conchuela Stink Bug, *Chlorochroa ligata* (Say), has been reported mainly from Mexico, Arizona, Texas, New Mexico, Colorado, Utah, and California.

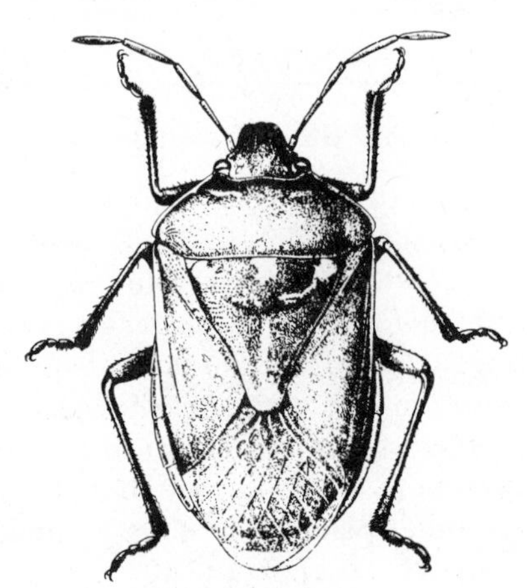

Figure 158. Say stink bug, *Chlorochroa sayi* Stål; 2 × (Caffrey and Barber, USDA)

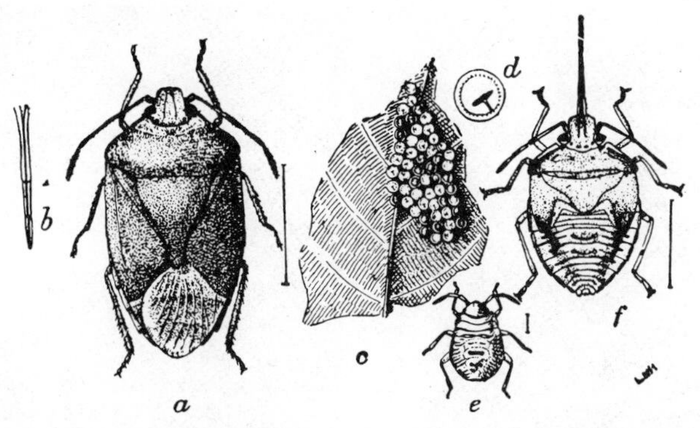

Figure 159. The green stink bug, *Acrosternum hilare* (Say), an occasional pest of cotton: *a*, adult; *b*, beak; *c*, eggs; *d*, end of egg more enlarged; *e*, young numph; *f*, last stage of nymph; 2 ×. (USDA)

The color varies from dull olive or ash gray to a green, purplish pink, or reddish brown. The most characteristic markings are an orange-red band along the lateral margins of the thorax and the costal margins of the wings and a spot of the same color on the tip of the scutellum.

Other species, at times injurious in various regions, are: brown stink bug, *Euschistus servus* (Say); consperse stink bug, *E. conspersus* Uhler; dusky stink bug, *E. tristigmus* (Say); and one-spot stink bug, *E. variolarius* (Pal. de Beau.) (Fig. 174).

The sceleonid egg parasites, *Telenomas podisi* Ashmead and *T. utahensis* Ashmead, are important in reducing stink bug populations.

Control measures consist of destroying hibernating quarters where practical; controlling weeds late in winter and in spring, in and surrounding the crop subject to attack; growing seed alfalfa as far removed as practical from sugar beets or small grains; planting and harvesting legume seed crops on uniform dates in all fields in a community or area; growing seed crops as rapidly as good agronomic practices permit; and using pesticides, where warranted.

REFERENCES: *Arizona Agr. Exp. Sta. Tech. Bul.* 140, 1960; *Bul.* A-23, 1962; *J. Kans. Ent. Soc.*, 34:151–157, 1961; *USDA Cir.* 903, 1952; *J. Econ. Ent.*, 45:254–257, 1952; 57:60–62, 1964; *Mo. Agr. Exp. Sta. Bul.* 805, 1962; *Ann. Ent. Soc. Amer.* 62:1246–1247, 1969.

MISCELLANEOUS LEGUME PESTS

Alfalfa Snout Beetle, *Otiorhynchus ligustici* (L.), is of central European origin and was first recorded in the vicinity of Oswego, N.Y., in 1933. However, checking of collections revealed speciments from that locality dated 1896. The known distribution is still confined to the region where the discovery of its presence was first made. Damage is caused by the larvae feeding on the roots

Figure 160. The alfalfa snout beetle, *Otiorhynchus ligustici* (L.) adult and larva; 3 ×. (Palm, Cornell Agr. Exp. Sta.)

and the adults feeding on the foliage of alfalfa, other legumes, and a great variety of plants.

All the beetles are females, and reproduction is therefore parthenogenetic. These flightless adults are nearly 12 mm in length, each having a large prominent snout (Fig. 160). Larvae are white and curved, resembling others in the family Curculionidae. The life cycle requires 2 years, perhaps 3 under unfavorable conditions. Hibernation occurs in the adult and half-grown larval stages.

Suggested control measures are short rotations with row crops, planting red clover in place of alfalfa, and poisoned baits.

REFERENCES: *Cornell Bul.* 629, 1935; 757, 1941; *J. Econ. Ent.*, 30:715, 1937; 45:298–302, 1952; 51:682–685, 1958; 54:601, 1961.

Alfalfa Caterpillar, *Colias eurytheme* Boisduval, is a green caterpillar covered with a fine pubescence, and approximately 35 mm long when fully developed (Fig. 161). It is the larva of 1 of the medium-sized yellow butterflies with black-bordered wings, common over the greater part of the country. Over much of its range this chewing caterpillar is of little consequence but in the southwestern states, especially in irrigated sections, it sometimes becomes so numerous as to destroy much alfalfa.

In the Southwest all stages may overwinter and there may be 7 generations per year, whereas in the North only pupae overwinter and the number of generations is less, depending on the latitude.

A nuclear polyhedrosis virus causing what is known as ''caterpillar wilt'' is perhaps the best controlling agent. The organisms causing this disease occur naturally in the soil, stubble, and crop remnants in fields, especially where alfalfa

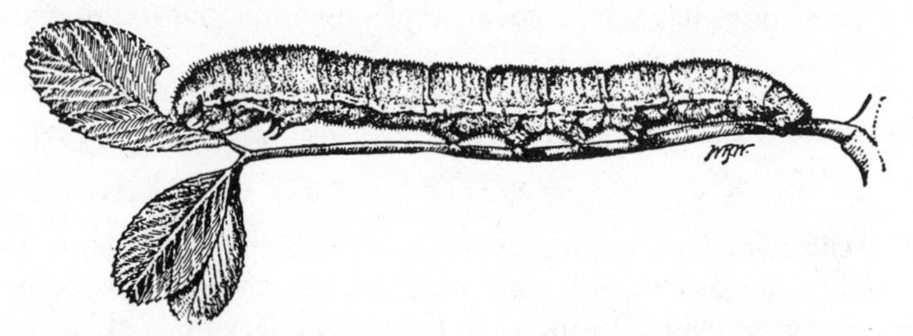

Figure 161. The alfalfa caterpillar, *Colias eurytheme* Boisduval; 2 ×. (Walton, USDA)

is grown. Both larvae and pupae are destroyed in great numbers. The California Agricultural Experiment Station has shown that a suspension of this virus applied as a spray to infested alfalfa gives more immediate control than dependence on the natural spread of the virus. If 15% of the caterpillars are killed by this disease when the plants are about ⅓ grown, pesticides may not be needed. *Bacillus thuringiensis* has been used alone or in conjunction with the polyhedrosis virus.

A braconid, *Apanteles medicaginis* Mues., a tachinid fly, *Euphorocera claripennis* (Macquart), and a chalcid, *Brachymeria ovata ovata* (Say), are important parasites that destroy great numbers of caterpillars, especially later summer generations.

Cutting an infested crop early, close, and clean will remove food and shelter for the caterpillars and butterflies, reducing their numbers. Flooding the closely mowed field after removal of the hay also contributes to control. To determine the need for chemical control measures, sweep the plants in different parts of the field with an insect net 15 inches in diameter. If 10 or more normal caterpillars are collected per sweep in an average stand a pesticide may be needed.

REFERENCES: *USDA Leaflet* 325, 1963; *Calif. Agr. Exp. Sta. Leaflets* 85 and 86, 1964.

Green Cloverworm, *Plathypena scabra* (F.), is universally distributed from the Mississippi Valley eastward, extending down from Canada to the Gulf. Damage is caused by the caterpillars devouring the leaves of clover, alfalfa, soybeans, cowpeas, and other leguminous plants. The slender green worms with faint white stripes may attain a length of 25 mm or more when fully grown. Adults are dark brown, almost black, with lighter markings, and a wing-spread of nearly 3 cm.

Winter is passed in the pupal or adult stage. Activity begans in the spring with the coming of warm weather, the eggs being deposited singly on the host plants. Larval development requires nearly a month and the pupal period from 2 to 3 weeks, this stage being passed in a silken cocoon in the litter on the ground. There are 3 to 4 generations in the South and 2 in the North. Control measures are seldom necessary because of numerous natural enemies, especially parasites.

REFERENCES: *USDA Dept. Bul.* 1336, 1925; *Bul. Ent. Soc. Amer.* 18:24–45, 1972; *Ann. Ent. Soc. Amer.* 64:647–651, 1971.

Alfalfa Webworm, *Loxostege commixtalis* (Walker), is the larva of a small moth not unlike the garden webworm pictured elsewhere in this book (p. 312), and has some resemblance to the corn borer. The larva is a slender, greenish yellow, striped caterpillar which reaches a length of about 25 mm. Pupation takes place in the soil with 3 generations occurring in the latitude of Colorado. There are many hosts in addition to alfalfa. When the food supply is nearly exhausted this species will sometimes migrate like the armyworm.

Harvesting infested crops for hay will greatly check this insect.

Garden Webworm, *Achyra rantalis* (Guenée), and the **Beet Webworm,** *Loxostege sticticalis* (L.), both damage alfalfa as well as their usual hosts in seasons when they are unusually abundant. Their larvae devour the leaves and stems and spin silken threads resulting in objectionable webbing on hay crops. The control measures are the same as those given for the alfalfa webworm.

REFERENCES: *USDA Agr. Handbook* 313, 1966; *J. Econ. Ent.*, 56:248–251, 1963; *Calif. Agr. Exp. Sta. Leaflet* 85, 1964.

Alfalfa Looper, *Autographa californica* (Speyer), is the larva of a noctuid moth. On several occasions this insect has done extensive damage in western Canada and from Montana southward in the United States. The looping larva devours the leaves of alfalfa, lettuce, and other crops. In California natural control by insect parasites, bacterial, viral, and fungus diseases has made the application of chemicals unnecessary.

REFERENCE: *J. Econ. Ent.*, 46:723, 1953.

Forage Looper, *Caenurgina erechtea* (Cramer), is considered a minor pest of clover and alfalfa in the plains region of Kansas and surrounding states. Overwintering occurs as a pupa enclosed in a cocoon in the soil near the base of the food plants. Moths appear early and deposit eggs which hatch into looping caterpillars; the latter devour the new growth of the hosts. There may be 3 or 4 generations per year in Kansas, and the damage done by the later generations is the most serious. Heavy parasitism of the larvae normally keeps the pest under control. A congener *C. crassiuscula* (Haworth) is called the **Clover Looper.**

REFERENCE: Kansas Bd. of Agr., *29th Ann. Rep.*, 1935.

Clover Head Caterpillar, *Grapholitha interstinctana* (Clemens), is a rather local and occasional pest, mainly on red clover. It is more abundant in the northeastern states and southern Canada. Damage is caused by the larvae feeding on the leaves near the crown and also in the developing clover heads. Overwintering occurs in the larval and pupal stages under crop residues in or near clover plantings. In the spring these larvae pupate and adult moths begin emerging from them as well as from the overwintering pupae, about the time red clover is beginning to bloom. The moths are nearly 12 mm in length, dark brown with white markings on the costal margins of the wings, and light dorsal spots forming a double crescent when the wings are folded. Eggs are laid on the leaves, stems, and heads, and upon hatching the larvae feed for a period of almost 5 weeks, then spin cocoons on the ground and change to pupae. After 2 weeks elapse the new adults begin emerging and soon afterwards lay eggs that form the second generation. A third generation may develop in more southern latitudes.

Harvesting the first crop while the worms are still partially developed usually prevents damage to the later crop grown for seed. There are several parasites known; since little mention has been made of this insect recently, it would seem that they must be rather effective in control.

Clover Seed Midge, *Dasineura leguminicola* (Lintner), is widely distributed in the United States and southern Canada. Red clover is the common host it attacks in injurious numbers. The midge passes the winter as a larva in the soil. Pupation occurs in early spring and the tiny mosquito-like adults (Fig. 162) emerge soon afterwards, laying their light yellow eggs in the young clover heads. In a few days hatching occurs and the yellow larvae feed, causing imperfect heads and preventing formation of seed. These heads often fall to the

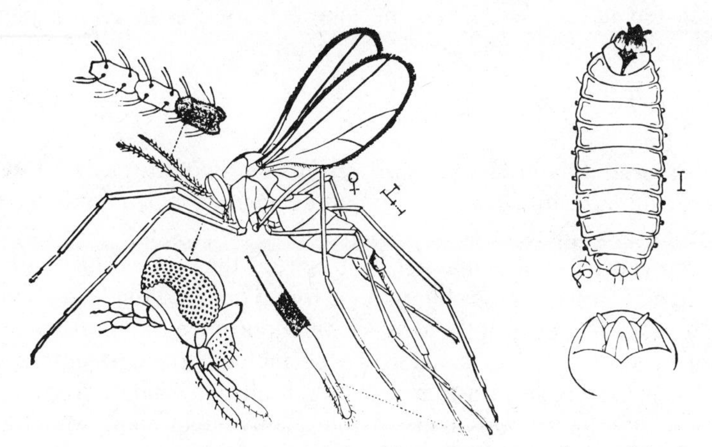

Figure 162. The clover seed midge, *Dasineura leguminicola* (Lintner). Actual size indicated by line beside figure; certain structures greatly enlarged. (from Riley)

ground and when the larvae in them have become fully grown they pupate in the soil and soon produce a second generation. Sometimes a third generation is produced in more southern areas of the country. If infested clover is cut before it comes into full bloom and is then removed from the field when cured, the young midge larvae are killed. To be effective this operation must be carried out before the larvae are fully developed. Pasturing red clover previous to starting a seed crop has given good control in the more northern latitudes. If neither pasture nor a hay crop is desired, clipping the field about May 20, leaving the clippings on the ground, will delay the production of heads until after the adults of the spring generation are gone. These heads will be advanced enough in development for seed production to be immune from attack by adults of the summer generation.

REFERENCES: *USDA Farmers' Bul.* 971, 1947; *Leaflet* 379, 1954.

Clover Stem Borer, *Languria mozardi* Latr., is a minor pest of red clover and alfalfa but has been found feeding on other clovers and a number of weeds. It is widely distributed throughout North America. The beetle is elongated, nearly 1 cm in length, with a red head and thorax and blue-black elytra. The pale yellow larva is very slender and tunnels in the stems of the hosts with pupation occurring in the same burrows. The average time required for the life cycle is about 50 days. There is a single generation in the East, but there may be three in the Southwest where the early broods develop in wild sweet clover. Overwintering occurs in the adult stage.

Several parasites attack this insect but the chalcid wasp, *Habrocytus languriae* Ashmead, is said to be the most important. Destruction of host plants or cutting infested hay crops before the larvae mature reduces the population, along with rotating crops and continual pasturing.

REFERENCE: *USDA Bul.* 889, 1920.

Three-Cornered Alfalfa Hopper, *Spissistilus festinus* (Say), is a tree-hopper that injures alfalfa, beans, cowpeas, tomatoes, melons, cotton, and other plants by puncturing the stems in sucking sap and in ovipositing. These punctures may be arranged in such a manner as to girdle the stem. The treehopper is injurious in the southern states but may be found north to the Canadian border. So far the insect has been only local in its importance and cultural measures have been proposed for its control. These include the destruction of weed patches and other vegetation that serve as food and shelter from which the hoppers may migrate to crop plants. A more convenient name sometimes used for this insect is alfalfa treehopper.

REFERENCES: *J. Agr. Res.*, 3:343–362, 1915; *J. Econ. Ent.*, 42:694, 1949; 52:428–432, 1959.

VELVETBEAN CATERPILLAR

Anticarsia gemmatalis (Hübner), Family Noctuidae

The velvetbean caterpillar is frequently a serious pest of soybeans, velvetbeans, cowpeas, peanuts, alfalfa, and related plants. It is the larva of a small noctuid moth, a tropical species found only in the Gulf states. Rarely does it survive the winter in continental United States except in the southern tip of Florida. However, it migrates northward each year and has been found as far as Delaware in late September. Adults usually appear in June or July and deposit tiny, round, white eggs on the undersides of the leaves of the host. These hatch in 4 days into slender green caterpillars with faint stripes, feed for about 3 weeks, then pupate less than 2 inches below the surface of the soil. Adults emerge approximately 10 days later, completing the cycle. As many as 3 generations occur per season and the most serious damage occurs in early autumn.

REFERENCES: *J. Econ. Ent.*, 41:803, 1948; *USDA Leaflet* 348, 1959.

PEA APHID

Acyrthosiphon pisum (Harris), Family Aphididae

The pea aphid, assumed to be of European or Asiatic origin, was first recorded in this country in 1879. Distribution is general throughout the United States and southern Canada where the host plants are grown. It attacks all kinds of peas, alfalfa, clover, and other leguminous crops. The resulting loss to the pea grower probably exceeds that caused by any other pest of the crop and the losses from forage crop attacks are often of considerable importance.

The pea aphid pierces the leaves, stems, blossoms, and pods of the plant and sucks the sap. This causes stunting of all parts, resulting in fewer and smaller pods, which are often only partly filled with peas. Pea aphid feeding causes some plants to wilt and turn yellow.

This aphid also transmits the causal organism of several virus diseases of pea plants. The diseases occur more frequently in the Pacific Northwest than in other sections of the country. One of the most important diseases is bean yellow mosaic, which is widespread in alfalfa. Another virus disease, known as enation mosaic, may appear late in the season. Infected pods become tough and difficult to shell, and the quality of the peas is lowered.

The adult aphid is nearly 4 mm long and ⅓ as wide; it is light to deep green in color, has red eyes and its legs and cornicles are usually tipped with yellow. Except for size, the young nymphs look like the adults (Fig. 163).

In the warmer areas this aphid remains active most of the winter and continues to reproduce with the egg stage entirely lacking. In cooler regions the winter eggs are laid on alfalfa and clover and the early generations of the fol-

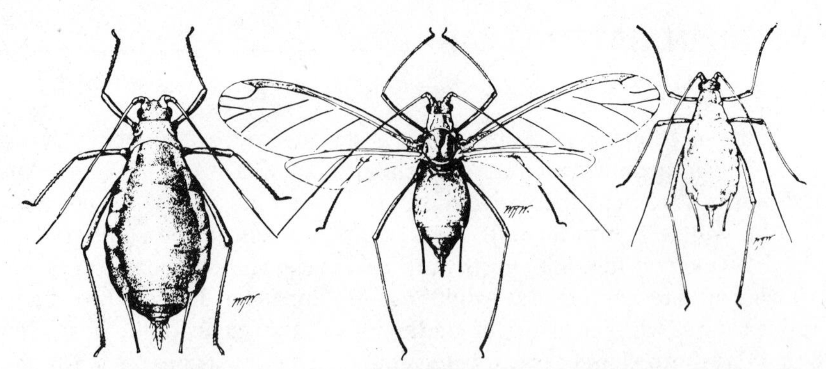

Figure 163. The pea aphid, *Acyrthosiphon pisum* (Harris); adult stem mother, winged ovoviviparous female and wingless ovoviviparous female; 8 ×. (USDA)

lowing year are present only on these plants. In April or May, depending on the latitude, the eggs hatch into wingless parthenogenetic females which, after reaching the adult stage, give birth to young nymphs, often 10 to 14 per day. Winged aphids appear at the second or third generation and fly to pea fields and other leguminous plants, often producing, under favorable conditions, 12 or more wingless generations in rapid succession throughout the summer. As peas and other hosts approach maturity and become less favorable for feeding, winged adults again appear. Many of these fly back to alfalfa and clover where males and egg-laying females are produced. The shining black eggs are deposited on the stems and leaves of alfalfa and clover, becoming the overwintering stage in the northern range of this insect.

Many natural controls often operate to keep the insect population below the point where commercial injury may result. There are 76 known parasites and predatory enemies of this aphid. The groups, in the order of their importance, are syrphid flies, lady beetles, fungus diseases, braconid parasites, and aphid-lions or lacewings. A preponderance of predators appears in the list; some beneficial forms themselves are hosts or prey to other parasites or predators.

Research in Wisconsin has shown that the pea varieties Pride, Yellow Admiral, and Onward were somewhat resistant to pea aphid attack. Some varieties of alfalfa also show resistance. If 30 to 50 aphids are obtained in one sweep of a standard insect net it is advisable to treat with a pesticide.

REFERENCES: *USDA Bul.* 276, 1915; *Tech. Bul.* 1287, 1963; *Leaflet* 529, 1964; *Farmers' Bul.* 1945, 1952; *Maine Bul.* 337, 1927; *J. Econ. Ent.*, 56:205–213, 1963; *Calif. Agr. Exp. Sta. Leaflet* 85, 1964; *Proc. 7th Calif. Alfalfa Symp.*, 1977.

PEA MOTH

Laspeyresia nigricana (Stephens), Family Olethreutidae

A native of Europe, the pea moth was introduced into Canada about 1893.

Since then it has become established in several northern states bordering Canada and some of the provinces of that country. It is sometimes considered an important pest.

The brown moths with a wingspread of 12 mm begin emerging from overwintering cocoons in July and are active until sometime in August; during this period the white flattened eggs are laid. Eggs may be placed on any part of the plant and, on hatching, the larvae bore into the pods. They feed on the developing seeds and spin a light web in the process. Several larvae may infest a single pod. Fungus growth often follows their work and spoils seeds that are undamaged by the worms. Larval development is completed in 3 or more weeks after which they bore out of the pods, enter the soil, and spin cocoons. A few of these larvae pupate and adults emerge to form a partial second generation, but most of them overwinter and pupate in late spring of the following year.

Control measures suggested are early planting and the selection of early-maturing varieties to produce the crop before the adults are present in the field. Rotation of crops and deep and clean plowing of infested fields in the fall or early spring are also of value in eliminating this pest.

REFERENCES: *Mich. Quart. Bul.* 14:87, 1931; *Wash. Agr. Exp. Sta. Bul.* 327, 1936; *Mich. Ext. Bul.* 312, 1952; *Wis. Agr. Exp. Sta. Bul.* 310, 1920.

PEA WEEVIL

Bruchus pisorum (L.), Family Bruchidae

So far as is known the pea weevil is a native of North America. It is widely distributed and attacks all varieties of edible and field peas whether grown for processing or for seed. Damage done by the larva is restricted to the seed which it first enters shortly after hatching. In the course of development the growing larva consumes the central portion of the pea, lowering or destroying the viability of the seed as well as rendering it unfit for human consumption and destroying its value for stock feed.

The adult insect is about 5 mm in length, dark brown or black, with light pubescence arranged in a characteristic pattern (Fig. 164). The white larva is thick-bodied, curved, and almost legless.

Adult beetles hibernate in a great variety of protected places but commonly in field crop remnants, stored peas, or pea hay. With the coming of warm weather they emerge and are attracted to the blooming peas upon which they feed and lay eggs. The orange oval eggs are cemented to the pods by a secretion. Hatching usually occurs in a week and the tiny larvae eat through the pod and enter the peas where development is completed at the end of about six weeks. Pupation takes place within the hollowed seed and after a period of a week or more the new adults emerge, completing the cycle which averages

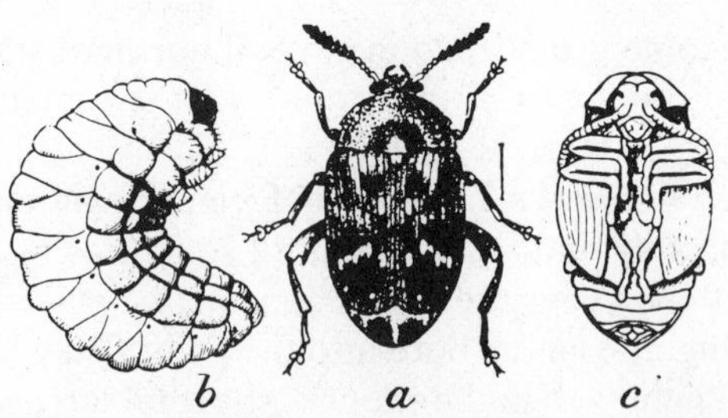

Figure 164. The pea weevil, *Bruchus pisorum* (L.): *a,* adult; *b,* larva; *c,* pupa; 5 ×. (Chittenden, USDA)

about two months in the northwestern states. There is only 1 generation per year.

The principal source of pea weevil infestation are peas shattered in the field, volunteer peas, pea hay containing weevil-infested peas, and weevil-infested seeds in storage. Any sanitation, mechanical, cultural, or chemical control measures limiting further development are of value in eliminating this pest. Fumigation or other treatment of the seed to kill the weevils is a very important practice in the control program (p. 501). Where infestation is heavy, even the greatest care in promoting these practices will not insure freedom from infestation, so that field treatment with a pesticide is imperative.

To determine the need for treatment use an insect net and sweep the upper part of the vines soon after the first blossoms appear. A population of 5 weevils in 50 sweeps results in an infestation at the canning stage of 1 to 2% in early varieties and 15 to 20% in later varieties. An infestation of 3 to 8% in harvested dry peas results from finding 1 weevil in 25 sweeps at blooming time.

REFERENCES: *USDA Tech. Bul.* 599, 1938; *Farmers' Bul.* 1971, 1952.

BEAN WEEVIL

Acanthoscelides obtectus (Say), Family Bruchidae

Although the bean weevil is well known in the North where it is a pest of stored beans only, in the South and Southwest it is a serious pest in the field as well as in storage. Once commonly considered native to this country, it is now thought to have been imported from Central or South America. It is widely distributed over the world. Food plants of importance are the different varieties of common beans, *Phaseolus vulgaris* L., and cowpeas, *Vigna sinensis* (L.). The weevils have been reared on many other kinds of garden beans and peas, as well as on seeds of plants in other groups. Damage consists of complete or partial destruction of the infested seeds. Under heavy infestations as many as a dozen or more weevils may develop from a single seed.

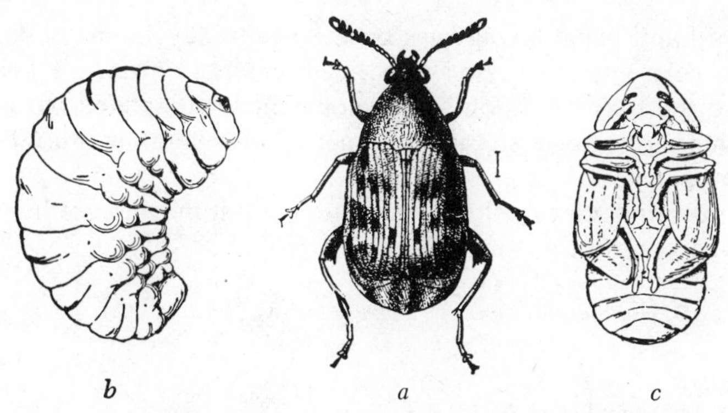

Figure 165. The bean weevil, *Acanthoscelides obtectus* (Say): *a*, beetle; *b*, larva; *c*, pupa; all greatly enlarged; 10 ×. (USDA)

Adult bean weevils are 3 mm in length, brownish black in color with lighter spots on the elytra (Fig. 165). The larvae are thick-bodied, curved, white, and footless except in the first instar. Larvae and pupae are found only in the seeds within which all development occurs. White oval eggs are laid on beans in storage and on or in pods in the field. Over 200 eggs have been recorded from one female. Hatching occurs in 3 to 9 days, larval development through 4 instars requires 12 days to 6 months, and the pupal period, 8 to 25 days. The prevailing temperatures greatly affect the speed of development, and frequently 6 or more generations are produced in a year. Usually only 1 or 2 of these are produced in the field, the others developing in stored beans.

Control of the bean weevil may be accomplished by thoroughly cleaning up and destroying all remnants of the crop after harvest. This may be done by plowing to bury all the vines and beans that have shattered. Storage places should be freed of weevils by sanitation and fumigation measures. All beans, whether for seed or food purposes, should be carefully inspected and, if any weevils are present, fumigated or given other treatments as recommended under control of stored product pests (p. 501). It is essential that all weevils in stored beans be eliminated in order to prevent field infestation in the spring or summer.

REFERENCE: *USDA Tech. Bul.* 593, 1938.

BROADBEAN WEEVIL

Bruchus rufimanus Boheman, Family Bruchidae

This weevil infests broad beans grown principally in California for human consumption or for stock feed. Damage is caused by the larva feeding and transforming to the adult within the beans. The insect resembles the pea weevil in habits and appearance but is smaller in size. Hibernating adults become active early in April and begin laying eggs on the newly developing pods. Approximately 19 weeks later the new adults begin emerging from the beans. Some adults leave the beans as soon as development is completed but others may remain within the seeds in storage for a period of several months. There is but 1 generation per year.

Delayed planting, clean culture, sanitation and fumigation of infested seeds are suggested control measures.

REFERENCES: *USDA Bul.* 807, 1920; *J. Econ. Ent.*, 44:240–243, 1951.

COWPEA CURCULIO

Chalcodermus aeneus Boheman, Family Curculionidae

One of the major insect pests of cowpeas, this curculio also attacks string beans, lima beans, peas, strawberries, young cotton seedlings, and several other plants. It is a native insect, most abundant in the South but found as far north as Iowa.

Damage to cowpeas results from the larvae feeding within the developing seeds. This destroys their value as human food and reduces their value as seeds for planting or for feeding to livestock. Early season damage to beans and peas may be severe and cotton seedlings are often destroyed, especially in plantings where cotton follows cowpeas in the crop rotation.

Overwintering adults emerge from hibernation in early summer and deposit their eggs in feeding punctures as the cowpeas begin to form. Hatching occurs in about 3 days and the white curved larvae feed and molt 3 times until fully developed, requiring almost a week, after which they leave the cowpeas and crawl in the soil. Pupation follows and the new adults emerge approximately 17 days later (Fig. 166). The entire life cycle from egg to adult may be completed in about 1 month. Two generations per year are produced in Alabama, only 1 in the latitude of Virginia.

Figure 166. The cowpea curculio; 7 ×. (Courtesy of F. S. Arant, Ala. Agr. Exp. Sta.)

Migration of the insect is primarily by crawling. Therefore rotation of crops and sanitation measures are of value in control. Chemical control is accomplished by spraying when the first pods are beginning to form and 2 to 4 applications may be necessary at approximately 5 day intervals.

REFERENCES: *Ala. Agr. Exp. Sta. Bul.* 246, 1938; *Va. Agr. Exp. Sta. Bul.* 409, 1947; *J. Econ. Ent.*, 42:856–857, 1949; 56:733–736, 1963; 65:778–781, 1679–1682, 1972.

COWPEA WEEVIL

Callosobruchus maculatus (Fabr.), Family Bruchidae

This insect is a primary pest of cowpeas but may also attack beans and seeds of other related plants. Development from larvae to adults takes place entirely within the seeds. Weevily cowpeas and beans are unfit for human food and if badly infested are unsuitable for planting. The cowpea weevil was probably imported from some part of the original range of its preferred food plant, the Old World subtropical regions. It occurs in the southern states and California, and there are scattered records from some northern states where it is found only in stored seeds.

Adults are a trifle larger than those of the bean weevil but very similar in other characteristics, except for a large lateral black spot on each wing cover and the darker posterior margins (Fig. 167). The whitish oval eggs are cemented in or on maturing pods in the field or on dry seeds in storage. The white larvae pass through 4 instars within the seed, requiring 14 days to 9 months, depending on the temperature. Pupation and adult emergence follow, with 6 or 7 gen-

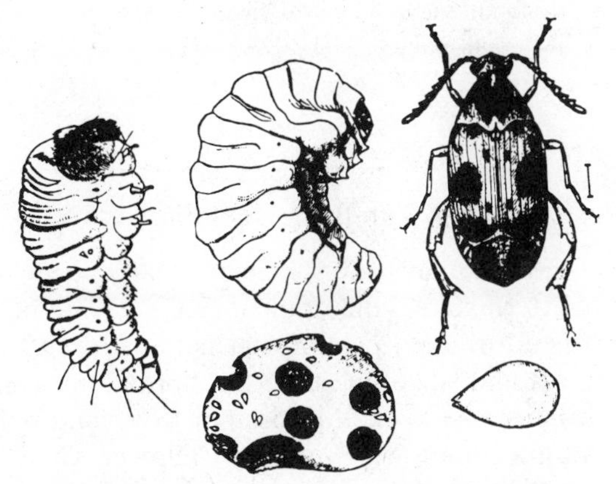

Figure 167. The cowpea weevil, *Callosobruchus maculatus* (Fabr.); adult, egg, larvae, and damaged seed; 9 ×. (Larson and Fisher, (USDA)

erations produced in California and nine in Texas. In more northern localities, the number of generations is greatly reduced.

A closely related tropicopolitan species is the southern cowpea weevil, *Callosobruchus chinensis* (L.). Its host list is extensive and includes various species of *Phaseolus, Cyamopsis, Dolichos, Vigna, Cajanus,* and *Vicia*. It infests beans in the field then becomes a storage pest in the granary, reinfesting time and again until the beans are reduced to fragments. The species is well known in India and Pakistan. It is a minor pest in the United States. The size and appearance is indicated in Fig. 168. Males have prominent pectinate antennae.

Control of these insects can be accomplished by the same measures recommended for the bean weevil.

REFERENCES: *Texas Bul.* 256, 1919; *USDA Tech. Bul.* 593, 1938; *J. Econ. Ent.*, 56:588–591, 1963.

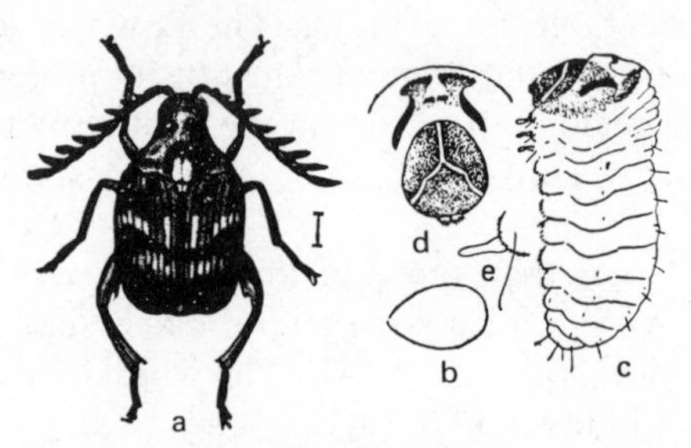

Figure 168. The southern cowpea weevil, *Callosobruchus chinensis* (L.): *a,* adult male; *b,* egg; *c,* young larva; *d,* front view of larval head; *e,* leg of larva; all enlarged. (USDA)

MEXICAN BEAN BEETLE

Epilachna varivestis Mulsant, Family Coccinellidae

This insect is one of the two plant-eating species of lady beetles. It is a native of the semiarid southwestern states, the original range having been in Mexico or in territory which once formed a part of that nation. In 1920 it was found in Birmingham, Alabama, and since that time has spread throughout the greater part of the United States east of the Mississippi River and bordering areas to the west as well. An infestation was found in California in 1946.

The Mexican bean beetle commonly attacks various varieties of bush, pole, and lima beans, the greatest damage usually occurring in the order named. Although it can reproduce on both cowpeas and soybeans, injury to soybeans is more common and in some regions is serious. It also attacks ladino clover and

Figure 169. Eggs, larvae, pupa, and adults of the Mexican bean beetle, *Epilachna varivestis* Mulsant; also leaf injury, 2 ×. (Original)

beggartick. The feeding by the larvae and adults, primarily on the lower surface of the leaves, results in skeletonized foliage.

The adult is yellow, coppery, or bronze, depending on its age, with 16 black spots on the wing covers. It is hemispherical in shape and about 8 mm long. The females deposit yellow eggs in masses of 40 to 60 on the undersides of the leaves. Over 1500 eggs may be deposited by a single female, but the average is about 460. The newly hatched spiny larvae are green, gradually becoming yellow as they near the fully developed stage, then changing to the broad yellow pupae that are attached to the plant by the gray-colored, last larval molt skin at the posterior end. All life stages are illustrated in Fig. 169.

Only the adult beetle overwinters, usually among plant remnants on the ground. The starting date of beetle activity in the spring is dependent on the prevailing temperature. In the North they are usually noticed feeding on bean foliage in late May and early June. Egg-laying soon follows with hatching taking place in about 7 days. There are 4 larval instars, each approximately 5 to 7 days apart, followed by a pupal period lasting a week. The total developmental period from egg to adult averages about 33 days in midsummer; this is greatly extended in cooler weather. There are 1 to 4 generations per year, depending on the latitude.

Some control occurs naturally by parasitic flies and wasps and predaceous bugs and lady beetles. Entomologists in Maryland have shown that the pest can be controlled biologically by the annual release of eulophid wasps reared for that purpose.

Plowing under infested crops after harvest destroys many life stages of the pest and is a recommended practice. Extremely hot, dry summers and cold winters also limit populations.

Several pesticides are available that give control if applied as the eggs are hatching and directed to the undersides of the leaves. They are rotenone, carbaryl, malathion, diazinon, methoxychlor, parathion, trithion, and naled. Great care should be exercised in handling the highly toxic pesticides. Home gardeners should use only rotenone, carbaryl, methoxychlor, or malathion.

REFERENCES: *USDA Farmers' Bul.* 1624, 1960; *Leaflet* 548, 1977; *Environ. Ent.* 4:947–957, 1975.

BEAN LEAF BEETLE

Cerotoma trifurcata (Förster), Family Chrysomelidae

The bean leaf beetle is widely distributed in the United States and attacks beans, peas, cowpeas, soybeans, and several other plants. Injury to the host plants is caused by the adults devouring the leaves and stems and the larvae eating the roots.

It resembles the better-known spotted cucumber beetle but is a trifle smaller (7 mm), more yellow, and marked differently (Fig. 170). Extreme variation in color pattern occurs.

Hibernation is in the adult stage in the soil or under crop remnants. With the coming of warm weather the beetles emerge and begin feeding and ovipositing.

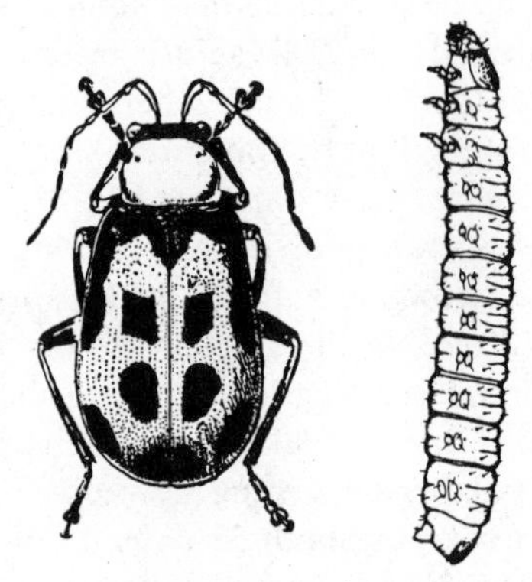

Figure 170. The bean leaf beetle, *Cerotoma trifurcata* (Förster), adult and larva, 5 ×. (USDA)

The orange eggs are placed in groups at the soil surface on or near the stems of the host plants. Hatching occurs in about 11 days, the slender white larvae then pass through three instars, transform to pupae, and emerge as adults. The complete life cycle requires 35 to 55 days. Only 1 generation occurs per year in the North but 2 and a partial third occur in the South.

The tachinid fly, *Celatoria diabroticae* (Shimer), often parasitizes about 20% of the adults. Cultural and mechanical control practices aid in checking this insect. Adjusting the planting date so that the young seedlings get started between the periods of high beetle population is of value in control. Resistant varieties of soybeans are recommended. When one or more pods per soybean plant show feeding, use a pesticide.

REFERENCES: *Ark. Bul.* 248, 1930; *S. C. Bul.* 265, 1930; *J. Econ. Ent.* 65:1669–1672, 1972; *Ohio Sta. Univ. Ext. Bul.* 545, 1977.

BEAN APHID

Aphis fabae Scopoli, Family Aphididae

This blue-black plant louse is often found on dock, nasturtium, and other plants, but occasionally becomes abundant on vegetable crops, such as beans, to such an extent that serious loss results. Continuous feeding causes the leaves to turn yellow and the plants to become dwarfed and malformed.

In the fall eggs of the species are placed on woody plants, *Euonymus* spp. and *Viburnum* spp. being the common winter hosts. In the spring these eggs hatch into wingless, parthenogenetic females that give birth to similar individuals. After a few generations, winged forms appear and fly to the summer hosts. Repeat generations occur throughout the summer. With the coming of cool fall weather winged forms again appear; they fly to the winter hosts where sexual forms are produced that lay the overwintering eggs.

Many natural enemies keep populations of this aphid checked.

REFERENCE: *Iowa Agr. Exp. Sta. Bul.* 23, 1894.

13

PESTS OF SOLANACEOUS CROPS

In addition to the pests covered in this chapter, damage is often serious from whiteflies, spider mites, lace bugs, European corn borer, and some of those discussed in Chapter 9.

TOBACCO AND TOMATO HORNWORMS

ORDER LEPIDOPTERA, FAMILY SPHINGIDAE

Hornworms are among the most destructive and widely distributed pests of tobacco and tomato plants. Even where they are not abundant these giant caterpillars may do a vast amount of damage, since each individual consumes a large quantity of food to reach full development.

Two species are found in most infestations and though they bear a close resemblance each has its distinguishing features. The tobacco hornworm, *Manduca sexta* (L.), has 7 diagonal stripes on each side of the body and the horn is curved and red, whereas the tomato hornworm, *M. quinquemaculata* (Haworth), has 8 curved stripes and the horn is straight and black (Fig. 171). These caterpillars are usually green but some brown or nearly black individuals occur; when fully grown they attain a length of almost 10 cm. The adults are large fast-flying hawk moths, which in flight are sometimes mistaken for humming-birds. They have a wingspread of about 12 cm, and they may be seen at dusk hovering over flowers sucking nectar.

Overwintering occurs in the soil as dark brown pupae, measuring nearly 7 cm. Adults emerge in late spring and deposit spherical green eggs on the undersides of the leaves. In 5 days hatching occurs and the larvae molt 4 or 5 times, reaching full development in 3 or 4 weeks. Pupation occurs in the soil and 2 to 4 weeks later adults emerge and lay eggs for a second generation. There may be 1 to 4 generations per year, depending on the latitude. The tobacco worm is gen-

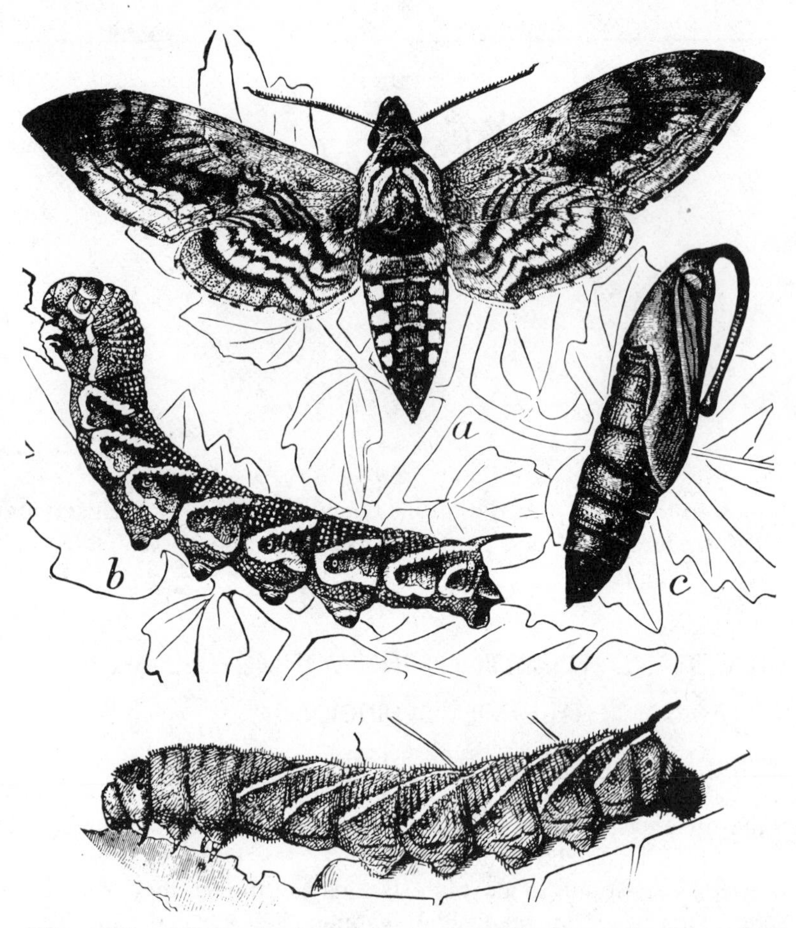

Figure 171. Above, the tomato hornworm, *Manduca quinquemaculata* (Haw.): *a*, moth, *b*, larva, and *c*, pupa. Below, larva of the tobacco hornworm, *M. sexta* (L.). (Howard USDA)

erally distributed, whereas the tomato worm is more southern in its range, occurring only sparingly in the middle states region.

As soon as harvesting is completed the elimination of crop residues by immediate plowing will reduce the overwintering population of hornworms. A braconid parasite, *Apanteles congregatus* (Say), is an important natural enemy. Its white cocoons, attached externally to the worms, are familiar to the layman. Numerous other natural enemies aid in control, particularly the spined stilt bug, *Jalysus spinosus* (Say), which attacks eggs.

Black light traps have reduced populations in the South. In gardens and small plantings the most economical control measure is often found to be handpicking. *Bacillus thuringiensis* has given good control in some regions, particularly against the smaller larvae. This holds true for other pesticides; therefore all applications should be made when the worms are small.

REFERENCES: *USDA Leaflet* 336, 1953; *Agr. Inf. Bul.* 380, 1975; *Ann. Ent. Soc. Amer.*, 52: 741–755, 1959; *J. Econ. Ent.* 61:352–356, 1968; *Ohio Sta. Univ. Ext. Bul.* 459, 1977.

TOBACCO BUDWORM

Heliothis virescens (Fabr.), Family Noctuidae

The budworm of tobacco is one of the more important pests with which the grower of that crop has to contend. In addition it may also attack cotton, tomatoes, garden peas, and other plants. This native insect of rather wide distribution is more prevalent in the South.

The injury done by this insect is destruction of the bud or growing tip of the plant and proves most serious when the plants are young. Where the bud is not entirely destroyed, the leaves growing from it are likely to be ragged and worthless. In many sections the damage from this insect exceeds that of all others.

The budworm overwinters as a pupa in the soil. Adult moths emerge in the spring and lay tiny white dome-shaped eggs on the undersides of host leaves. Hatching occurs in approximately 5 days. The caterpillars, varying from light to dark green or even brown, with paler stripes running lengthwise of the body, begin devouring the host plants. Development is complete in 18 to 31 days, after which they leave the plants, crawl in the soil, and pupate. Six to 12 days are spent in the brown pupal stage; then the new adults begin emerging. The adult moths have a wingspread of 4 cm; the green fore wings are obliquely crossed with 3 lighter stripes; the hind wings are silvery and are bordered with brown hairs (Fig. 172). About 3 generations develop on tobacco, and 2 late-season generations follow on other hosts. The adults and larvae of the corn earworm closely resemble these stages of the tobacco budworm.

Some natural control results from predatory spiders and wasps, and, most important, the parasitic braconid wasp, *Cardiochiles nigriceps* Viereck. Important pathogens are nuclear polyhedrosis viruses and *Bacillus thuringiensis*.

Cultural practices that contribute to budworm control are topping plants, plowing tobacco fields in the fall or winter, and destroying all plants in the field after harvesting is completed and all plants in seedbeds as soon as the beds are abandoned.

Chemical control measures should begin when 5 young budworm larvae per 100 plants are noticed, and repeated applications made at weekly intervals as needed. Usually 2 to 4 applications are sufficient.

REFERENCES: *USDA Farmers' Bul.* 2174, 1962; *Tech. Bul.* 1454, 1972; *Ann. Ent. Soc. Amer.* 63:67–70, 1970.

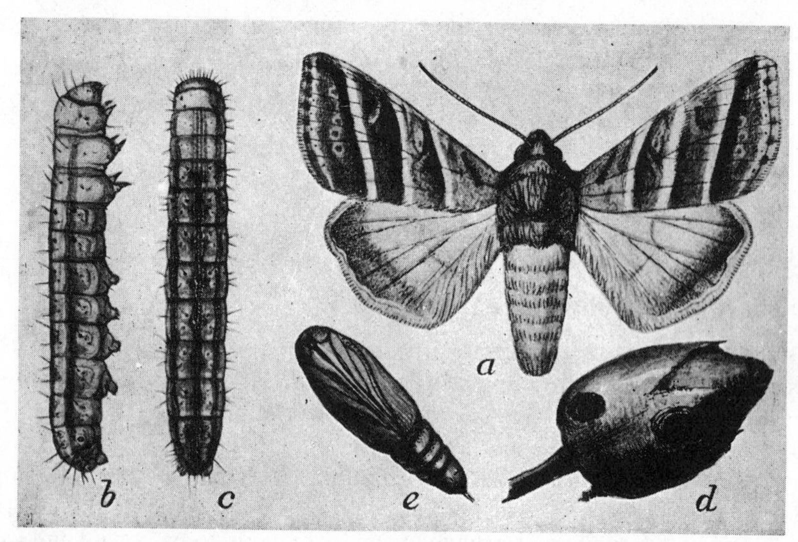

Figure 172. The tobacco budworm, *Heliothis virescens* (Fabr.); *a,* adult moth; *b,* and *c,* side and top view of larva; *d,* injured bud; and *e,* pupa; 1½ ×. (USDA)

OTHER TOBACCO PESTS

Corn Earworm, *Heliothis zea* (Boddie), is sometimes called the false budworm. It devours tobacco in much the same way as the budworm and each year may cause widespread damage late in the season to flue-cured varieties. Damage is most likely to occur where growers allow hairy vetch to mature near tobacco plantings. Control measures are the same as for the tobacco budworm.

Tobacco Thrips, *Frankliniella fusca* (Hinds), (Fig. 173) feed on many plants, including wheat and cotton, but are likely to damage tobacco the most, especially in dry years. Adults and nymphs produce gray or silvery feeding marks, principally along the veins of the lower leaves, with their rasping-sucking mouthparts. Heavy infestations cause the foliage to turn yellow. Injury may be more pronounced when the leaves are cured. Eggs of thrips are deposited in the leaf tissues; the young reach maturity in a little more than a week, and there are several overlapping generations each year. Adult thrips pass the winter inside stems of grasses and other plants, on or near the ground.

REFERENCES: *Conn. Cir.* 179, 1950; *Bul.* 379, 1935.

Suckfly, *Cyrtopeltis notatus* (Distant), a dark green plant bug, attacks tobacco grown for flue-curing in many parts of the South. Periodically it becomes abundant enough to cause serious damage in late-planted fields. Its feeding may

Figure 173. The tobacco thrips, *Frankliniella fusca* (Hinds); 20 ×. (USDA)

reduce the coloration, weight, and thickness of the cured leaves, as well as lower the quality because of specks of excrement on the undersides.

One-spot Stink Bug, *Euschistus variolarius* (P. de B.), and occasionally other species of stink bugs sometimes attack tobacco and tomatoes. Removal of plant sap by the piercing-sucking nymphs and adults (Fig. 174), near the point where the leaf petiole joins the main stalk, results in wilting. This bug is also said to be predaceous on other insects.

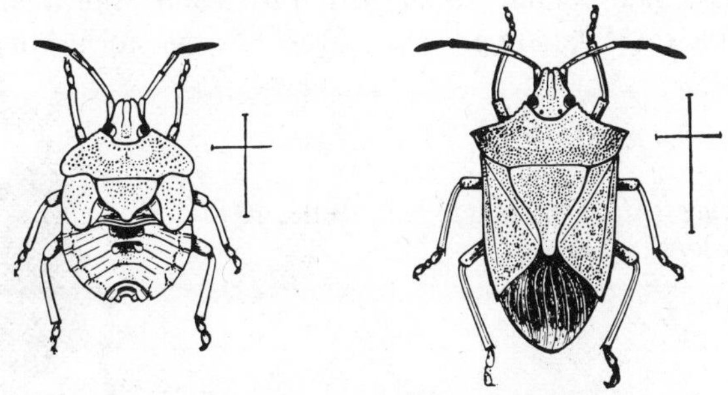

Figure 174. One-spot stink bug, *Euschistus variolarius* (P. de B.): nymph (left), adult (right); enlarged. (USDA)

Green Peach Aphid, *Myzus persicae* (Sulz.), attacks tobacco wherever it is grown, particularly where the crop is shade-grown. The areas most seriously affected are in Georgia, Florida, and the Connecticut Valley. Aphid feeding causes curled, stunted, distorted leaves that become contaminated with cast molt skins, honeydew, and sooty fungus. This damage quickly ruins a cigar-wrapper crop. In addition to the damage inflicted by sucking plant sap, aphids are important in the transmission of mosaic viruses causing disease in tobacco. The life cycle of this insect is covered on p. 427.

REFERENCES: *USDA Leaflet* 405, 1957; *Tech. Bul.* 1175, 1958; *Ohio Agr. Exp. Sta. Res. Cir.* 50, 1958; *J. Econ. Ent.* 62:593–596, 1969.

Grasshoppers, cutworms, stalk borers, plant bugs, wireworms, and seed corn maggots may attack tobacco. Discussion and control of these pests are found elsewhere in this book.

FLEA BEETLES

ORDER COLEOPTERA, FAMILY CHRYSOMELIDAE

Serious injury to tobacco, potato, and eggplant results from the feeding of flea beetles and their larvae; occasional injury is observed on tomato, pepper, and related plants, but it is usually not as severe. Adult beetles chew small holes in the leaves, giving them a sieve-like appearance; the larvae feed on the underground parts of their host plants. Some species, particularly those of the tuber flea beetle, scar the surface of potato tubers or bore into them and cause discoloration, resulting in waste when the potatoes are pared. Potato flea beetles are also known to be vectors of the organisms causing spindle tuber, blight, brown rot, and potato scab. Tobacco is seriously damaged both in the seed bed as well as the field. Adult feeding substantially reduces the market value of tobacco grown for cigar wrappers.

Important species are the potato flea beetle, *Epitrix cucumeris* (Harris) (Fig. 175), the eggplant flea beetle, *Epitrix fuscula* Crotch, the tobacco flea beetle, *Epitrix hirtipennis* (Melsheimer) (Fig. 34), the western potato flea beetle,

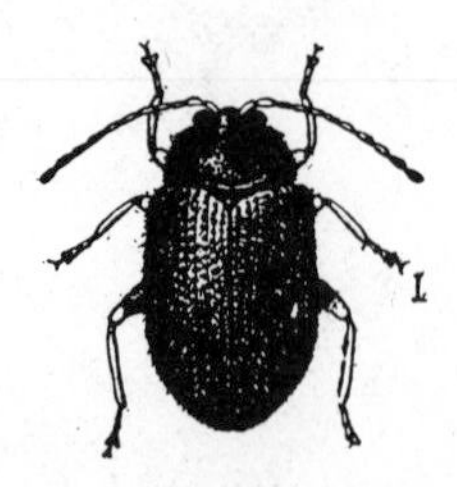

Figure 175. The potato flea beetle; 10 ×. (USDA)

Epitrix subcrinita LeConte, and the tuber flea beetle, *Epitrix tuberis* Gentner. The potato flea beetle is generally distributed; the tobacco and eggplant flea beetles range over a large area but are usually more southern. Tuber flea beetles are potato pests primarily in Washington, Oregon, Colorado, Nebraska, British Columbia, and Vancouver Island. Little damage to other crops is caused by this species. The western potato flea beetle is found primarily in western United States.

Adults of the various flea beetles are approximately 2 mm in length. Some are entirely black, others brown-black with faint lighter markings. These hibernate in the soil, crop remnants, or other vegetation and become active in the spring, feeding on the host plants as new growth appears. Consequently, early injury is likely to be severe. Eggs are laid on or in the soil near the base of the plant. These hatch in about a week and the slender white larvae feed on the plant roots or tubers for a period of 2 or 3 weeks, after which pupation occurs, followed by emergence of the new adults. The entire life cycle from egg to adult may be completed in 6 weeks or less. One to 4 generations develop each year, depending on the species and on the region of the country in which they occur. Feeding by the adults may extend over a period of 2 months.

Natural control of flea beetles results primarily from the climatic factors that limit their numbers and distribution. Screening tobacco seed beds with cloth furnishes excellent protection to young seedlings.

Pesticides are necessary to combat these insects and the one selected depends on the crop to be protected.

REFERENCES: *Can. Dept. Agr. Pub.* 94, 1949; 96, 1951; *Va. Agr. Exp. Sta. Bul.* 355, 1943; *Conn. Agr. Exp. Sta. Cir.* 179, 1950; *USDA Farmers' Bul.* 2168, 1965; *Agr. Handbook* 264, 1964; *Ohio Sta. Univ. Ext. Bul.* 459, 1977.

POTATO LEAFHOPPER

Empoasca fabae (Harris), Family Cicadellidae

The potato leafhopper is considered the most important of all the insect pests of this crop in the United States. Besides potatoes it also seriously damages beans, alfalfa, clover, cotton, and deciduous nursery stock. It is also known to feed on nearly 200 other kinds of plants. This hopper occurs widely in North America, with the possible exception of the Northwest, but is most abundant east of the Rocky Mountains. From western Texas to the Pacific Coast, the closely related species, *Empoasca abrupta* DeL., *E. arida* DeL., *E. solana* DeL., and *E. filamenta* DeL., are more common.

Feeding by this leafhopper on potatoes, eggplant, rhubarb, and dahlia causes curling, stunting, and dwarfing, accompanied by a yellowing, browning, or blighting of the foilage known as hopperburn or tipburn. This is caused by the injection of saliva into the phloem during feeding which results in a physiological disturbance producing diseaselike manifestations. On beans and other hosts, a marked curling-under of the leaf edges is produced, and a crinkling effect of the upper surface along with the usual stunting effects already described.

The adult is pale green, somewhat wedge-shaped, about 3 mm long with inconspicuous white spots on the head and pronotum. Adults are very active, jumping or flying when disturbed. Both nymphs and adults can run backwards or sideways as rapidly as they move forward. The females deposit slender white eggs within the stems and larger veins of the leaves. Hatching occurs in 6 to 9 days during the summer, and the pale green nymphs molt 5 times before

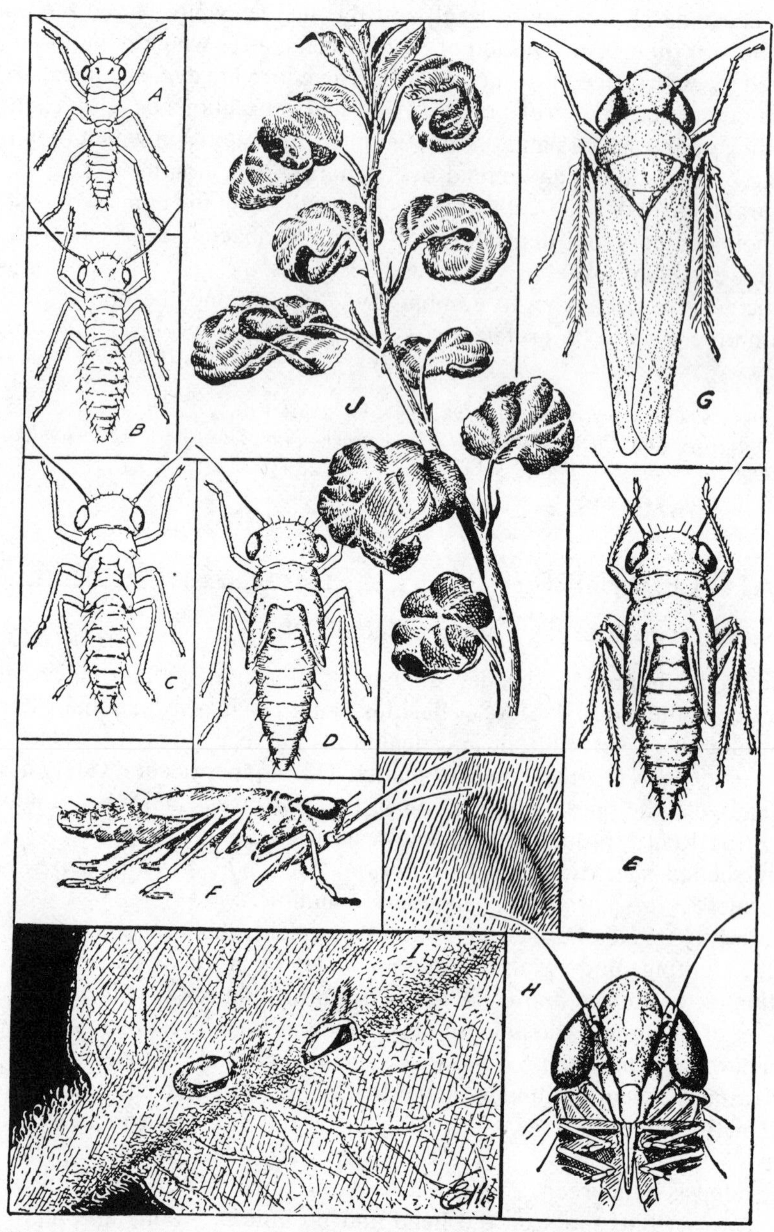

Figure 176. The potato leafhopper, *Empoasca fabae* (Harris); *A,* first nymphal stage; *B,* second stage; *C,* third stage; *D,* fourth stage; *E,* fifth stage; *F,* side view of fifth stage; *G,* adult; *H,* front view of head of adult; *I,* eggs in tissue on underside of apple leaf; *J,* curled terminal leaves owing to attack by the leafhopper; all enlarged. (Ackerman, USDA)

they become fully grown and transform to winged adults (Fig. 176). Shortly after the adults appear, mating takes place, followed by oviposition. The period from egg to adult is about 3 weeks during warm weather; several overlapping generations develop each season. The potato leafhopper has not been found overwintering north of the Gulf states where it breeds throughout the year. Migration northward with the warm spring winds occurs annually. In Ohio the first adults usually appear in mid May. Adults have been collected in late April as far north as Wisconsin.

Leafhopper populations are limited by many factors in the environment. The major natural enemies are the mymarid egg parasite, *Anagrus epos* Girault, and dryinid wasps. Pesticide applications are necessary to control potato leafhoppers, especially on beans, alfalfa, and potatoes.

REFERENCES: *USDA Tech. Bul.* 618, 1938; *Farmers' Bul.* 2168, 1965; *Leaflet* 229, 1952; 521, 1963; *Agr. Handbook* 264, 1964; *Ann. Ent. Soc. Amer.*, 57:588–591, 1964; *Ohio Sta. Univ. Ext. Bul.* 459, 1977.

COLORADO POTATO BEETLE

Leptinotarsa decemlineata (Say), Family Chrysomelidae

Universally known among growers as the potato bug, this insect was long considered the most dangerous enemy of potatoes and is still capable of doing much damage. In commercial potato-growing areas, where spraying for disease and leafhopper control is a regular practice, pesticides have so reduced the population that it is no longer a serious problem. How great a part natural control has played in bringing about this condition has not been determined. Nevertheless, in some places the control of this insect still demands attention.

The Colorado potato beetle is an invader from the semiarid regions of eastern and southern Colorado. It fed originally on a species of *Solanum,* called the sandbur. When the early settlers planted potatoes within the range of the insect, it promptly attacked them and thrived so well that it multiplied far beyond the original numbers; it then migrated eastward, following the potato plantings. This was before pesticides and spraying methods had been developed. The absolute necessity for control led to the first use of arsenicals, such as paris green and London purple. Chemical control today is still necessary in regions where populations become great.

The adults are stout, oval, strongly convex beetles, about 1 cm long, with black and yellow stripes running lengthwise along the wing covers (Fig. 177). The 10 black stripes suggested the specific name *decemlineata*. Overwintering

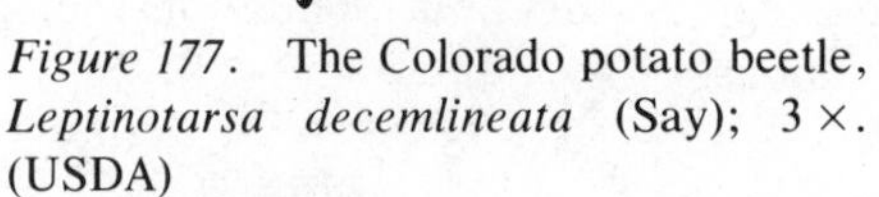

Figure 177. The Colorado potato beetle, *Leptinotarsa decemlineata* (Say); 3 ×. (USDA)

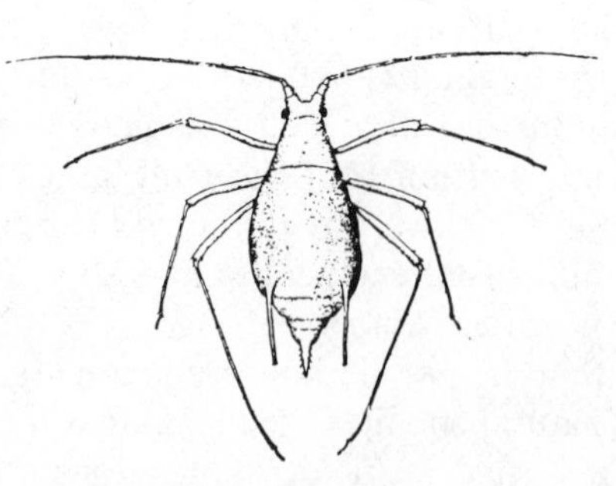

Figure 178. The potato aphid, wingless summer form; 5 ×. (Ohio Agr. Exp. Sta.)

beetles hibernate in the soil, emerging in the spring about the time that potatoes come through the ground. They lay orange-yellow eggs in groups of a dozen or more on the undersides of the leaves, each female depositing approximately 300 over a 5-week period. In a few days hatching occurs and the dark red larvae devour the foliage, becoming orange-colored as maturity approaches. There are 2 rows of conspicuous black dots on the sides of their bodies. The larvae are often called soft bugs or grubs by growers of potatoes. When mature they leave the plant, enter the soil and pupate, emerging as adults several days later. The life cycle requires about a month, and there are 1 to 3 generations per year, depending on the latitude. Injury is due to actual consumption of foliage and stems by adults and larvae. The adult is said to be one of the agents in the spread of organisms causing the diseases spindle tuber, bacterial wilt, and ring rot of potatoes.

REFERENCES: *USDA Farmers' Bul.* 2168, 1965; *Agr. Handbook* 264, 1964; *Ohio Sta. Univ. Ext. Bul.* 459, 1977.

POTATO APHID

Macrosiphum euphorbiae (Thomas), Family Aphididae

This widely distributed insect is also called the pink and green potato aphid because it occurs in two color phases. Damage is caused by both nymphs and adults sucking plant sap from the foliage, especially the terminal growth. In addition, this aphid may carry and transmit the viruses causing leaf roll, mild mosaic, rugose mosaic, spindle tuber, and unmottled curly dwarf from diseased to healthy plants. Besides potato this aphid also attacks tomato, eggplant, pepper, sunflower, jimson weeds, and many other plants. It is found throughout the United States and southern Canada.

In the North, winter is passed in the egg stage, principally on wild rose bushes. These black eggs hatch in the spring into wingless females which give birth to young. As the season progresses both winged and wingless individuals develop, the winged ones dispersing to various hosts including potatoes. Here generations occur throughout the summer. With the coming of cool weather in autumn, winged individuals fly to the winter hosts and produce oviparous females which mate and deposit overwintering eggs. Many generations occur throughout the summer in the North, and in the South the number is even larger with no sexual generations developing at all. The pink or green wingless adult (Fig. 178) is about 5 mm in length with long legs and prominent cornicles.

Other aphids attacking potatoes, peppers, and related hosts are the green peach aphid, *Myzus persicae* (Sulzer), the buckthorn aphid, *Aphis nasturtii* Kltb., and the foxglove aphid, *Acyrthosiphon solani* (Kaltenbach). They have life cycles similar to that of the potato aphid.

Aphids have many natural enemies that often keep populations checked. Important ones are aphidlions, syrphid larvae, lady beetles, and parasitic wasps.

REFERENCES: *USDA Farmers' Bul.* 2168; 1965; *Agr. Handbook* 264, 1964; *Tech. Bul.* 1338, 1965; *Ohio Sta. Univ. Ext. Bul.* 459, 1977.

POTATO TUBERWORM

Phthorimaea operculella (Zeller), Family Gelechiidae

The tuberworm is of cosmopolitan distribution, occurring in most areas where Irish potatoes or other solanaceous plants are grown or shipped. Damage consists of foliage injury caused by the larvae mining between the leaf surfaces and in the stems. On tobacco this injury is known as "splitworm" damage. Severe loss of potato tubers also results, both in the field and in storage, owing to the larval tunnels which are contaminated with excrement and permit the entrance of decay organisms (Fig. 179).

The adult is a small gray moth with a wing expanse of about 13 mm. The wings are narrow, fringed with hairs, and mottled with black and brown spots (Fig. 180). Larvae are slender caterpillars with dark heads and white bodies which are sometimes tinged with pink or green. The winter is passed as larvae or pupae in the soil, where the weather makes hibernation necessary; in California, all stages are said to be present at all seasons in the infested districts. Moths emerge with the coming of warm weather and begin laying white eggs on potato, tobacco, eggplant, and related weeds. These hatch in a few days and the larvae pass through 4 instars in reaching full development, then pupate in the soil and emerge a few days later. The average time required to produce a generation is 25 to 30 days. Several generations are produced each season,

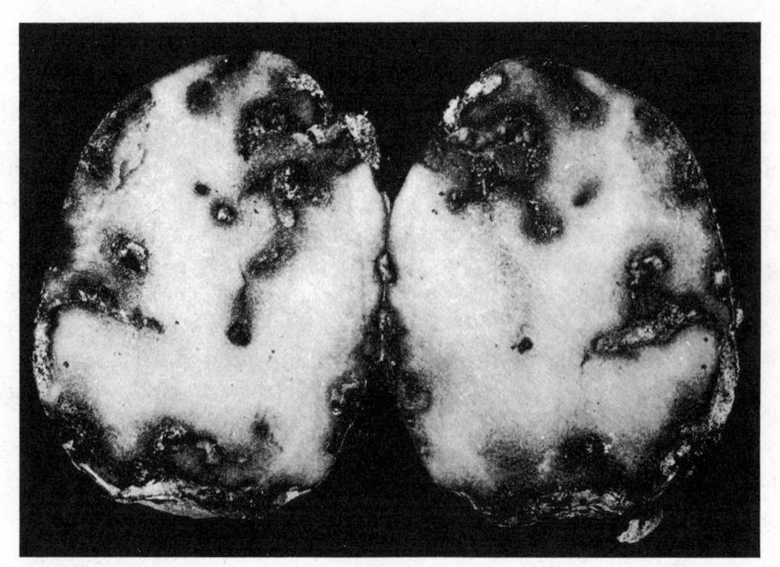

Figure 179. Potato section showing injury by potato tuberworm. (USDA)

the later ones attacking tubers which are exposed or only lightly covered with soil. In storage the insect may continue to breed, provided that the temperatures are approximately 50 F (10 C) or above.

Cultural practices that are of value in controlling the tuberworm are: plant the spring crop early, keep the developing tubers covered with at least 2 inches of soil, harvest as early as possible, avoid leaving exposed potatoes in the field overnight, destroy all infested or discarded potatoes which may serve as breeding material, avoid planting the fall crop adjacent to spring plantings, and arrange for prompt marketing.

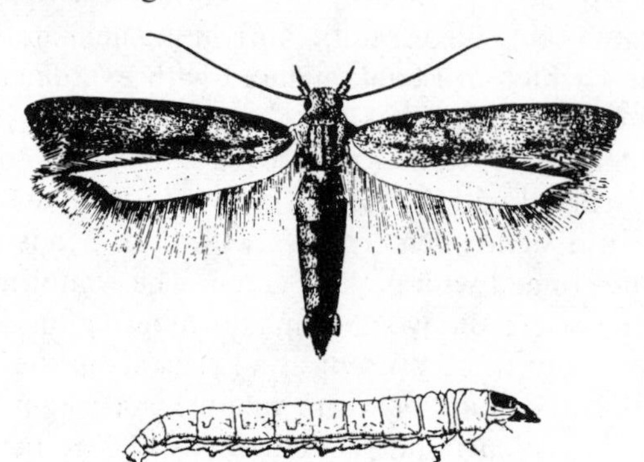

Figure 180. The potato tuberworm, *Phthorimaea operculella* (Zeller); moth and larva; 5 ×. (USDA)

Potatoes stored at 50 F (10 C) or lower will not be damaged by tuberworms, but development is only arrested and will resume when higher temperatures prevail. The braconid wasps, *Orgilus lepidus* Muesebeck and *Bracon gelechiae* Ashmead, parasitize the larvae. A three-hour exposure to the vapors of methyl bromide at a concentration of 2 pounds per 1000 cubic feet of space will eliminate an established infestation in storage.

REFERENCES: *Va. Bul.* 251, 1926; *Va. Truck Exp. Sta. Bul.* 61, 1927; 111, 1949; *USDA Farmers' Bul.* 2168, 1965; *Agr. Handbook* 264, 1964; *Ann. Ent. Soc. Amer.* 62:1407–1414, 1969; *Ohio Sta. Univ. Ext. Bul.* 459, 1977.

MISCELLANEOUS PESTS

Blister Beetles in the family Meloidae are often injurious to plants. Some of the most common species are the margined blister beetle, *Epicauta pestifera* Werner, striped blister beetle, *E. vittata* (F.) (Fig. 181), threestriped blister beetle, *E. lemniscata* (Fabr.), ebony blister beetle, *E. funebris* Horn, black blister beetle, *E. pennsylvanica* (DeG.), spotted blister beetle, *E. maculata* (Say), clematis blister beetle, *E. cinerea* (Först.), and ash gray blister beetle, *E. fabricii* (LeC.). Among crop plants alfalfa suffers much damage at times; sugar beets may be destroyed; garden vegetables in wide variety and many flowers and other ornamentals are often seriously injured. Some blister beetles are known as "old-fashioned potato beetles" when they feed on that host. Adults have been known to infect potatoes with organisms causing brown rot. The larvae are beneficial since they feed on grasshopper egg masses.

Adults of the several species have similar habits. They appear when summer is well advanced, the entire population emerging in a very short period and

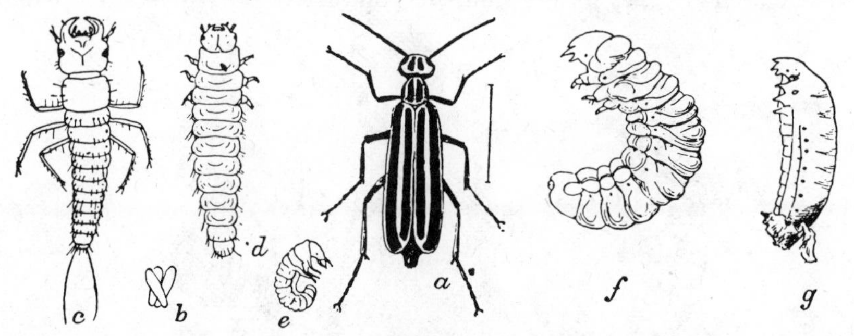

Figure 181. The striped blister beetle, *Epicauta vittata* (Fabr.), which destroys eggs of grasshoppers in the larval stages: *a,* adult; *b,* eggs; *c,* 1st instar larva; *d,* 2nd instar larva; *e,* same as *f,* as doubled up in a grasshopper egg pod; *f,* 3rd instar larva; *g,* pupa. All but *e* enlarged; actual size of adult indicated by the line. This insect is said to undergo hypermetamorphosis. (USDA)

likely doing much damage before they are noticed. They feed on the foliage, usually in large numbers, and after defoliating a plant will migrate to others. Usually only 1 generation is produced each year.

REFERENCES: *U.S. Ent. Comm. 1st Rept.*, 1878; *Iowa Bul.* 155, 1915, *USDA Dept. Bul.* 967, 1921.

Wireworms have been discussed as enemies of other plants (p. 121). As pests of solanaceous crops, particularly tobacco and potatoes, these insects often are of special importance. Several species may be responsible for causing damage, the relative abundance of each depending on the locality. On tobacco, the newly set plants show the most noticeable injury. Their roots and crowns are destroyed and the worms may even burrow up into the stems. On potatoes, damage to seed pieces soon after planting may be so serious as to reduce the stand. Later the wireworms chew deep pits and tunnels in the developing tubers, decreasing the market value of the crop. This injury also favors the spread of rhizoctonia or other diseases affecting potato tubers.

White Grubs likewise are pests of other plants as well as potatoes. Damage to potatoes is often the result of planting in grub-infested soil, the injury consisting of large shallow holes made in the tubers by the larvae. The affected plants do not reveal the injury. The description of the common species and their life cycles is given in Chapter 9.

Potato Stalk Borer, *Trichobaris trinotata* (Say), feeds in the stems of several solanaceous plants and is occasionally destructive to eggplant and potato plantings. The adults are snout beetles, 5 mm in length, black with gray pubescence over the body except for 3 spots at the base of the wing covers and the head (Fig. 182). Overwintering adults emerge from hibernation in the spring, feed on the foliage, and lay eggs on the stems of potatoes or other hosts. On hatching

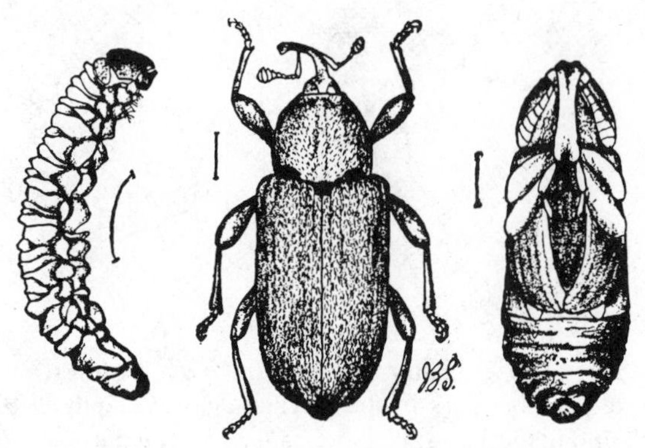

Figure 182. Potato stalk borer, *Trichobaris trinotata* (Say); larva, pupa, and adult; 8 ×. (J. B. Smith.)

the pale yellow larvae with brown heads burrow into the stalks, causing the greatest damage. By late July pupation occurs and transformation to adults takes place within the burrows where they normally remain until the following spring. Destruction of vines after harvest is of value in control. Potatoes receiving a regular spray program are not damaged by this pest.

Green June Beetle, *Cotinis nitida* (L.), is a large green species well known in the southern states. Adults devour the foliage and fruits of a number of trees, shrubs, and small fruit crops. The large larvae, somewhat like ordinary white grubs (Fig. 183), loosen the soil about plants and may feed on their roots. Grasses, legumes, tobacco in plant beds, and other plants growing in soils very high in decaying organic matter are most often injured. Seedling plants may be

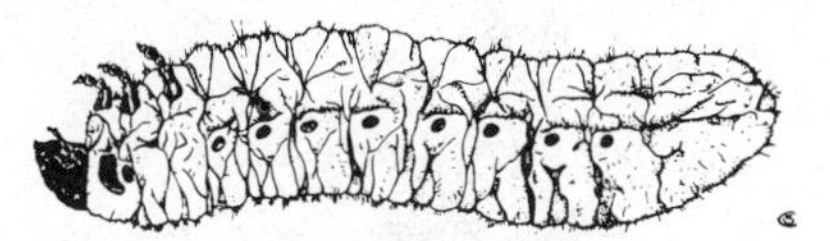

Figure 183. Larva of the green June beetle, *Cotinis nitida* (L.); about natural size. (USDA)

damaged so severely as to reduce the stand. The larvae have the habit of crawling on their backs. Hibernating larvae pass the winter deep in the soil, then in the spring tunnel near the surface where feeding takes place. Pupation occurs in late spring and adults begin emerging in June. Adult abundance is greatest in July and August. Eggs are deposited in the soil, and the young grubs hatching from them feed until cold weather comes. One generation develops each year.

Control measures are directed toward killing the larvae in the soil. Tobacco beds may be fumigated, drenched, or baited. Methyl bromide is an effective fumigant. Probably the easiest way to control larvae is to drench the infested area with a pesticide.

REFERENCES: *N.C. Agr. Exp. Sta. Bul.* 242, 1921; *Va. Agr. Exp. Sta. Bul.* 454, 1952; *J. Econ. Ent.*, 46:705–706, 766–771, 1953; 50:96–100, 1957; *USDA Leaflet* 504, 1962.

Potato Psyllid, *Paratrioza cockerelli* (Sulc), occurs in most of the western states except Oregon and Washington, and is an important pest of potato and tomato in Colorado, Utah, Nebraska, and New Mexico. Other solanaceous plants may also serve as hosts. The removal of plant sap by the nymphs and adults results in curled yellow leaves known as psyllid yellows. The set of tubers is increased, but they are usually too small to be marketable. Aerial tubers sometimes form in the leaf axils.

Adult psyllids pass the winter in Texas and New Mexico, with development occurring on weeds in the nightshade family. In the spring, migration northward takes place as the season advances, adults appearing in Colorado and Utah in May and June.

The light yellow eggs are spindle-shaped and suspend from the leaves on short stalks. These hatch in 3 to 8 days and the flattened, scalelike, pale yellow-green nymphs pass through 5 instars in 15 to 20 days before transforming to adults. The adult resembles a tiny cicada (Fig. 184), about 3 mm long with 4

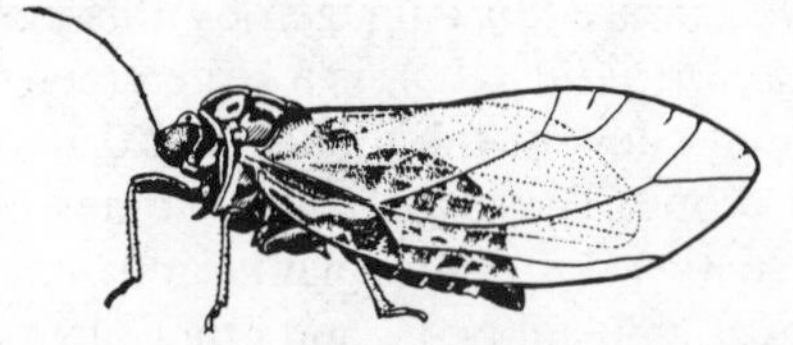

Figure 184. The potato psyllid, *Paratrioza cockerelli* (Sulc), 14 ×. (Knowlton, Utah. Agr. Exp. Sta.)

membranous wings held rooflike over the body. There may be 10 or more generations per year.

When one or more adult psyllids are caught in 100 sweeps of an insect net, it is advisable to begin treatment with a pesticide. Four to 5 applications about 2 weeks apart are required where infestations are heavy.

REFERENCES: *Colo. Bul.* 454, 1939; *Nebr. Bul.* 327, 1940; *USDA Farmers' Bul.* 2168, 1965; *Tech. Bul.* 1107, 1955; *Agr. Handbook* 264, 1964; *Mont. Agr. Exp. Sta. Tech. Bul.* 446, 1947.

Potato Scab Gnat, *Pnyxia scabiei* (Hopkins), is a sciarid fly or fungus gnat, the larva of which is capable of causing serious injury to potatoes. Egg-laying adult females are attracted to decaying plant materials and the wounds in potato tubers that may have been caused by other insects or disease organisms. The white maggots attack seed pieces or growing tubers and continue to develop after harvest in the stored potatoes. The result is a condition that is described as scabby. Adult females are wingless (Fig. 185); the males are winged. Overwintering gnats occur in stored potatoes and in the field. De-

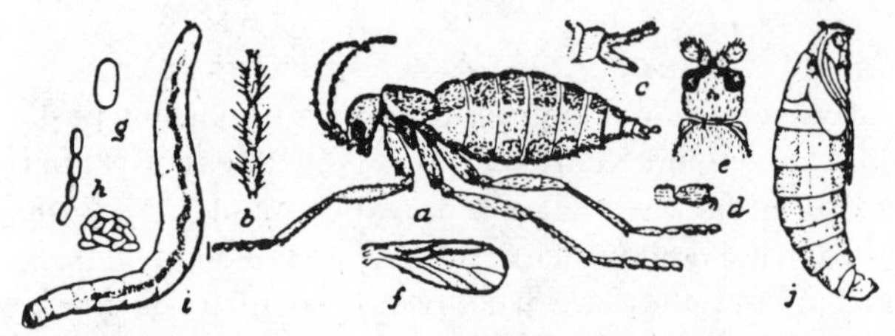

Figure 185. Potato scab gnat, *Pnyxia scabiei* (Hopkins): *a,* female fly; *b,* antennal segments; *c,* tip of abdomen; *d,* tip of hind leg; *e,* head, dorsal view; *f,* wing of male; *g,* egg; *h,* egg mass; *i,* larva; *j,* pupa; much enlarged. (Hopkins, W. Va. Agr. Exp. Sta.)

velopment is rapid in warm weather; the entire life cycle may be completed in 15 days. Although this pest is rather widely distributed in the northeastern and north central states, it is not a common pest of potatoes.

Disposing of infested tubers before planting time, treating seed with hot formalin, rotating potato plantings with less susceptible crops, where possible growing the crop in acid soil (about pH 5), selection of certified seed, preventing heating and sweating of tubers in bins, are all suggested control measures.

REFERENCES: *Ohio Agr. Exp. Sta. Bul.* 524, 1933; *USDA Farmers' Bul.* 1881, 1948; 2168, 1965; *Agr. Handbook* 264, 1964.

Tomato Fruitworm, *Heliothis zea* (Boddie), is merely the corn earworm attacking tomatoes. This insect feeds on tomato foliage and fruits, especially the green fruits which are often damaged so much that they are unmarketable. Treat with a pesticide if 5 adults per black light trap are caught 2 or 3 consecutive days and wormy fruits are evident.

REFERENCES: *USDA Tech. Bul.* 1147, 1956; *Leaflet* 367, 1976; *J. Econ. Ent.*, 56:813–817, 1963.

Tomato Russet Mite, *Aculops lycopersici* (Massee), is a pest of tomatoes but may feed on other solanaceous hosts in the subtropical or tropical regions of the world. When feeding on tomatoes the surface of the leaves and stems are russeted or bronzed in appearance. After 3 or 4 weeks of attack the main stalk of the plant cracks or checks and the leaves turn brown. If no control measures are initiated, the entire plant turns brown and dies.

Adult mites are orange-yellow, wedge-shaped, tapering posteriorly, about 200 microns in length, 50 microns in thickness, with a short piercing-sucking beak. At 70 F (21 C) development from egg to adult occurs in 7 days. Reproduction is continuous where environmental conditions permit but slows down with cooler temperatures. All stages are killed by prolonged temperatures of freezing or below.

REFERENCES: *J. Econ. Ent.*, 36:706–712, 1943; 46:502–504, 1953; *USDA* E-876, 1954; *Ann. Ent. Soc.*, 55:431–435, 1962.

Tomato Pinworm, *Keiferia lycopersicella* (Walsingham), is primarily a pest in warmer climates but may infest tomatoes grown in glasshouses in more northern areas. Damage is done by the larvae feeding as leaf miners, which occasionally invade the stems and fruits. In addition to the United States, pinworms occur in Mexico, Hawaii, Central and South America, and some of the West Indian islands.

The adults are small gray moths 6 mm in length. They are more active at dusk and deposit their tiny oval eggs on any part of the plant but usually on the lower surface of the leaves. Hatching occurs in a week and the young larvae are light orange at first, becoming purplish black as maturity is reached when they attain a length of 6 mm. Pupation may occur in the soil, in the folded leaves, or in tomatoes. Adults emerge and soon begin laying eggs, the total length of the life cycle varying from 3 to 6 weeks, depending on the temperature. Checking for infestation on transplants and plowing under crop residues at the end of the growing season are recommended practices. Azinphosmethyl and methomyl are recommended pesticides.

REFERENCES: *USDA Cir.* 440, 1937; *Ohio Res. Bul.* 702, 1950; *J. Econ. Ent.*, 52:155–158, 1963.

Pepper Weevil, *Anthonomus eugenii* Cano, is a Mexican insect that is established in Texas, New Mexico, California, Arizona, Georgia, and Florida. The adult is a snout beetle about 3 mm in length, black in color but sparsely covered with gray or tan hairs. It lays its eggs in the buds or young pods where the white larvae with brown heads feed and develop, rendering the peppers unmarketable. Pupation also occurs inside the peppers and adults emerge and begin the cycle anew. There may be 5 to 8 generations per year, depending largely on the prevailing temperatures.

REFERENCES: *USDA Leaflet* 226, 1942; *Farmers' Bul.* 2051, 1959; *Home & Garden Bul.* 46, 1963.

Pepper Maggot, *Zonosemata electa* (Say), is the larva of a tephritid fly, which was first observed in New Jersey in 1915. It is now found from New York and Massachusetts to Kansas and southward to Florida and Texas. The natural food of the insect is said to be the horse nettle, *Solanum carolinense*. Eggplants are sometimes attacked but serious injury has been done only to pepper. The slender white maggots feed within the pods, a single maggot rendering the peppers worthless for marketing.

The adults are 2-winged, yellow-striped flies, about 7 mm long. They deposit eggs beneath the skin of the peppers, and all larval development is completed inside. Pupation takes place in the soil at a depth of 2 to 4 inches, but the adults do not emerge until the following summer, starting sometime in July.

REFERENCES: *N.J. Bul.* 373, 1923; *USDA Farmers' Bul.* 2051, 1959.

14

PESTS OF CUCURBIT AND CRUCIFEROUS CROPS

These crops may also be damaged by some of the pests discussed in Chapters 9 and 16.

STRIPED CUCUMBER BEETLE

Acalymma vittatum (Fabr.), Family Chrysomelidae

This native insect is found from Mexico to Canada, primarily east of the Rockies. As soon as cucumber, squash, pumpkin, melons, and related plant seedlings push through the soil the beetles attack them, eating off the stems and cotyledons. If the plants survive, later feeding on the leaves, vines, and fruits by adults, along with larval mining of the roots, is often serious. In addition, the beetles are vectors of the organisms causing bacterial wilt and mosaic. Infection by these disease organisms is often more damaging than the insects.

The cucumber beetle is 6 mm in length, yellow-green, with 3 black longitudinal stripes (Fig. 186). This stage passes the winter in the shelter of plant remnants, becomes active early in the spring, and feeds on the blossoms and leaves of many other cultivated or wild plants. When favored host plants appear, migration to these areas takes place. Pale orange-yellow eggs are placed in the soil near the base of the plants. In about 10 days hatching occurs and the slender white larvae, darker at both ends, with 3 pairs of short legs, reach a length of almost 9 mm in 3 or more weeks, then pupate in the soil, and 10 days later emerge as adults of the next generation. Only 1 generation is produced in the northern range of the insect; farther south there are 2 and may be more in the Gulf states.

Parasitic enemies include a tachina fly, *Celatoria setosa* (Coq.), a braconid wasp, *Syrrhizus diabroticae* Gahan, and a nematode, *Howardula benigna*

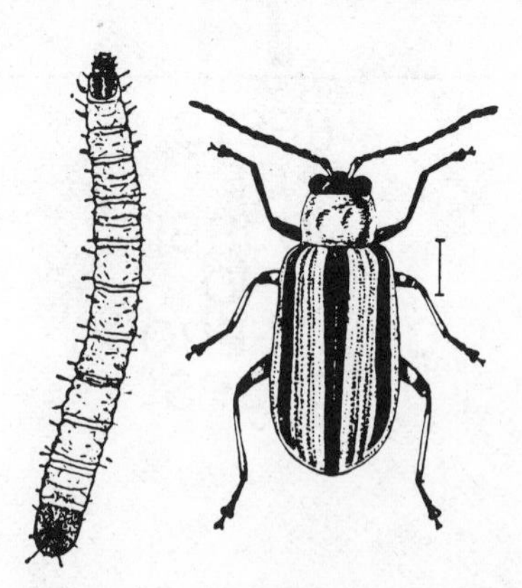

Figure 186. Larva and adult of the striped cucumber beetle. Actual size indicated by line. (USDA)

Cobb. Of the several predators a soldier beetle, *Chauliognathus pennsylvanicus* DeGeer, is considered the most important.

Although mechanical barriers of screening material and the use of trap crops are of value in protecting plants, pesticides are necessary. Applications should be made early and repeated at weekly intervals or after each rain in order to kill all beetles carrying the bacterial wilt organisms. There is no control for the disease once infection has set in.

REFERENCES: *Ohio Agr. Exp. Sta. Bul.* 388, 1925; *Purdue Agr. Exp. Sta. Bul.* 490, 1944; *USDA Agr. Handbook* 290, 1965; *Ohio Sta. Univ. Ext. Bul.* 459, 1977.

OTHER CUCUMBER BEETLES

ORDER COLEOPTERA, FAMILY CHRYSOMELIDAE

Other species of beetles attack cucumbers and related plants and cause the same type of injury. For the most part they have similar life cycles.

Spotted Cucumber Beetle (Fig. 105), *Diabrotica undecimpunctata howardi* Barber, feeds extensively on all cucurbits but is also a general feeder on many other plants. The larva of this species is known as the southern corn rootworm. Further discussion of this insect will be found under the heading "Corn Rootworms" (p. 168).

Western Spotted Cucumber Beetle, *Diabrotica undecimpunctata* Mann., is the common species west of the Rocky Mountains. Both larvae and adults attack cucurbits as well as several other species of plants.

Western Striped Cucumber Beetle, *Acalymma trivittatum* (Mann.), also found in the West, resembles the eastern species in appearance and habits, but it has a somewhat more general feeding range.

Banded Cucumber Beetle, *Diabrotica balteata* LeC., is a southern species with rather general feeding habits; sweetpotatoes are often damaged.

Control measures suggested for the striped cucumber beetle are applicable for all species.

SQUASH VINE BORER

Melittia satyriniformis (Hübner), Family Aegeriidae or Sesiidae

The vine borer ranges from Canada to Argentina and is the most serious enemy of squashes and gourds. It causes much trouble where only a few of them are grown in gardens. It rarely attacks cucumbers and melons, and great variation exists in the susceptibility of squash and pumpkin varieties. Hubbard squash is highly susceptible.

Damage is caused by the larvae tunnelling into the stems, often killing the plants, especially when they are working in the basal portions of the vines. Sometimes the fruits are also attacked. Sudden wilting of a vine and the presence of sawdustlike excrement coming from holes in the stem are evidences of attack.

The adult is one of the moths known as "clear wings" because the hind wings are almost without scales. It is 37 mm in wing expanse, of metallic green-black color; the hind legs are fringed with black and orange hairs, and markings of similar color occur over much of the abdomen. The moths are day fliers and are often mistaken for wasps (Fig. 187). Larvae are white, heavy-bodied, and considerably over 25 mm long when fully grown.

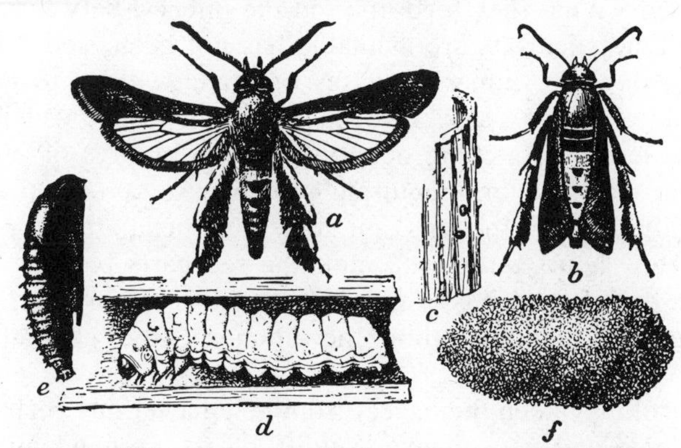

Figure 187. The squash vine borer: *a,* male moth; *b,* female with wings folded at rest; *c,* eggs shown on bit of stem; *d,* full-grown larva in vine; *e,* pupa; *f,* pupal cell; all one-third larger than natural size. (USDA)

The insect winters in the soil as a larva or pupa enclosed in a cocoon. Moths emerge in early summer and lay eggs on the stems of the plants, usually during April and May in the South, and June and July in the North. On hatching, the larvae bore into the vines and complete their development in 4 or more weeks, then leave the plant, crawl into the soil, spin a cocoon, and transform to pupae in areas where 2 generations develop. There is 1 generation in the North, 2 in the South, and a partial second in intermediate regions.

Control can be achieved by repeatedly treating the base of the plants with a pesticide in order to kill the young larvae before they enter the vines. Begin treatment when the eggs appear (usually the last week in June for Ohio), and repeat 5 times at weekly intervals. Pheromone baited traps are helpful in timing pesticide applications.

REFERENCES: *Conn. Agr. Exp. Sta. Bul.* 328, 1931; *USDA Agr. Inf. Bul.* 380, 1975; *Ohio Sta. Univ. Ext. Bul.* 459, 1977.

SQUASH BUG

Anasa tristis (DeGeer), Family Coreidae

The squash bug is well known and widely distributed in North America. Injury is caused by both nymphs and adults sucking sap from the leaves and stems of squash and pumpkin, but related plants may also be attacked. This feeding causes wilting and when severe the leaves become black and crisp.

The winged adult is gray-black and nearly 25 mm in length (Fig. 188). Only adult bugs live through the winter. They are found in all kinds of protected places, both out-of-doors and in buildings, long before the cultivated food plants are present. What they feed upon in the interval between spring emergence and the time cucurbits are available has not been recorded. When the plants begin to develop runners, masses of orange-yellow to bronze-brown eggs, each containing about a dozen or more, are deposited usually on the undersides of the leaves. Hatching occurs in about 10 days or more, and the nymphs pass through 5 instars, requiring 4 to 6 weeks to reach maturity. Early instars are highly colored; the legs, antennae, and head are red and the abdomen green. A few hours after each molt the red parts become black. Later instars are of a dark greenish gray color. Because of the protracted egg-laying period all stages are found throughout the summer. Only 1 generation develops each year.

Natural control may keep this insect from being a serious pest. Winter mortality is often high and parasitism of adults by the tachinid fly, *Trichopoda pennipes* (Fabr.), sometimes reaches 32%.

If only a few vines are involved, the easiest control method is collection of the bugs and eggs by hand. Burying infested crop remnants by plowing after

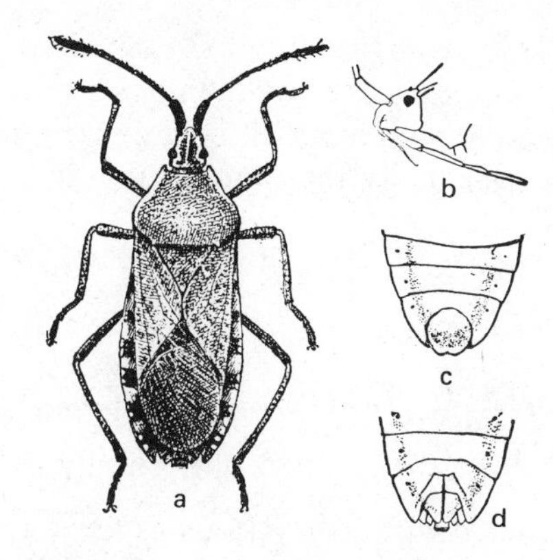

Figure 188. The squash bug; *a*, mature female; *b*, side view of head showing beak; *c*, abdominal segments of male; *d*, same of female; *a*, twice natural size; *b*, *c*, *d*, more enlarged. (After Chittenden, USDA)

harvest reduces a population. When bugs appear in large plantings, application of a pesticide is necessary.

Selecting varieties of squash resistant to the squash bug is a good practice. The following varieties are listed according to degree of resistance from high to low: Butternut, Royal Acorn, Sweet Cheese, Green Striped Cushaw, Pink Banana, and Black Zucchini.

REFERENCES: *Conn. Agr. Exp. Sta. Bul.* 440, 1940; *Utah Agr. Ext. Cir.* 214, 1954; *J. Econ. Ent.*, 55:912–919, 1962; *Ohio Ext. Bul.* 459, 1977.

SQUASH BEETLE

Epilachna borealis (Fabr.), Family Coccinellidae

This is the second of the 2 plant-feeding lady beetles that attack crops in this country. All life stages resemble those of the Mexican bean beetle, except the markings of the adult which are apparent in the illustration (Fig. 189). The in-

Figure 189. The squash beetle, *Epilachna borealis* (Fabr.); natural size. (USDA)

sect is generally distributed east of the Rocky Mountains but is comparatively rare as a pest. The common hosts are squash and pumpkins. Adults hibernate during the winter and 2 or 3 generations develop each season, the life cycle being essentially the same as that of the bean beetle.

REFERENCE: *Va. Agr. Exp. Sta. Bul.* 232, 1923.

PICKLEWORM

Diaphania nitidalis (Stoll), Family Pyralidae

The pickleworm is injurious as far north as the central states only at comparatively rare intervals; in the South Atlantic and Gulf Coast states it is likely to be a serious pest every year. Summer squash is the favored host, but cucumber and muskmelons are also attacked. The larvae feed on the blossom and vines and mine into the developing fruits, usually the undersides, entirely ruining them for market purposes.

The moth has dark brown wing margins with central areas of light yellow and a wingspread of slightly more than 25 mm. The tip of the abdomen is tufted with hairs (Fig. 190). New adults emerge from the overwintering pupae and are active at night, laying tiny eggs in small clusters on the leaves, buds, vines, and fruits. Hatching occurs in a few days and the black dotted light green larvae feed for nearly 2 weeks or more until fully grown, then pupate inside silken cocoons on the leaves. In 5 or more days the adults appear and lay eggs which develop into the second generation. Two generations are often produced in its northern range and 5 or more in the South. Activity is continuous through the winter in southern Florida.

Figure 190. The melonworm, (left) *Diaphania hyalinata* (L.), and the pickleworm, (right) *D. nitidalis* (Stoll); 2 ×. (Quaintance.)

Severe injury develops relatively late in the season; this may be avoided, to some extent, by early planting. Cucumbers and muskmelons can be protected by planting summer squash as a trap crop. Deep plowing to bury crop remnants containing pupae and other life stages reduces the infestation.

When the pickleworm first appears start applying a pesticide at weekly intervals.

REFERENCES: *J. Econ. Ent.*, 41:334–335, 1948; 44:817–818, 1951; *USDA Yearbook of Agriculture*, Plate 51, 1952; E–856, 1953; *Leaflet* 455, 1961.

MELONWORM

Diaphania hyalinata (L.), Family Pyralidae

The light-colored areas in the wings of the moth of this species are of a pearly, iridescent whiteness, and the margins are velvety black. The arrangement of these colors is shown in Fig. 190. The moth is a day flier. Its habits are similar to those of the pickleworm, which it greatly resembles in the main features of its biology. The melonworm is not likely to occur in the central states, but injury has been noted as far north as northern Kansas. The young larvae of the melonworm feed on foliage rather than blossoms; later they mine into the stems and the fruits. They are distinguished by the 2 dorsal white stripes along the length of the body.

Control of this species is easy owing to the foilage-feeding habits of the young larvae. They may be poisoned readily with the same pesticides used for the pickleworm.

MELON APHID

Aphis gossypii Glover, Family Aphididae

The melon aphid is widely distributed. It is more likely to injure melons and cucumbers, less frequently squashes and pumpkins. It has been recorded as feeding on 64 different plant species in Florida. Damage is caused by the nymphs and adults removing plant sap, which results in stunting of growth, curling of leaves, or death of the plant. In addition, this insect may transmit viruses causing mosaic diseases in plants, maize dwarf mosaic of corn is an example. The aphids vary from pale yellow to light or dark green, with black leg joints, cornicles, and eyes (Fig. 191). A more detailed description of this insect and its life cycle will be found under cotton aphid (p. 203), another common name for the melon aphid.

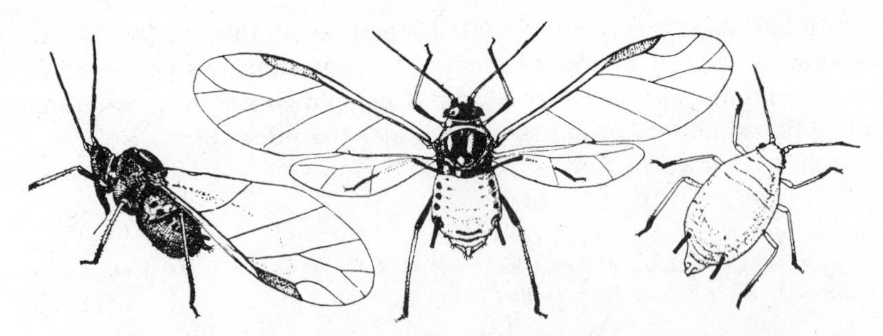

Figure 191. The melon aphid, *Aphis gossypii* Glover; 6 ×. (USDA)

Natural enemies are numerous, the most common being lady beetles, syrphid fly larvae, aphidlions, and the parasite, *Aphidius testaceipes* (Cresson).

REFERENCES: *Ohio Agr. Exp. Sta. Pub. Ser.* 11, 1964; *USDA Agr. Handbook* 290, 1965; *Farmers' Bul.* 2148, 1977; *Leaflet* 389, 1964; *Ohio Sta. Univ. Ext. Bul.* 459, 1977.

CABBAGE MAGGOT

Hylemya brassicae (Wiedemann), Family Anthomyiidae

The most destructive early-season pest of cabbage and cauliflower is the maggot. It is also a serious pest of radishes and often attacks broccoli, brussels sprouts, turnips, celery, and beets. Evident symptoms of infested cabbage and cauliflower are that the lower leaves become tinged with yellow and young plants fail to grow, may even wither and die. Injury results from the maggots feeding on the surface of the roots or tunnelling through them. Fleshy parts of radishes and turnips become brown-streaked with these tunnels, reducing their market value. Cool moist weather favors the development of this insect, and northern United States and southern Canada are the commonly infested areas of North America.

The winters are usually passed as brown puparia in the soil. Adult flies, slightly smaller than the house fly, begin emerging in April and soon afterwards lay white eggs at the bases of newly set plants or infest them while in the seed beds. Hatching occurs several days later and the white tapered maggots feed on the roots. Development continues for 20 to 30 days, after which transformation to puparia takes place in the surrounding soil. In a few more weeks emerging adult flies deposit eggs that develop into the second generation. Two to 4 generations occur annually, the number depending on the locality. Later generations injure late cabbage, turnips, and radishes.

Small plantings of radishes and turnips can be protected from ovipositing flies by screen-covered frames.

Recommended chemical control measures consist of one or more of the following operations: surface application in a narrow band over the row immediately after transplanting in the field, adding pesticide to the transplant water, treating the furrow at the time of field seeding, or broadcast treatment before planting.

REFERENCES: *N.Y. Agr. Exp. Sta. Cir.* 164, 1937; *Ohio Agr. Exp. Sta. Pub. Ser.* 11, 1964; *USDA Agr. Inf. Bul.* 380, 1975; *J. Econ. Ent.*, 55:160–164, 1962; *Can. Dept. Agr. Pub.* 1027, 1957.

SEED CORN MAGGOT

Hylemya platura (Meigen), Family Anthomyiidae

Supposedly of European origin, the seed corn maggot has been known in this country for many years and is widely distributed in both the United States and southern Canada.

The sprouting seeds of several vegetable crops are attacked by this white maggot, resulting in weakened plants and poor stands. It occasionally injures sprouting corn but is much more likely to damage beans and peas. Cabbage, turnip, radish, onion, beet, spinach, and sprouting potatoes are other plants which may be attacked. Cool wet periods favor the development of the maggots.

The adult is a small fly, not more than 7 mm long (Fig. 192). It is a near relative of several other root-infesting maggots which attack vegetable crops. Since the adult appears in early spring it is thought that the puparium in the soil is the

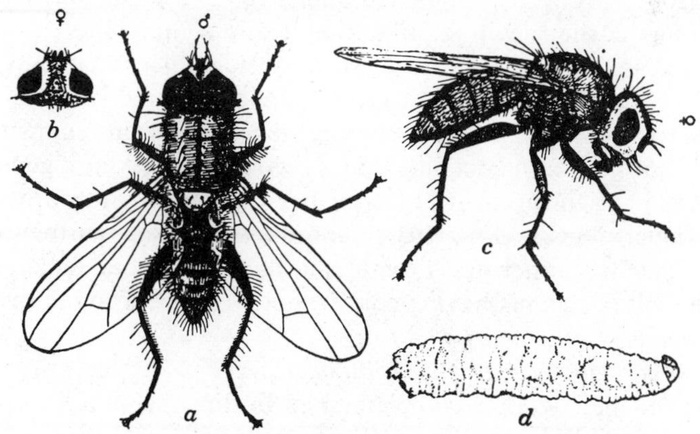

Figure 192. The seed corn maggot, *Hylemya platura* (Meigen): *a*, adult male; *b*, head of female; *c*, side view of female; *d*, larva or maggot, 5 ×. (USDA)

overwintering stage, but this has not been definitely determined. All life stages have been found during the winter in North and South Carolina. Eggs are laid in moist soil where there is an abundance of decaying vegetation. Newly hatched larvae may feed as scavengers but eventually crawl to the roots of plants and complete their feeding in 2 or 3 weeks. Pupation occurs in the soil and in a week or more the new adults emerge and repeat the cycle. Injury by later generations is usually of less importance.

Damage may be prevented by delaying planting until the maggots of the first generation have become fully grown and are changing to puparia. This date varies with the locality but is approximately June 10 for the state of New York. Avoid organic fertilizer in the seeded row, if possible. Thorough incorporation of organic matter in the soil, preparation of the surface layers of the soil for rapid germination, and shallow planting reduce damage since the insect is attracted by humus and moisture.

For protection of early planted seeds it is best to treat with one of the commercial seed protectants according to label directions.

REFERENCES: *USDA Tech. Bul.* 723, 1940; *Leaflet* 370, 1962; *J. Econ. Ent.*, 51:704–707, 1958; *Can. Ent.* 92:210–221, 1960; *N.Y. Agr. Exp. Sta. Bul.* 752, 1952; *Ohio Sta. Univ. Ext. Bul.* 545, 1977; 459, 1977.

CABBAGE APHID

Brevicoryne brassicae (L.), Family Aphididae

Aphids on cabbage and related plants are often of this species, imported presumably from Europe. Nymphs and adults remove plant sap, causing distortion, stunting, curling, wilting, and often death of the plants.

The aphid is a green species with a considerable amount of gray, waxy "bloom" on the surface, which gives heavily infested plants a whitish appearance. The winter is passed in the egg stage usually on remnants of the host plants, but in the South breeding may be continuous. Many generations may be produced in a season, over 21 having been recorded. Both winged and wingless individuals occur (Fig. 193), reproduction being parthenogenetic and ovoviviparous in warmer areas, with males and oviparous females developing only in the fall in regions having cold winter weather. The life cycle is typical of most aphids.

Braconid parasites, especially *Aphidius testaceipes* (Cresson), are numerous and effective; their work is supplemented by the predatory attacks of many lady beetles, aphidlions, and syrphid fly larvae.

Control operations involve burying crop remnants of the previous season to destroy eggs in the North and all stages in the South; use insect-free trans-

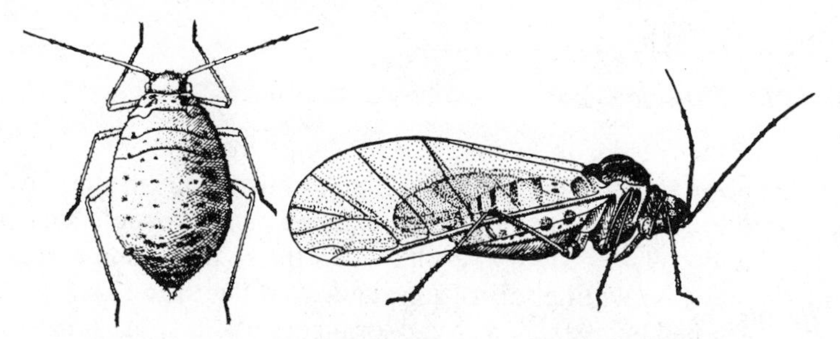

Figure 193. The cabbage aphid, *Brevicoryne brassicae* (L.), 9 ×. (Knowlton, Utah, Agr. Exp. Sta.)

plants; apply a pesticide only when necessary and before the leaves become cupped and distorted.

REFERENCES: *Ohio Sta. Univ. Ext. Bul.* 459, 1977; *USDA Farmers' Bul.* 2148, 1977.

TURNIP APHID

Hyadaphis erysimi (Kaltenbach), Family Aphididae

This widely distributed species was not known to be distinct from the cabbage aphid until 1914. Its host plants, habits, and damage are similar to those of the cabbage aphid. The most serious damage from this pest occurs in late summer or fall in the southern states where breeding may be continuous.

Wingless turnip aphids are pale green and almost 2 mm in length when fully grown (Fig. 194). Winged forms are pale green with black body markings and black heads. The biology is essentially similar to that of the cabbage aphid. In the Gulf states from 15 to 45 generations occur each year.

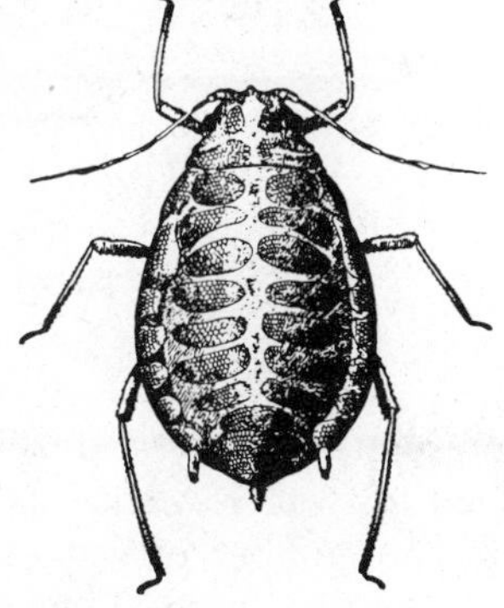

Figure 194. The turnip aphid; wingless ovoviviparous female, 15 ×. (Paddock.)

This aphid is attacked by the same natural enemies and controlled by the same pesticides as the cabbage aphid.

REFERENCES: *USDA Farmers' Bul.* 2148, 1977; *Ohio Sta. Univ. Ext. Bul.* 459, 1977.

HARLEQUIN BUG

Murgantia histrionica (Hahn), Family Pentatomidae

The striking colors of this insect suggest its common name. Essentially southern in distribution, it is seldom injurious much farther north than 39° N. latitude. Favorite food plants are cole crops like cabbage, broccoli, turnip, horseradish, kale, and collards. Injury results when the nymphs and adults remove the plant sap causing wilting, distortion, and even death.

The shield-shaped black bugs with red, orange, or yellow markings are nearly 1 cm in length (Fig. 195). In its northern range the adult hibernates; in the Gulf states activity may continue all year. Highly ornamented keg-shaped eggs are

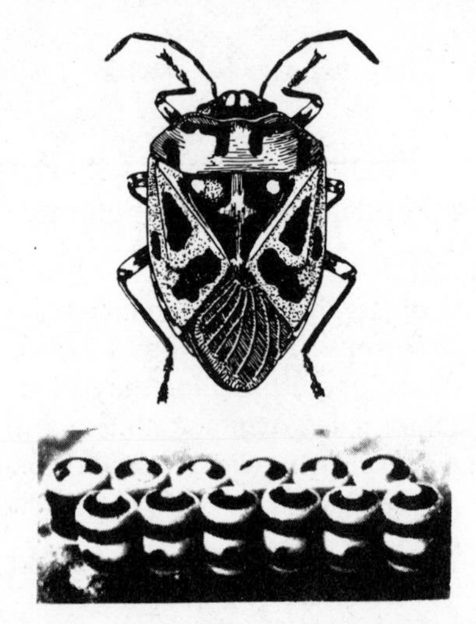

Figure 195. The harlequin bug, *Murgantia histrionica* (Hahn); adult, and group of eggs; 3 ×. (USDA)

laid in groups of about a dozen. These hatch in a week or more into nymphs that feed for 6 to 8 weeks and, after passing through 5 instars, transform to adults. The number of generations depends on the climate, at least 4 being produced in the South, 2 being maximum north of Virginia.

Plowing under crop remnants, preferably after cold weather comes, is of value and the benefits are increased if the practice is community-wide. Trap crops and hand-picking have also been practiced as control measures.

REFERENCES: *USDA Farmers' Bul.* 1712, 1933; *Agr. Inf. Bul.* 380, 1975; *J. Econ. Ent.*, 41: 808–809, 1948.

IMPORTED CABBAGE BUTTERFLY

Pieris rapae (L.), Family Pieridae

The cabbage butterfly is the adult stage of the imported cabbage worm, which is widely distributed over much of the world. Cabbage, cauliflower, broccoli, kale, mustard, and related plants, as well as nasturtiums and lettuce are common hosts. Damage is caused by the larvae devouring the leaves and sometimes boring into the heads of cabbage (Fig. 196). A related species, mostly southern in distribution, is the southern cabbage worm, *Pieris protodice*. This species has a similar life cycle.

In the northern states the imported cabbage worm passes the winter in the chrysalis or pupal stage. White butterflies, tinged with yellow and having several black spots on the wings, emerge with the coming of warm weather in early spring. They lay tiny yellow eggs, usually on the undersides of the leaves. These hatch in 7 or more days into velvety green pubescent caterpillars, each having a narrow orange stripe dorsally and paler broken stripes laterally. They have 5 pairs of prolegs. After feeding for about 15 days they attain a length of almost 3 cm, then change to pupae on the plants or nearby objects, and in about 10 days new adults emerge, completing the cycle. There are 2 or 3 generations each year in the North and several in the more southern latitudes.

Known predators are stink bugs and *Polistes* wasps. Probably the most common and effective parasite is the braconid, *Apanteles glomeratus* (L.), (Fig. 197), but the chalcid, *Pteromalus puparum* (L.), and the tachinid, *Phryxe vulgaris* (Fallen), are of importance in some seasons.

Dusting or spraying with pesticides is necessary to control cabbage worms.

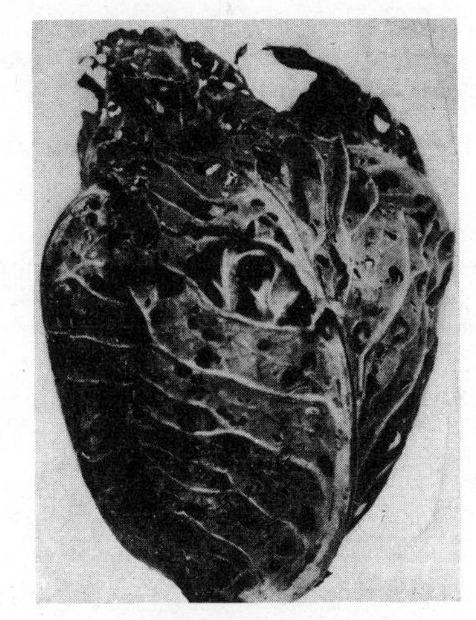

Figure 196. Life stages of the imported cabbage butterfly, *Pieris rapae* (L.); (right) cabbage showing typical injury. (Pettit, Mich. Agr. Exp. Sta.)

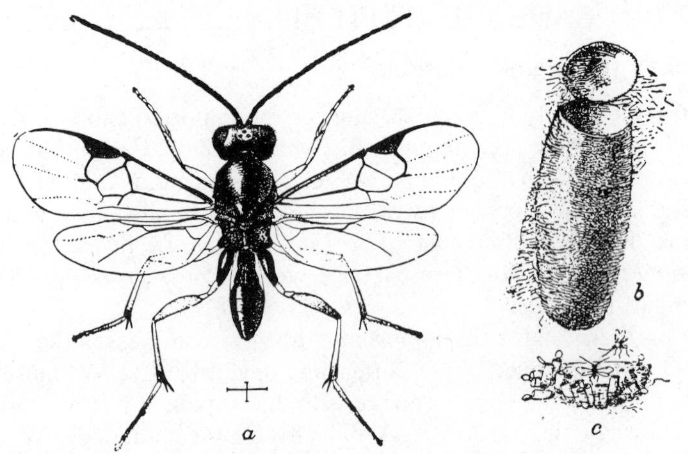

Figure 197. *Apanteles glomeratus* (L.), a parasite of the cabbage worm: *a*, adult; *b*, cocoon; *c*, adults escaping from cocoons natural size; *a*, *b*, highly magnified. (USDA)

Begin the applications when the caterpillars appear, and repeat at 7- to 10-day intervals until the infestation is checked. The microbial pesticides *Bacillus thuringiensis* and granulosis viruses have given good control.

REFERENCES: *USDA Cir.* 615, 1942; *Farmers' Bul.* 2099, 1960; *Agr. Handbook* 290, 1965; *Can. Dept. Agr. Pub.* 97, 1953; *Proc. Ent. Soc. Ont.* 93:85–87, 1963; *Ohio Ext. Bul.* 459, 1977.

CABBAGE LOOPER

Trichoplusia ni (Hbn.), Family Noctuidae

Widely distributed in North America, this light green caterpillar, with a few white or pale yellow stripes and only 3 pairs of prolegs, is often found feeding on cabbage and related plants (Fig. 198). In addition, beans and lettuce are sometimes seriously damaged as well as some glasshouse crops. Its work is often mistaken for that of the butterfly larva, and the two may be found on the same plants. The adult of the looper has dark brown mottled fore wings, each having a small silvery spot resembling a figure 8 near its center; the hind wings are almost uniformly light brown. The moths have a wingspread of slightly more than 4 cm.

The loopers winter in the pupal stage, the pupae being enclosed in flimsy silken cocoons attached to the food plants or to nearby objects. Moths emerge in the spring and deposit dome-shaped, pale green eggs on the host plants, chiefly at night. After hatching, the destructive larval stage reaches full development in 2 to 4 weeks; pupation then occurs and in almost 10 days the new adults emerge. Three or more generations are produced each season, depending on the latitude.

There is a high degree of natural control of this insect. Several parasites are usually numerous, a common species being the encyrtid, *Copidosoma truncatellum* (Dalm.). Predators attack it freely and there is often a high mortality

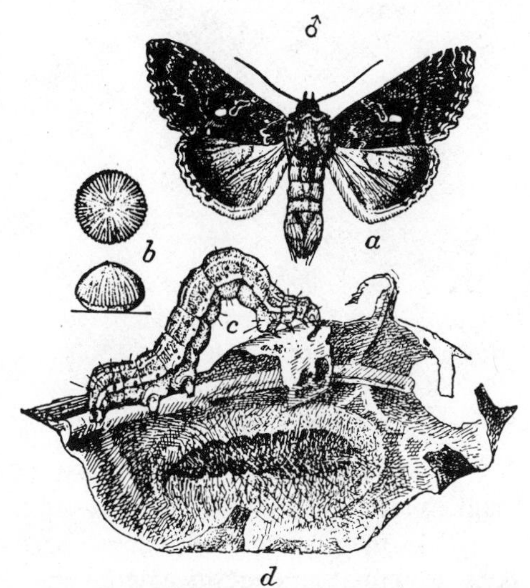

Figure 198. The cabbage looper: *a,* male moth; *b,* egg from above and from side; *c,* full-grown larva in natural feeding position; *d,* pupa in cocoon; *b,* greatly enlarged. (USDA)

from a polyhedral virus disease. Plowing in the spring to bury the overwintering pupae on crop remnants before emergence of adults is advisable. Pesticidal control is secured by the chemicals recommended for the imported cabbage worm.

REFERENCES: *USDA Tech. Bul.* 846, 1944; *Farmers' Bul.* 2099, 1960; *Agr. Handbook* 290, 1965; *Proc. Ent. Soc. Ont.*, 93:61–75, 1963; *Ohio Sta. Univ. Ext. Bul.* 459, 1977.

DIAMONDBACK MOTH

Plutella xylostella (L.), Family Yponomeutidae

An introduced species, this moth is widespread in North America but does commercial damage only in limited areas; during periods of several years it may attract no attention as a pest. Figure 199 shows the life stages of the moth. When the moth is in its normal resting position it is about 8 mm long and the light-colored areas that show as anal margins of the fore wings in the picture fit together to form diamond-shaped spots, the basis for the common name.

The diamondback winters as a moth, hibernating for a short period or remaining active through the winter, depending on the prevailing weather. Tiny

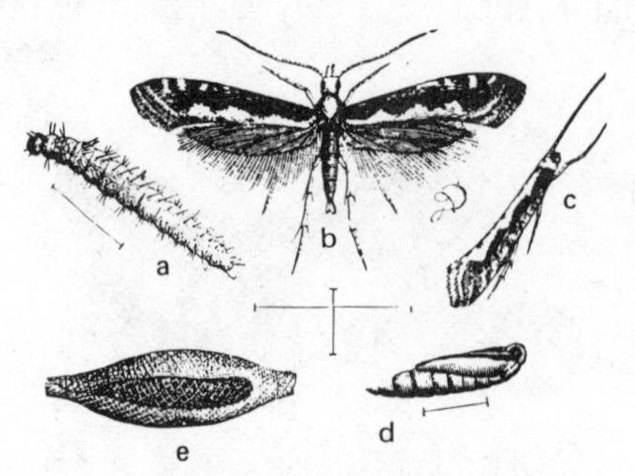

Figure 199. The diamondback moth: *a,* larva; *b,* adult; *c,* adult with wings folded; *d,* pupa; *e,* cocoon. (After Riley, (USDA)

eggs are laid on the leaves of cabbage and related hosts. On hatching the pale green larvae chew small cavities and holes in the leaves, feeding mainly on the undersides. In about 2 weeks they become fully developed, spin a loose mesh silken cocoon and change to pupae, the moths emerging a week or more later. As many as 7 generations have been observed in regions where there is a hibernating period.

A small ichneumon wasp, *Horogenes insularis* (Cresson), has been recorded as parasitizing as many as 95% of these insects. Other species parasitic on the diamondback larva are *H. plutellae* (Viereck) and *Diadromus plutellae* (Ashmead).

The control practices recommended for suppression of other worms on cabbage and related plants are effective for diamondback larvae.

REFERENCES: *J. Agr. Res.*, 10:1–10, 1917; *USDA Farmers' Bul.* 2099, 1960; *Proc. Ent. Soc. Ont.*, 93:61–75, 1963; *Can. Ent.* 89:554–564, 1957; *Ann. Ent. Soc. Amer.* 64:651–655, 1971; *Ohio Ext. Bul.* 459, 1977.

FLEA BEETLES

ORDER COLEOPTERA, FAMILY CHRYSOMELIDAE

Flea beetles are general feeders and are frequently found on the foliage of vegetable and flower garden plants, where they chew small holes through the leaves from the underside, producing a shot-hole or sievelike appearance. The larvae of many species feed on the underground parts of their host plants;

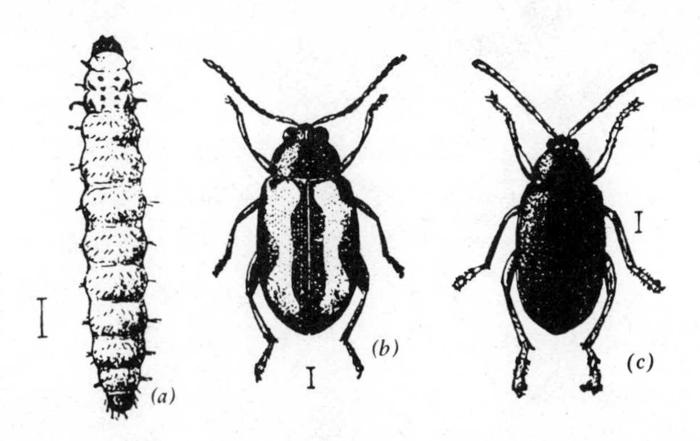

Figure 200. The striped flea beetle, *Phyllotreta striolata* (F.): *a*, larva; *b*, adult; *c*, adult of the western black flea beetle, *P. pusilla* Horn. (USDA)

others feed on the stems and foliage. Some species may also be vectors of organisms causing plant diseases. These insects derive their common name from the fact that they are provided with well-developed hind legs and, when disturbed, jump in a manner resembling fleas.

The horseradish flea beetle, *Phyllotreta armoraciae* (Koch), is most commonly found on horseradish and wild mustard. Larvae mine in stems and leaf veins. Adults are almost 3 mm long and black, with a yellow stripe on each wing cover.

The striped flea beetle, *P. striolata* (F.), is very common and widely distributed, attacking cabbage, turnips, radishes, and related plants (Fig. 200).

The western striped flea beetle, *P. ramosa* (Crotch) is mainly southwestern in distribution and feeds on crucifers, weeds, and other plants.

The western black flea beetle, *P. pusilla* Horn, is found mostly from Canada to Texas westward to the coast. It is shiny, bronze-black, and attacks flowers and weeds, as well as crucifers (Fig. 200).

All species have very similar life cycles. Usually two or more generations are produced annually, depending on the latitude, and the adults pass the winter hidden under plant remnants. Other species of flea beetles are discussed in detail in Chapters 10, 13, 15, and 16.

15

PESTS OF OTHER VEGETABLE CROPS

Occasionally crops in this chapter are damaged by pests discussed in Chapters 9 and 16.

ASPARAGUS BEETLE

Criocerus asparagi (L.), Family Chrysomelidae

Introduced from Europe in 1860 this beetle has spread west over most of the areas of the United States where asparagus is grown.

The adult is 6 mm in length, blue-black, with a reddish prothorax; the wings are marked with a characteristic pattern of creamy yellow spots with red borders. Dark gray, fleshy, soft-bodied larvae with black heads are about 9 mm in length when fully grown. Eggs are slate black in color, elongate oval, and attached by one end to the stems of the host. The appearance of these stages is shown in Fig. 201.

Adult beetles hibernate and emerge about the time asparagus is cut for market. They chew portions of the green shoots, causing blemishes. In addition the eggs are placed in abundance on the growing tips, making them unfit for market. Hatching may occur in a week, the larvae causing additional damage by devouring leaves and stems. In 2 or more weeks they become fully grown, drop to the ground, and enter the soil where pupation takes place. About 10 days later the new adults emerge. There may be 2 or more generations each season.

To avoid poisonous residues during the cutting season only rotenone, carbaryl, or malathion are recommended. For post-harvest treatments these same pesticides may be used to eliminate infestations and greatly reduce the populations of beetles that eventually go into hibernation.

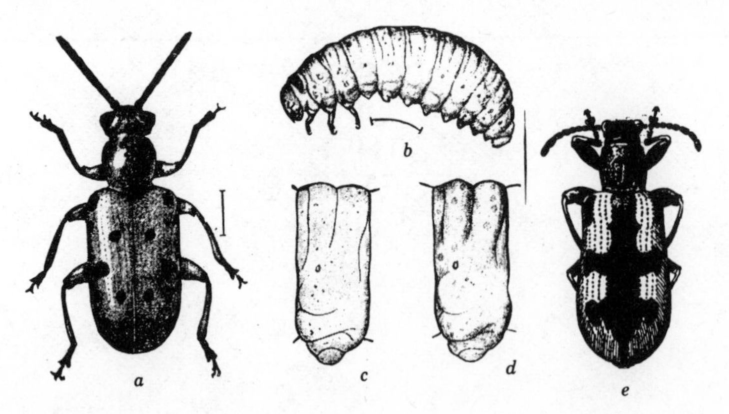

Figure 201. Asparagus beetles: *a, Criocerus duodecimpunctata* (L.), adult; *b,* larva; *c,* second abdominal segment of larva; *d,* same of *C. asparagi*; *e, C. asparagi* (L) adult; *a, b, e,* 5 ×; *c, d,* more enlarged. (USDA)

SPOTTED ASPARAGUS BEETLE

Criocerus duodecimpunctata (L.), Family Chrysomelidae

Although this insect was introduced from Europe only a few years after the advent of *C. asparagi,* it has not become so widely distributed, and is found primarily east of the Mississippi River. Except for the fact that the larvae of this species feed on or in the berries, there is little difference in the habits and life cycles of the 2 species. The adult beetle is orange-red with 12 black spots and of a slightly heavier form than its congener (Fig. 201). The larvae are orange-colored. Control is achieved with the same pesticides recommended for *C. asparagi.*

REFERENCES: *USDA Farmers' Bul.* 837, 1917; *Agr. Inf. Bul.* 380, 1975; *N.Y. Agr. Exp. Sta. Cir.* 171, 1937; *Ohio. Ext. Bul.* 459, 1977.

ASPARAGUS MINER

Ophiomyia simplex (Loew), Family Agromyzidae

Occasionally injury by the small larvae of this fly has been noted. These mine just beneath the surface of asparagus stems which weakens or sometimes kills that portion of the plant by girdling. Injury is most likely to occur in beds of seedlings or newly set plants but may also occur in older plants. The black-bodied fly is 4 mm long and the white larva is about 5 mm long. Puparia pass the winter in stalks of asparagus. Adults appear in May and again in late summer. Eggs are laid near the base of the stalk and the larvae tend to work upward. Unless the insect becomes more abundant and destructive than it hitherto has been, the only control needed is the destruction of stalks which show evidence of the presence of maggots.

REFERENCE: *Cornell Agr. Exp. Sta. Bul.* 331, 1913.

BEET LEAFHOPPER

Circulifer tenellus (Baker), Family Cicadellidae

Occurring in the arid and semiarid regions of western United States, northern Mexico, and southwestern Canada, this insect attacks red beets, sugar beets, beans, tomatoes, melons, spinach, ornamental plants, and many weeds. Some damage is caused by the piercing-sucking nymphs and adults feeding on plant sap, causing curling, stunting, and distortion. The most serious damage results from a plant disease known as "curly top," the virus organisms causing it being carried only by the beet leafhopper. The organisms survive the winter in the bodies of the leafhoppers, which then become the source of infection the following year. Leafhoppers transmit the virus during feeding. Citrus stubborn disease organism, *Spiroplasma citri,* is also transmitted by this leafhopper.

The beet leafhopper is 3 mm in length, pale green to tan, sometimes with darker markings (Fig. 202). This stage overwinters chiefly in uncultivated and overgrazed areas where mustards, desert plantains, flixweed, pepperweed, and other weeds grow. Activity and feeding continue during the winter whenever temperature permits. Oviposition takes place when the plants begin spring growth, the eggs being inserted inside the tissues of leaves and stems. Each

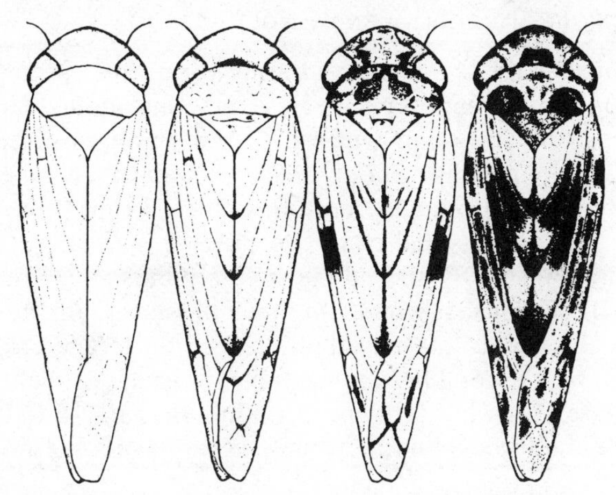

Figure 202. The beet leafhopper, *Circulifer tenellus* (Baker), showing variation in color patterns; 18 ×. (Knowlton, Utah Agr. Exp. Sta.)

female deposits 300 to 400 eggs. Hatching occurs in 5 to 40 days depending on the temperature, and the tiny nymphs feed and molt 5 times before reaching the adult stage. Development from egg to adult requires 1 to 2 months. Generations overlap considerably, 3 being produced in northern areas and 5 or more in Arizona and California. The first generation is on spring weed hosts, chiefly mustards; later generations are on Russian thistle, many other weeds, and cultivated crop plants.

Natural control results from many parasites and predators. The chief parasites are the Strepsiptera, dryinid wasps, and the big-headed fly, *Tomosvaryella subvirescens* (Loew) (Fig. 56). Important predators are a lygaeid bug, *Geocorus pallens* Stål, a damsel bug, *Nabis alternatus* Parshley, spiders, lizards, and birds.

Growing curly-top resistant varieties, early planting, controlling weed hosts in major breeding areas by proper land management, replacement of weed hosts with perennial grasses that are not leafhopper-breeding hosts, and guarding against overgrazing are cultural control measures of some practical value. Chemicals are used especially during the spring migration period to reduce leafhopper populations and curly-top infection.

REFERENCES: *Utah Tech. Bul.* 234, 1932; *USDA Tech. Bul.* 607, 1938; *Cir.* 518, 1939; *Tech. Bul.* 848, 1943; 855, 1943; 897, 1945; *Yearbook of Agriculture*, pp. 544–550, 1952; *Tech. Bul.* 1155, 1957, *Prod. Res. Rept.* 18, 1958; *Leaflet* 389, 1964; *J. Econ. Ent.*, 57:85–89, 1964; *Calif. Agr.* (6)30:15, 1976.

BEET ARMYWORM

Spodoptera exigua (Hübner), Family Noctuidae

A somewhat smaller relative of the fall armyworm, this species has a general resemblance to the more important species. Its forewings are dark with mottled lighter markings, and the hind wings are very thinly covered with whitish scales. The insect ranges throughout most of southern United States. Besides beets, it also feeds on asparagus, cotton, lettuce, beans, cole crops, alfalfa, peas, peppers, other crops, and many weeds.

The larvae are indistinctly striped green caterpillars which reach a length of 30 mm (Fig. 203). The insect winters in the adult stage, and the early brood of larvae that usually do not attract attention feed on forage crops and weeds. This is followed by a second brood which may invade beet fields and defoliate thousands of acres. In Florida there is no hibernation, and all stages of the insect may be found throughout the year. The economic threshold is 15 larvae/net sweep.

REFERENCES: *Fla. Agr. Exp. Sta. Bul.* 271, 1934; *J. Econ. Ent.*, 54:192–193, 1961; *USDA Agr. Handbook* 253, 1963; *Farmers' Bul.* 2219, 1976; *Tech. Bul.* 1454, 1972.

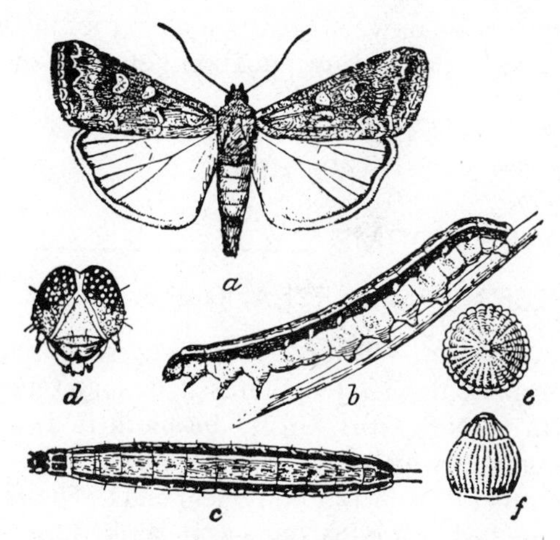

Figure 203. The beet armyworm, *Spodoptera exigua* (Hübner): *a,* moth; *b,* larva, side view; *c,* larva, dorsal view; *d,* head of larva; *e,* egg from above; *f,* egg from side; all enlarged. (Chittenden, USDA)

BEET WEBWORM

Loxostege sticticalis (L.), Family Pyralidae

Although this species is designated the beet webworm, sometimes the sugar-beet webworm, it has as much claim to the name garden webworm as the insect to which the latter name is applied. The beet webworm is an important pest of sugar beets, and attacks table beets, mangel-wurzels, other garden or field crops, as well as weeds. Damage is caused by the webbing larvae devouring the foliage. It is distributed from Canada to Texas westward as a pest of importance, but it is also present in some eastern states.

The brown moths have wings marked with darker and lighter spots, and a wing expanse of about 25 mm. Larvae are rather slender, green to yellow with a dorsal black stripe, and attain a length of 25 mm. The larvae hibernate in the soil in long silken tubes where they pupate in the spring, emerging as moths about May. Eggs are laid in rows usually on the undersides of leaves. Newly hatched larvae feed largely on alfalfa and weeds. There is some tendency for these caterpillars to migrate from one field to another or from weed patches into cultivated crops. When fully developed, pupation takes place a few inches below the soil surface, and new adults emerge in 2 or more weeks. These appear in July and a third generation sometimes develops in August. There may be a fourth generation in more southern areas.

Clean culture, destruction of weed hosts which harbor the worms, and, when possible, deep plowing at the time of pupation help prevent injury.

REFERENCES: *Colo. Ent. Cir.* 58, 1933; *Mont. Agr. Exp. Sta. Bul.* 389, 1941; *USDA Agr. Inf. Bul.* 380, 1975; *Agr. Handbook* 290, 1965; *Farmers' Bul.* 2219, 1976.

OTHER BEET WEBWORMS

ORDER LEPIDOPTERA, FAMILY PYRALIDAE

The spotted beet webworm, *Hymenia perspectalis* (Hübner), is widely distributed in the world. In the United States it is sometimes injurious in late autumn as far north as New York City. The moth is described as coppery-brown, the wings marked with white. Larvae are green with purple spots on the head and dark tubercles on the body segments. The host plants and the damage caused by the larvae are the same as described for the beet webworm.

The Hawaiian beet webworm, *Hymenia recurvalis* (F.), is similar to its congener in distribution, size, and habits, but it is darker in color with much more conspicuous white markings. Damage by the larvae is similar to that of the beet webworm. There are several generations per year in warmer climates.

REFERENCES: *USDA Bul.* 109, 1911; *Bul.* 127, 1913.

SUGAR-BEET ROOT APHID

Pemphigus populivenae Fitch, Family Aphididae

This aphid is usually of little importance, but in certain years it has been the most serious enemy of commercial beets in some of the western states. The aphid passes the winter in the egg stage on cottonwood. Early generations cause galls to form on the petioles of these trees; migration to beets takes place early in the season, and feeding on the root sap of beets and related plants continues through the summer. In the fall, winged migrants fly back to cottonwood and produce sexual stages which mate and lay the overwintering eggs. Excess moisture is unfavorable to the root aphid and, where irrigation is possible, injury can be prevented.

REFERENCES: *Utah Ext. Cir.* 214, 1954; *USDA Farmers' Bul.* 2219, 1976.

SUGAR-BEET ROOT MAGGOT

Tetanops myopaeformis (Röder), Family Otitidae

This maggot is an occasional pest of sugar beets in Utah and nearby states. It causes severe injury in some parts of individual fields, particularly in dry sandy

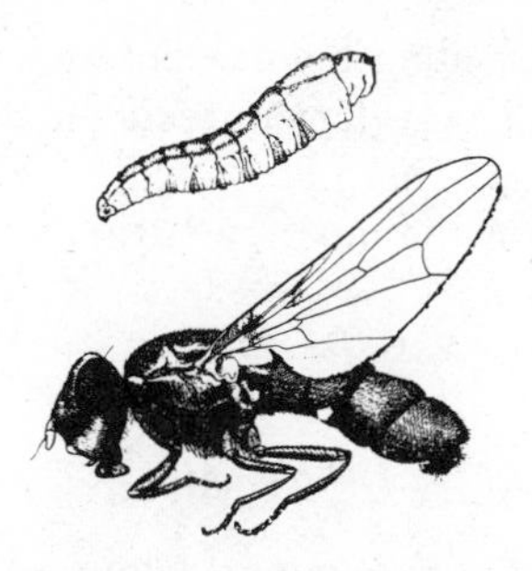

Figure 204. Larva and adult of the sugar-beet root maggot; 6 ×. (Knowlton, Utah Agr. Exp. Sta.)

soil. The maggots feed on the taproot of sugar beets, sometimes cutting it off, resulting in death of the plant. Adult flies are 6 mm long and black-bodied. The larvae are white tapered maggots (Fig. 204). Winter is passed in the soil, and 1 or 2 generations are produced each season.

Where possible irrigation, to keep the soil-moisture content up, will result in the maggots feeding high enough to prevent injury to the taproot.

REFERENCES: *Utah Leaflet* 22, 1934; *USDA Farmers' Bul.* 1903, 1942; 2219, 1976; *Agr. Handbook* 290, 1965.

SPINACH FLEA BEETLE

Disonycha xanthomelas (Dalm.), Family Chrysomelidae

Most of the garden flea beetles discussed previously (pp. 257, 286) attack red beets, sugar beets, and spinach, but this species is somewhat more specific to these hosts. Both adults and larvae devour the foliage. The adult is nearly 6 mm long, the head is black, the wing covers of a blue-green luster, the prothorax and abdomen orange-yellow, and the appendages black (Fig. 205).

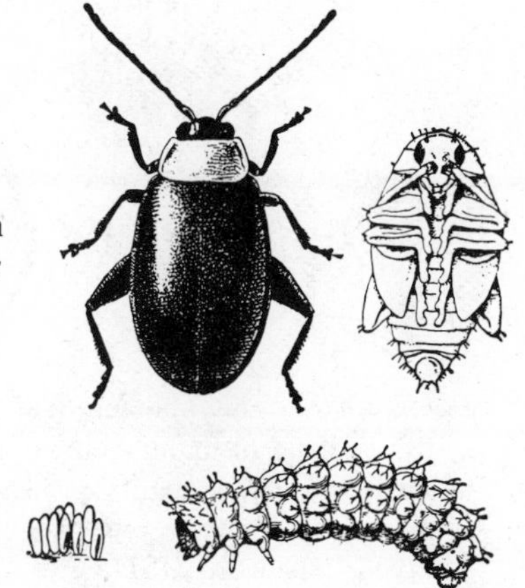

Figure 205. Life stages of the spinach flea beetle, *Disonycha xanthomelas* (Dalm.); 4 ×. (USDA)

Hibernating beetles emerge and lay eggs on the host plants. On hatching, the dull gray larvae feed on the leaves and reach a length of 8 mm before pupating. There are usually 2 generations, the earlier one developing on weed hosts for the most part, the latter one on beets and spinach.

REFERENCE: *USDA Farmers' Bul.* 1371, 1934.

SPINACH LEAFMINER

Pegomya hyoscyami (Panzer), Family Anthomyiidae

Beets, spinach, chard, and related plants are damaged by larvae of a small fly, mining between the upper and lower leaf surfaces. Leaf crops are made unfit for food, and the photosynthetic surface is reduced for proper development of root crops. Most of the damage is noticed late in the growing season. The winter is passed as puparia in the soil. Adult flies, 6 mm long (Fig. 206), emerge in the spring and lay eggs on the leaves of the host plants. After hatching the tiny maggots mine into the leaves and, when fully grown, drop to the ground and change into puparia. In a few weeks new adults emerge and soon lay eggs of the next generation. There are 3 or more generations annually. This insect is also called the beet leafminer.

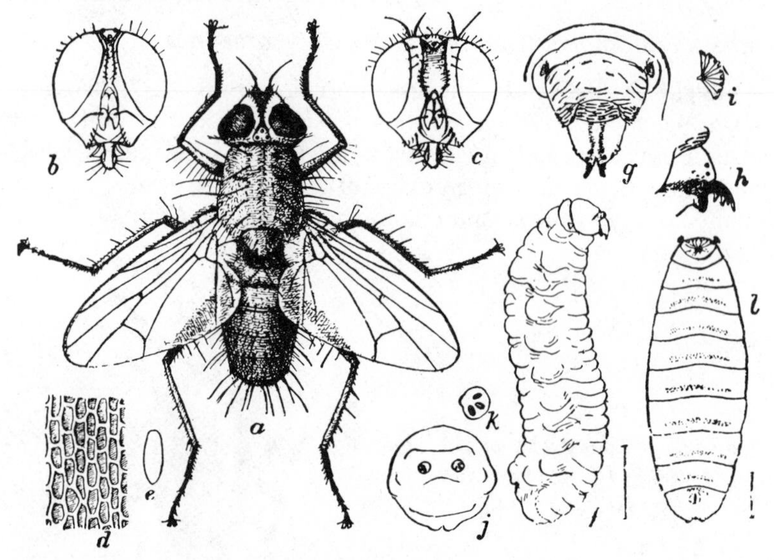

Figure 206. The spinach leaf miner, *Pegomya hyoscyami* (Panzer): *a,* fly; *b,* head of male; *c,* head of female; *d,* surface of egg, highly magnified; *e,* egg; *f,* maggot; *g,* head of same; *h,* cephalic hooks of larva; *i,* thoracic spiracle; *j,* anal segment; *k,* anal spiracles; *l,* puparium; all enlarged. (Howard, USDA)

Deep plowing in early spring, destruction of wild hosts, as well as old crop remnants, are suggested cultural control measures. Screen-covered frames over small plantings of spinach, chard, and beets gives protection from ovipositing flies.

REFERENCE: *Ohio Sta. Univ. Ext. Bul.* 459, 1977.

CARROT RUST FLY

Psila rosae (Fabr.), Family Psilidae

The rust fly is a native of Europe and was found in Canada in 1885. It has since spread to most of southern Canada and northern United States. Besides carrots this insect also attacks parsnips, celery, parsley, and related plants. Damage is caused by the maggots burrowing into the roots, killing young seedlings or impairing the market value of older roots by their rust-colored tunnels. Heavy maggot feeding is indicated by drooping, discolored foliage.

The fly is shiny black, less than 5 mm in length, with a pale yellow head, legs, and wings (Fig. 207). They begin emerging from the overwintering puparia in the soil in early May and soon deposit tiny white eggs at the base of the host plants. On hatching, the slender white maggots feed, becoming fully developed in 3 or 4 weeks, then leave the roots and change to puparia in the soil. Second-generation adults appear from early July until late August. In some areas a third generation develops; these adults appear in September.

Delaying seeding until mid-May and harvesting early in July before the second-generation flies appear are means of avoiding injury. Rotation of crops,

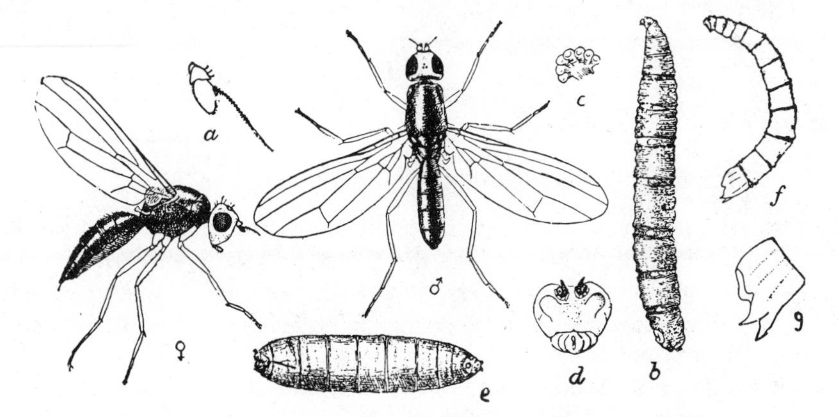

Figure 207. The carrot rust fly, *Psila rosae* (F.): *a*, antenna of male; *b*, fully grown larva; *c*, enlargement of larval spiracle; *d*, anal extremity of larvae; *e*, puparium; *f*, young larva; *g*, enlargement of anal segment from side; male and female adults; 5 ×. (USDA)

destroying wild hosts, and deep plowing in fall or spring are other control measures.

REFERENCES: *Mass. Agr. Exp. Sta. Bul.* 352, 1938; *Can. Dept. Agr. Pub.* 91, 1951; 939, 1955; *Ohio Sta. Univ. Ext. Bul.* 459, 1977.

CARROT WEEVIL

Listronotus oregonensis (LeC.), Family Curculionidae

This weevil is known to exist in the states from Virginia to New York and Connecticut; it has also been found in Illinois and Iowa. Carrots are the plants most often attacked, but it may feed on celery, parsley, parsnips, dill, wild carrot, plantain, and dock.

The adult beetle is nearly 6 mm long, dark brown, with typical chewing mouthparts of weevils. The larvae are white, legless, curved grubs which cause most damage by boring into roots and stems. Eggs are usually laid in stems, the larvae boring downward and doing their most destructive work to the roots of carrots; in celery the damage may be largely to the stalks. Hibernating adults become active in May, or earlier, and by July beetles of the first generation are mature. It is reported that there may be a partial second generation.

The beetles fly, but apparently they do not travel far. Consequently, the removal of plantings of susceptible crops to uninfested ground is an effective control measure.

REFERENCES: *Cornell Ext. Bul.* 206, 1931; *N. J. Agr. Exp. Sta. Bul.* 693, 1942; *J. Econ. Ent.*, 50:183–184, 797–799, 1957.

CARROT BEETLE

Ligyrus gibbosus (DeGeer), Family Scarabaeidae

Widely distributed over most of the United States except the most northern states, the adult carrot beetle attacks the roots and foilage of carrots, celery, corn, potatoes, parsnips, sugar beets, and sweet potatoes. The larvae are root feeders as well as scavengers, and soils high in organic matter seem to favor the development of populations.

Carrot beetles are broad, stout, red-brown, and about 13 mm long. They hibernate deeply in the soil and appear in the spring. Eggs are laid in the soil, and the hatching larvae resemble white grubs, being curved and white, often with a blue tinge. When fully grown they are over an inch long. There is but 1 generation a year.

Plowing in the fall reduces populations of the carrot beetle. Crop rotation is also recommended.

REFERENCES: *USDA Farmer's Bul.* 2168, 1963; *Agr. Handbook* 264, 1964.

CELERY LOOPER

Syngrapha falcifera (Kirby), Family Noctuidae

Widely distributed in southern Canada and throughout the United States, this looping caterpillar devours the leaves of beets, beans, celery, lettuce, and other plants. The adult is a moth resembling the adults of the cabbage looper and several of the cutworms. Its markings are shown in Fig. 208. Larvae are pale green with rather faint stripes, the body tapering toward the head, with 3 pairs of abdominal prolegs. Along the sides of the abdomen is a row of small black spots, each spot having a white dot in the center. These mark the spiracles. Winter is passed in the partly grown larval stage, and development is completed in the spring; there are 3 generations through the central part of the range of this pest. *Bacillus thuringiensis* is now widely recommended for control.

REFERENCES: *USDA Farmers' Bul.* 1269, 1944; *Tech. Bul.* 1567, 1978; *Ohio Sta. Univ. Ext. Bul.* 459, 1977.

Figure 208. The celery looper, *Syngrapha falcifera* (Kirby): male moth and larva. (Chittenden, USDA)

CELERY LEAFTIER

Udea rubigalis (Guenée), Family Pyralidae

This insect is also known as the greenhouse leaftier because it often becomes a serious pest of glasshouse crops. Besides celery it also attacks cabbage, beets, spinach, tobacco, many flower garden plants, and weeds. In the gardens of the North it usually is not a problem pest; in Florida it is one of the major insects attacking celery.

The larva is light green, with dorsal lighter and darker stripes. Fully grown it is about 2 cm long. It feeds on the inner surface of the folded leaves, which are webbed with silk, but may also eat into buds and flowers. Pupation takes place in a silken cocoon inside the rolled edge of a leaf. New tan-colored moths with darker irregular markings and a wingspread of nearly 2 cm emerge in about 10 days. Tiny pale green eggs are soon deposited on the leaves, and the life cycle is repeated. There are several generations each year. In glasshouses a generation may be completed in 35 to 40 days.

REFERENCES: *Mich. Quart. Bul.*, 14:91, 1931; *Fla. Bul.* 251, 1932; *J. Agr. Res.*, 29:137–158, 1924; *USDA Tech. Bul.* 463, 1935.

SWEETPOTATO WEEVIL

Cylas formicarius elegantulus (Sum.), Family Curculionidae

This weevil is the worst pest of sweetpotatoes in the areas where it is now found. Of tropical origin it is distributed in much of Louisiana and Florida, many counties of southern and eastern Texas, the coastal counties of Mississippi, Alabama, Georgia, and South Carolina. It is also present in Hawaii. Other hosts are morning-glories and related species.

The beetle is 6 mm long, dark blue on the head, snout, and back, and brick red on the prothorax and legs. It resembles a large ant (Fig. 209). Eggs are laid in the plants near the surface of the soil. In warm weather hatching occurs in about a week, and the white legless larvae with pale brown heads feed and develop, attaining a length of 9 mm. These larvae or grubs bore into the vines and through the roots and tubers, resulting in death of the plants and unmarketable tubers. After a week or more in the pupal stage inside the tubers the adults emerge and feed on the leaves, vines, and roots, soon laying eggs that develop into the next generation. In the Gulf states there are as many as 8 generations each year. Weevils continue to feed and breed in sweetpotatoes in storage, although development is retarded, especially where cool temperatures occur.

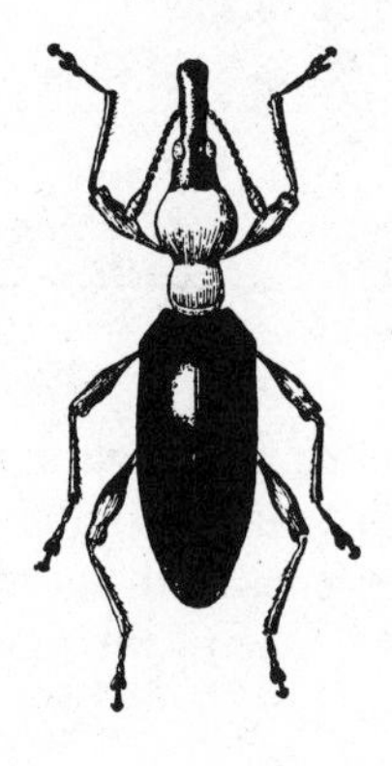

Figure 209. The sweetpotato weevil; 6 ×. (USDA)

The control program for the weevil includes several operations. At least one month before the new crop is planted there should be a thorough cleaning of all sweetpotato storage places and complete destruction of all infested potatoes. Then dust or spray the storage place with an approved pesticide. Use only state-certified seed sweet potatoes. If seed is selected at harvest, treat it with an approved pesticide. Destroy all the vines and sweetpotatoes in the seed beds as soon as sufficient plants have been obtained for the new crop. Place, where possible, in fields which were not planted with sweet potatoes the season before and as far from such fields as feasible. Set the plants deeply, and mound up the soil to protect the stems and make it difficult for the grubs to reach the roots. Keep volunteer plants destroyed and harvest promptly and cleanly, leaving no culls in the field. Turning hogs into the field after the crop is harvested assists in cleaning up vines and small potatoes. Fumigating infested sweet potatoes with methyl bromide at the rate of 1 pound per 1000 cubic feet of storage space prevents development of the weevil in storage and insures insect-free seed. Quarantines are maintained in the principal sweetpotato-growing states to prevent introduction and spread of the weevil and to assist in eradication campaigns.

REFERENCES: *USDA Leaflets* 121, 1954; 431, 1960; *Agr. Handbook* 329, 1967; *PA* 874, 1968; *La. Sta. Univ. Tech. Bul.* 483, 1954.

SWEETPOTATO LEAF BEETLE

Typophorus nigritus viridicyaneus (Crotch), Family Chrysomelidae

This is a minor pest of the sweetpotato, widely distributed throughout the United States from the Ohio River southward. The adults feed on foliage and the larvae bore into the roots and vines, damaging both.

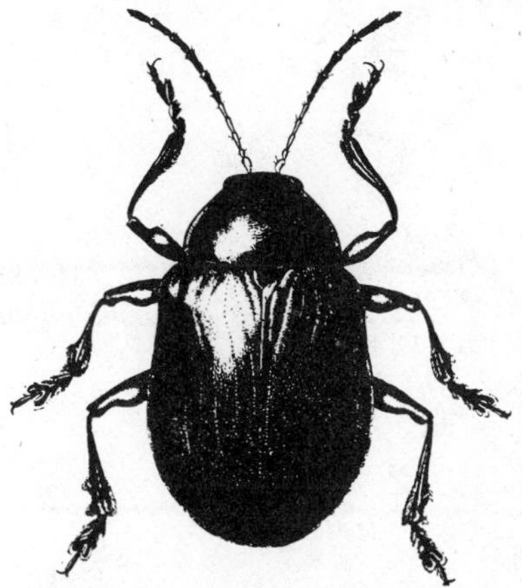

Figure 210. The sweetpotato leaf beetle; 6 ×. (USDA)

These beetles are shiny blue-green, 6 mm in length (Fig. 210). They begin to emerge the latter part of May and are present until July. After a preoviposition period of 2 weeks each female deposits small groups of lemon-yellow eggs in the soil near the base of the plants. Hatching occurs in 9 days and the pale yellow larvae feed until fully developed, which is usually before harvest of the crop. They leave the roots, migrate downward in the soil, and pass the winter. Pupation begins in early May of the following year, the entire life cycle requiring almost a year.

Since most adults emerge from the soil within a period of about a week late in May or early in June, and there is a preoviposition period of about 14 days, pesticides should be applied during this period to kill adults before eggs are deposited.

REFERENCES: *USDA Cir.* 495, 1938; *Agr. Handbook* 329, 1967.

SWEETPOTATO FLEA BEETLE

Chaetocnema confinis Crotch, Family Chrysomelidae

The sweetpotato flea beetle is a widely distributed species east of the Rocky Mountains. Feeding by this beetle is somewhat unlike that of other flea beetles. It cuts shallow irregular channels in the surfaces of the leaf, but it does not puncture the leaves. When these beetles (Fig. 211) attack very young plants serious damage results, often requiring replanting. Adults are black and about 3 mm long. They pass the winter in sheltered places and move to sweetpotato fields soon after planting has been completed. Eggs are laid in the soil and a

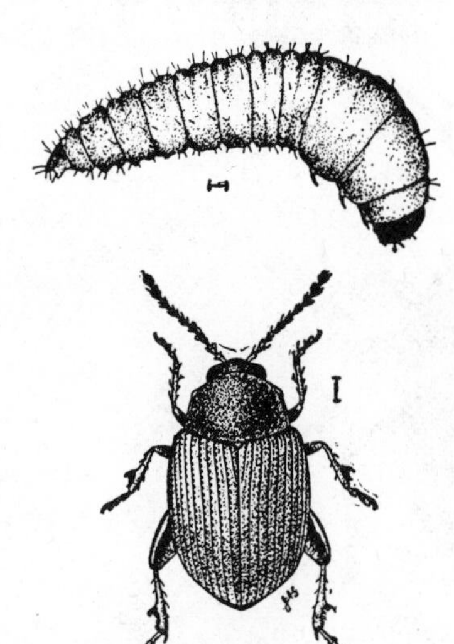

Figure 211. The sweetpotato flea beetle, *Chaetocnema confinis* Crotch; 9 ×. (J. B. Smith.)

few days later they hatch into white slender larvae which, when fully grown, are about 5 mm long. The larvae make small winding tunnels just under the skin of sweetpotato roots. These are nearly invisible at first but soon darken and can be seen through the skin. As the roots grow, the skin over the tunnels splits away leaving shallow scars. Although large larval populations are often found in sweetpotato plantings, they cause little injury to the harvested roots. Pupation occurs in the soil and new adults appear. In the warmer months the period from egg to adult is about 30 days. A succession of generations develops from spring to autumn. Adults occur almost everywhere sweetpotatoes are grown. They also live on bindweed and have been found feeding on morning-glory, corn, wheat, oats, clover, and sugar beets. Sweetpotato varieties differ widely in their susceptibility to damage from this insect.

REFERENCES: *N.J. Agr. Exp. Sta. Bul.* 229, 1910; *J. Econ. Ent.* 58:581–583, 1965; *USDA Handbook* 329, 1967.

TORTOISE BEETLES

ORDER COLEOPTERA, FAMILY CHRYSOMELIDAE

A group of leaf beetles, sometimes called tortoise beetles because the heads are concealed under the margin of the thorax, attack moonflower, Chinese lantern, sweetpotato, morning-glory, and bindweed, often with serious injury caused by the larvae and adults devouring the foliage.

Some common species are the argus tortoise beetle, *Chelymorpha cassidea* (Fabr.); the black-legged tortoise beetle *Jonthonota nigripes* (Oliv.); the mottled tortoise beetle, *Deloyala guttata* (Oliv.); the golden tortoise beetle, *Metriona bicolor* (Fabr.) (Fig. 212); and the striped tortoise beetle, *Agrioconota bivittata* (Say).

The adults are oval flattened beetles, nearly 6 mm in length, with a color pattern indicated by the common name, except for the argus tortoise, which is orange-red with black spots and about 9 mm long.

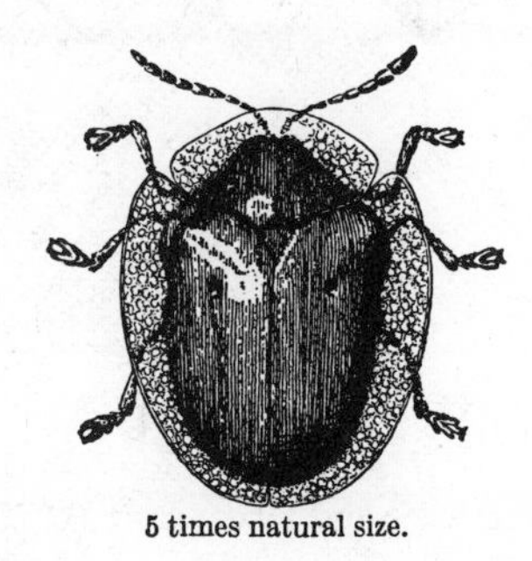

Figure 212. A tortoise beetle. (USDA)

Larvae are short, flattened, spiny-margined, with a forked posterior appendage bent forward over the body to which is attached a mass of cast skins and excreta.

All species have similar life cycles. Adult beetles hibernate in the winter, become active in the spring, and feed on morning-glory or bindweed plants until sweetpotatoes are planted. Eggs are laid about June and the larvae feed on the foliage, later pupating on the leaves. New adults appear by midsummer and feed until fall. In more southern areas some species may produce several generations during the summer.

REFERENCES: *J. Agr. Research*, 27:43–53, 1924; *USDA Farmers' Bul.* 1371, 1934.

ONION MAGGOT

Hylemya antiqua (Meigen), Family Anthomyiidae

This insect, generally considered the most serious pest of onions, is a native of Europe. It is found in this country in the northern states and is also established in Canada. The flies are slender-bodied, long-legged, and resemble the house fly (Fig. 213) except that they are only half as large. The larvae are typical white maggots that burrow into the developing bulbs of the onion, causing the plants to turn yellow and die. Cool wet weather favors the development of serious infestations.

Hibernation takes place as puparia concealed in the soil or in the shelter of weeds or crop remnants. Emerging adults appear in the spring and lay their

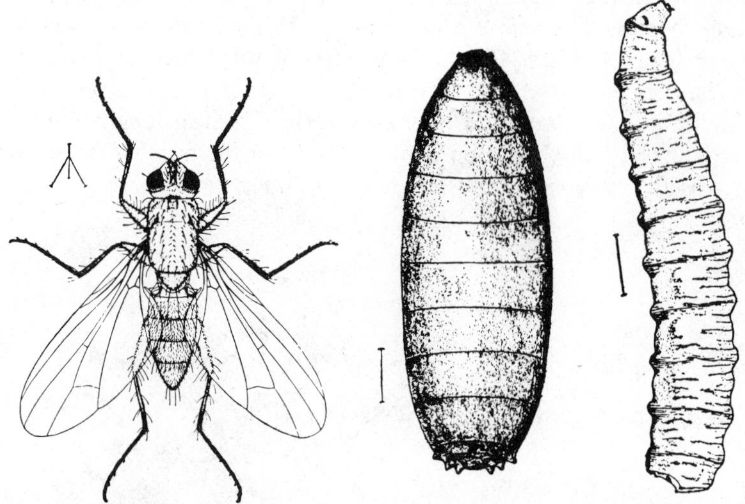

Figure 213. The onion maggot, *Hylemya antiqua* (Meigen); adult, puparium, and larva; 8 ×. (USDA)

white eggs in the axils of the leaves and in the soil near the plant. When the onions are small and growing close together a maggot may move from one to another and destroy several plants in the course of its development. When fully grown they enter the soil, change to puparia, and in a few weeks emerge as adults, producing a second generation. In some areas a third generation often develops in onions before harvest.

Treatment of the furrow has become a widely adopted practice. At planting time introduce an approved chemical into the seed furrow either as a drench or as granules. An additional control measure consists of surface treatments of the growing onion plants with a pesticide. This is directed at killing all the newly emerging adults before they lay eggs, therefore repeated applications will be required. Screen-covered frames over the rows of small plantings give protection from this pest.

REFERENCES: *Can. Dept. Agr. Bul.* 161, 1932; *Pub.* 135, 1961; *Ohio Ext. Bul.* 459, 1977; *J. Econ. Ent.*, 47:852–859, 1954; 56:580–584, 1963.

ONION THRIPS

Thrips tabaci Lindeman, Family Thripidae

Thrips are serious pests in practically all onion-growing areas of the United States and Canada. Besides onions many other vegetable and flower garden plants, as well as weeds, are hosts. Damage is caused by the nymphs and adults sucking the plant sap with their rasping-sucking mouthparts. This injury has the appearance of tiny white spots on the leaves, and when severe the entire plant may wilt and die.

In the North both adults and nymphs pass the winter concealed in grass stems or other plant remnants, whereas activity may continue the year around in warmer climates. Adult thrips are under 2 mm in length, dark-bodied, with 4 wings fringed with hairs (Fig. 214). The females deposit eggs in tender plant

Figure 214. The onion thrips, *Thrips tabaci* Lindeman; 15 ×. (USDA)

tissues. The eggs hatch in a few days into tiny, wingless, almost white nymphs that feed and molt 4 times before reaching the adult stage. Development from egg to adult may not require more than 2 weeks, resulting in several overlapping generations per season. These insects may feed between the leaves, well down toward the base of the plant where it is difficult to reach them with pesticides.

Several species of lady beetles and the minute pirate bug, *Orius tristicolor* (White), are predators of thrips. Some varieties of sweet Spanish onions show resistance to thrips' attack.

Pesticide applications are the common and effective control measures. Two or 3 applications about a week apart are usually sufficient to check an infestation.

REFERENCES: *USDA Leaflet* 372, 1960; *Agr. Handbook* 313, 1966; *Ohio Sta. Univ. Ext. Bul.* 459, 1977.

RHUBARB CURCULIO

Lixus concavus Say, Family Curculionidae

The adult rhubarb curculio is one of the largest snout beetles common to the eastern three-fourths of North America. It is 12 mm in average length and nearly black in ground color (Fig. 215). Fresh specimens are so densely covered with pollen that they appear reddish. The beetle bores into the stalks of rhubarb with its chewing mouthparts and then lays eggs in the punctures, but the larvae seldom develop in that plant. The injury is merely the result of the punctures and the small amount of feeding by the adults on rhubarb foliage. Larvae develop in common curled dock, and on that plant most adult feeding occurs. The curculio is not considered a serious pest, but elimination of dock from the vicinity of rhubarb should be of value as a control measure.

REFERENCE: *USDA Home and Garden Bul.* 23, 1953.

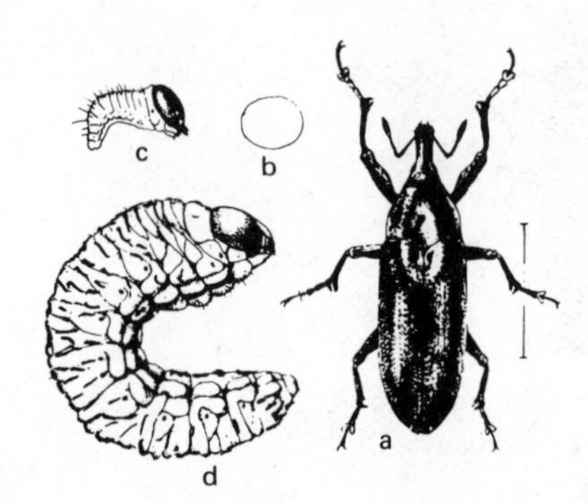

Figure 215. The rhubarb curculio: *a*, beetle; *b*, egg; *c*, newly hatched larva; *d*, full-grown larva; 2×. (USDA)

16

PESTS OF GLASSHOUSE AND GARDEN PLANTS

This chapter deals with some of the major and minor pests that attack a wide variety of glasshouse and garden plants. Other pests that may attack the same plants are discussed in Chapters 9, 13, 17, and 22.

The name glasshouse (common terminology in England) is used because it is descriptive of the structure: a house made of glass except for the supporting structures. Glasshouse pests are the same as those that occur outdoors; therefore, one of the best ways to avoid pest problems in glasshouses is to practice sanitation. Before planting a crop, be sure the house is pest-free. This can be accomplished by cleaning thoroughly and fumigating with HCN. Then to keep it that way see that all soil used in the house is steam sterilized, all plants brought into the house are pest-free, all vents are screened to prevent flying pests from entering, and the outdoor area surrounding the house is free of weeds and other plants that may harbor pests. Be alert to any developing pest problems in the house and practice sanitation and/or use a pesticide to solve the problem.

TWO-SPOTTED SPIDER MITE

Tetranychus urticae Koch, Family Tetranychidae

This widely distributed species is probably the most troublesome of all the spider mites. It feeds on many hosts including field, garden, glasshouse, nursery, and ornamental plants as well as weeds. All active stages remove plant sap, usually from the undersurface of the leaves, resulting in tiny light spots. This gives the foliage a speckled appearance, and with severe infestation the entire plant may be killed.

The spider mite is less than 0.5 mm in length and variable in color, usually with shades of green, yellow, or red, and two darker dorsal pigmented spots

(Fig. 216). The first instar is pale yellow and has 3 pairs of legs. Later instars have 4 pairs of legs, the mature females being the larger of the 2 sexes and showing more pigmentation. Males are recognized by their greater activity, smaller size, and narrow more pointed abdomen. Tiny clear to pale green spherical eggs are usually deposited on the undersides of leaves where feeding and spinning of delicate webs takes place. Hot dry weather favors rapid development of spider mites, whereas low temperature, high humidity, and excess moisture are unfavorable. Under ideal conditions eggs may hatch in 3 days and after 3 molts the adult stage may be reached 5 days later. Ten generations develop out-of-doors each year at Blacksburg, Virginia, and these overlap so much that all life stages are always present. In glasshouses reproduction is continuous throughout the year, with many generations developing. Out-of-doors in colder climates the hibernating orange-colored females and sometimes even the normal crawling stages remain throughout the winter. Each female mite may deposit nearly 200 eggs and may live nearly 70 days, but on the average about 70 eggs are deposited and longevity is 30 days. Unfertilized eggs develop into male mites.

Predaceous mites in the genera *Amblyseius, Phytoseiulus,* and *Metaseiulus* are very important in keeping mite populations checked. Other predators of importance are the lady beetles, *Stethorus punctum* (LeC.) and *S. picipes* Casey; the minute pirate bug, *Orius tristicolor* (White); the thrips, *Leptothrips mali* Fitch, *Haplothrips faurei* Hood, and *Scolothrips sexmaculatus* (Pergande); the chrysopids, *Chrysopa carnea* Stephens, *C. oculata* Say, *C. nigricornis* (Burmeister), *C. rufilabris* Burmeister; and some mirid species. Certain pesticides are highly toxic to the natural predators, especially predaceous mites. Following their use populations of this mite and other destructive species (p. 204) often increase in numbers. Then miticides are necessary. Every effort should be made to integrate biological control into your management program to get maximum value from the natural enemies with minimal usage of costly miticides. Frequent syringing of plants with a strong stream of water is

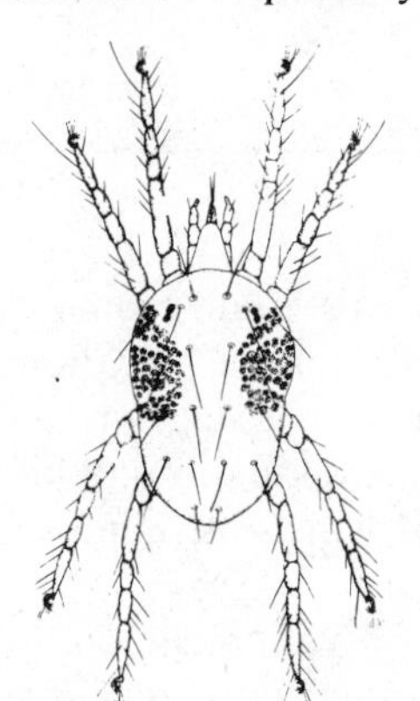

Figure 216. Adult female two-spotted spider mite; 45 ×. (Cagle, Va. Agr. Exp. Sta.)

of value in spider mite control, particularly on plants that can stand this treatment.

Dusting sulfur is effective as a miticide, especially during hot weather. A wide variety of miticides can now be obtained. For proper dosages follow the directions of the manufacturer. Some possess residual toxicity whereas others are primarily contact in action, requiring repeated applications. Control in glasshouses may be accomplished by spray, dust, or aerosol formulations of these chemicals. Continuous use of one kind of miticide should be avoided to prevent the development of strains of mites resistant to that chemical.

REFERENCES: *USDA Bul.* 416, 1917; *Va. Agr. Exp. Sta. Tech. Bul.* 113, 1949; *Ohio Sta. Univ. Ext. Bul.* 459, 1977; 538, 1977; 504, 1977.

SPRUCE SPIDER MITE

Oligonychus ununguis (Jacobi), Family Tetranychidae

This mite is widely distributed in the United States and southern Canada. It feeds on a number of conifers, especially spruce, cedar, arborvitae, juniper, hemlock, and some pines. Damage results from their sucking plant sap from the needles, producing a yellow mottled effect. Many needles turn brown and drop. Populations greatly increase in periods of hot dry weather. Outbreaks have followed applications of certain pesticides that evidently eliminated the natural enemies.

Resembling the 2-spotted species in many ways, this mite produces abundant webbing, the young are pale green and the adults are dark green to brown with pigmented areas; it differs by overwintering as spherical eggs at the base of the needles. Overlapping generations occur about every 18 days throughout the summer.

REFERENCES: *USDA Misc. Pub.* 626, 1948; 1175, 1972; *Ohio Sta. Univ. Ext. Bul.* 504, 1977.

CYCLAMEN MITE

Steneotarsonemus pallidus (Banks), Family Tarsonemidae

This mite is chiefly a pest of glasshouse plants, especially cyclamen, gerbera, snapdragon, and African violets. At times it seriously injures garden plants such as delphinium, dahlia, and strawberry. Symptoms of its presence, caused by removal of plant sap, are distortion and stunting of flowers and foliage, with or without blackening of the areas fed upon. Attacks are confined primarily to the younger leaves.

Adult mites are amber- or caramel-colored; the younger stages are white (Fig. 217). They are too small to be seen without magnification. Tiny pearl-

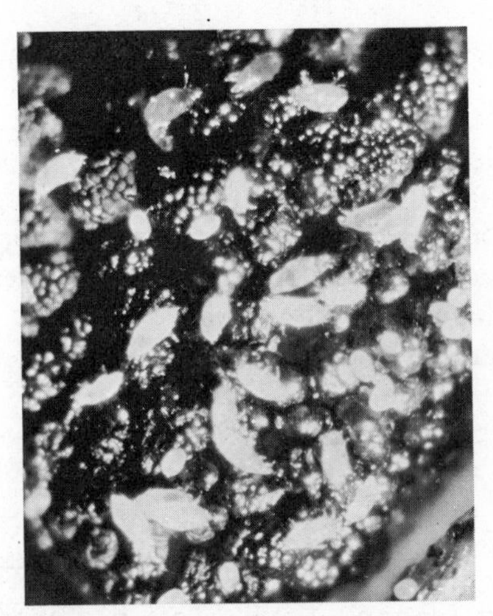

Figure 217. Various stages of the cyclamen mite; greatly magnified. (USDA)

like eggs hatch in a week or more. The first instar larva has only 3 pairs of legs, subsequent instars, 4 pairs. Breeding is continuous in glasshouses, and a generation may be completed in about 3 weeks; outdoors this may be greatly extended, depending on the temperature. Outside glasshouses, the mites winter in the crowns of the host plants.

Considerable natural control results from predators, especially the mites *Amblyseius aurescens* Athias-Henriot, and *A. cucumeris* (Oud). (p. 386). Sanitation measures should be practiced where infested plants are handled. Fumigation with methyl bromide, 2 pounds per 1000 cubic feet for 2 hours at 75 F (24 C) is an effective method of ridding plants of the mites if the plants are tolerant to this fumigant. Immersing house plants for 15 minutes in water held at 110 F (44 C) controls the mite.

REFERENCES: *J. Agr. Res.*, 10:373–390, 1917; *Calif. Agr. Exp. Sta. Bul.* 713, 1949; *J. Econ. Ent.*, 46:555–560, 707–708, 1953; 50:648–652, 1957; 56:565–571, 1963; *Ohio Sta. Univ. Ext. Bul.* 506, 538, 1977.

GREENHOUSE WHITEFLY

Trialeurodes vaporariorum (Westwood), Family Aleyrodidae

The widely distributed greenhouse whitefly feeds in both the immature and adult stages by sucking plant juices from the under leaf surface. Heavy feeding gives the infested leaves a mottled appearance or causes them to turn yellow and die. The sticky honeydew excreted by the insect often glazes the lower leaves and permits the development of black sooty mold on the surface, thus detracting from the beauty of the plants.

This insect is chiefly a pest in glasshouses but at times it attacks outdoor plants such as cucumber, tomato, potato, tobacco, strawberry, grape, ageratum, aster, calendula, goldenglow, and lantana.

The adult is about 2 mm long, yellow-bodied, with 4 wings that, along with the dorsal part of the body, are covered with a white waxy powder. Except for the first instar, which is a crawler much like that same stage of a scale insect, the light green oval flattened nymphs are about the size of a small pinhead and attached to the leaf surface until mature. Their bodies are covered with radiating long and short waxy filamentous threads. They resemble young soft-scale insects. The length of the nymphal period is almost a month, the last instar being more elevated and slightly segmented. Emerging adults soon mate and begin depositing their elongated yellow eggs, attaching them to the host by a short stalk. Before hatching the eggs darken. Several overlapping generations may develop each year in glasshouses or out-of-doors in the South. This species does not survive the winters in the North except in glasshouses. During the summer infestations on outdoor plants in the North result from infested transplants or adults escaping from infested glasshouses.

Other species that are similar in appearance and habits are: azalea whitefly, *Pealius azaleae* (B. and M.); avocado whitefly, *Trialeurodes floridensis* (Quaintance); strawberry whitefly, *T. packardi* (Morrill); grape whitefly, *T. vittata* (Quaintance); rhododendron whitefly, *Dialeurodes chittendeni* Laing, and citrus whitefly, *D. citri* (Ashmead) (p. 476).

Eulophid parasites are common and, at times, will keep populations under satisfactory control. The important species attacking greenhouse whitefly is *Encarsia formosa* Gahan. Other species found attacking whiteflies are *E. luteola* Howard, *E. meritoria* Gahan, and *E. pergandiella* Howard. In any glasshouse management program every effort should be made to utilize the natural enemies. The juvenile hormone preparation, kinoprene, is very effective in controlling whiteflies in glasshouses.

REFERENCES: *Conn. Agr. Exp. Sta. Bul.* 140, 1902; *USDA Misc. Pub.* 626, 1948; *Farmers' Bul.* 1306, 1923; *Calif. Agr. Exp. Sta. Bul.* 713, 1949; *Ohio Sta. Univ. Ext. Bul.* 538, 1977.

THRIPS

ORDER THYSANOPTERA, FAMILY THRIPIDAE

Several species of thrips may seriously damage garden flowers and vegetable crops, the more common ones being flower thrips, *Frankliniella tritici* (Fitch) (Fig. 23); tobacco thrips, *F. fusca* (Hinds) (Fig. 173); bean thrips, *Caliothrips fasciatus* (Perg.); greenhouse thrips, *Heliothrips haemorrhoidalis* (Bouché);

chrysanthemum thrips, *Thrips nigropilosus* Uzel; onion thrips, *T. tabaci* Lind. (Fig. 214); gladiolus thrips, *Thrips simplex* (Mor.); western flower thrips, *Frankliniella occidentalis* (Pergande); and iris thrips, *Iridothrips iridis* (Watson).

Plant injury is caused by both nymphs and adults rasping the bud, flower, and leaf tissues of the host plants and then sucking the exuding sap. This causes distorted and discolored flowers and buds, and gray or silvery speckled areas on the leaves. Gladiolus thrips also feed on the corms in storage, causing russeted areas and lowered vigor, which results in retarded growth and smaller flowers.

Adult thrips are generally less than 2 mm in length, usually tan-to-dark brown-bodied, with 4 featherlike wings. Nymphs are creamy white and wingless. Eggs are deposited in the plant tissues, the young developing to maturity in approximately 2 or more weeks. The number of generations produced each year depends primarily on the species, the temperature, and other climatic factors. Most species produce many generations in a season. Females may lay fertilized or unfertilized eggs, the latter developing into males only.

Control for garden crops is accomplished by pesticide sprays or dusts. Two or 3 applications at 7 to 10-day intervals are generally needed to control an infestation. Aerosols of these chemicals are commonly employed for thrips control in glasshouses. Migrating or wind-borne thrips can be prevented from entering glasshouses by installing cheesecloth coverings impregnated with a pesticide over the vents.

REFERENCES: *Calif. Cir.* 337, 1935; *Bul.* 609, 1973; *Fla. Agr. Exp. Sta. Bul.* 357, 1941; *USDA Cir.* 445, 1937; *Can. Dept. Agr. Pub.* 69, 1952; *J. Econ. Ent.*, 57:357–360, 1964; *Ohio Ext. Buls.* 459, 538, 1977.

GARDEN WEBWORM

Achyra rantalis (Guenée), Family Pyralidae

The webworm is the larva of a small buff yellow moth with a wingspread of about 25 mm. The wings have both lighter and darker markings; their arrangement and the shape of the insect are illustrated in Fig. 218. The larval stage is a

Figure 218. The garden webworm; 2½ ×. (USDA)

slender caterpillar, usually green in color but often varying to different shades of yellow, with black dots on each segment. It has the habit of spinning a silken thread wherever it crawls, and as a result there is considerable webbing of the plants on which it feeds.

This species is a general feeder, devouring the foliage of both garden and field crops, especially beets, legumes, and cotton. Several weeds are also hosts, pigweed being the most common. Though widely distributed in North America, it is more of a pest in the Middle West and Southwest.

Development may continue throughout the year in the extreme South, whereas in the North the winter is passed as a pupa. Between these 2 areas overwintering as larvae also occurs. When moths appear in the spring, eggs are deposited in groups consisting of a few to almost 50. On hatching, the larvae feed for 3 or more weeks then construct silken-lined burrows on or in the soil in which pupation occurs. As many as 5 generations are produced in Texas, a reduced number in the more northern latitudes where the insect is found.

Harvesting infested hay crops checks this insect but migrating larvae often seriously damage adjacent garden or field crops. The larvae are easier to kill in the early instars before much webbing has occurred. Chemicals have given satisfactory control and selection of the proper one will depend on the residues involved on a particular crop.

REFERENCES: *USDA Farmers' Bul.* 944, 1918; *Leaflet* 304, 1951.

GARDEN FLEAHOPPER

Halticus bracteatus (Say), Family Miridae

Fleahoppers, probably American in origin, are found from Maine to Florida and west through the central states to Utah. The appearance of the insects, except for size, is shown in Fig. 219, which presents a winged male and short-winged female. Long-winged females also occur, and resemble the males. The large hind legs enable them to hop actively. Maximum length of the winged forms is about 3 mm.

Fleahoppers hibernate as adults. The number of generations produced varies in different latitudes, 5 being recorded for some southern states. Feeding is accomplished by sucking plant sap, which makes small discolored areas on the foliage. These areas often become numerous enough to cover leaves completely and to cause their death, injuring the plant seriously. Clover, cowpeas, alfalfa, many weeds, and garden crops are attacked by fleahoppers. Natural enemies have been observed but are said to have little effect on the population of the hoppers. Climatic factors probably keep them under control in the North.

REFERENCES: *Va. Agr. Exp. Sta. Tech. Bul.* 101, 1946; 107, 1947; *J. Econ. Ent.*, 40:675–679, 1947; *USDA Agr. Handbook* 290, 1965.

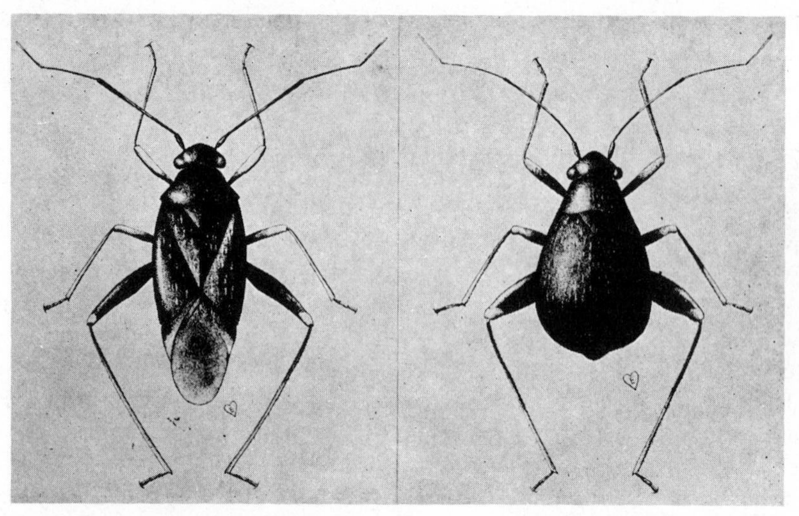

Figure 219. The garden fleahopper, *Halticus bracteatus* (Say): (left) male; (right) short-winged female. The long-winged female resembles the male; 14 ×. (Beyer, USDA)

OMNIVOROUS LEAF ROLLER

Platynota stultana Walsingham, Family Tortricidae

This insect is becoming increasingly important in North America as a pest of a wide variety of crops, especially in areas where the year-round growing season provides a sequence of crop plants on which to feed. Cotton and alfalfa are favored hosts but celery, lettuce, red pepper, tomato, sugar beet, citrus, rose, grape, carnation, chrysanthemum, aster, and over thirty other plants, in glasshouses and out-of-doors, have also been attacked. It was described from Sonora, Mexico in 1884.

Adults are typical tortricid moths. The female is lighter colored than the male and slightly larger (♂ 7.5 mm; ♀ 10 mm long). Each female may lay over 300 eggs during her life span. Pale green eggs are deposited in masses, usually on the upper leaf surface and primarily during the hours of twilight and dawn. Individual eggs overlap each other like fish scales and the entire mass is cemented together and to the plant with a secretion from the colleterial glands. Hatching occurs in about 7 days and the black-headed pale yellow larvae begin making silken ladders on which they move about and feed. The larval habit of concealing itself by rolling a leaf usually begins in the second instar which measures 7 mm long. It takes approximately 5 minutes to complete the leafrolled nest. Some larvae complete their development in a single nest; others may build 6 or more. After a total of 5 larval instars (now 14 mm long) and about 25 days, pupa-

Figure 220. The omnivorous leaf roller, *Platynota stultana* Walsingham; adult female, egg mass, and second instar larva; 7 ×. (Courtesy of E. L. Atkins, Univ. of Calif.)

tion takes place within the rolled leaf; a week later adults emerge. This rate of development is based on 80 F (27 C); at higher temperatures the life cycle is completed in less time (see Fig. 220).

Four parasitic wasps and 7 predators were found attacking this pest in California but none was of sufficient importance to effect practical control.

REFERENCES: *J. Econ. Ent.*, 29:306–312, 1936; 50:59–64, 1957; *Ann. Ent. Soc. Amer.*, 50:251–259, 1957; *Calif. Agr. Exp. Sta. Leaflets* 83, 86, 1964; *Down to Earth*, 20:18–22, 1965.

WHITE-FRINGED BEETLES

ORDER COLEOPTERA, FAMILY CURCULIONIDAE

Originally from South America and discovered in Okaloosa County, Florida, in 1936, these beetles were collected later in the same year in Alabama and in other localities of Florida. Since that time they have been found in Mississippi, Louisiana, Georgia, Maryland, North Carolina, South Carolina, Virginia, Arkansas, and Tennessee. By the end of 1949, 3 species of white-fringed beetles were described as occurring in the United States: *Graphognathus leucoloma* (Boheman), *G. minor* (Buch.), and *G. peregrinus* (Buch.). *G. leucoloma* is represented by 5 races: *dubious, pilosis, imitator, striatus,* and *fecundis*. Information on the life history and habits of the original infestation, which consisted of *G. leucoloma fecundis,* is presented here.

The larvae and adults have been observed to feed on 385 species of plants, of which cotton, peanuts, okra, velvetbeans, soybeans, cowpeas, sweet potato, beans, and peas comprise the most important garden and field crops commonly attacked. Damage is done to the roots of the food plants by the larvae, and to the foliage and other plant parts by the adults.

The adult is a large, dark gray snout beetle, just under 13 mm in length, with a lighter band along the outer wing margins (Fig. 221). All the beetles are females reproducing parthenogenetically and incapable of flight. The legless white larvae are typical of the weevils.

Winter is usually passed as larvae in the soil, but the egg stage may also survive in well-protected situations such as haystacks or among unshelled peanuts. The first adults generally appear in May and emergence continues into August in most latitudes. Egg-laying begins in 5 to 25 days and continues for 30 to 60 days. Masses containing up to 60 or more white eggs are most often deposited on plants or other objects at or near the point of contact with the soil. Hatching occurs in about 17 days in midsummer, and the larvae feed in the soil until the following spring when they pupate and transform to adults. There is only 1 generation per year, but some larvae from eggs deposited late in the year do not reach maturity until the second summer.

Quarantines have been established to prevent the movement of soil or plant materials likely to be infested, since spread from one locality to another takes place primarily in this manner. Fumigation of such plant materials is a recommended procedure.

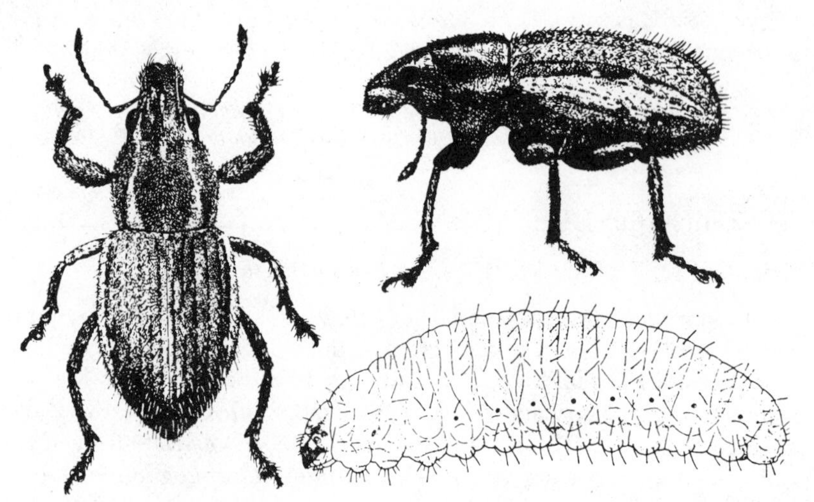

Figure 221. The white-fringed beetle, *Graphognathus leucoloma* (Boheman); 4 ×. (USDA)

Cultural control practices suppress white-fringed beetle populations. These consist of planting oats or other small grains on the heavily infested portion of the farm and following a rotation in which peanuts, soybeans, velvetbeans, and other summer legumes are only planted in those areas once every 3 or 4 years. Treating infested soil with a pesticide has given control of the larvae.

REFERENCES: *USDA Cir.* 850, 1950; *Leaflet* 550, 1972; *J. Econ. Ent.*, 45:475–461, 1952.

MOLE CRICKETS

ORDER ORTHOPTERA, FAMILY GRYLLATALPIDAE

Outbreaks of these insects have occurred at various times and places in the South Atlantic and Gulf Coast states from North Carolina to Texas. Garden vegetables, tobacco, peanuts, strawberries, and grasses are often damaged by the nymphs and adults that feed underground at or near the soil surface. They chew the roots, tubers, and underground stems, and also attack strawberries and other fruits that touch the ground.

Four kinds of mole crickets occur in the infested area mentioned. The most common and abundant species is the southern mole cricket, *Scapteriscus acletus* Rehn and Heb. which in some localities is accompanied by the changa or Puerto Rican mole cricket, *S. vicinus* Scudder. These 2 species are most often found in sandy soils. The short-winged mole cricket, *S. abbreviatus* Scudder, is found only in the southern third of Florida, and the northern mole cricket, *Neocurtilla hexadactyla* (Perty) (Fig. 222), is rather scarce except in a few localities with wet heavy soils.

All species closely resemble each other except for size. Fully grown mole crickets are 38 mm long, brown in color, with beadlike eyes, and short stout front legs fitted for digging. Winter is passed as nymphs or adults in the soil, migration downward occurring during cold weather. In the spring and early summer the eggs are laid in cells constructed in the soil by the female. Ordinarily about 35 eggs are placed in each cell. Hatching occurs in 10 to 40 days de-

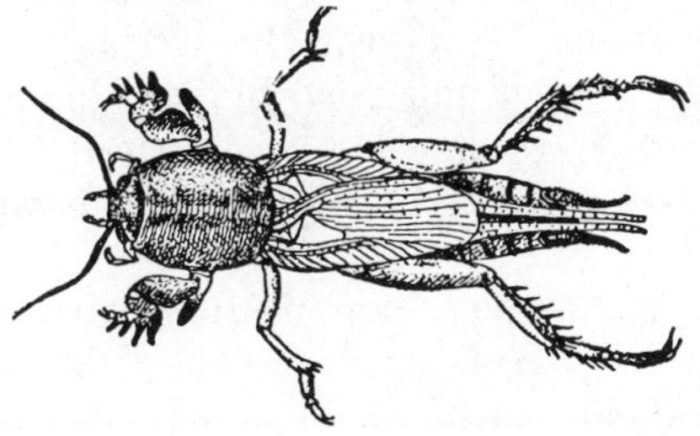

Figure 222. The northern mole cricket, *Neocurtilla hexadactyla* (Perty), natural size. (USDA)

pending on the prevailing temperatures. The young crickets grow rapidly during the summer and early fall, most of the species becoming adults before midwinter. Only 1 generation occurs each year. If the soil is flooded, mole crickets leave their burrows and swim about in an effort to locate dry land.

Soil chemicals and poisoned baits have been used to control these insects. Increased effectiveness can be expected when recently plowed and disk-harrowed fields are baited, dusted, or sprayed after a good rain or irrigation.

REFERENCE: *USDA Leaflet* 237, 1953.

SERPENTINE LEAFMINER

Liriomyza brassicae (Riley), Family Agromyzidae

The serpentine leafminer is found in America and Europe, being more prevalent in warmer areas. It causes damage in the larval stage by mining between the upper and lower surfaces of the leaves of many plants. These mines are light in color, narrow, and winding. Besides rendering edible portions of vegetable crops unsalable, the leafminer interferes with the normal photosynthesis of the plant.

Shiny black flies with yellow markings deposit pale white oval eggs into plant tissues. Inserted from the bottom of the leaf, the eggs are placed against the upper epidermis. The average hatching period is 4 days; the white to pale yellow larvae with black mouthparts mine the leaves for about 5 days in the summer months, but for longer periods when temperatures are lower. Normally the larvae change into shiny brown puparia in the soil, but in arid and semi-arid areas they have been observed changing to this stage in the leaf mines. In about 10 days the flies, 1 to 1.75 mm in length, emerge and start the cycle anew. The average period for the entire life cycle is 23 days, and several overlapping generations occur each year. In the northern areas of distribution overwintering is in the puparium stage. Other species of leafminers cause similar damage and have similar life cycles. They are the pea leafminer, *L. huidobrensis* (Blanchard), and vegetable leafminer, *L. sativae* Blanchard.

Many natural enemies attack this pest. Some eulophid parasites of importance are *Solenotus begini* (Ashmead), *S. websteri* (Crawford), *Derostenus pictipes* Crawford, *D. arizonensis* Crawford, *Chrysocharis ainsliei* Crawford, and *C. parksi* Crawford.

Sprays containing diazinon, parathion, or trichlorfon have given satisfactory control. Consideration must be given to the possibility of poisonous residues when selecting the pesticide for protecting a particular crop.

REFERENCES: *J. Agr. Res.*, 1:59–88, 1913; *Fla. Agr. Exp. Sta. Press Bul.* 639, 1947; *J. Econ. Ent.*, 51:357–359, 1958; *Ohio Sta. Univ. Ext. Bul.* 459, 1977.

ARBORVITAE LEAFMINER

Argyresthia thuiella (Packard), Family Yponomeutidae

In the eastern half of North America the tips of arborvitae twigs may become discolored and gradually turn brown because the interior of the needles is eaten by arborvitae leafminers (Fig. 223). The damage is done by small green caterpillars, about 3 mm long, red-tinged with shiny black heads. Winter is passed as larvae in the mines, with pupation occurring the following spring and adults emerging in late May and June. Adults are tiny gray moths with small conspicuous silver markings on the wings which when spread, measure 8 mm. They lay their greenish eggs between the tip of one leaf scale and the base of an adjoining one. On hatching, the larvae enter the twigs. Five miners in a 3½ inch twig is considered a heavy infestation. One generation occurs each year.

Cutting off and burning infested twigs in autumn or winter is an effective nonchemical control method. Several chemicals have been effective in controlling this pest if applied in May or June.

REFERENCES: *Conn. Agr. Exp. Sta. Bul.* 348, 1933; 693, 1971; *Can. Ent.* 89 (4):171–182, 1957.

AZALEA LEAFMINER

Caloptilia azaleella (Brants), Family Gracillariidae

Native to Asia this pest was introduced into North America from Europe before 1910. The foliage of both evergreen and some deciduous varieties of azaleas

Figure 223. Typical damage caused by arborvitae leafminer. (Courtesy of D. G. Nielson, OARDC)

may be mined and skeletonized, both outdoors or in glasshouses. Damaged leaves turn yellow and drop.

The damaging stage is a yellowish caterpillar about 12 mm long when fully grown. Until nearly half-grown it mines inside the leaves; then it comes out, folds the tip or margin of the leaf under (held in place by silken strands), where skeletonizing occurs. Pupation takes place within the folded leaves. Emerging adults are moths with about an 8 mm wingspread; the fore wings are yellow with purplish markings. Creamy white eggs are placed on the underside of leaves; these hatch in about 6 days. Most of the miners overwinter as mature larvae. Adults of the first generation appear in May and June. There are 3 complete generations outdoors, whereas in a glasshouse reproduction is continuous.

REFERENCES: *USDA Misc. Pub.* 626, 1948; *Conn. Agr. Exp. Sta. Bul.* 693, 1971; *Ohio Agr. Res.* and *Dev. Center Res. Bul.* 983, 1966.

BIRCH LEAFMINER

Fenusa pusilla (Lepeletier), Family Tenthredinidae

This sawfly, an introduced species first discovered in North America in Connecticut in 1923, now occurs in southeastern Canada and from Maine to New Jersey, westward to Ohio and the Lake States. It is a major pest of gray, white, paper, and cutleaf varieties of European white birch, with occasional damage to red or river birch.

Winter is passed as fully grown larvae in cocoons in earthen cells 1 to 2 inches below the soil surface. Pupation occurs in the spring and adults begin to appear about mid-May and deposit eggs in the new leaves. The larvae feed on the tissue between the leaf surfaces producing blisterlike areas; these wrinkle and turn brown. Fully grown larvae are flattened, yellowish-tinged, and about 6 mm long. They chew out of the leaves and drop to the ground where they pupate in cocoons. There are 3 or 4 generations per year in the southern infested regions. The later generations are less important.

Outbreaks occur frequently in the northeastern states and some tree killing may occur, but the greatest damage is weakening the trees which leads to attack by other insects, particularly borers.

REFERENCES: *Conn. Agr. Exp. Sta. Bul.* 348, 1933; *Cir.* 185, 1954; 693, 1971.

BOXWOOD LEAFMINER

Monarthropalpus buxi (Laboulbène), Family Cecidomyiidae

The adult is a very small orange-yellow fly about 4 mm long, the larvae or maggots of which feed between the leaf surfaces of boxwood. The mines produced by this feeding appear as blotches or blisters on the lower leaf surface and when numerous they kill the leaves and disfigure the plant. The tiny yellowish maggots require a year for development, from time of hatching in late spring until the following spring, when they transform to pupae within the mines. Gnatlike flies emerge over a period of 2 weeks, usually starting the first or second week of May in the latitude of Washington, D.C. This occurs earlier farther south and later farther north. The tail end of the empty pupal case usually remains attached to the leaf. The adults live only a few days so ovipositing in new leaf tissue begins very soon after emergence.

All varieties of boxwood may be attacked, but some are more seriously injured than others. Infestations spread rapidly through all boxwood of a given region and control is, to a considerable extent, a community problem.

Systemic pesticides have proven satisfactory when larvae are present in the mines and a number of other chemicals are effective when applied during the period adults are emerging.

REFERENCES: *USDA Misc. Pub.* 626, 1948; *Farmer's Bul.* 1855, 1940; *Cir.* 305, 1934; *Conn. Agr. Exp. Sta. Bul.* 565, 1953; *Bul.* 693, 1971; *Ohio Ext. Bul.* 504, 1977.

BOXWOOD PSYLLID

Psylla buxi (L.), Family Psyllidae

A characteristic cupping, mainly of terminal boxwood leaves, results from the feeding of this piercing-sucking insect. It is gray-green in color and about 3 mm long. The nymphs are covered with a whitish waxy secretion. Overwintering occurs as first instar nymphs protected by the outer scales at the base of buds. Adults begin to appear in May and become more abundant as the season advances. Eggs are laid at the base of the buds. Only one generation occurs each year. Severe infestations retard growth.

Chemicals are usually more effective in checking this pest if applied in May.

REFERENCES: *USDA Misc. Pub.* 626, 1948; *Farmer's Bul.* 1855, 1940; *Conn. Agr. Exp. Sta. Bul.* 565, 1953.

HOLLY LEAFMINER

Phytomyza ilicis Curtis, Family Agromyzidae

This European insect is a primary pest of American or Christmas holly. The adults are tiny black flies about 1 mm in length. They begin emerging from leafmines, usually after mid-May in many areas and continue to do so into June. Eggs are deposited in punctures made in the undersides of newly developing leaves, often near the midrib. The larvae or maggots are pale yellow to white and about 3 mm long when fully grown. They mine just under the upper leaf surface. At first the mines are threadlike and inconspicuous, but by late autumn they widen, because of larval growth, into blotches or blisters. The larvae pass the winter in the mines and change to brown pupae during April. Only 1 generation occurs each year.

The native holly leafminer, *P. ilicicola* Loew, feeds only on *Ilex opaca* and its cultivars. It has a life cycle similar to *P. ilicis*. However, the adult female causes additional injury by puncturing leaves with the ovipositor from which oozes sap, imbibed by both sexes. There are 4 more described species of leafminers on holly some of which are quite host specific.

Picking off and destroying infested leaves in autumn is the best nonchemical control. Applying a nonsystemic pesticide during the adult emergence period or a systemic chemical in early April or in June to kill the maggots are the recommended control measures.

REFERENCES: *USDA Misc. Pub.* 626, 1948; *Ohio Res. and Dev. Center Bul.* 983, 1966; *Md. Agr. Exp. Sta. Bul.* A-55:4, 1968; *Conn. Agr. Exp. Sta. Bul.* 568, 1953; *Bul.* 693, 1971.

LOCUST LEAFMINER

Odontota dorsalis (Thunberg), Family Chrysomelidae

This insect occurs throughout the eastern half of North America. Damage is caused by the adults skeletonizing the leaves of black locust, elm, beech, oak, dogwood, cherry, hawthorn, wisteria, and often other herbaceous plants, especially soybeans grown in the vicinity of black locust, and also by the larvae mining the leaves of black locust.

The black bodied beetles are about 6 mm long with reddish brown elytra and a median black dorsal stripe. They hibernate in sheltered areas during winter, emerge in the spring and begin feeding on developing foliage of black locust. In a short time flat oval eggs are deposited on the underside of leaves. These soon hatch and the flattened yellow-tinged larvae eat into the inner layers of the leaf forming mines resembling white stellate blisters, which later turn brown. The head, legs, thoracic, and anal shield of the larvae are black. Pupation occurs within the mines and first generation adults appear in July. These scatter over the area and lay eggs that produce a second generation. Affected black locust trees have the appearance of having been swept by fire. Complete defoliation 2 successive years where 2 sets of leaves are produced in 1 summer will kill the tree.

Natural enemies of the locust leaf miner are the parasites, *Closterocerus tricinctus* (Ashmead), *Trichogramma odontotae* Howard, *Spilochalcis odontotae* Howard, and the predator *Arilus cristatus* (L.).

REFERENCES: *Ohio Agr. Exp. Sta. Bul.* 322, 1918; *Mass. Agr. Exp. Sta. Bul.* 353, 1938; *J. Econ. Ent.* 33:742–745, 1941; *USDA Misc. Pub.* 657, 1950.

JUNIPER WEBWORM

Dichomeris marginella (Fabr.), Family Gelechiidae

This pest of juniper is of European origin, first reported in the United States in 1910. It is found in North America from Maine to North Carolina and west to Missouri and Michigan; also in California and southern Canada. The reddish brown caterpillars with lighter stripes feed on the needles and spin silken threads, thus webbing them together. The webbed portions die and turn brown. Partly grown larvae spend the winter in the webbed nests and complete their feeding the following spring, attaining a length of 14 mm. Pupation occurs in the webs and the brown moths with a wing span of 15 mm emerge in late spring or early summer and lay eggs. These hatch in 9–16 days, depending on the temperature. One generation develops each year.

Natural enemies seem to control this pest in forested areas, the most important being the parasites *Itoplectis conquisitor* (Say), *Bracon aequalis* Provancher, *Bracon gelechiae* Ashmead, *Catolaccus aeneoviridis* (Gir.), and *Tetrastichus* spp. A pesticide is often needed for ornamentals around the home. Cutting out and destroying webbed masses is a helpful control practice.

REFERENCES: *USDA Misc. Pub.* 626, 1948; *Ann. Ent. Soc. Amer.* 62:287–292, 1969; *OARDC Res. Bull.* 983, 1966.

SPRUCE BUDWORM

Choristoneura fumiferana (Clemens), Family Tortricidae

This pest occurs in the region of North America extending from Virginia to Labrador and westward to the MacKenzie River Valley, Yukon Territory, Canada. Until recently it had been considered present throughout the ranges of spruce and fir, but forms occurring in western Canada and western U.S. are now considered different species (*C. occidentalis* Freeman is one). Balsam fir is the preferred host but feeding also occurs on white, red, and blue spruce, and on larch, pine, and hemlock.

Spruce budworm adults are dull gray with brown or red markings on the forewings, with a wingspread of 25 to 30 mm. They are active from late June to early August, depending on location. The females deposit their pale green flattened oval eggs in overlapping elongate masses, like shingles on a roof. Each mass may contain nearly 60 eggs. Hatching occurs in 8 to 12 days and the young larvae disperse throughout the trees and spin hibernaculae in suitable sites and molt to the second instar. This is the overwintering stage. The following spring the tiny caterpillars emerge and mine the old needles or feed in the expanding flowers, buds, and new needles. As the new growth elongates, the larvae tie the tips of the twigs together with silk, forming nests in which they feed. By late

June the red-brown larvae covered with yellow warts and with black heads are fully grown, reaching a length of 25 mm. Pupation occurs and new adults emerge about 10 days later. There is 1 generation each year.

In large persistent populations all the new foliage may be consumed for successive years. Top killing occurs in about 3 years, and tree mortality after about 5 years. Weakened trees may die as a result of subsequent attacks by inner-bark borers such as bark beetles. The magnitude of these losses is very great to the pulpwood industry.

The spruce budworm has many natural enemies, including parasites, predators, and pathogenic organisms.

Silvicultural practices are thought, by those who have studied the problem, to be the best answer to spruce budworm control. For shade or ornamental trees a pesticide may be necessary to prevent damage. An average of 10 or more caterpillars per 40 cm branch indicates a heavy population.

REFERENCES: *Can. Ent.* 91:758–783, 1959; 92:384–396, 1960; 97:129–136, 1965; 99:456–463, 1967; *USDA Leaflet* 242, 1944; *Misc. Pub.* 626, 1948; 657, 1950; *Can. Dept. Agr. Pub.* 1035, 1946; *U.S. Forest Serv. Res. paper* RM-44, 1968.

SPRUCE GALL APHIDS OR ADELGIDS

ORDER HOMOPTERA, FAMILY PHYLLOXERIDAE

Galls that resemble miniature pineapples (25 mm long) (Fig. 224) on spruce at the base of the current season's growth are caused by the eastern spruce gall aphid, *Adelges abietis* (L.). This species is found primarily in southeastern Canada and northeastern and north central United States. Norway spruce is damaged most but the insect may also occur on white, black, and red spruce.

Figure 224. Galls of spruce gall aphids or adelgids: *left,* eastern; *right,* Cooley; about natural size. (Original)

Overwintering blue-gray nymphs, located mainly at the base of spruce buds, mature in the spring and lay eggs in masses covered with waxy threads. These hatch about the time the shucks break away from the buds. The new aphids crawl into the mass of new needles to feed which causes the plant to develop the galls. In late July or August the galls turn brown, open, and the enclosed aphids escape. These soon develop wings and lay eggs from which the overwintering nymphs hatch.

The Cooley spruce gall aphid, *A. cooleyi* Gillette, causes galls on Colorado blue, Sitka, Engelmann, white, and big-cone spruce. This species occurs from coast to coast in northern United States and southern Canada, wherever the hosts are present. The galls vary from 25 to 75 mm in length, are light green to dark purple, and occur at the tips of twigs (Fig. 224). The life cycle of this aphid is the same as that of *A. abietis* when it remains on spruce. However, some of the aphids emerging from galls in July or August fly to Douglas fir to lay eggs. The eggs hatch and the nymphs live during the winter on fir, mature in the spring and become covered with a white cottony excretion. These produce a summer generation and may cause some injury from sap-sucking but they do not cause gall formation on fir. Some of these aphids develop wings and fly to spruce to deposit eggs; others are wingless and remain on fir where they deposit eggs that develop into the overwintering nymphs.

If feasible, removing and destroying the galls before they open in July will aid in control of these pests. Plantings of Douglas fir and spruces together should be avoided, where practical. Applying an approved pesticide in early April before the shucks break away from the buds, and/or in early autumn when the nymphs are on the foliage, is sometimes recommended. These aphids are more troublesome in Christmas tree plantings.

REFERENCES: *Can. Ent.* 91:601–617, 1959; *Conn. Agr. Exp. Sta. Bul.* 566, 1953; *Mich. Univ. Sch. Forestry Circ.* 2, 1937; *Mich. Sta. Univ. Ext. Bul.* 353, 1962; *USDA Misc. Pub.* 626, 1948, 1175, 1972; *OARDC Res. Bul.* 983, 1966; *Ohio Agr. Ext. Bul.* 504, 1976.

IRIS BORER

Macronoctua onusta Grote, Family Noctuidae

This is the most troublesome insect attacking iris. Overwintering eggs on the old iris leaves hatch in early spring, and the tiny larvae make slender feeding channels into new leaves somewhat resembling the burrows of leafminers. As the white to pink larvae grow they feed behind leaf sheaths, in stems and flower buds, gradually moving downward into the rhizomes where they complete their growth, pupate, and emerge as dusky brown moths by late summer and early autumn. Fully grown larvae are nearly 5 cm in length and have a brown head; moths have a wingspread of 5 cm.

Cleaning up and destroying old iris leaves, stems, and debris in late autumn or early winter will eliminate overwintering eggs. Destroy larvae and infested rhizomes when re-setting iris roots. Sanitation can be supplemented with properly timed pesticide sprays or dusts. These should be applied soon after the eggs hatch. Malathion, diazinon, or dimethoate have been used with success.

REFERENCES: *Ohio Agr. Exp. Sta. Res. Bul.* 892, 1961; *USDA Misc. Pub.* 626, 1948; *Agr. Inf. Bul.* 237, 1962; *Home and Garden Bul.* 66, 1966.

ASTER LEAFHOPPER

Macrosteles fascifrons (Ståhl), Family Cicadellidae

This insect, also known as the six-spotted leafhopper, sucks sap from the leaves of many plants. Besides aster, it attacks other flowering plants, weeds, and several vegetable and field crops. This feeding, when severe, distorts the plants, which often turn brown and die. In addition, this leafhopper transmits a virus from infected to healthy plants, causing a disease known as "aster yellows." The insect is widely distributed in North America and is a major vector of the virus. In western regions there are other strains of aster yellows transmitted by several species of leafhoppers. Strains affecting celery are said to be carried by 22 leafhopper species.

The adult is light green-yellow, the head marked with black dots or spots arranged in pairs (Fig. 225). The body dorsally is black, the border of the abdomen yellow and the basal portion yellow with black. The nymphs have similar markings but are usually darker green. The elytra are nearly transparent with the apex somewhat smoky. Total length is about 4 mm.

Winter is passed as adults in milder regions with migration northward each spring. There are 5 nymphal instars and several overlapping generations per year.

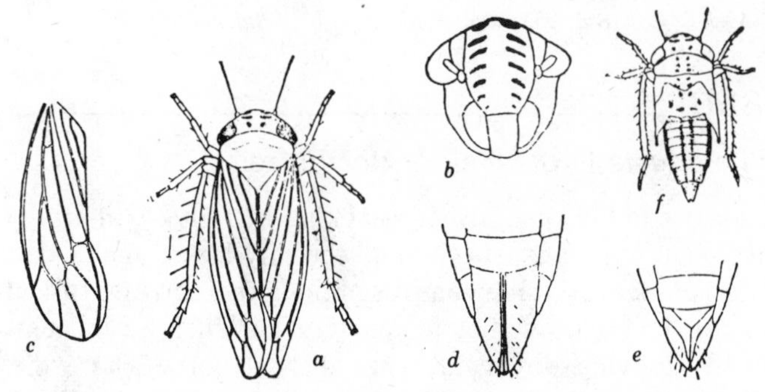

Figure 225. *Macrosteles fascifrons* (Ståhl): *a,* adult; *b,* face; *c,* wing elytron; *d,* ventral view of female abdomen; *e,* ventral view of male abdomen; *f,* nymph. All enlarged. (From Osborn, USDA)

Control of the disease aster yellows depends on control of the leafhoppers. Chemicals are usually necessary.

REFERENCES: *Me. Agr. Exp. Sta. Bul.* 248, 1916; *Ent. Soc. Wash. Memoir No. 3,* 1949; *Ohio St. Univ. Ext. Bul.* 459, 1978.

MISCELLANEOUS PESTS

False Chinch Bug, *Nysius raphanus* Howard, resembles the chinch bug for which it is often mistaken, but is slightly shorter and more slender. It is a pest of beets, potatoes, cabbage, and other crucifers and is sometimes found on grape, corn and sorghums. It is one of the more important sugar-beet pests. Although widely distributed in the United States, it causes the most damage in the semiarid regions west of the Mississippi River. The bugs increase in numbers during the early spring and summer, chiefly among weeds and other uncultivated plants. When these plants become less succulent, migration to garden crops takes place, and the removal of sap by the piercing-sucking bugs causes the leaves to wilt and die.

The adults are about 3 mm long, narrow-bodied, and gray-brown. Both nymphs and adults overwinter among plant remnants and become active early in the spring. Crescent-shaped pink eggs, laid in the soil and on foliage and flowers of low-growing plants, hatch in 4 days, and the red-brown nymphs feed for about 3 weeks before transforming to adults. There are 4 or 5 generations each year in the North.

REFERENCES: *USDA Farmers' Bul.* 762, 1916; 2168, 1965; *J. Agr. Res.* 13:571–578, 1918.

Zebra Caterpillar, *Ceramica picta* (Harris), is velvety black with prominent yellow stripes and reaches a length of 4 cm when fully grown. The moth lacks easily described markings but is of a general rusty brown color and has a wingspread of almost 5 cm. Eggs are laid in masses and the hatching larvae devour the foliage of a wide variety of flowers and vegetable garden plants in mid and late summer. Cabbage and other cruciferous plants are often damaged. Distribution of the insect is general. Spot treatment with a pesticide is generally sufficient to eliminate the caterpillars.

Vegetable Weevil, *Listroderes costirostris obliquus* (Klug), is a South American beetle that has been known in this country since 1922. It now occurs in the Gulf Coast states and in California where it attacks potatoes, tomatoes, turnips, carrots, and many other vegetable crops, as well as weeds. Both larvae and adults feed, principally at night, on the buds, foliage, and roots of such vegetables as turnips and carrots. Stems of plants may be cut off at ground level, the injury resembling that of cutworms.

The adult weevil is about 1 cm long, dull gray-brown with a pale gray mark near the posterior end of each wing cover, forming an inconspicuous V-shaped spot when the wings are at rest (Fig. 226). In common with most weevils, the adults have the habit of feigning death when disturbed. In the southern states they are active throughout the fall, winter, and spring, and estivate during the

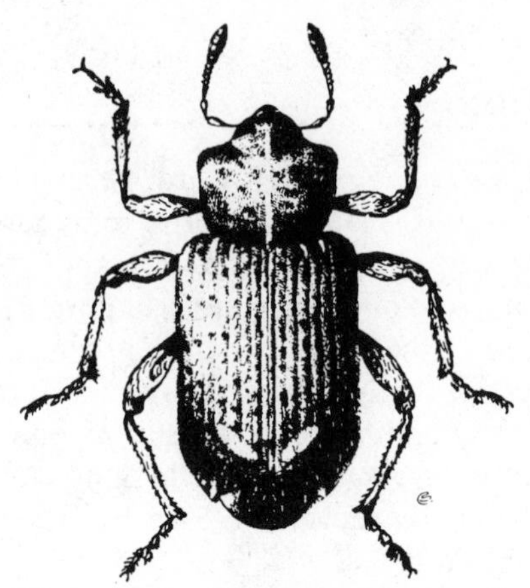

Figure 226. The vegetable weevil, *Listroderes costirostris obliquus* (Klug); 4 ×. (USDA)

summer months. Some individuals live almost 2 years. Only female weevils are known; therefore reproduction is parthenogenetic. Oviposition begins in the autumn, and the eggs are placed on the plants or soil near-by. Hatching occurs after an incubation period of 2 or more weeks and the pale green legless larvae begin feeding, becoming fully grown in 23 to 45 days. Pupation takes place in the soil, and transformation to adults may occur over a period of a few days to 2 weeks depending on the temperature. Completion of the life cycle varies from a little over a month to nearly 4 months. There is only 1 generation per year.

REFERENCES: *Calif. Agr. Exp. Sta. Bul.* 546, 1932; *USDA Cir.* 530, 1939; *Farmers' Bul.* 2168, 1965.

Pale-Striped Flea Beetle, *Systena blanda* Melsheimer, is one of the larger species. It is almost 4 mm in length, brown-black with a median cream-colored stripe on each wing cover (Fig. 227). Corn, tomatoes, peas, beans, peanuts, oats, cotton, sweet potatoes, and strawberries are some of the many hosts on which it feeds.

The life cycle has not been definitely recorded, but adults appear in the spring and continue to be numerous until midsummer, some still being found during the remainder of the summer and in the fall. Larvae injure roots and seeds, and adults eat the leaves of food plants, typical of most flea beetles. Two generations occur each year in the South.

REFERENCES: *N.Y. Agr. Exp. Sta. Memoir* 55, 1922; *Tenn. Agr. Exp. Sta. Bul.* 262, 1946; *USDA Agr. Handbook* 329, 1967.

European Earwig, *Forficula auricularia* L., a native of Europe, was first reported in Seattle, Washington, in 1907. It has since become a pest in many areas from the west to the east coast of the United States and southern Canada. The earwig is found in gardens where it eats flowers and foliage of many plants such as seedling vegetables, dahlias, carnations, chrysanthemums, lettuce, celery, marigolds, and potatoes. Another objection to this pest is its habit of entering houses.

The adult female appears as shown in Fig. 228. This stage passes the winter in the soil. White eggs are laid in underground nests during January and February, hatching occurs in April and new adults appear in July. Some of these adults enter the soil, lay eggs, and produce a second generation by August.

Since earwigs seldom fly, a favorite control method is spreading granules or dusts of diazinon or carbaryl where they are apt to crawl. Make 1 or 2 applications in late spring about a week apart and treat the soil around the foundations of houses, along walks, fences, and around trees.

REFERENCES: *USDA Tech. Bul.* 766, 1941; *Agr. Inf. Bul.* 237, 1962; *Home & Garden Bul.* 75, 1977; *Can. Dept. Agr. Pub.* 21, 1953; *J. Econ. Ent.*, 56:29–31, 1963; 62:686–689, 1969.

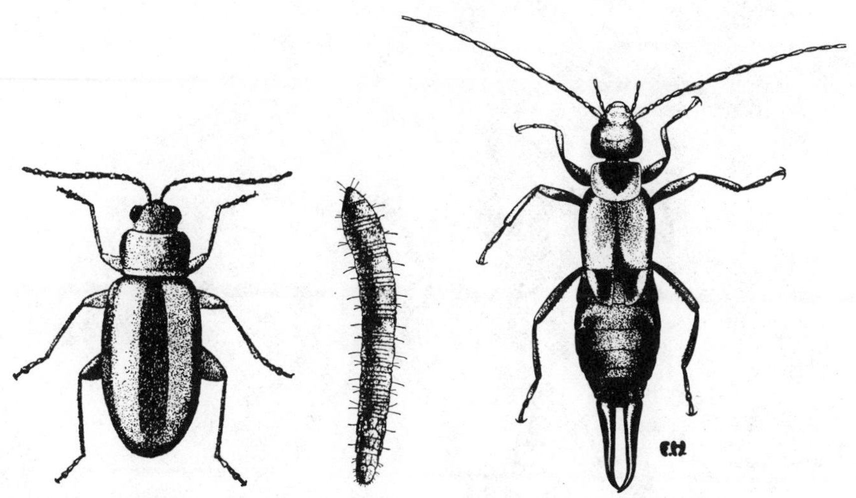

Figure 227. Adult and larva of the pale-striped flea beetle, *Systena blanda* Melsheimer; 8 ×. (USDA)

Figure 228. A female European earwig; enlarged 4 ×. (Can Dept. of Agr.)

Slugs and Snails are distributed throughout the world, frequently causing damage to glasshouse and garden plants (Fig. 229). Some slugs are especially injurious in mushroom houses. Young seedlings and the more succulent parts of plants are devoured by these pests. They leave a trail of mucus on the surfaces on which they crawl, and, on drying, silvery marks result which are objectionable, especially on floral or ornamental plants. Only very humid environments favor the development of slugs and snails.

Thirty-two species of slugs have been recorded in the United States. Those of most importance as pests are the spotted garden slug, *Limax maximus* L. (Fig. 230), the tawny garden slug, *L. flavus* L., the greenhouse slug, *Milax gagates* (Draparnaud), and the gray garden slug, *Agriolimax reticulatus* (Müller). All slugs are similar in general appearance but vary in coloration and length, depending on the species. The spotted garden slug may reach a length of 20 cm, whereas the true garden slug averages about 25 mm in length. Tawny slugs rarely attain a length of 10 cm.

Only a few of the several hundred species of snails recorded for the United States are of importance as pests. These are the brown garden snail, *Helix aspersa* Müller, the banded wood snail, *Cepaea nemoralis* (L.), the bush snail, *Zonitoides arboreus* Say, the white garden snail, *Theba pisana* (Müller), the subulina snail, *Subulina octona* (Brug.), and four species of cellar snails, *Oxychilus cellarius* (Müller), *draparnaldi* (Beck), *helveticus* (Blum), and *alliarius,* Müller. They are usually of some shade of gray, but their shells vary from

Figure 229. Garden slugs, and damage to beans. (USDA)

Figure 230. The spotted garden slug, *Limax maximus* L. About one-half natural size. (USDA)

nearly white through brown to black and are often ornamented with stripes or mottlings of contrasting colors (Fig. 231). Size varies with the species, bush snails being about 5 mm long; other species are larger.

Slugs and snails pass the winter in sheltered situations out-of-doors in colder regions but continue activity throughout the year in glasshouses or in warmer regions. Snails seem to be more winter-hardy. Most slug species overwinter as eggs. Both animals deposit eggs in moist habitats with development to maturity requiring a year or more depending on the species. Common garden snails may survive in captivity for 9 years.

Figure 231. A fully grown snail; about natural size. (USDA)

In California the rove beetle, *Ocypus olens* Müller, is a promising predator of the brown garden snail. Presumably it might attack other species. Metaldehyde, either as a bait or as a 15% dust placed where slugs crawl, has been widely used in control. It is not only toxic but also induces excessive mucus secretion which leads to death by desiccation. Methiocarb is a highly recommended effective chemical. For small gardens a nonhazardous material is beer poured into shallow pans and placed at ground level in the evening; it attracts slugs and they fall into the liquid and are killed. Under no-tillage or reduced-tillage crop management, slugs are now becoming a most serious problem.

REFERENCES: *Conn. Agr. Exp. Sta. Cir.* 203, 1958; *USDA Farmers' Bul.* 1895, 1953; *Agr. Inf. Bul.* 380, 1975; *Pflanzenschutz-Nachrichten* 22(2) 205–243, 1969; *Calif. Agr.* 30(3) 20–21, 1976; *Bul. Ent. Soc. Amer.* 22:302–304, 1976; *N.Y. Food and Life Sci.* 5 (3):16–19, 1972.

Garden Symphylan, *Scutigerella immaculata* (Newport), or centipede, is not a true insect but a near relative. It is generally distributed and feeds on the roots of many plants both in the field and in glasshouses.

Adults are white narrow-bodied creatures, almost 1 cm in length, with a pair of long antennae and 12 pairs of short legs (Fig. 232). Egg-laying begins in the spring and continues throughout the summer. Masses containing up to 20 eggs are deposited and hatching takes place in about 10 days. The symphylans molt six times and are fully developed in 45 to 60 days, depending on temperature,

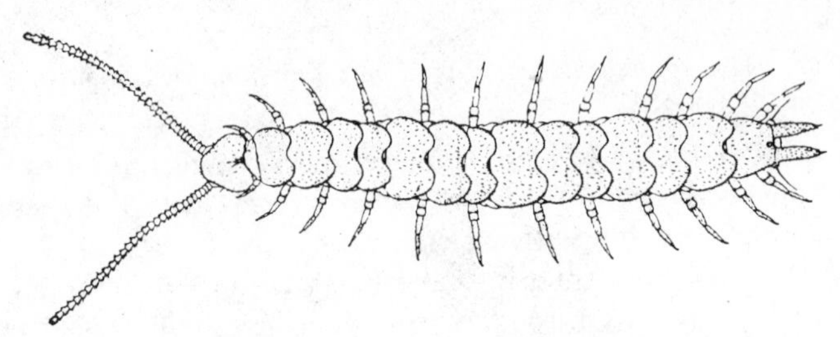

Figure 232. Adult garden symphylan; 6 ×. (Drawing by A. E. Michelbacher.)

with all life stages occurring in the soil. During very cold or hot dry weather they migrate into the subsoil. Life span is said to be 4 to 5 years.

Where irrigation water is available, continuous submergence of fields for 2 weeks in the summer or 1 month in the winter will rid the soil of symphylans for 2 years. Fumigation or steam sterilization of soils have also been employed with success.

REFERENCES: *Ohio Agr. Exp. Sta. Bul.* 486, 1931; *Conn. Agr. Exp. Sta. Cir.* 203, 1958; *USDA Agr. Handbook* 264, 1964; *J. Econ. Ent.*, 48:246–250, 1955; 52:666–683, 1959; 57:360–363, 525–527, 1964; *Can. Ent.* 99:696–702, 1967; 100:772–778, 1968; *Hilgardia* 11:1–248, 1938.

Millipedes eat the roots of various plants, including garden and glasshouse vegetables and flowers. They are usually abundant in moist soils high in organic matter. They resemble but can be distinguished from wireworms by their many body segments and numerous legs. Millipedes have 2 pairs of legs per segment except for the first 6, which have only 1 pair per segment, and wireworms have 3 pairs of legs located on the 3 segments posterior to the head. Life cycles of millipedes have not been thoroughly studied. It is known that their eggs are laid on or in the soil, usually in clusters of 20 to 100. These hatch in about 3 weeks, and the young feed on decaying vegetable matter and grow rather slowly. Adult millipedes are said to overwinter in the soil and other secluded places that are moist (see Fig. 16).

Nematodes of various species attack a wide variety of garden and glasshouse plants. Probably the most destructive species feed on the roots or tubers, causing stunting or development of knots or galls, an example being the root-knot nematodes. Some species are leaf or bud feeders and are called foliar nematodes; still others are known as cyst nematodes, the sugar beet cyst and soybean cyst nematodes being examples. These pests are not insects but tiny roundworms. The root-knot species may complete their life cycle in one month. Control is difficult, but sanitation measures, resistant varieties, crop rotation, fumigation, or steam sterilization of soils are the commonly recommended practices. Foliar nematodes are controlled by sprays of demeton.

Soil fumigation can be done at any time, but the treatment appears to be more effective in the fall when most nematodes are nearer the soil surface. Application of the fumigant and fall plowing can be combined into one operation, thus saving labor costs. Chemicals that have been used in nematode control are D-D mixture, Nemagon, ethylene dibromide, CH_3Br, chloropicrin, Vapam, Nemacide, Telone, Zinophos, Temik, and Furadan.

REFERENCES: *USDA Agr. Inf. Bul.* 380, 1975; *Ohio Agr. Exp. Bul.* 451, 1930; *Tenn. Agr. Exp. Sta. Cir.* 54, 1935; *Ariz. Agr. Exp. Sta. Bul.* 212, 1948; *Calif. Agr.* 29(12):6–7, 1975; 30(1):4–7, 1976; *Ohio Sta. Univ. Ext. Bul.* 538, 1977; *Conn. Agr. Exp. Sta. Bul.* 768, 1977.

17

PESTS OF TREES AND ORNAMENTAL PLANTS

Fruit growing is a very important industry in the United States and Canada, and in many areas of these countries it is impossible to produce a quality product unless both insects and diseases are controlled. In order to do this as economically as possible, both insecticides and fungicides are combined into a spray program. Such a program must be developed for each region, must include compatible chemicals that are effective in control, and must be the least hazardous from the standpoint of application and residues on the marketed product. Since many of these programs are complex and are prepared primarily for the experienced commercial grower many persons who normally enjoy growing small quantities of different fruits as a hobby are often denied this pleasure. Some all-purpose spray formulas have been developed for the amateur fruit grower that are relatively safe and effective if thoroughly applied. The following is a recommended mixture*:

Materials		Amount Used for	
		100 gal.	10 gal.
Methoxychlor	50% wettable powder	2½ lb.	4 oz.
Malathion	25% wettable powder	3 lb.	5 oz.
Captan		2 lb.	3 oz.

These ingredients may be purchased separately and mixed before application. Commercially blended mixtures may also be available.

This formula is recommended for peaches, plums, sweet and sour cherries, grapes, red raspberries, blackberries, pears, quinces, and apples. Proper timing of the applications depends on the insects and diseases present in a given re-

* C. R. Cutright, *Ohio Farm and Home Research,* 39:26–27, 1954.

gion. To determine the insects present and to learn the timing of spray applications for their control, reference should be made to this and the following five chapters. Generally, the first spray should be applied on stone fruits and berries when 95 to 100% of the peach petals have fallen, and should be repeated in 10 days. After this, 2 more applications should be made at 2-week intervals. Apples, cherries, pears, and quinces may be in full bloom when the first spray is needed on other fruits; in that event they should not be sprayed until their petals have dropped. Do not spray when any of the fruits are in blossom.

ARMORED SCALES

SAN JOSE SCALE

Quadraspidiotus perniciosus (Comstock), Family Diaspididae

The San Jose scale was discovered in this country in the vicinity of San Jose, California, before 1880. It was brought to this country from Japan, but its original habitat seems to have been northern China. From California it was transported to the Atlantic seaboard on nursery stock, and from centers of infestation thus established it spread rapidly and now may be found throughout most of the United States and in Canada. In recent years it has not been a serious problem because of natural enemies and effective pesticides properly applied.

San Jose scale attacks most cultivated fruits and a large number of ornamental shrubs and trees. Osage orange is often heavily infested and serves as a reservoir for reinfestation. Several forest trees maintain the insect but injury is usually not serious. Vigor is reduced and fruits are blemished by the nymphs and adults removing sap from any part of the plant but especially the wood. Diagnosis can be easily made by the encrusted scales on the branches and the tiny red circles with white centers on apple and pear fruits. Unchecked heavy infestation may kill trees.

The fully grown female scale is circular, about the size of a pinhead, dark brown to black, with a raised dull yellow center (Fig. 233). Young scales are very light colored but soon become sooty black, often ashy in appearance. The male scales are smaller than the females and of oval form. Near the center of the scale cover is a depressed ring surrounding a raised center. Where this shows clearly it serves as a convenient means of identification.

Winter is passed in the partly grown nymphal stage under the scale coverings on the host plants. Development continues in the spring, and maturity is reached usually in May or June when minute, 2-winged adult males (Fig. 234) appear which mate with the females under the edge of their scale coverings and die soon afterwards. The female produces young which crawl from under the edge of her covering. Newly born lemon-yellow nymphs resemble mites, but they can be distinguished from them by the presence of 3 pairs of legs and a pair of antennae. They are called "crawlers" and migrate over the host for a few hours before inserting their mouthparts into the bark, leaves, or fruit to feed. After the first molt the legs and antennae are shed; these are incorporated in the scalelike covering that increases in size as the nymphs continue to grow. Underneath these scalelike coverings may be found the yellow nearly circular saclike bodies of the insects (Fig. 235). Growth is completed in about 6 weeks, and 2 or more overlapping generations are produced each season.

Local spread is accomplished by the crawlers. Many of these are doubtless carried from place to place on the feet of birds and on other insects. Some may be carried by the wind. Long-distance dispersal is largely through transportation of infested plants by man.

Natural control of this scale is exercised partly through climatic factors, extreme cold being fatal. Several parasitic eulophid wasps, principally *Aphytis mytilaspidis* (LeBaron), *A. proclia* (Walker), *Prospaltella aurantii* (Howard) (Fig. 236), *P. perniciosi* Tower, *Aspidiotiphagus citrinus citrinus* (Craw), and *Coccophagus lycimnia* (Walker), as well as predatory lady beetles and mites, are very important in keeping populations checked.

An early effective pesticide was dormant-strength liquid lime-sulfur, and later wettable sulfur, applied in the spring before the buds burst. Dormant sprays containing 2% superior-type emulsifiable oils have largely replaced lime-sulfur for scale control because they are more effective and kill other insect life stages as well. Superior oil plus ethion is recommended in the delayed dormant period in some regions. Other organic phosphorous compounds are useful in control of this scale.

REFERENCES: *Farmers' Bul.* 650, 1915; *Cir.* 263, 1923; *Ohio Sta. Univ. Ext. Bul.* 506, 1977.

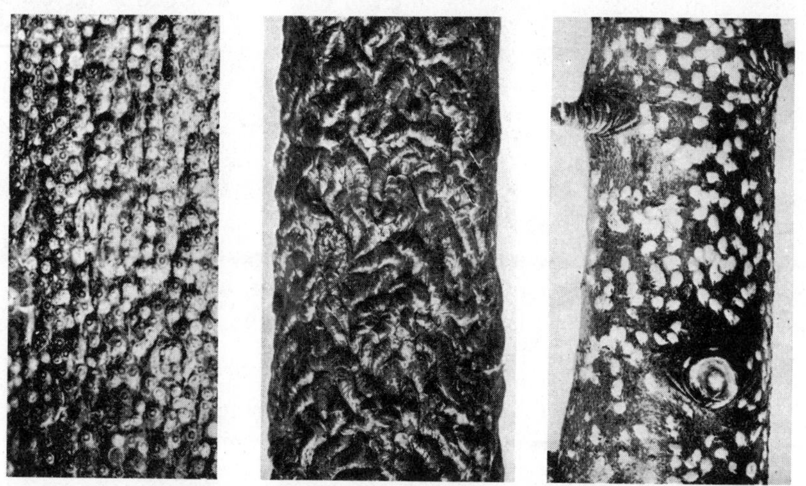

Figure 233. San Jose scale (Left), oystershell scale (center), scurfy scale (right); about natural size. (Pettit, Mich. Agr. Exp. Sta.)

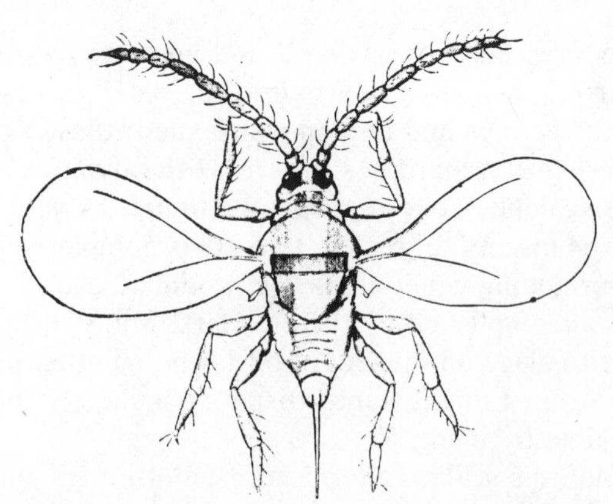

Figure 234. Adult male San Jose scale, greatly enlarged. (Howard and Marlatt.)

OYSTERSHELL SCALE

Lepidosaphes ulmi (L.), Family Diaspididae

World-wide in distribution oystershell scale is a pest of apple, pear, occasionally other fruit trees, and many shade and ornamental plants. Lilac and ash trees are often seriously attacked. Heavy infestations of these piercing-sucking insects greatly reduce plant vigor and often kill the host if no control measures are initiated.

The scale covering is brown, sometimes appearing gray, and shaped somewhat like oystershells (Fig. 233). The covering of the female is nearly 3 mm long when completed, narrow at the anterior end and widened at the rounded posterior end. The covering of the males is smaller and more oval-shaped.

Winter is passed as minute white eggs beneath the scale of the female insect which produced them. When spring is well advanced the eggs hatch, and the tiny nymphs crawl out and migrate over the plant; in a few hours they settle down, insert their mouthparts, and begin sucking sap. At the first ecdysis the legs and antennae are shed, and the molt skin is incorporated in the scale covering formed over the body. By mid-August nymphal development is completed; the 2-winged males emerge and mate with the females, which soon deposit the overwintering eggs. The shrivelled body of the female may be found beneath the anterior end of the scale. One generation is produced each year in the North, and 2 in the southern half of the country.

Natural control results mainly from the same parasitic wasps, predatory lady beetles, and mites as listed under San Jose scale. Where feasible, pruning infested plant parts and destroying them in the winter will contribute to control.

REFERENCES: *Ohio Biol. Surv. Bul.*, Vol. II, No. 2, 1963; *Ohio Agr. Exp. Sta. Bul.* 332, 1918; *USDA Farmers' Bul.* 723, 1916; 1270, 1922; *Ohio Sta. Univ. Ext. Bul.* 506, 1977.

SCURFY SCALE

Chionaspis furfura (Fitch), Family Diaspididae

This native insect is widely distributed in North America. It may attack a great number of shade trees, bush fruits, apple, pear, and quince, but it is not considered a serious pest. Damage from these piercing-sucking insects results in blemished fruits, reduced vigor of the plant, and sometimes death of heavily infested branches.

Except for the brown anterior tip, the scale covering of the female is white in sheltered situations, gray where exposed, more or less pear-shaped, and nearly 2.5 mm in length (Fig. 237). The males are also white, much smaller, and parallel-sided with 3 dorsal ridges. The insect beneath the scale is yellow and rounded or elongate according to the sex.

Winter is passed as red-purple eggs under the scale coverings of females. Hatching takes place in April or May depending on the latitude, and in a few hours the crawlers settle and begin to feed. Like other armored scales, they lose their legs and antennae after the first molt and soon secrete the scale covering. When fully grown the 2-winged males emerge, mate with the females, and die shortly afterwards. The females deposit their eggs under the scale coverings and then die. In the more northern range there is only 1 generation each year, but in North Carolina there are 2 complete generations.

Many of the parasites and predators that attack San Jose and oystershell scales will check scurfy scale. Usually this insect is not a problem in orchards.

REFERENCES: *J. Econ. Ent.*, 46:969–972, 1953; *Va. Poly. Inst. Tech. Bul.* 119, 1952.

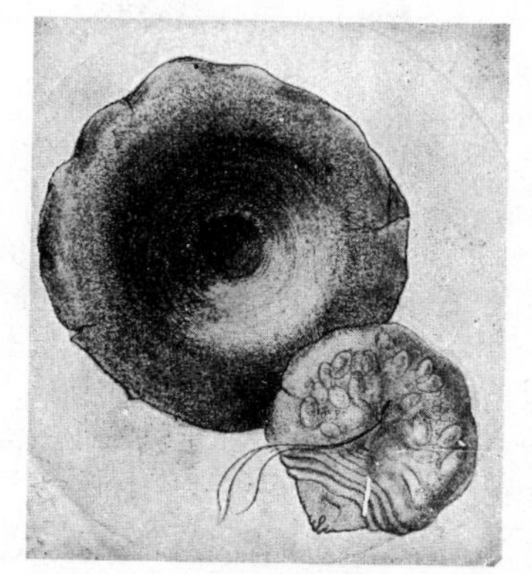

Figure 235. Adult female San Jose scale, with scale removed to expose the insect; greatly enlarged. (Alwood.)

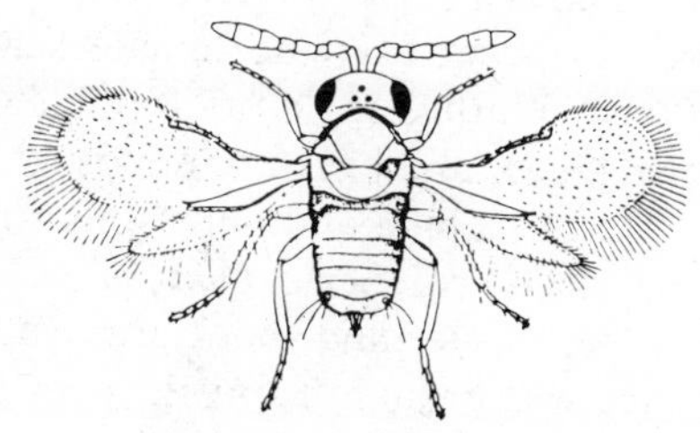

Figure 236. *Prospaltella aurantii* (Howard), a minute parasite of scale insects; greatly enlarged. (USDA)

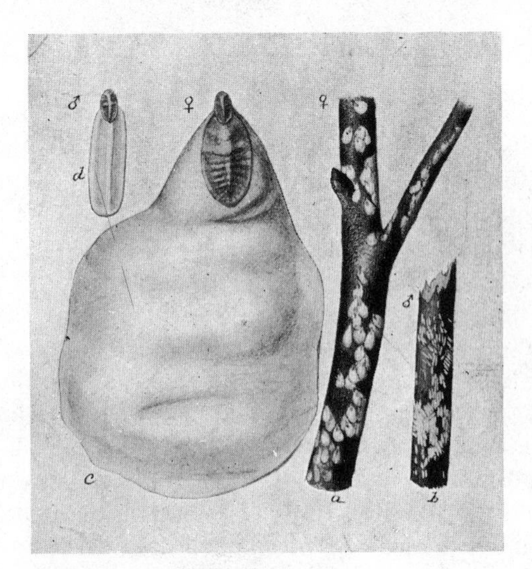

Figure 237. The scurfy scale: *a, c,* females; *b, d,* males; *c, d,* greatly enlarged. (Howard, USDA)

OTHER ARMORED SCALES*

ORDER HOMOPTERA, FAMILY DIASPIDIDAE

Forbes Scale, *Quadraspidiotus forbesi* (Johnson), attacks cherry, apple, pear, peach, plum, quince, and currants. It is similar to San Jose scale but is a trifle larger and brown instead of dull black in the half-grown stage. The raised central point or nipple of the mature scale cover is orange instead of pale yellow, characteristic of San Jose scale. It winters as a partly grown scale in the North and as an adult female in the South. In the North reproduction is said to be both oviparous and ovoviviparous; nymphs appear in May and June. In North Carolina the young are born starting the last of May. There are 1 to 3 generations each year depending on the locality.

Putnam Scale, *Diaspidiotus ancylus* (Putnam), resembles San Jose scale very closely, except that the female scale covering is a trifle larger and slightly darker, with the nipple brick red and off center. Winter is passed in the partly grown stage, and eggs are deposited in late spring. There is only 1 generation per season. Apple and numerous shade and forest trees may be attacked.

European Fruit Scale, *Quadraspidiotus ostreaeformis* (Curtis), is very similar to Forbes scale, and the 2 are usually confused. As the name suggests, this species was brought from Europe. It has been established in many localities, mainly in the northeastern parts of the country.

* Kosztarab, M. "The armored scale insects of Ohio," *Ohio Biol. Surv. Bul.*, N.S. Vol. II, No. 2, 1963.

Walnut Scale, *Quadraspidiotus juglansregiae* Comstock, is rather widely distributed and attacks many kinds of fruit and shade trees as well as ornamental plants. It is more abundant in the South than in northern regions. It is slightly larger than San Jose and Putnam scales, which it closely resembles in shape. Winter is passed as adult females beneath the gray-brown scale covering. The eggs are deposited in the spring. There are usually 2 generations each year.

Greedy Scale, *Hemiberlesia rapax* (Comstock), is a common species that attacks a wide range of hosts including citrus, deciduous fruits, and many ornamental trees and shrubs. It is generally more troublesome in Florida and California.

Oleander Scale, *Aspidiotus nerii* Bouché, also called ivy scale, encrusts the leaves of English ivy, oleander, palms, and other semitropical plants. It is found in glasshouses in the North. The scale coverings are usually circular, the females being light buff with a tinge of purple and about three times the size of the pure white males.

Obscure Scale, *Melanaspis obscura* (Comstock), is a serious pest of pecans in the South. It also occurs on forest and shade trees in other parts of the Pecan Belt. The scales are somewhat circular and grayish; they occur on the branches, and under heavy infestations those less than 3 inches in diameter are killed. One generation occurs each year and the crawler stage is present from May until early August. A close relative that is common on red maple trees is **Gloomy Scale,** *M. tenebricosa* (Comstock).

Euonymus Scale, *Unaspis euonymi* (Comstock), attacks bittersweet, lilac, orange, and pachysandra, in addition to euonymus. The female scale resembles the female oystershell scale, and the male is white and narrow like the male of scurfy scale. Winter is passed as fully grown females under the scale covering. Pale yellow crawlers appear in May and June. There is a second and sometimes a third generation in some areas each season.

White Peach Scale, *Pseudaulacaspis pentagona* (Targ.), has been found in a number of localities but chiefly in the South. On peach trees it is sometimes as injurious as San Jose scale. It also attacks other stone fruits and is often found on privet, lilac, and other shrubs. The females are circular and white tinged with brown. The males, which are usually found in groups, are elongated and pure white. There are 4 or 5 generations annually in the South.

Tea Scale, *Fiorinia theae* Green, is a common species on camellia and tea. It is more prevalent on the undersides of leaves. The female scales are about 2 mm long, elongate oval, and dark brown to black. The males are snow white with a ridge down the middle; they give the undersides of the leaves a frosty appearance. The life cycle from egg to adult is completed in 60 to 70 days and overlapping generations occur from March to November in most areas. Reproduction is continuous in glasshouses.

Camellia Scale, *Lepidosaphes camelliae* Hoke, is an occasional pest of that host. The females of this species resemble oystershell scale, being somewhat pear-shaped, flattened, brown, and about 3 mm long. Overlapping generations occur outdoors from March to November and 60 to 70 days are required to complete a generation.

Peony Scale, *Pseudaonidia paeoniae* Cockerell, is a common pest of azaleas in the South. It also attacks camellia and a few other woody plants. On azalea the thin bark usually grows over the insects leaving small bumps on the bark. If these bumps or swellings are opened the circular convex gray-brown scale, about 2 mm in diameter, may be found. When the insect is removed a thin layer of whitish wax usually remains, giving the appearance of a white scar. The young scales attach to the bark and remain exposed about 4 weeks before the bark covers them. This is the critical period for successful control with chemicals. Winter is passed as eggs under the female scales. One generation develops each year.

Juniper Scale, *Carulaspis juniperi* (Bouché), is a major pest of junipers but arborvitae, incense cedar, and cypress may be attacked. Winter is passed as fertilized females that mature by May and lay eggs with crawlers appearing in June. The scale covering the female is white, round, convex, with a yellow center, and about 2 mm in diameter; the male scale is also white but smaller and narrow with a ridge down the center. One generation develops each year.

Pine Needle Scale, *Chionaspis pinifoliae* (Fitch), is troublesome on ornamental plantings of most species of pine and spruce, and may attack Douglas fir and hemlock. Pink eggs under old female scales survive the winter and hatch into crawlers in May. These develop scale coverings that are white and 3 mm long for females (Fig. 30) and about half this size for males. Maturity is reached by July and a second generation develops that matures by autumn and lays the overwintering eggs.

All these armored scale species have many natural enemies in addition to the parasites and predators listed under San Jose scale. Sanitation measures are of value in control, and chemicals are more effective when crawler stages are present.

UNARMORED SCALES

TERRAPIN SCALE

Lecanium nigrofasciatum Pergande, Family Coccidae

Terrapin scale is a soft or unarmored species; the protective covering is the thickened body wall itself rather than a separate structure. Widely distributed over eastern United States and southern Canada, it is one of the largest native scales commonly found in orchards; many shade trees are also hosts of this species.

Adults are dark red-brown, sometimes with black banding and mottling,

Figure 238. Terrapin scale, *Lecanium nigrofasciatum* Pergande; 6×. (Courtesy of M. Kosztarab and VPI)

2 mm in length, very convex, hemispherical, and fluted or crimped near the edges of the body (Fig. 238). Early instars are much more flattened and lighter in color.

Only fertilized ovoviviparous females survive the winter. Crawlers appear in spring and migrate to the leaves where they suck sap for nearly 6 weeks. The females then return to the twigs and branches where they continue their feeding and growth. The tiny 2-winged males are rarely seen but appear in late August and early September, effect fertilization, and die. The females continue feeding until cold weather; only 1 generation occurs each year. During feeding, especially in the later period, the scales excrete quantities of honeydew, which covers foliage and fruit, collects dust, and serves as a medium for the growth of sooty fungus. The presence of honeydew and attendant fungus growth on fruits greatly reduces its value.

Terrapin scale is partly controlled by its many natural enemies, some of them of unusual interest. One is the larva of a pyralid moth, *Laetilia coccidivora* Comstock, which feeds as a predator on this and other scales. Another predator is the lady beetle, *Hyperaspus binotata* Say, the larvae of which live for a time in the cavity underneath the upper covering of the scale insect, later attacking and killing the developing young. Other species of lady beetles, aphidlions, and hemerobiids are also active predators. The most important parasites include *Aphycus stomachosus* Girault and several species of *Coccophagus,* especially *C. lycimnia* (Walker).

REFERENCES: *Conn. Agr. Exp. Sta. Bul.* 575, 1953; *USDA Farmers' Bul.* 1861, 1954. *Ohio Sta. Univ. Ext. Bul.* 504, 1977; 619, 1977.

OTHER UNARMORED SCALES*

ORDER HOMOPTERA, FAMILY COCCIDAE

European Fruit Lecanium, *Lecanium corni* Bouché, is widely distributed over the entire United States and southern Canada. It attacks numerous shade and

* M. L. Williams and M. Kosztarab, "Coccidae of Virginia," *Res. Bul.* 74, Va. Poly. Inst. and Sta. Univ., 1972.

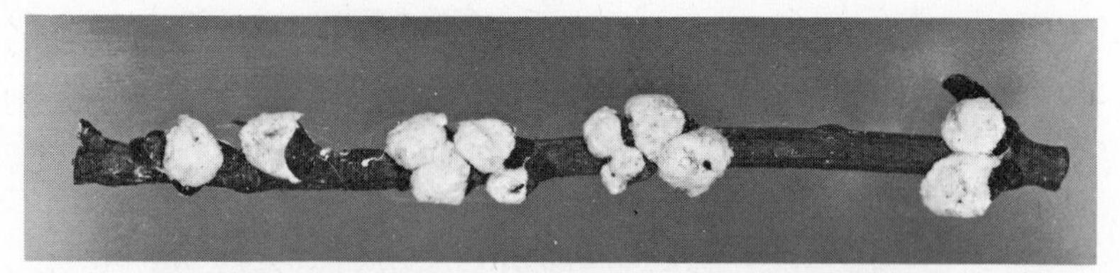

Figure 239. Cottony maple scale. (Original.)

forest trees as well as many fruit trees. The female is oviparous, somewhat larger than terrapin scale, and overwinters as a nymph. Maturity is reached in early spring with white eggs being laid in mid-May and hatching occurring in June. Crawlers migrate to the leaves and feed, and by late summer move back to the bark of branches where they complete their growth by spring. One or 2 generations develop each year depending on latitude. On some hosts the scale is covered by a white powder.

Fletcher Scale, *Lecanium fletcheri* Cockerell, is also widely distributed in the United States and southern Canada. Sometimes called the *Taxus* lecanium, it is often a pest of yew and it also occurs on arborvitae and juniper. Sooty fungus often develops on the copious quantities of honeydew excretions. The mature female is yellowish-brown, hemispherical, and 4 mm in diameter. Eggs are laid in late spring and crawlers appear by July, feeding first on the foliage and later on the bark. Winter is passed as a nymph and 1 generation develops each year. This species may be synonymous with *L. corni*.

Cottony Peach Scale, *Pulvinaria amygdali* Cockerell, has been found as a pest in orchards in New York and Ontario; it apparently is of little importance elsewhere. It is a large species resembling those occurring on maple and grape. The eggs are deposited in a large white cottony mass which protrudes from under the posterior part of the female scale insect. These eggs hatch into the crawler stage just before midsummer.

Cottony Maple Scale, *Pulvinaria innumerabilis* (Rathvon), is a common and quite conspicuous pest of soft maple (Fig. 239). At times it is found on other kinds of maples as well as a variety of other trees and shrubs. Heavy infestations excrete great quantities of honeydew, contaminating the foliage and objects below. Sooty fungus often develops on the honeydew covered leaves and branches.

Partly grown, brown, oval, female scales overwinter on the twigs and branches. Growth is rapid in the spring; this is followed by egg-laying in large cottony masses resembling popped corn. Hatching occurs in June and July, and the crawlers migrate to the leaves and suck sap. Maturity is reached in August; after mating the females crawl back to the twigs. Only one generation develops each year. Superior type oil emulsions in the dormant period and malathion or parathion summer sprays applied as the eggs hatch are the commonly recommended control chemicals.

Tuliptree Scale, *Toumeyella liriodendri* (Gmeln), is hemispherical and gray-green mottled with black, the females being nearly 8 mm in diameter. This scale is primarily a pest of tulip poplar but has been recorded on magnolia and yellow poplar. Second instar nymphs overwinter on the bark and are fully grown in August when the ovoviviparous females give birth to the crawlers; some females produce over 3000 during their life span. One generation develops each year.

Pine Tortoise Scale, *Toumeyella parvicornis* (Cockerell), is primarily a pest of Scotch, Austrian, red, and mugho pines. Females are red-brown, oval, and very convex; they are 6 mm long when fully grown. Immature females survive the winter, reach maturity, and begin egg-laying in June. These eggs soon hatch into crawlers that develop throughout the year, there being 1 generation annually in the North and sometimes 2 in the South.

Magnolia Scale, *Neolecanium cornuparvum* (Thro), is primarily a pest of magnolia spp. and is one of the largest scale insects occurring in North America. Mature females are 12.5 mm in length, elliptical, convex, smooth, dark shiny brown, and covered with a white waxy bloom. Winter is passed as first instar nymphs on the newer wood; these become fully grown by August and the ovoviviparous females give birth to crawlers, mostly through September. One generation develops each year. Heavily infested trees are not only weakened but are rendered unsightly by sooty fungus developing in the copious quantities of honeydew excretions.

OTHER SCALES

ORDER HOMOPTERA, FAMILY ERIOCOCCIDAE

Azalea Bark Scale, *Eriococcus azaleae* Comstock, has the general appearance of a mealybug because of the cottony masses found on the stems, especially in the axils of the plant and in the matted filaments covering the egg sac and the spent female. This insect winters mainly as a nymph in the South and as eggs in the North.

European Elm Scale, *Gossyparia spuria* (Modeer), is destructive at times to all species of elms. The adult female is oval, red-brown about 4 mm long, with a white waxy fringe along the body margin. Eggs are deposited in late spring and young begin appearing in June. Most of the young become established on the lower leaf surfaces but later migrate to the trunk and branches where they live during the winter as nymphs. One generation develops each year.

REFERENCES: *Ala. Agr. Exp. Sta. Cir.* 84, 1940; *USDA Misc. Pub.* 626, 1948; 1175, 1972; *Univ. Md. Ext. Bul.* 154, 1955; *Ohio Agr. Res. and Dev. Ctr. Res. Bul.* 983, 1966; *Ohio Sta. Univ. Ext. Bul.* 504, 1977; *Ill. Nat. Hist. Surv. Cir.* 47, 1958.

PERIODICAL CICADAS

Magicicada spp., Family Cicadidae

Periodical cicadas are one of the best-known insects within their range of distribution. They closely resemble the common or annual species, some of which are seen every year in July and August. The annual species are nearly 6 cm long with brown-black bodies often ornamented with green and with green wing veins. Their life cycles are not accurately known. Periodical cicadas are 3 cm in length, black-bodied, with orange wing veins and bright red eyes, and occur in abundance. There are 6 species of periodical cicadas in the United States, 3 with a 17-year cycle and 3 with a 13-year cycle. They are listed with distinguishing characteristics in the following table taken from Alexander and Moore, *Univ. of Michigan Museum of Zoology Misc. Pub.* 121, 1962:

Characteristics	17-Year Cycle	13-Year Cycle
Body length, 27–33 mm Propleura and lateral extensions of pronotum between eyes and wing bases red Abdominal sterna primarily red-brown or yellow Song: "Phaaaaaroah," a low buzz, 1–3 sec in length, with a drop in pitch at the end	Linnaeus' 17-year cicada, *M. septéndecim* (L.)	Riley's 13-year cicada, *M. trédecim* Walsh and Riley
Body length, 20–28 mm Propleura and lateral extensions of pronotum between eyes and wing bases black Abdominal sterna all black, or a few with a narrow band of red-brown or yellow on apical third, this band often constricted or interrupted medially Last tarsal segment with apical half or more black Song: 2–3 sec of ticks alternating with 1–3 sec buzzes that rise and then fall in pitch and intensity	Cassin's 17-year cicada, *M. cássini* (Fisher)	Cassin's 13-year cicada, *M. tredecássini* Alexander and Moore

Body length, 19–27 mm Propleura and lateral extensions of pronotum between eyes and wing bases black Abdominal sterna black basally, with a broad apical band of red-yellow or brown on posterior half of each sternum, this band not interrupted medially Last tarsal segment entirely brown or yellow, or at most, the apical third black Song: 20–40 short high-pitched phrases, each like a short buzz and tick delivered together, at the rate of 3–5 per sec, the final phrases shorter and lacking the short buzz	The Little 17-year cicada: *M. septendécula* Alexander and Moore	The Little 13-year cicada: *M. tredécula* Alexander and Moore

Damage is caused by the females ovipositing in the twigs and branches of trees and shrubs. Branches having a diameter slightly larger than that of a pencil are most often attacked. A series of these oviposition wounds, often 3 to 4 inches in length, results in the weakening of twigs that may break off or, if this does not happen, are permanently scarred and abnormal (Fig. 240). Addi-

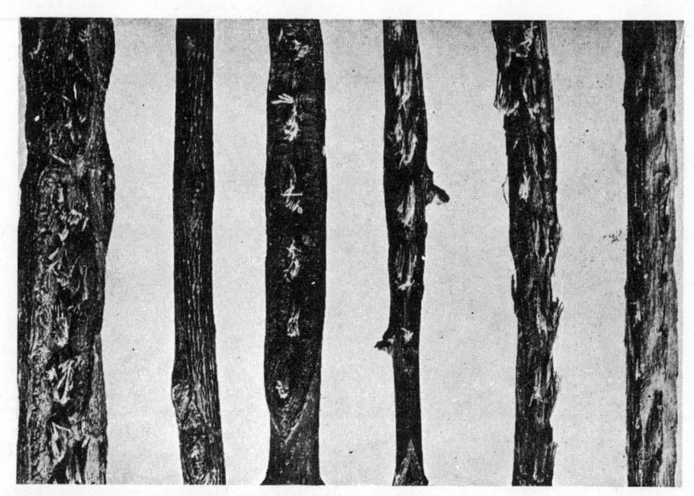

Figure 240. Typical cicada oviposition injury. (W. E. Rumsey.)

tional injury is caused by the removal of plant sap from the tree roots by the developing nymphal cicadas in the soil.

Adult periodical cicadas begin emerging from the soil, usually late in May, and continue activity through a period of about 6 weeks. Individuals live almost a month, during which time very little nourishment is taken. Egg-laying begins approximately 2 weeks after emergence, and each female may deposit over 500 eggs. These are placed in the twigs in groups of 15 or more. Hatching takes place nearly 7 weeks later, and the nymphs drop to the ground, enter the soil, and begin feeding on root sap. This feeding continues for 17 years in the North and 13 years in the South. At the end of this prolonged growth period the nymphs (Fig. 241) make their way to the surface and emerge through holes almost 13 mm in diameter.

In moist soil the nymphs, before emerging, construct mud tubes extending up to 7 cm above the ground level. Great numbers emerge at the same time, starting at dusk. They crawl up on tree trunks or other objects in the vicinity and anchor their bodies. Several hours later the new adults emerge, expand their wings, and begin the short period of adult existence. The familiar song or call of the cicada is made only by the males and is produced by a pair of drumlike organs on the basal segments of the abdomen.

There are several broods of the periodical cicada. For convenience they have been designated by Roman numerals. The numerals I through XVII are assigned to the 17-year broods, and XVIII through XXX to the 13-year broods. The largest of all is brood X, which last occurred in 1970 and is due to appear again in 1987. Many broods overlap in distribution. Those requiring only 13 years for development are found primarily in southeastern United States, but

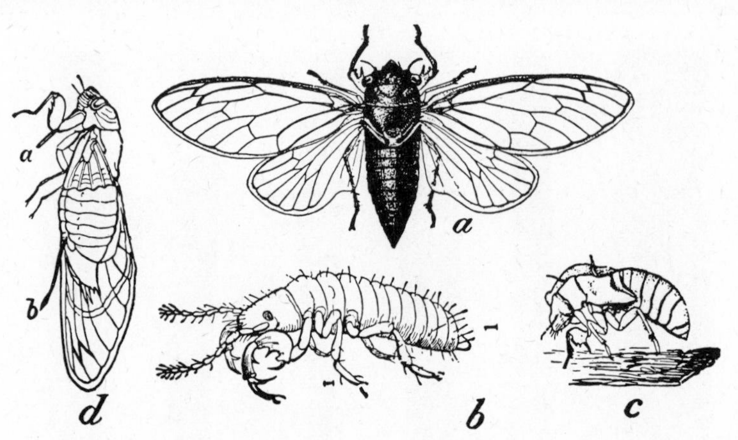

Figure 241. Stages of the periodical cicada, *Magicicada septendecim* (L): *a,* adult; *b,* young nymph, enlarged; *c,* cast skin of full-grown nymph; *d,* adult female, showing ovipositor at *b,* and beak at *a* (natural size). (Marlatt.)

one brood extends into Illinois. The latter is generally distributed in the South and will appear again in 1985.

When a large brood of cicadas is due to appear in any area it is well to defer planting young trees and shrubs until after this period is passed. It is also well to postpone pruning trees during the year before an outbreak so that injury may be distributed on a maximum number of twigs. The worst of these may be removed by pruning the following winter, and enough may escape serious injury so that the tree will not be permanently deformed. Where feasible, valuable trees and shrubs may be protected by covering them with cheesecloth or similar material during the oviposition period.

Natural enemies include birds, a fungus disease that kills some adults, and some insects and mites that attack the eggs.

REFERENCES: *USDA Bul.* 71, 1907; *Leaflet* 540, 1971; *Ohio Agr. Exp. Sta. Bul.* 311, 1917; *J. Econ. Ent.*, 42:359–362, 1949; 46:385, 1953; 50:713–715, 1957; 57:295, 1964.

TREEHOPPERS

ORDER HOMOPTERA, FAMILY MEMBRACIDAE

Treehoppers of many species attack fruit and ornamental trees, as well as other woody plants. Some of those most frequently mentioned are buffalo treehopper, *Stictocephala bisonia* (Kopp & Yonke) (Fig. 242); green clover treehopper, *Stictocephala inermis* F.; dark-colored treehopper, *Ceresa basalis* Walker; quince treehopper, *Glossonotus crataegi* (Fitch); and twomarked treehopper, *Enchenopa binotata* (Say). There are doubtless more important species found in other localities. The range of treehoppers is throughout the country, but certain species may be localized in given regions.

Injury is due almost entirely to the twig punctures that the females make in ovipositing. These may be severe to young trees and to new growth in older

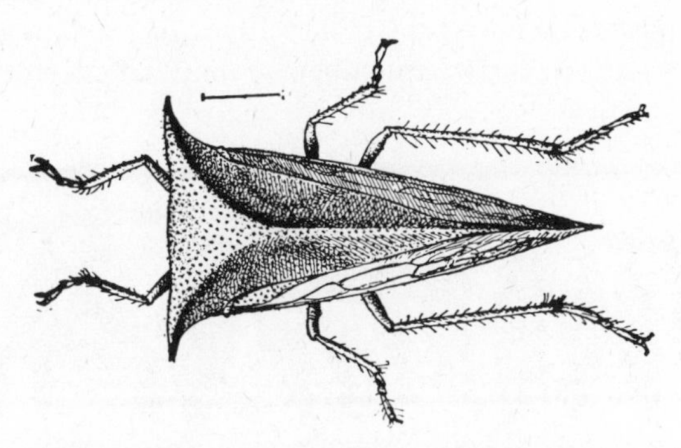

Figure 242. The buffalo treehopper; 6×. (USDA)

trees. Treehoppers, especially the nymphs, also cause damage to clovers, alfalfa, corn, and other plants by sucking sap. They are most likely to be injurious to trees where the orchard is adjacent to a legume crop or has a legume ground cover.

Since most treehoppers are considered secondary pests and since the injury from various species is similar in character, no attempt will be made to give separate life cycles. The winter is passed as eggs in the twigs from which nymphs hatch in April or May. These drop to the ground and feed on the cover crop, becoming adults in July. Oviposition takes place in August, usually in the current and previous season's wood.

Injury can be largely prevented by clean cultivation, but this is often undesirable especially in older orchards. Avoiding cover crops of alfalfa or clovers is also a desirable practice where treehoppers are causing damage. A thorough application of 4% dormant oil kills most of the eggs in twigs.

REFERENCES: *USDA Cir.* 270, 1950; *Agr. Handbook* 290, 1965; *Conn. Agr. Exp. Sta. Bul.* 552, 1952.

BAGWORM

Thyridopteryx ephemeraeformis (Haworth), Family Psychidae

Bagworms produce conspicuous spindle-shaped cocoons or cases which are familiar objects on trees and shrubs throughout the United States east of the Rocky Mountains (Fig. 243). These caterpillars devour the foliage of a large variety of fruit and shade trees as well as ornamental shrubs, particularly junipers and arborvitae; then they are often called evergreen bagworms.

Overwintering light tan eggs inside the bags begin hatching, when proper temperatures prevail, into dark brown larvae, which start feeding and immediately begin constructing a new case or bag of silk plus foliage of the host plant. This bag is carried around by the larva and enlarged as development proceeds; when full growth is attained it is attached to a twig or small branch with a band of silk. Pupation occurs in early September; several days later the adult winged

Figure 243. Cocoons of the bagworm; slightly reduced.

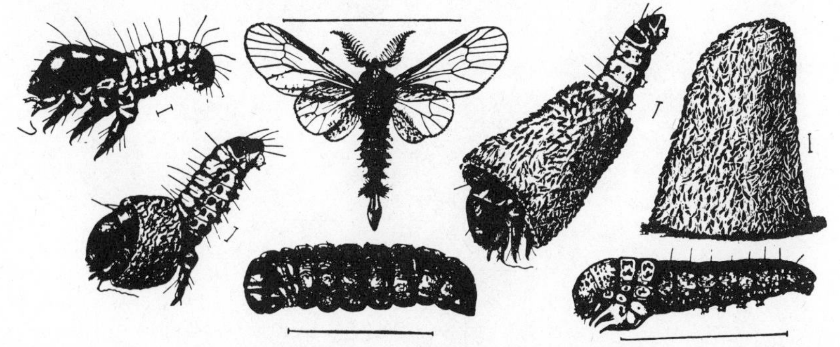

Figure 244. The bagworm. Note early stages in formation of the bag, the winged male, and wingless female; actual size indicated by lines. (Riley.)

males (Fig. 244) emerge, fly to the bags containing wingless females, and mate with them at the lower tip of the case. The male moth is almost black, with four transparent wings practically free of scales. It is seldom seen during its short life in late summer. After mating, the female moths deposit their eggs in the old pupal case and die. There is usually only one generation each year.

Populations of this insect vary greatly from year to year because there is an unusually high degree of natural control from parasites. Some common species are the ichneumons, *Itoplectis conquisitor* (Say), *Phobetes thyridopteryx* (Riley), *Scambus hispae* (Harris), and the chalcids, *Brachymeria ovata ovata* (Say), and *Spilochalcis maria* (Riley).

Choice plants of juniper and arborvitae should have all bagworm cases removed during the winter including the silken band around the twig, to prevent disfiguration by girdling. Clipping the bags from all small shrubs and destroying them gives complete control. However, many immature parasites are also in the bags and holding them in a screened or fine mesh container until adult parasites emerge and are released enhances biological control. Pesticides are more effective if applied soon after eggs hatch. Select one that does the least damage to the natural enemies.

REFERENCES: *USDA Farmers' Bul.* 701, 1916; *Misc. Pub.* 657, 1950; *Tex. Agr. Exp. Bul.* 382, 1928; *J. Econ. Ent.* 58:863–866, 1965; *Fla. Ent.* 52(2):62–66, 1969; *Ohio Sta. Univ. Ext. Bul.* 504, 1979.

CANKERWORMS

ORDER LEPIDOPTERA, FAMILY GEOMETRIDAE

Cankerworms devour the foliage of many forest, shade, and fruit trees. Apple and elm species are the most common hosts. Repeated defoliation weakens the trees, making them susceptible to attacks of other insect pests. The larvae are known as measuring worms, inchworms, or loopers because of their habit

Figure 245. The fall cankerworm, *Alsophila pometaria* (Harris), showing several developmental stages: *A,* full-grown larva (2½×); *B,* egg mass (7×); *C.* pupae (2¼×); *D,* male moth (3×); *E,* female moth laying eggs (3×); *F,* pupae in cocoons (1¾×); *G,* female moths (2¼×). (Porter and Alden, USDA)

of arching their bodies high in the center while in the act of crawling. Sometimes they spin silken threads on which they may be seen crawling or swinging.

The fall cankerworm, *Alsophila pometaria* (Harris), occurs in the area from Nova Scotia to North Carolina, west to Missouri, northwest to Montana and Manitoba. It is also found in Colorado, Utah, and California. Most adult moths emerge in November and early December, although some continue to appear during warm periods in the winter. Each female deposits gray-brown eggs in compact masses of about 100 on twigs and branches of trees. Hatching occurs as soon as the foliage begins to appear in the spring, and the young larvae feed on the leaves for a period of 3 or 4 weeks, then drop to the ground and pupate in silken cocoons formed in the soil. There is 1 generation per year.

The larvae are slender, striped, green, looping caterpillars, about 2.5 cm long, with 3 pairs of abdominal prolegs. The male moth is ash gray, with faint markings of both black and white, and a wingspread of 2.5 cm. The female is entirely wingless. The appearance of each stage is illustrated (Fig. 245).

The spring cankerworm, *Paleacrita vernata* (Peck.), has the same general distribution as that of the fall species, except that its range extends farther south, especially in the plains region, where it reaches northern Texas. Adult moths emerge in February, March, or April, and deposit their dull-colored oval eggs in loose irregular clusters of about 50. The eggs are usually laid under bark scales, crevices, and other secluded places. Hatching and larval development occur at the same time as for the fall species, but no cocoon is formed by the spring species before pupation in the soil. One generation develops each season.

The slender larvae are light green to brown or black, with white lines down the back. Only 2 pairs of abdominal prolegs are present. Adults are very similar, except that the female of the spring species has double rows of red spines across the abdominal segments (Fig. 246), whereas the female of the fall species is spineless.

Cankerworms are normally kept under control by natural forces, consisting of many predatory and parasitic enemies and climatic factors.

Sticky tree bands placed on tree trunks in early October capture many adult female fall cankerworms. These bands must be renewed in late February to capture the adult female spring cankerworms. The bands are constructed of a strip of cheap cotton batting 2 inches wide, placed around the tree trunk and covered with a strip of single-ply tarred paper 5 inches wide, which is tacked securely. The tarred paper is then covered with the sticky tree-banding substance.

Cankerworms are not pests of fruit trees where a spraying program is followed. Therefore they are a problem only on ornamental trees or in forested areas. Control may be readily accomplished by spraying a pesticide soon after the eggs hatch.

REFERENCES: *USDA Bul.* 1238, 1924; *Leaflet* 183, 1953; *Conn. Agr. Exp. Sta. Cir.* 214, 1960.

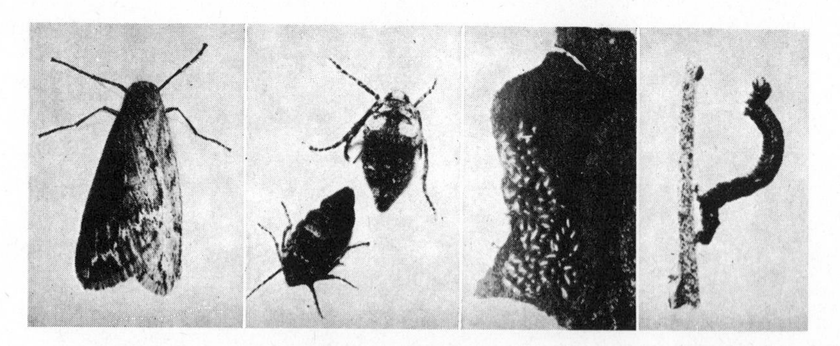

Figure 246. The spring cankerworm, *Paleacrita vernata* (Peck); male and female moths, eggs, and a larva. (Porter and Alden, 2×. USDA)

WINTER MOTH

Operophtera brumata (L.), Family Geometridae

For many years the larval stage of this insect has been a major defoliator of fruit and forest trees in European and some Asian countries. At times it also injures the fruit. The first known infested area in North America was in Nova Scotia, Canada, where it was discovered in 1949. In 1962 it was found in New Brunswick and Prince Edward Island, and in 1978 Washington and Oregon.

Fully grown larvae are green with dark-brown heads, and about 25 mm long; they have a dark mid-dorsal longitudinal line and 3 yellow stripes along each side of the body. They crawl in the same manner as all so-called "measuring worms." The male moth has fully developed wings and flies. Fore wings are light brown-gray; hind wings are paler, the wingspan being nearly 25 mm. The female adult is almost wingless (only stubs), grayish-brown, and moves only by crawling. Eggs are laid on tree trunks and limbs in late autumn; these hatch in early spring, and the newly hatched larvae crawl or are carried to the foliage by air currents. If this occurs after the buds open larval survival is high (about 60%). But if buds are still closed most of the larvae reaching the branch tips starve. Pupation occurs in June and July and adults appear in late October. One generation develops each year.

Two parasites, introduced into Canada from Europe, are now established and are controlling this pest. They are, *Cyzenis albicans* (Fall.), a tachinid fly, and *Agrypon flaveolatum* (Grav.), an ichneumonid wasp.

REFERENCES: *USDA PA* 580, 1963; *Can. Ent.* 92:862–864, 1960; 98:1159–1168, 1966; 97:631–639, 1965; *Entomophaga* 5:111–129, 1960.

HEMLOCK LOOPER

Lambdina fiscellaria (Guenée), Family Geometridae

This species defoliates eastern hemlock throughout its range from southern Canada to Georgia and west to Wisconsin, mainly in mountain regions. Other hosts, often damaged, especially where mixed planting with hemlock occur, are Douglas and balsam fir, cedar, sitka, white, red, and black spruce.

Moths are light tan, slender, with a wingspread of about 4 cm; each fore wing is crossed with 2 thin dark wavy lines, whereas each hind wing has only one such line. The moths are present from late August to early October and soon after emerging begin depositing blue to brown colored eggs singly or in groups of 2 or 3 on twigs, branches, and tree trunks. The eggs overwinter and hatch during May and early June into typical looper larvae having 2 pairs of prolegs. The larvae vary in coloration from yellowish green to brown, the head and body flecked with black dots; when fully grown they are 30 mm in length. By August larval development is complete and pupation occurs in bark crevices or in litter. There is only 1 generation per year.

Epidemics of the western hemlock looper, *L. fiscellaria lugubrosa* (Hulst) have occurred mainly in the coastal areas of the Pacific Northwest of North America; it is estimated that this subspecies has killed three-fourths million board feet of timber. The appearance and life cycle are very similar to the eastern species, except the larvae of *lugubrosa* have diamond-shaped markings along the back.

The most important natural controlling agents are a virus disease of larvae and heavy rains occurring during moth flight.

REFERENCES: *Can. Ent.* 88:587–599, 1956; *Can. Pub.* 1180, 1967; *USDA Misc. Pub.* 1175, 1972.

EUROPEAN PINE SHOOT MOTH

Rhyacionia buolina, (Schiffermüller), Family, Olethreutidae

This introduced species was first recorded in North America on Long Island, N.Y., in 1914. It is now widely distributed from coast to coast in southern Canada and northern United States. Scotch, red, mugho, and Austrian pines are the hosts most attacked in the eastern regions, but jack, eastern white, pitch, longleaf, ponderosa, Virginia and other pine species may also be injured.

The moth is rusty orange-red and has a wingspan of 18 mm. The fore wings are marked with several forked silvery cross lines; the hind wings are dark brown, and the legs white. Larvae are brown and when fully grown are nearly 18 mm long.

Adults appear in late spring and most activity is at night. They lay eggs singly or in groups on the needles, twigs, and shoots. Hatching occurs in 10 to 20 days, depending on the temperature, and the brown larvae feed at the base of the needles under tentlike webs for a short time, then move to the buds. Their feeding here causes the exudation of resin which, along with webbing, frass and other debris, accumulates on the side of a bud where it hardens to form a yellow-white patch mass. Feeding ceases in August and winter is passed in the feeding tunnels of buds. When activity resumes in mid-April the larvae move to other buds and continue to feed in new shoots until fully grown, then pupate, with moth emergence occurring primarily in June.

Repeated killing of terminal buds results in stunted and bushy trees. Injury is most severe in young forest plantings.

Native and introduced parasites destroy about 10% of the population; the braconid, *Orgilus obscurator* (Nees), appears to be the most promising of the introduced species.

Since the larvae are only exposed during two periods in their life cycle any pesticide application should be made soon after the eggs hatch or in the spring when they leave their hibernating quarters to search for developing buds. Pruning infested shoots and burning them before July 1 will aid in control.

REFERENCES: *Ohio Agr. Exp. Sta. Res. Bul.* 760, 1955; *Res. Bul.* 983, 1966; *Mich. State Univ. Ext. Bul.* 353, 1962; *USDA Misc. Pub.* 1175, p. 361, 1972; *Proc. Ent. Soc. Ont.* 92:58–69, 1962; *Yale Univ. School of Forestry Bul.* 37, 1933.

EASTERN TENT CATERPILLAR

Malacosoma americanum (F.), Family Lasiocampidae

Widely distributed throughout the United States east of the Rocky Mountains, and similar to other defoliating caterpillars, this insect may become a problem in neglected orchards and on shade or forest trees, but it is unimportant where spraying is regularly practiced. Apple and wild cherry are the most common hosts. Other closely related species are: western tent caterpillar, *M. californicum* (Packard); Pacific tent caterpillar, *M. constrictum* (Edwards); southwestern tent caterpillar, *M. incurvum* (Edwards); and Sonoran tent caterpillar, *M. tigris* (Dyar).

Dark brown masses of eggs that encircle the twigs are the overwintering stage. Each mass may contain 150 to 350 eggs. Embryonic development starts immediately after deposition and within a month the eggs contain fully formed larvae. Hatching takes place about the time that new leaves on wild cherry trees begin to appear in the spring. The larvae are gregarious and soon construct a tent of silk in a crotch of the tree, enlarging it as they grow. During cloudy and rainy weather the larvae remain within the tent, but when the weather is favorable they emerge onto the foliage and feed at regular intervals, spinning threads of silk wherever they crawl. About 6 weeks after hatching, the characteristically marked larvae become fully grown, reaching a length of 5 cm.

Pupation occurs in silken cocoons which are found on tree trunks, fences, leaf litter, or debris on the ground. About 2 weeks later emergence of the red-brown moths with diagonal white stripes takes place and deposition of the overwintering eggs soon follows, after which the moths die. (See Figs. 247, 248, 249, 251, 252.) One generation occurs each year.

Tent caterpillar populations fluctuate greatly because of control by natural enemies, some of which are birds, predaceous and parasitic insects, and disease organisms. Parasitism in some areas may be nearly 80%. Twenty-four species have been reared from the larvae or pupae; of these the ichneumons predomi-

Figure 247. Adults of the tent caterpillar, *Malacosoma americanum* (F.), slightly enlarged. (USDA)

Figure 248. Tent caterpillars, *Malacosoma americanum* (F.); 1.5×. (USDA)

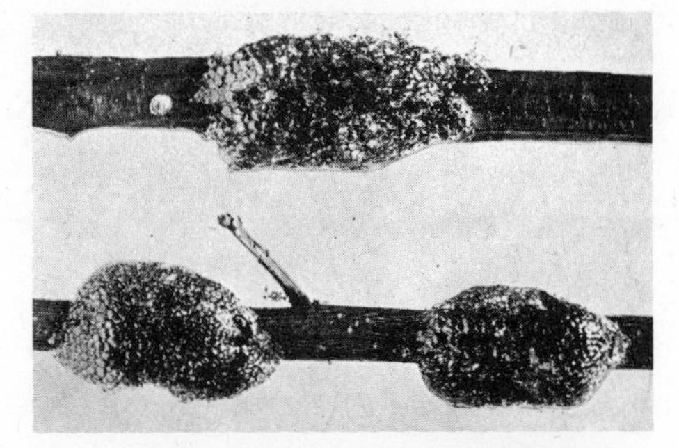

Figure 249. Eggs of the tent caterpillar.

Figure 250. Forest tent caterpillars; 0.5×. (Britton.)

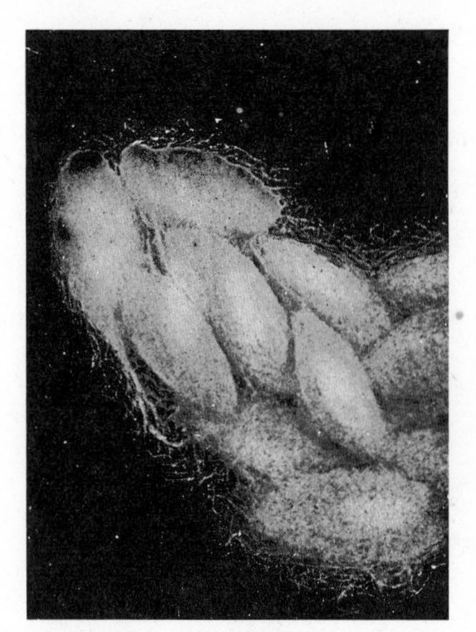

Figure 251. Cocoons of the tent caterpillar; natural size. (Lowe.)

Figure 252. Newly formed tent of eastern tent caterpillar; reduced. (USDA)

nate, followed by the braconids and chalcids. Unfavorable weather also acts as a natural check on this insect.

Collecting and destroying egg masses during the winter is sometimes suggested as a means of control on a small scale.

REFERENCES: *USDA* G-178, 1973; *Misc. Pub.* 657–417, 1950; 1175:328–329, 1972; *Agr. Handbook* 290, 1965; *Conn. Agr. Exp. Sta. Bul.* 378, 1935.

FOREST TENT CATERPILLAR

Malacosoma disstria Hübner, Family Lasiocampidae

This species is a widespread defoliator of deciduous forest and shade trees throughout Canada and the United States from Minnesota eastward. Serious outbreaks often persist for three or more years before natural factors bring the pest under control. Heavy damage has been wrought in Canada, the New England states, New York, Minnesota, Michigan, Mississippi, and Louisiana.

The life cycle is essentially the same as that of the eastern tent caterpillar. Adult moths are light buff-brown, with a wing expanse of 25 to 35 mm, the fore wings having 2 darker, oblique lines near the middle. The egg mass encircles the twig and contains up to 350 eggs cemented together with a dark varnishlike substance. Fully grown larvae are about 25 mm long, dark in color with a row of white spots along the back (Fig. 250). These larvae are gregarious until nearly full size, and instead of a tent they make a silken mat on the trunk or branch where they congregate after feeding or during molting. Pupation takes place inside silken cocoons on tree trunks or other objects near-by; approximately 2 weeks later the adults emerge. After mating, the eggs are deposited which become the overwintering life stage. Embryonic development begins immediately after egg deposition and in about 30 days they contain fully formed larvae. Only 1 generation develops each year.

Many of the same natural enemies mentioned for eastern tent caterpillar also attack this species.

REFERENCES: *USDA Misc. Pub.* 657:418–419, 1950; 1175:330–333, 1972; *J. Econ. Ent.*, 57: 157–160, 1964.

FALL WEBWORM

Hyphantria cunea (Drury), Family Arctiidae

Distributed throughout the greater part of the United States and Canada, this caterpillar is a general feeder on deciduous trees and shrubs. It is often seen on roadside trees and in neglected orchards. Besides defoliation the larvae produce large unsightly webs on the branches (Fig. 253). There are two distinct forms or races of this insect in Louisiana.

Figure 253. Web of the fall webworm, *Hyphantria cunea* (Drury).

The moth is pure white, with a wing expanse of 3 cm, the fore wings sometimes marked with black dots. These adults emerge from overwintering pupae in May to July, depending on the climate, and lay masses of 400 to 900 white eggs usually on the undersides of the leaves. In a few days hatching occurs and the larvae begin to spin silken webs over the foliage, extending them as the enclosed leaves are devoured. The hairy light tan larvae are gregarious until the last instar, after which they may be found crawling any place on the host plants. They are nearly 25 mm in length at this stage of development. Pupation takes place in a silken cocoon among debris on the ground. Generally, the larvae of the first generation are active in June and July, the second generation in August and September. Where a single generation occurs larval activity is usually from July to September.

A number of natural enemies keep the fall webworm in check. Two important parasites are the ichneumon wasp, *Itoplectis conquistor* (Say) (Fig. 54), and the braconid wasp, *Meteorus hyphantriae* Riley. Pruning and destroying the larval webs as soon as they are discovered are often practical control measures. Orchards in which a spray schedule is followed are not troubled by fall webworms.

REFERENCES: *USDA Misc. Pub.* 657:387–388, 1950; *Conn. Agr. Exp. Sta. Bul.* 203, 1918; *Ann. Ent. Soc. Amer.*, 57:192–194, 1964; *J. Econ. Ent.*, 57:314–318, 1964.

WHITE-MARKED TUSSOCK MOTH

Orgyia leucostigma (Smith), Family Lymantriidae or Liparidae

This species occurs in eastern United States and Canada, and west into Colorado and British Columbia. It is a general feeder on foliage of deciduous trees and shrubs, particularly apple, basswood, elm, poplars, rose, sycamore, willow, wisteria, and Norway, silver, and sycamore maples.

The male moth is ash gray with feathery antennae and a wingspread of about 3 cm. The fore wings have wavy bands of a darker shade. All females are wingless and white- or gray-colored. They deposit their eggs in white masses cemented to the surface of the old cocoons; the winter is passed in this way. Hatching takes place between April and June, depending on the latitude, and the larvae become fully grown in 30 to 40 days. At that time they are nearly 3 cm long, slender, cream yellow, with a broad black longitudinal stripe on the back and a broader gray one on each side. The head is red and the body is ornamented by 2 tufts of long black hairs on the prothorax, another black tuft on the eighth abdominal segment, 4 brushlike tufts of light tan hairs on each of the first 4 abdominal segments, and red dots on the sixth and seventh abdominal segments (Fig. 254). Pupation takes place in cocoons made of silk and old larval hairs. These cocoons may be found on the tree trunk, on branches, or on other

Figure 254. The white-marked tussock larva, *Orgyia leucostigma* (Smith). (Ohio Agr. Exp. Sta.)

objects near infested trees. In 2 weeks the new adults emerge, mate, and soon afterwards deposit eggs. There are 3 generations per year in the latitude of Washington, D.C., and one in the latitude of northern New York.

In some areas, other species of tussock moths may be more abundant than the one under consideration, viz., western tussock moth, *O. vetusta* (Boisduval), and Douglas-fir tussock moth, *O. pseudotsugata* (McDunnough).

Tussock moth larvae and eggs are attacked by many natural enemies. Those of greatest importance are insect parasites and predators, birds, and wilt disease. Daubing the egg masses during the winter with creosote or oil is sometimes suggested as a control measure.

REFERENCES: *Ohio Agr. Exp. Sta. Bul.* 332, 1918; *Ill. Nat. Hist. Sur. Cir.* 47, 1958; *Bul. Ent. Soc. Amer.* 23:167–180, 1977.

GYPSY MOTH

Lymantria dispar (L.), Family Lymantriidae or Liparidae

The gypsy moth was brought from Europe to Massachusetts in 1869 for experiments in silk production. It was accidentally liberated and has since spread to most of the New England states and southeastern Canada, many of the North Central and South Central states, and in 1976 was discovered in San Jose, California.

Damage results from the caterpillars devouring the foliage of shrubs, fruit, shade, and forest trees, including conifers. It is more important as a pest of forests than of orchards, but it may be very destructive in neglected orchards. Repeated defoliations can kill most hardwoods and a single defoliation can kill white pine, spruce, and hemlock.

The adults are shown in Fig. 255; the female is light in color with a wingspread of 50 mm, whereas the male is darker with a wingspread of 37 mm. Although both sexes are winged, only the male can fly. After emergence and

Figure 255. The gypsy moth, *Lymantria dispar* (L.); (left) female, (right) male; about natural size. (USDA)

mating, the female crawls up some nearby object, usually the tree trunk or branches, and deposits masses of eggs, each containing 100 to 1000 or more (Fig. 256). The eggs are packed with tan hairs from her body. Mature larvae are nearly 6 cm long; the head has yellow markings and the body is dusky and hairy, with a double row of five pairs of blue tubercles followed by 6 pairs of red tubercles on the dorsal side. The diagrammatic representation of the life cycle gives all the essential information concerning developmental activity (Fig. 257). There is only 1 generation per year.

They are spread from one area to another primarily by shipment or movement of infested nursery stock or any object on which eggs may be deposited. When the first instar larvae spin down on silken threads they may be transported by man or borne long distances by wind.

Natural and biological control agents check the ravages of this pest. One predator, a ground beetle closely related to several native species, *Calosoma sycophanta* (L.), was brought from Europe and has become permanently established and of definite value. Larval and pupal parasites of proven worth are the tachinid flies, *Exorista larvarum* (L.) (Fig. 55), *Compsilura concinnata* (Meigen) (Fig. 55), and *Blepharipa scutellata* (Rob.-Desv.), and the braconid wasp,

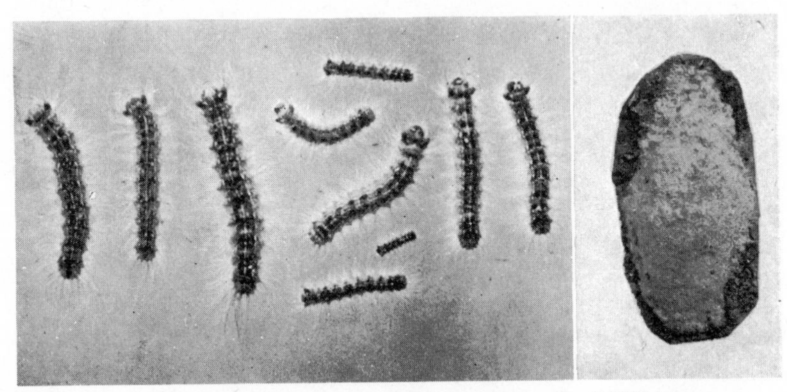

Figure 256. Larvae and egg mass of the gypsy moth. (USDA)

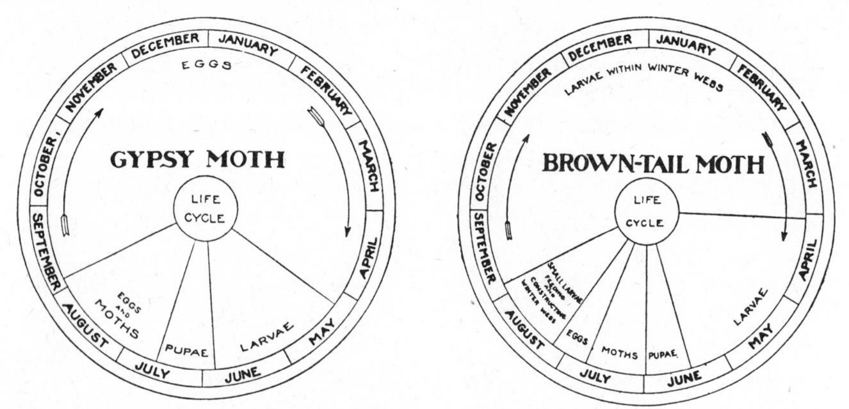

Figure 257. Diagrams showing life cycles of the gypsy and brown-tail moths. (USDA)

Apanteles melanoscelus (Ratz.). Egg parasites of importance are *Anastatus disparis* Ruschka and *Ooencyrtus kuwanai* (Howard). Other established parasites of some value in control are *Apanteles lacteicolor* Vier., *Phobocampe disparis* (Vier.), *Monodontomeris aereus* Walker, and *Parasetigena agilis* (Rob.-Desv.). A "wilt" or nuclear polyhedral virus disease as well as climatic factors and insectivorous birds also help to prevent serious outbreaks.

Developing stands of trees somewhat resistant to gypsy moth attack, by reducing the proportion of favored host species such as basswood, hawthorn, oaks, poplars, willows, and gray and river birch, is of value in forested areas. Egg clusters can be destroyed by saturating them with a commercial creosote mixture. On occasion, particularly where a few ornamental shrubs or shade trees are concerned, a band of burlap may be tied around the tree trunk and folded down at the middle. This affords shelter for the larvae in the daytime. These bands should be examined frequently and the larvae destroyed. Gyplure, the synthetic form of the sex lure found in unmated female gypsy moths, may find a place in the control program. By using this lure all males might be caught in traps or poisoned by baits and reproduction of the species halted because all eggs are sterile.

REFERENCES: *USDA Cir.* 464, 1938; E–726, 1947; *Misc. Pub.*, 657:408–412, 1950; *Conn. Cir.* 186, 1954; 231, 1969; *Bul.* 655, 1963; 735, 1973; *Pa. Dept. Agr. Misc. Bul.* 4404, 1962; *Bul. Ent. Soc. Amer.* 19:15–19, 1973; *Calif. Agr.* 31(3):23–24, 1977.

BROWN-TAIL MOTH

Euproctis chrysorrhoea (L.), Family Lymantriidae or Liparidae

The first record of the brown-tail moth in the United States was in Somerville, Massachusetts, in 1897, but actual introduction probably took place several years earlier. Later it became established in the eastern part of Connecticut and Vermont, in New Hampshire, Maine, and the provinces of New Brunswick and Nova Scotia. In recent years the most serious infestations have been confined to the southeastern part of Maine, the southern half of New Hampshire, and the eastern part of Massachusetts.

The larvae most often devour the foliage of pear, apple, plum, cherry, hawthorn, white oak, rose, and willow, but occasionally other deciduous trees are attacked.

As illustrated (Fig. 258) the moths are white with the abdomen densely clothed with brown hairs, especially at the tip. The females have a wing expanse of about 37 mm; the males are slightly smaller. Eggs are usually deposited on the underside of a leaf in masses, each containing almost 300. They are closely packed and covered with brown hairs (Fig. 259). Fully grown larvae are nearly 37 mm long, the head is light brown, the body dark brown to black, with a broken white line on each side and 2 conspicuous red spots near the posterior end. The larvae possess numerous hairs that are irritating to the skin of persons on whom they might chance to lodge, either from handling or from those floating in the air during the larval molting periods. Avoid breathing the hairs into the lungs.

There is 1 generation annually in the New England states. The essential difference in the life cycle of this insect and of the gypsy moth is that the eggs of the brown-tail moth hatch the same summer they are laid, the larvae feed for a short time in August or early September, then form shelters of silk spun about leaf clusters at the tips of twigs, in which they hibernate during the winter. The complete life cycle is shown (Fig. 257) for comparison with that of the gypsy moth.

Federal quarantines regulating the shipments of nursery stock, eradication programs carried on by the federal, state, and municipal governments, and the

Figure 258. Male and female brown-tail moths. (USDA)

Figure 259. Larvae of brown-tail moth and egg mass on leaf. (USDA)

introduction and establishment of parasites have greatly reduced infestations of the brown-tail moth and thus prevented widespread dispersal.

Low winter temperatures play an important part in the control of this pest. In addition, the parasites *Compsilura concinnata* (Meigen), *Townsendiellomyia nidicola* (Townsend), *Carcelia laxifrons* Vill., *Meteorus versicolor* (Wesm.), and *Apanteles lacteicolor* Vier. are also effective enemies along with the predator *Calosoma sycophanta* (L.) A fungus disease, *Entomophthora aulicae* (Reich.), greatly reduces a larval population when weather conditions favor it.

REFERENCES: *USDA Cir.* 464, 1938; *Misc. Pub.*, 657:412–414, 1950; *PA* 282, 1956.

RED-HUMPED CATERPILLAR

Schizura concinna (Smith), Family Notodontidae

This species is widely distributed from Canada to the Gulf and from the Atlantic to the Pacific. The larvae commonly feed on apple, elm, poplar, rose, plum, willow, walnut, wild black cherry, and other deciduous trees. Small larvae skeletonize the leaves; the larger ones devour all but the midrib. They are locally abundant in unsprayed apple orchards, nurseries, or on ornamental shade trees.

The adult is a gray-brown moth with a wing expanse of slightly more than 26 mm. Fully grown larvae are over 26 mm in length and strikingly colored. The head is coral red, the body marked with black and yellow lines, and on the first abdominal segment is a conspicuous red hump (Fig. 260). The tip of the abdomen is usually held in an elevated position.

Winter is passed as fully grown larvae inside silken cocoons in the duff on the ground. Pupation takes place in the spring and adults emerge from late May

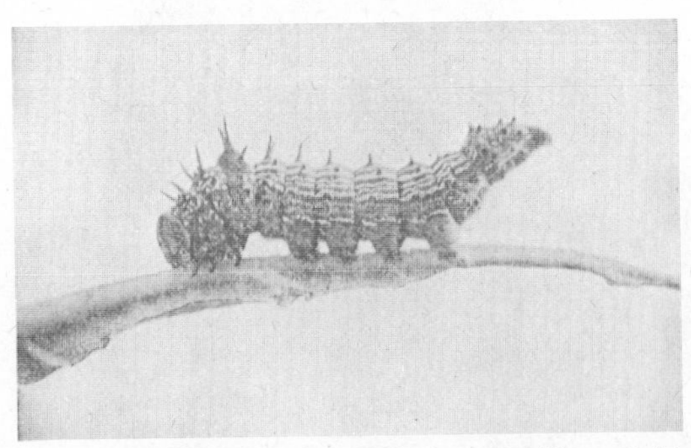

Figure 260. The red-humped caterpillar, *Schizura concinna,* 1.5× (Smith).

to July. The white eggs are deposited on the undersides of the leaves in masses of 100 or less. On hatching, the caterpillars feed in groups, often defoliating a single branch. Full growth is attained in 3 or more weeks, and transformation to pupae takes place, with new adults appearing in July and August. Two or more generations develop in warmed areas; only 1 occurs in their northern range.

Clipping off and destroying infested branches controls the insect and is a recommended practice where only a few plants are involved. Apple orchards receiving a regular spray program are not troubled with this insect.

REFERENCES: *Ohio Agr. Exp. Sta. Bul.* 332, 1918; *USDA Misc. Pub.* 1175:311, 1972.

YELLOW-NECKED CATERPILLAR

Datana ministra (Drury), Family Notodontidae

This species occurs throughout most of the United States and southern Canada. Its food plants include apple, birch, blueberry, basswood, cherry, elm, hawthorn, oak, and other deciduous trees. Often it is locally abundant, particularly on unsprayed apple trees. Injury results when the larvae devour the foliage.

The yellow-and-black-striped caterpillars, which have a prominent yellow spot just back of the head, reach a length of 5 cm when fully grown. When disturbed they raise their head and tip of abdomen in an upright position and cling to the twig or branch by 4 of their 5 pairs of abdominal prolegs (Fig. 261). They are gregarious in habit and tend to congregate in crotches of larger branches at the molting periods (Fig. 262). The adults have cinnamon brown front wings, each crossed by four darker lines; the hind wings are light buff (Fig. 261). They have a wing expanse of nearly 5 cm.

Winter is passed as brown pupae in the soil. Adults appear during June and

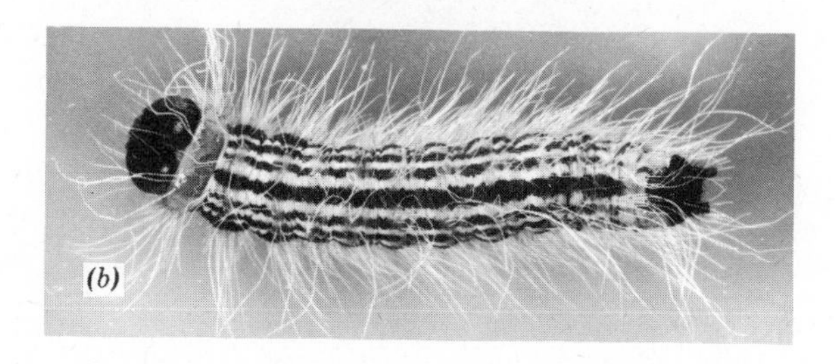

Figure 261. The yellownecked caterpillar. (*a*) Larva in alarm position, (*b*) larva, dorsal view, (*c*) adult, 2.2×. (Courtesy of David G. Nielsen, OARDC)

Figure 262. Characteristic cluster of yellow-necked caterpillars.

July, and lay white eggs on the leaves in masses containing up to 100. The caterpillars may be found from late June to October in some regions, but development is normally completed in 3 or 4 weeks when they enter the soil and change to pupae. One generation per year is the usual number.

The simplest method of control in light infestations is by removing and destroying each colony of caterpillars. Apple orchards receiving a regular spray program are never troubled by this insect. The tachinid parasites, *Compsilura concinnata* (Meigen), and *Winthemia datanae* (Townsend), are important natural enemies.

REFERENCES: *Ohio Agr. Exp. Sta. Bul.* 332, 1918; *USDA Misc. Pub.* 1175:311, 1972.

WALNUT CATERPILLAR

Datana integerrima G. and R., Family Notodontidae

A close relative of *D. ministra,* this caterpillar is found throughout most of eastern United States west to Kansas. Its food plants are butternut, walnut, hickory, and pecan. When this insect becomes abundant the trees may be completely defoliated.

The fully grown larvae are almost 5 cm long, black, and clothed in long gray hairs. Younger instars are nearly naked and vary from brick red to dark red-brown with pale yellow-to-gray longitudinal stripes. When disturbed they raise their head and tip of abdomen in an upright position and hold to the twig or branch with 4 of their 5 pairs of abdominal prolegs. They are also gregarious in habit, feeding together and congregating in large masses on the tree trunk during the molting periods. The adults closely resemble *D. ministra*.

The life cycle is also the same as that of the yellow-necked caterpillar. However, in the South there are 2 generations, the first brood of larvae appearing in June and July, and the second in August and September.

Light infestations on small trees can be eliminated by crushing or burning the masses of larvae. The tachinid, *Archytas metallicus* (Rob.-Des.), is one of the natural enemies.

REFERENCES: *Ohio Agr. Exp. Sta. Bul.* 332, 1918; *USDA Farmers' Bul.* 1829, 1954; *Agr. Handbook* 240, 1963.

ELM LEAF BEETLE

Pyrrhalta luteola (Müller). Family Chrysomelidae

The elm leaf beetle may be found nearly every place that elm trees grow in North America. Larvae devour the lower surface of the leaves causing them to die and drop prematurely. Adults eat small holes in the leaves. Heavy defoliation weakens the trees and makes them subject to the attacks of bark beetles and borers, as well as disease organisms. Elms growing in urban areas are most heavily attacked.

The adult beetle is 6 mm long, varying from yellow when young to olive green when aged, with a black stripe along the margin of each wing cover. The eyes are black; the legs and antennae are yellow. These beetles pass the winter in secluded places, often in buildings and under plant debris. They emerge in the spring and fly to elm trees with unfolding leaves where feeding takes place, followed by oviposition. The yellow eggs are placed in groups of 5 to 25 on the undersides of the leaves, each female laying over 400. Hatching occurs in about a week, and the dull yellow larvae with 2 dark stripes and black tubercles skeletonize the leaves. In almost 3 weeks they are fully grown, attaining a length of 12 mm. Pupation then follows in crotches or crevices in the bark or on the ground. In approximately 10 days the new adults (Fig. 263) emerge from the bright yellow pupae. There are 2 generations per year in the latitude of Columbus, Ohio, but in more southern areas there may be a partial third generation.

Natural enemies include birds, toads, diseases, and predaceous and parasitic insects. A chalcid wasp, *Tetrastichus brevistigma* Gahan, frequently kills many pupae. The fungus, *Beauveria bassiana* (Bals.), also kills pupae and adults late in the summer, especially in humid seasons.

Chemical control is accomplished by spraying when the eggs are hatching, usually the latter part of May. If a thorough job is done at that time usually nothing need be applied for later generations.

REFERENCES: *USDA Leaflet* 184, 1952; *Ill. Nat. Hist. Sur. Cir.* 47, 1958.

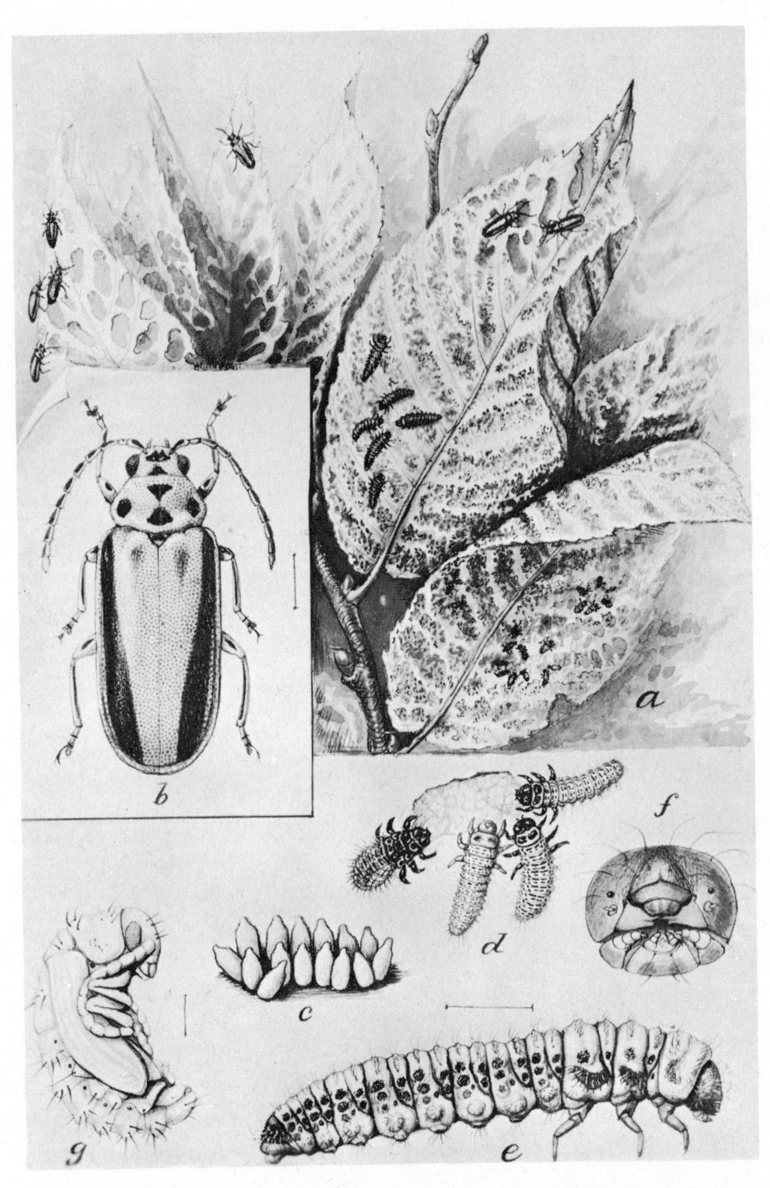

Figure 263. The elm leaf beetle: *a*, adult and larval feeding damage; *b*, adult beetle; *c*, eggs; *d*, young larvae; *e*, mature larva; *f*, mouthparts of larva; *g*, pupa; all figures enlarged except *a*. (USDA)

ELM BARK BEETLES

ORDER COLEOPTERA, FAMILY SCOLYTIDAE

Two kinds of bark beetles commonly attack elm trees in Canada and the United States. One is the smaller European elm bark beetle, *Scolytus multistriatus* (Marsham), and the other the native elm bark beetle, *Hylurgopinus rufipes* (Eichh.). So far as is known, these bark beetles attack only elm and are important pests principally because they spread the fungus, *Ceratocystis ulmi,* which causes what is commonly known as Dutch elm disease. The disease is present throughout most of the United States, found in California in 1975 and since in Oregon.

Native to Europe, the smaller European elm bark beetle was found in the vicinity of Boston, Massachusetts, as early as 1904. Since then it has been found in most of the states east of the Rocky Mountains, in California, Washington, Oregon, and the bordering provinces of Canada from Ontario to Quebec. It undoubtedly will spread throughout the natural range of the elm tree.

The adult is a shiny, dark red-brown beetle, approximately 3 mm long; the underside of the posterior end of the body is concave with a prominent spine. These adults appear in the spring about the middle of May, emerging from holes they make in the bark. They feed in the crotches of living elm twigs and, if carrying the Dutch elm disease fungus, may introduce these organisms into healthy trees. Later they bore through the bark of recently cut, dead, or dying trees, forming galleries by grooving the surfaces of both the wood and the inner bark. These galleries are parallel with the grain of the wood, and in niches along the sides are placed the globular, pearly white eggs (Fig. 264). From these eggs hatch the legless white brownheaded larvae which feed in the cambium region, constructing galleries across the grain of the branches. As the larvae increase in size the galleries are also widened, and, when fully developed, pupation occurs in the enlarged gallery tip (Fig. 265). In the latitude of New Jersey the length of the life cycle ranges from 45 to 60 days under favorable conditions, and 2 full generations and a partial third are produced each year. The overwintering larvae develop from the eggs laid late in the summer or early fall.

The native elm bark beetle presumably originated in the United States. It occurs in most of the eastern states from Maine to Virginia, in Mississippi, Kansas to Minnesota, and also in eastern Canada. It may be present throughout the natural range of the American elm, or approximately the eastern ⅔ of the country. It feeds on various species of elm and is recorded as feeding on basswood.

The adult beetle is dull brown, a little less than 3 mm long, and without the concave underside of the abdomen or prominent spine evident in *S. multistriatus*. The egg, larva, and pupa closely resemble the European elm bark beetle in color and appearance, and the life cycle is also very similar; the winter, however, is passed in both the larval and adult stages. Only diseased,

Figure 264. The smaller European elm bark beetle; *a,* adult; *b,* larva; *c,* eggs; *d,* pupa; all greatly enlarged. (USDA)

dying, or recently cut elm trees are attacked, and the parent or egg galleries consist of 2 branches diverging from the point where the beetles enter the bark. These galleries extend across the grain, and the larval galleries almost always run with the grain.

Unless elm bark beetles are associated with Dutch elm disease, there is little need for control measures, since healthy elm trees are not subject to serious damage by the beetles. When elm leaves on a branch or two suddenly wilt, become yellow or dry, then drop off, the tree may have Dutch elm disease. Cut off several small branches showing these symptoms and examine for a brown discoloration in one or more of the annual growth rings. If present, the tree is probably infected with Dutch elm disease. To be positive a laboratory diagnosis is necessary.

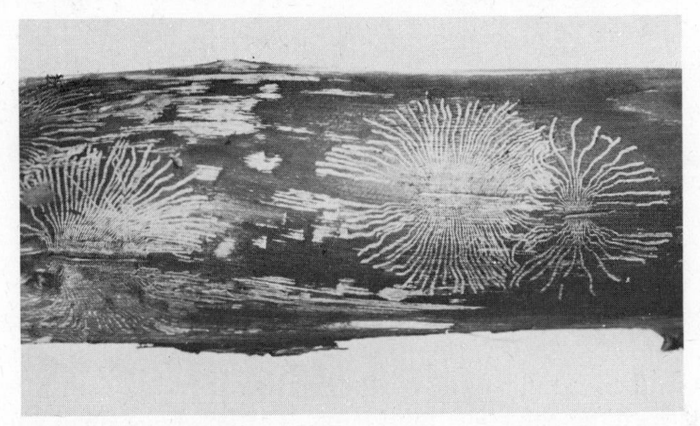

Figure 265. Larval and adult galleries of the smaller European elm bark beetle. (USDA)

Dutch elm disease can be checked by community-wide effort by killing the bark beetles that carry the organisms. The most recent approved way to try to save trees is to pressure inject the chemical Lignasan, systemic in action, into the tree trunk or roots. Another promising control measure is luring the beetles into traps by a combination of female sex attractant, other beetle odors, and the odor of decaying elm trees. Scientists are also attempting to develop an elm resistant to the disease.

Other helpful control measures are the removal and burning of all dead, dying, or diseased elm trees in a relatively large area where both the beetles and the disease are present, and proper pruning, fertilizing, and watering to keep remaining trees as vigorous as possible.

REFERENCES: *Conn. Agr. Exp. Sta. Bul.* 420, 1939; *USDA Agr. Inf. Bul.* 193, 1964; *Leaflet* 185, 1966.

SHOT-HOLE BORER

Scolytus rugulosus (Ratz.), Family Scolytidae

Introduced from Europe in 1878, this bark beetle has spread over most of the United States and southern Canada. Common hosts include peach, plum, cherry, apple, pear, and related plants. Its presence is indicated by the numerous small holes in the bark of twigs and branches (Fig. 266). These holes are about the diameter of small lead shot and are often filled with gummy exudates, especially on peach and other stone fruits. They are not serious enemies of

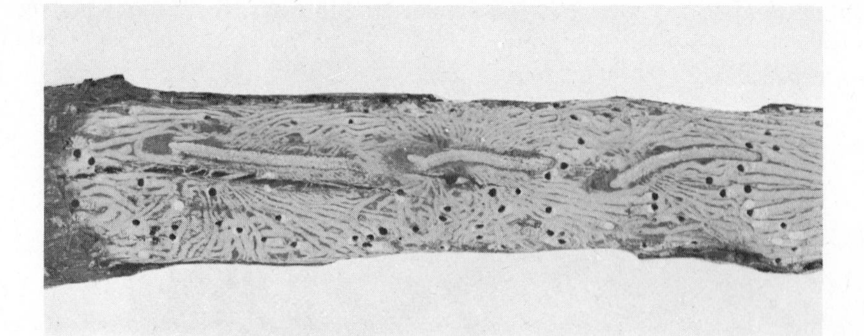

Figure 266. Feeding damage by the shot-hole borer, about natural size.

healthy trees, but occasionally they may be found infesting broken, dead, or dying branches.

The adult beetle is 2.5 mm or less in length, brown-black, with a short, stubby snout in which are the chewing mouthparts. The larva is a small white footless grub, slightly larger at its anterior end and about 4 mm in length (Fig. 267).

Larvae of these beetles hibernate in their burrows beneath the bark, complete their development, and emerge as adults in the spring or early summer. These adults bore into the twigs and branches, and make tunnels more or less parallel with the grain of the wood just under the bark where they deposit their eggs. The larvae extend their tunnels somewhat at right angles to the parent gallery, which may so injure the cambium layer that the branch is killed. When grown, the larvae pupate, transform to adults, and emerge from holes which they chew in the bark. These exit and entrance holes give the shot-hole effect. There are 1 or 2 generations per year in the North and 3 or more in the South.

The shot-hole borer is said to have several natural enemies, the most important being parasitic chalcid wasps.

Since healthy, vigorous, well-cared-for trees are less subject to attack by shot-hole borers, any operation that contributes to this end is of value in control. Removal and burning of infested or diseased branches or trees during the

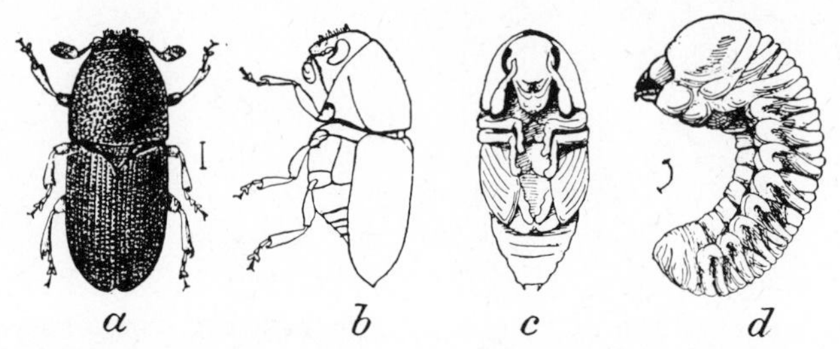

Figure 267. The shot-hole borer, *Scolytus rugulosus* (Ratz.): *a, b,* adult; *c,* pupa; *d,* larva; enlarged. (USDA)

winter period destroy the insects while still in their burrows. This helps prevent the increase of beetles to the point where they might attack healthy wood. Orchards receiving a regular spray program are not troubled by this insect.

REFERENCES: *USDA Cir.* 270, 1950; *Agr. Handbook* 290, 1965.

FLATHEADED APPLE TREE BORER

Chrysobothris femorata (Olivier), Family Buprestidae

This flatheaded borer attacks practically all fruit, shade, or forest trees. It is widely distributed throughout the United States and sections of Canada, but on the Pacific Coast it is largely supplanted by a related species, *C. mali* Horn, the Pacific flatheaded borer. Both species cause similar damage by tunnelling under the bark in the larval stage and feeding on the foliage as adults. Young trees are often attacked, as well as older trees that are unthrifty because of drought, defoliation, sun scald, or other factors.

The beetles are flat, bronzy-black, and nearly 12 mm in length. The larvae are white, legless, and about 25 mm long when fully grown, with the fore part of the body broad and flattened (Fig. 268). The adults are more prevalent in sunny locations.

Partly grown larvae hibernate in the burrows. In the spring before pupation they tunnel a cavity almost at a right angle to the bark surface. Adults begin emerging in May and this continues until midsummer. Eggs are laid beneath scales of bark and on hatching the larvae burrow under the bark. Development is usually completed with 1 year, but sometimes 2 years are required.

Figure 268. (Left) the flatheaded borer, *Chrysobothris femorata* (Oliv.); (center) the Pacific flatheaded borer, *C. mali* Horn; and, (right) larva in burrow; 3×. (USDA)

Maintenance of tree vigor is considered of utmost importance. Shading the trunks of newly set trees by wrapping with paper or burlap impregnated with a residual pesticide inhibits egg deposition and kills newly hatched larvae. Applying external white latex paint to the trunks of young trees before borer eggs are laid prevents sunburn and reduces borer attacks. Removing and destroying infested wood during winter are recommended practices.

REFERENCES: *USDA Farmers' Bul.* 1065, 1919; *Tech. Bul.* 83, 1929; *Leaflet* 274, 1965; *Cir.* 270, 1950; *Ill. Nat. Hist. Cir.* 47, 1958; *Okla. Bul.* C-259, 1942; *Calif. Agr.* 22(4):6–7, 1968.

BRONZE BIRCH BORER

Agrilus anxius Gory, Family Buprestidae

This native insect apparently occurs throughout the range of birch in North America, with white and paper birches, grown as shade and ornamental plants, most seriously injured. Trees that have been weakened or injured in some way are most often damaged.

Creamy white slender flattened larvae about 25 mm long pass the winter in galleries under the bark of the host. In late April or May pupation occurs and green-bronze adults begin emerging by late May or early June, depending on the locality. The adults are about 13 mm in length, feed on foliage during June, July, and August, and oviposit in bark crevices, mostly on unshaded parts of the tree. Young hatching larvae bore into the cambium region, making long winding galleries. The injury often appears first in the top of the tree and, if the infestation is not checked, gradually moves downward with death the final result. In its northern range 2 years are required for a generation; in the South 1 year.

Some natural control results from woodpeckers and a chalcid wasp, *Phasgonophora sulcata* Westwood. Keeping trees vigorous is recommended. At times, a pesticide may be needed. It should be applied to kill adults or young hatching larvae before they enter the bark.

REFERENCES: *Cornell Agr. Exp. Sta. Bul.* 234, 1906; *USDA Misc. Pub.* 657, 1950; 1175, 1972; *Can. Ent.* 89:12–36, 1957; *OARDC Res. Bul.* 983, 1966.

ROUNDHEADED APPLE TREE BORER

Saperda candida Fabr., Family Cerambycidae

This borer is a native North American insect that attacks shadbush, wild crabs, mountain ash, and hawthorn, the importance being approximately in the order named. In addition, apple, pear, quince, plum, and cherry may also be injured, particularly when some of the preferred hosts are in the vicinity. Distribution is rather general throughout the area from Maine and southern Canada to North Dakota, south to Texas and central Georgia. Greatest damage is caused by the larvae feeding in the cambium region near the ground level which results in girdling, and in the heartwood which weakens the trees so much that they are easily broken by winds. Young trees are especially vulnerable to attack if they are in the vicinity of shadbush or service berry. Adults feed on foliage and fruits but it is considered unimportant.

The appearance of the insect is shown in Fig. 269. It is necessary to add that the adult ground color is light olive brown with conspicuous white stripes, and the larva is creamy white with darker mouthparts and head capsule.

The life cycle requires 2 to 4 years; 3 is the usual period. Pupation occurs in the spring of the last year of larval life and takes place in the galleries excavated by the larvae. Beetles are active from late April to September depending on the latitude; however, the greatest numbers are to be found in June. Less than a week after the adults emerge egg-laying begins; the eggs are deposited in small cavities in the tree trunk, usually close to the ground. On hatching, the larvae bore into the tree and feed. In large trees the first year is spent in the sapwood, subsequent feeding being in the heartwood; in young trees they may go directly to the heartwood, and in such trees the work of a single larva may be enough to kill the tree. Much fibrous frass exuding from holes in the bark is an indication of larval activity.

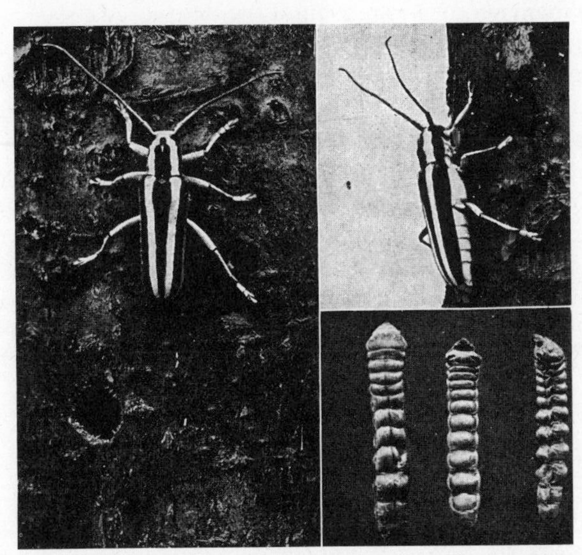

Figure 269. The roundheaded apple tree borer, *Saperda candida* Fabr.; larvae, adults, and exit hole; natural size. (After Rumsey and Brooks.)

Eradication of wild host plants from the vicinity of the orchard is advised. Removing the larvae by hand is feasible if only a few trees are involved. Good control of larvae in trees already infested has been obtained by injecting fumigants into the tunnels. Bearing orchards receiving a regular spray program are not troubled by this insect.

REFERENCES: *USDA Dept. Bul.* 847, 1920; *Leaflet* 274, 1965; *N.Y. Agr. Exp. Sta. Bul.* 688, 1940; *Ohio Agr. Exp. Sta. Res. Bul.* 930, 1963.

PHILOEM NECROSIS LEAFHOPPER

Scaphoideus luteolus, V.D., Family Cicadellidae

This white-banded leafhopper, a common pest of elms, is important primarily because it is the known vector of the mycoplasma producing phloem necrosis of the American and winged elms. Once a tree is affected, death is certain though not immediate. The present known area of infection by this disease is from Ohio to Wyoming, south to Texas, east to South Carolina, and north to Lake Erie.

When elm leaves turn yellow, wilt, and drop off, the tree may be infected. Cut through the bark at several places around the lower trunk or on buttress roots and expose the inner bark. If the inner bark that lies next to the wood is a yellow or butterscotch color, sometimes flecked with brown, remove a strip of the discolored layer and place it in a stoppered vial for a few minutes in a warm place. Phloem necrosis is present if the discolored tissue gives off a faint odor of wintergreen.

The leafhopper eggs are deposited in the outer bark of living elm trees and the winter is passed in this stage. Hatching occurs in the spring soon after elm foliage appears. The nymphs pass through 5 instars, and approximately 50 or more days later the adult stage is reached. The adults are brown with faint mottling, and nearly 5 mm in length. They are present throughout the remainder of the summer, during which they deposit the overwintering eggs.

To prevent the spread of the disease it is necessary to control the leafhoppers. Spraying thoroughly with methoxychlor has been most effective if applied when elm leaves are fully grown, usually in June, followed by a second application in mid-August.

REFERENCES: *J. Econ. Ent.,* 42:729–732, 1949; *USDA Leaflet* 329, 1952.

ROSE SLUGS OR SAWFLIES

ORDER HYMENOPTERA, FAMILY TENTHREDINIDAE

The larvae of three species of sawflies injure roses by skeletonizing the foliage or chewing large ragged holes in the leaves. These larvae (Fig. 270) are re-

Figure 270. Rose slug, *Endelomyia aethiops* (F.); 2×. (USDA)

ferred to as "slugs" because of their resemblance to those animals, and false caterpillars because of their resemblance to the larvae of moths and butterflies. The adults are black-bodied, 4-winged insects about the size of the house fly. They insert their eggs into the leaf tissue.

The bristly rose slug, *Cladius difformis* (Panzer), is yellow-green with a darker green line down the back, and has 7 pairs of abdominal prolegs. The body is covered with stiff hairs from which it derives its common name. When fully grown it is about 12 mm in length. It skeletonizes the leaves when young but may devour the entire leaf in later instars. Pupation takes place inside a thin membranous cocoon attached to objects above or on the ground. There are 6 overlapping generations a year in the vicinity of Washington, D.C.

The rose slug, *Endelomyia aethiops* (F.), formerly called the European rose slug, skeletonizes the leaves usually from the upper surface. The pale green larva is nearly 12 mm long when fully grown, with 8 pairs of abdominal prolegs. This species has only one generation per year. It appears in early spring, completing its development and entering the soil where it remains inactive inside an earthern cell until the following spring.

The curled rose sawfly, *Allantus cinctus* (L.), sometimes referred to as the coiled rose worm, eats the entire leaf tissue, usually from a curled position along the leaf edges. The larva has 8 pairs of abdominal prolegs. It is green with gray sides and legs, and can be distinguished from the other species by its yellow-brown head marked with a black spot. Pupation occurs inside cells bored in the pith of soft wood or pruned rose stems. There are 2 generations each season and the winter is passed in the pupal stage.

Dusts containing either diazinon, malathion, or rotenone easily control these insects. The all-purpose rose dusts are also quite satisfactory if they contain at least one of these pesticides.

REFERENCES: *USDA Misc. Pub.* 626, 1948; *Agr. Inf. Bul.* 237, 1962.

MIMOSA WEBWORM

Homadaula anisocentra Meyrick, Family Glyphipterygidae

This web-forming defoliator that attacks mimosa and honey locust may be found in almost all areas of the eastern half of the United States where these hosts abound. At times damage is severe and the trees greatly disfigured. The species was discovered in the vicinity of Washington, D.C. in 1940 and is thought to be native to the Indo-Australian region.

Adults of mimosa webworm have a wingspread of about 25 mm. The forewings are silvery gray with about 20 conspicuous black dots; the hindwings are brownish gray. In Ohio, moths have been observed from mid-May to mid-October.

The pearly gray oval eggs are deposited on the flowers and leaves of the hosts. The color soon changes to a shade of pink, and hatching occurs within a few days. Larvae spin webs and pull leaflets together into tunnellike masses where they feed on the inner surfaces. Damaged leaves turn brown and often absciss. When disturbed, larvae will twist violently. Mature larvae are nearly 25 mm long, vary from pale green to dark brown, and have 5 longitudinal white stripes. Fully grown larvae often descend to the ground on silken threads where they pupate within silken cocoons in the litter, bark crevices, or other objects; others pupate in webbed foliage. Winter is passed in the pupal stage and 4 generations per year have been observed. Foliar sprays or systemic organophosphorus compounds have been successful in control.

REFERENCES: *Proc. U.S. Nat. Mus.*, 93:205–208, 1943; *J. Econ. Ent.*, 40:546–553, 1947; *Arborist's News*, 29:17–20, 1964; *Ohio Res. Bul.* 983, 1966; *Ohio Sta. Univ. Ext. Bul.* 504, 1977.

CONIFER SAWFLIES

ORDER HYMENOPTERA, FAMILY DIPRIONIDAE

Colonies of caterpillars devouring the needles of conifers are usually the larvae of sawflies. Sometimes they become so numerous that complete defoliation of the tree results. Repeated yearly attacks will cause death of the tree. The caterpillars are 25 mm or less in length with 6 to 8 pairs of abdominal prolegs without crochets. Adults are typical sawflies with 4 membranous wings and a broad attachment of abdomen to thorax. There are several injurious species in North America.

The European pine sawfly, *Neodiprion sertifer* (Geoffroy) overwinters as eggs in slits in the needles. These hatch the following April or May and the larvae feed on the old needles, becoming fully grown by late May or early June. The black-headed larvae are gray-green with a longitudinal dorsal line of lighter shade; laterally there are 2 white lines bordering a stripe of intense green to black. Pupation occurs in a brown cocoon, mainly in the duff beneath the trees, but occasionally some cocoons are found on twigs. Adults emerge in autumn and lay their eggs. Only 1 generation develops each year. The favored hosts are red, Scotch, jack, Japanese red, Swiss mountain, and mugho pines. This species is often a problem in Christmas tree plantations found in the eastern half of North America.

The red-headed pine sawfly, *N. lecontei* (Fitch) (Fig. 271), frequently attacks young plantings of a wide variety of hard pines and other conifers. Young larvae are white and unspotted with a brown head; later, after a series of molts, they become yellow with 6 rows of black dots on the body, and the head becomes red. Winter is passed as brown cocoons in the duff under the trees. One to 5 generations develop each year, depending on latitude.

The white-pine sawfly, *N. pinetum* (Norton), is mainly a white pine pest but it occasionally attacks other species. It resembles the red-headed species but has only 4 rows of black dots on the body, and the head is black. The life cycle is also similar to that of the red-headed pine sawfly.

Other sawfly species of importance at times are: red-pine sawfly, *N. nanulus nanulus* Schedl; Swaine jack-pine sawfly, *N. swainei* Middleton; jack-pine sawfly, *N. pratti banksianae* Rohwer; balsam-fir sawfly, *N. abietis* (Harris); lodgepole sawfly, *N. burkei* Middleton; hemlock sawfly, *N. tsugae* Middleton; Virginia pine sawfly, *N. pratti pratti* (Dyar); loblolly pine sawfly, *N. taedae linearis* Ross; blackheaded pine sawfly, *N. excitans* Rohwer; European spruce sawfly, *Gilpinia hercyniae* (Hartig); and introduced pine sawfly, *Diprion similis* (Hartig).

Natural enemies include rodents, predaceous beetles, a few insect parasites, and virus diseases.

REFERENCES: *J. Econ. Ent.*, 32:887–888, 1939; *USDA Misc. Pub.* 657, 1950; *Tech. Bul.* 1118, 1955; *Ill. Nat. Hist. Survey Cir.* 47, 1958; *Ohio Res. Bul.* 983, 1966.

Figure 271. Red-headed pine sawfly, *Neodiprion lecontei* (Fitch); 2×. (Courtesy of R. E. Treece, OARDC)

BOXELDER BUGS

Leptocoris trivittatus (Say), Family Rhopalidae

Widely distributed in North America, boxelder bugs are found almost everywhere boxelder trees grow. They feed primarily on the pistillate or seed-bearing trees by sucking sap from the leaves, tender twigs, and developing seeds. Occasionally they have been observed feeding on maple and ash. The plant damage is not considered important and the bugs are much better known as a source of annoyance to home owners because of their habit of congregating in large numbers in or on the outside of houses in autumn, when adults are seeking hibernation quarters. If they gain entrance to houses they do not harm clothing, foods, or other household articles.

The adult bug (Fig. 272) is about 13 mm long, brown-black, with 3 longitudinal red stripes on the thorax and red margins on the basal half of the wings. Under the wings the abdomen is bright red.

Overwintering adults leave their hibernating quarters with the coming of warm spring weather and begin laying red eggs in crevices of the bark and on other objects in the vicinity of the host plant. Hatching occurs in about 14 days and new adults appear in July and begin laying eggs that result in a second generation by early autumn.

The western boxelder bug, *L. rubrolineatus* Barber, is the western North American species, and, in addition to feeding on boxelder, frequently damages pear fruits of orchards growing near that host.

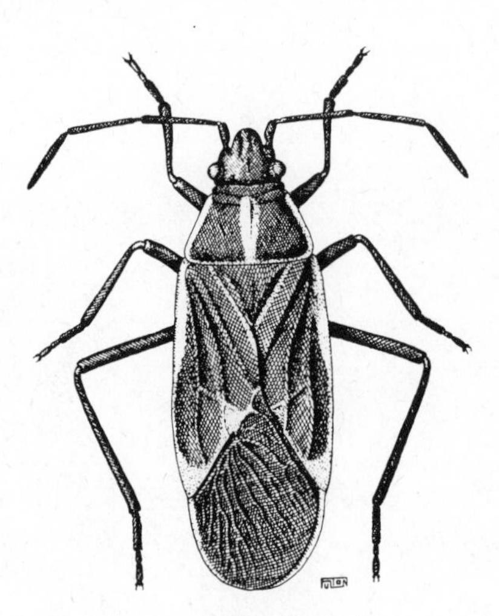

Figure 272. The boxelder bug; 4×. (Courtesy of Knowlton, Utah Agr. Exp. Sta.)

Spraying infested trees in June while nymphs are present will reduce bug populations in the vicinity of homes and orchards. Endosulfan, diazinon, malathion, and carbaryl have been employed with success. Practical control of these insects inside the home consists of collecting them with a tank-type vacuum cleaner and destroying them. Silvicultural control is having nurserymen propagate boxelder trees only from staminate cuttings. To eliminate the bug problem all pistillate boxelder trees should be removed from a given area and replaced with tree species that have less insect pests and are of higher ornamental value. Boxelder trees should never be allowed to grow near homes or orchards.

REFERENCES: *USDA Misc. Pub.*, 657:118–119, 1950; *Wis. State Dept. Agr. Bul.*, 330:22, 1955; *Proc. Ent. Soc. Ontario,* 92:202–203, 1961; *Rutgers Univ. Ext. Leaflet* 371, 1963; *Calif. Agr.* 30(10):23, 1976.

MAPLE BLADDERGALL MITE

Vasates quadripedes Shimer, Family Eriophyidae

Globular galls, about 3 mm in diameter, occurring on the upper leaf surface of silver maple trees, are evidence of this pest. When first formed the galls may be pink or green but they often become bright red and later almost black. The cause of the gall formation is a microscopic mite that lives through the winter in bark crevices of the host, then crawls to new leaf growth in the spring. Mites feed on the lower leaf surface by means of their sucking beaks which causes the galls to develop on the upper surface. As the galls enlarge, a cavity forms within with an opening to the lower leaf surface; the mites continue to feed and reproduce inside until autumn when they migrate to the bark before the leaves absciss. They cause no appreciable damage to the tree except to reduce its ornamental value. Silver maple trees are usually not recommended for planting on streets or near homes.

REFERENCES: *OARDC Res. Bul.* 983, 1966; *USDA Misc. Pub.* 1175, 1972.

18

PESTS OF POME FRUITS

Many pests discussed in previous chapters may also injure pome fruits. The seriousness of the pests depends greatly on the regions concerned and the management practices followed.

EUROPEAN RED MITE

Panonychus ulmi (Koch), Family Tetranychidae

Introduced from Europe some time previous to 1911, this spider mite has become one of the most important pests of fruit trees in northern United States and adjacent regions of Canada. It has not been reported on deciduous trees south of 34° N. latitude. The mite attacks elm, apple, pear, peach, plum, and prune to an injurious extent, and may be found on other deciduous trees and shrubs as well.

Sap removal by the piercing-sucking active stages results in bronzing and off-colored foliage and, under severe infestations, defoliation, and undersized, poorly colored fruits.

The mites are usually rusty in color, but newly emerged females are bright velvety red, changing in time to dark red-brown. Males are dull green to fulvous. Prominent, curved, dorsal white spines on the body help distinguish the red mite from red individuals of the two-spotted spider mite. The red-orange eggs of the European red mite are easily recognized by the hairlike projection on the upper side.

Winter is passed in the egg stage, usually on twigs and branches of the hosts. Hatching takes place at or before the pink stage. The first instar has 6 legs, and, after molting, succeeding instars (called protonymphs, deutonymphs, and adults) all have 8 legs. Females are globular-shaped; the males are narrower with a more pointed abdomen (Fig. 273). Development from egg to adult is

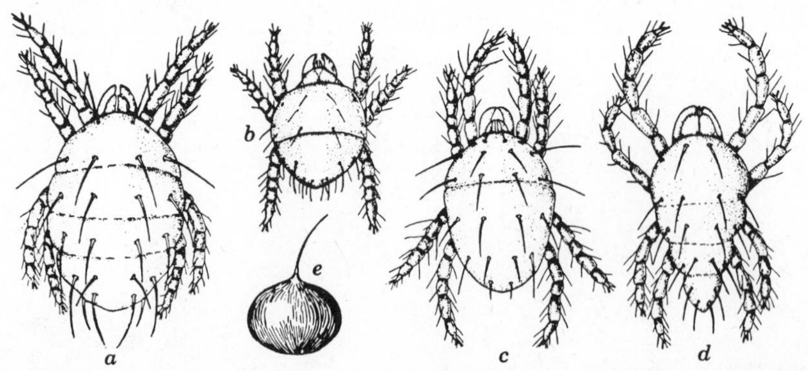

Figure 273. The European red mite, *Panonychus ulmi* (Koch): *a,* adult female; *b,* first instar; *c,* deutonymph; *d,* adult male; *e,* egg; 60×. (Newcomer & Yothers, USDA)

affected by temperature but generally requires 1 to 3 weeks. Fertilized eggs develop into males and females, but unfertilized eggs develop only into males. There may be as many as 6 to 8 overlapping generations each season. Summer eggs are laid on foliage. Winter eggs are laid on twigs, their production is influenced by photoperiod, temperature, and nutritional factors in the leaves, and deposition usually begins in mid-August and continues until cold weather.

Several predators attack this mite as well as other orchard mites. A lady beetle, *Stethorus picipes* Casey, is an important predator in the Northwest, *S. punctum* (LeConte) in the East and Midwest; *S. punctillum* Weise in various regions of Canada. The stigmaeid mites, *Zetzellia mali* (Ewing), and *Agistemus fleschneri* Summers, and species of phytoseiid mites in the genera *Amblyseius, Phytoseiulus,* and *Metaseiulus,* are all very important in checking plant-feeding mite populations. Other valuable predators are the minute pirate bug, *Orius tristicolor* (White); the thrips, *Leptothrips mali* Fitch, *Haplothrips faurei* Hood, and *Scolothrips sexmaculatus* (Pergande); the chrysopids, *Chrysopa carnea* Stephens, *C. oculata* Say, *C. nigricornis* (Burmeister), and *C. rufilabris* Burmeister; the plant bugs (Miridae), *Hyaloides vitipennis* (Say); *H. harti* Knight; *Deraeocoris nebulosus* (Uhler); *D. brevis* Uhler; and *Plagiognathus politus* Uhler. Frequently this group of predators is so effective that no miticides are necessary.

A few mites must be on the leaves to support the appetites of the predators, otherwise they starve. It has been noted* that as many as 20 mites/leaf are no cause for alarm if they do not persist at this level more than 10 to 15 days. An ideal orchard management program attempts to maintain a balance between predators and their prey while using necessary pesticides. For killing overwin-

* R. P. Holdsworth, "Major predators of the European red mite in Ohio," *Ohio Res. and Dev. Center Res. Cir.* 192, 1972.

tering European red mite eggs, a delayed dormant (½ inch green tip stage) application of a 2% superior-type oil emulsion, or an oil-ethion combination is still a recommended practice. The eggs become increasingly susceptible to pesticides as hatching time approaches. Should emergency mite control arise in mid-summer use a miticide least damaging to the predators. Also keep in mind that continuous use of the same miticide favors the development of resistant strains of mites. To avoid or delay this development growers are urged to rotate the use of different classes of miticides. For example, apply oil in early season each year and in mid-summer the first year use an organophosphate, the second year a chlorinated hydrocarbon, the third year a sulfur-based compound, and the fourth year begin this sequence of sprays again.

Selection of the proper chemical depends on such factors as cost, availability, toxicity to other pests and beneficial insects, presence of resistance, phytotoxicity, and hazards in handling.

REFERENCES: *USDA Tech. Bul.* 89, 1929; *Va. Agr. Exp. Sta. Tech. Bul.* 98, 1946; *J. Econ. Ent.*, 46:112–115, 894–896, 1085–1086, 1953; 57:35–37, 1964; *Hilgardia*, 21:253–287, 1952; *Ohio Agr. Exp. Sta. Cir.* 37, 1956; *Ann. Ent. Soc. Amer.* 62:1261–1267, 1969; *Bul. Ent. Soc. Amer.* 17:89–91, 1971; *Ohio Sta. Univ. Ext. Bul.* 506, 1979.

OTHER SPIDER MITES

ORDER ACARI, FAMILY TETRANYCHIDAE

In addition to the European red mite and the two-spotted spider mite (p. 307), there are many other species that attack a wide variety of trees, shrubs, flowers, and agricultural crops. A few of the more common ones are mentioned here with some comment on their habits, appearance, life cycle, distribution, and control. They all cause injury by removal of plant sap with their piercing-sucking mouthparts, and are less than 0.5 mm long.

Clover Mite, *Bryobia praetiosa* Koch, is a worldwide pest of fruit trees, clover, and other legumes, as well as a number of garden and field crops, including cotton. It may also become a nuisance by invading houses, particularly in the spring and occasionally in autumn. It is seldom found on the aerial portions of trees, does not web leaves with silk, and is more tolerant of cold than the brown mite. Winter is passed principally as smooth spherical cherry red eggs and as flattened, dark brown to dull green adults with rather long front legs (Fig. 274). The winter eggs hatch in early spring into scarlet nymphs which, on reaching the adult stage lay the summer eggs that usually estivate until September. Occasionally some of these eggs hatch, giving rise to a succession of summer generations. Only one generation in the fall develops from the hatching estivating eggs. Keeping all turf plant growth 18 inches from the foundation of homes prevents these mites from entering.

Brown Mite, *Bryobia rubrioculus* (Scheuten), is of economic importance in many countries, living in the aerial portions of orchard trees, for example,

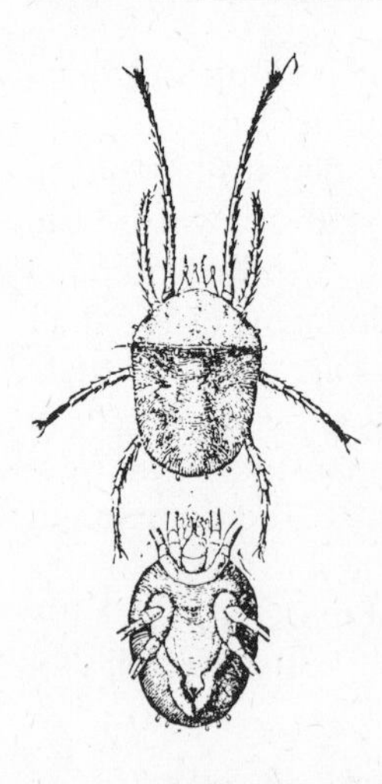

Figure 274. Dorsal and ventral (legs removed) views of the clover mite, *Bryobia praetiosa* Koch; 30×. (USDA)

apple, almond, apricot, cherry, peach, pear, and plum. It closely resembles the clover mite with which it has been confused. Winter is passed as spherical red eggs on twigs and branches; these begin hatching in the late delayed dormant to early pink periods of apple tree development and this continues until full bloom is reached. No webbing of silk is associated with their foliage feeding. With the approach of autumn the brown mite migrates from the leaves back to the twigs and branches where overwintering eggs are deposited. The maximum number of generations per year in Canada is four.

Pacific Spider Mite, *Tetranychus pacificus* McGregor (Fig. 136) is one of the most serious pests of deciduous fruits in western United States. It also attacks many small fruit, legume, and vegetable crops, as well as ornamental plants and weeds. Adult females are pale green with dorsal dark spots medially and a pair of dark spots near the posterior end of the body. Overwintering females are often bright orange and are found under the bark of trees or in the duff on the ground. Feeding is general over the undersurface of the leaves, and dense webbing is produced. The eggs are spherical and pearly white. Treat with a miticide if 8 mites per leaf are found on grape.

Schoene Spider Mite, *Tetranychus schoenei* McGregor, is widely distributed throughout the southeastern states, attacking deciduous fruit and ornamental trees and shrubs, as well as field crops. Adult females are pale green with four dorsal darkened spots. Overwintering females are bright orange. The development, webbing, and feeding habits are similar to those of the Pacific mite.

Four-Spotted Spider Mite, *Tetranychus canadensis* (McGregor) (Fig. 275), has been found in many areas of the eastern half of the United States and in southern Canada. Specimens have been taken from apple, cotton, rose, elm, linden, plum, horse chestnut, and osage orange. Adult females resemble those of the Schoene mite. The biology of this species is the same as for *T. schoenei*.

Strawberry Spider Mite, *Tetranychus turkestani* Ugar. and Nik., is found

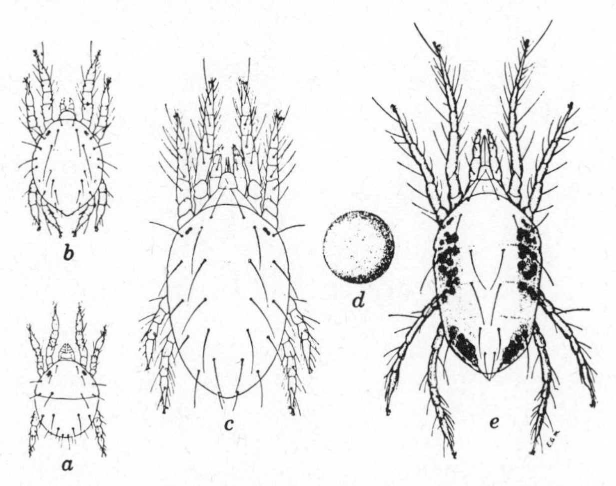

Figure 275. The four-spotted spider mite, *Tetranychus canadensis* (McGregor): *a,* first instar; *b,* second instar or protonymph; *c,* third instar or deutonymph; *d,* egg; *e,* adult female; 60×. (McGregor and McDonough)

throughout the United States. It commonly feeds on cotton, strawberries, legumes, ornamental shrubs, and fruit trees. The appearance and life cycle are similar to the two-spotted mite, *T. urticae* Koch (see p. 307).

Other species are: McDaniel spider mite, *T. mcdanieli* McGregor; carmine spider mite, *T. cinnabarinus* (Boisduval); yellow spider mite, *E. carpini borealis* (Ewing); Yuma spider mite, *E. yumensis* (McGregor); avocado brown mite, *Oligonychus punicae* (Hirst); southern red mite, *O. ilicis* (McGregor); Banks grass mite, *O. pratensis* (Banks); and avocado red mite, *O. yothersi* (McGregor). These species resemble many of the mites already described in greater detail. Often a trained acarologist may have to make the final determination of species. References cited should be of value to those interested in further details concerning spider mites.

REFERENCES: *Conn. Agr. Exp. Sta. Cir.* 180, 1951; *Bul.* 552, 1952; *Hilgardia,* 21(9), 1952; 22(7), 1953; *Va. Agr. Exp. Sta. Tech. Bul.* 87, 1943; 124, 1956; *J. Econ. Ent.,* 50:135–141, 1957; *Can. Ent.,* 90:23–42, 1958; *Ann. Ent. Soc. Amer.* 57:220–226, 1964; *Cornell Agr. Exp. Sta. Memoir* 380, 1963; *USDA Home and Gard. Bul.* 134, 1967; *Ann. Rev. Ent.* 14:125–174, 1969.

APPLE APHIDS

ORDER HOMOPTERA, FAMILY APHIDIDAE

Three species of aphids are usually found on apple and related hosts every year, sometimes in destructive numbers. They are known as the apple aphid, *Aphis pomi* DeGeer (Fig. 276), the rosy apple aphid, *Dysaphis plantaginea* (Passe-

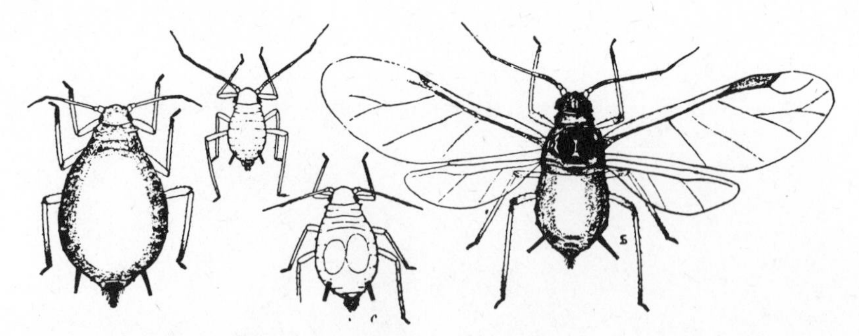

Figure 276. The apple aphid, *Aphis pomi* DeGeer; ovoviviparous female, young nymph, oviparous female, and spring migrant; much enlarged. (USDA)

rini) (Fig. 277), and the apple grain aphid, *Rhopalosiphum fitchii* (Sanderson) (p. 181). All species are widely distributed in the apple-growing areas of North America. Both foliage and fruits are injured by the piercing-sucking nymphs and adults. Evidence of their activity is indicated by curled, twisted, stunted leaves, especially the tender terminal growth of new shoots, and stunted malformed clusters of fruits later in the season. Rosy aphids are considered the most destructive species, with *A. pomi* sometimes becoming a problem in some areas.

All three species have very similar life cycles. Winter is passed in the egg stage on twigs, usually around buds or in crevices in the bark (Fig. 278). Eggs

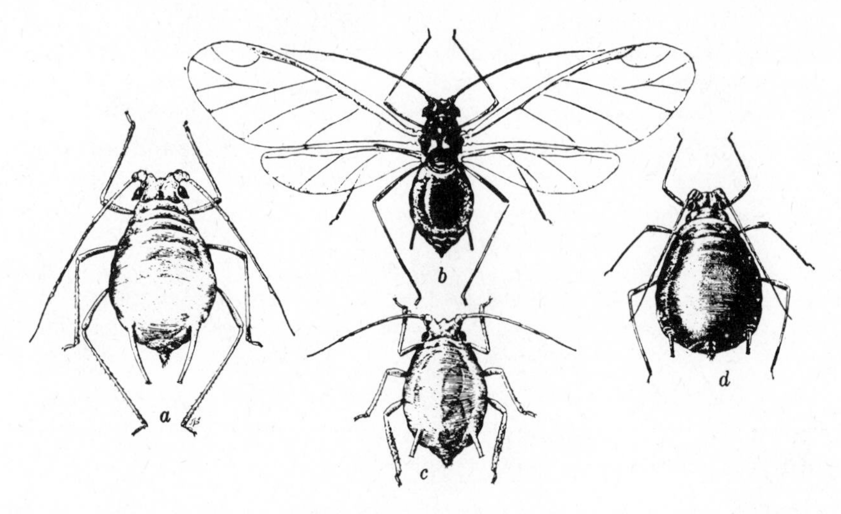

Figure 277. The rosy apple aphid, *Dysaphis plantaginea* (Passerini): *a,* ovoviviparous female; *b,* fall migrant; *c,* oviparous female, and *d,* spring wingless form; much enlarged. (USDA)

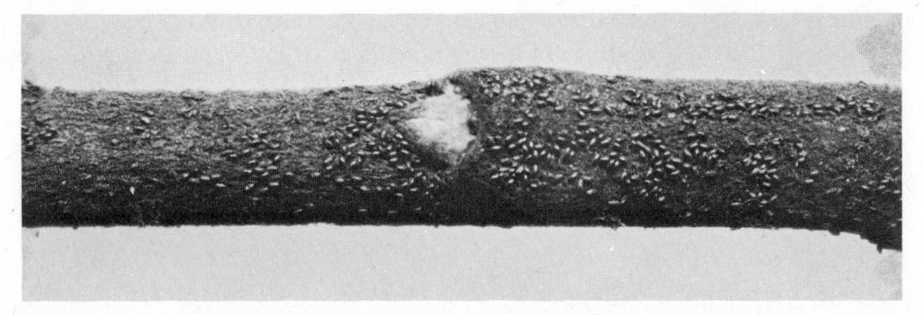

Figure 278. Eggs of apple aphids. (Courtesy of Cutright, Ohio Agr. Exp. Sta.)

begin hatching in the spring just as the buds are in the green tip stage. Newly hatched nymphs are all wingless females and when fully grown are called stem mothers. These parthenogenetic females give birth to young, with repeated generations occurring about every two weeks. Early in the summer winged forms are produced which migrate to new host plants of the same species or to plants of a different species. These alternate hosts are often called the secondary hosts. Rosy aphids may remain on apple throughout the summer, but they usually migrate to narrow leaf plantain; apple grain aphids remain on apple a shorter period and migrate to grains and grasses, usually before they have done any damage to apple; *Aphis pomi* is usually a permanent resident of apple and related trees.

Reproduction continues throughout the summer, and in autumn winged forms develop on the secondary hosts that return to the primary host. Here are born nymphs that develop into true sexual forms; these mate, the females depositing the overwintering eggs. The eggs are green when first laid but soon turn shiny black.

Color and cornicles serve as useful characters to distinguish newly hatched nymphs. Rosy aphids are very dark purple, covered lightly with gray powder, and the cornicle length greatly exceeds the diameter; apple grain aphids are dark green, and the cornicles scarcely project from the abdomen; the apple aphid is green, with the cornicle length about equal to the diameter. Stem mothers of the rosy aphid are slate-colored, some rosy or pink, all with dark appendages; those of apple grain aphids are light green, having a median dorsal dark stripe with cross-bars of the same color and pale yellow appendages with darkened tips; those of apple aphids are yellow-green with black cornicles and darkened tips on the antennae, tarsi, and tibiae. Adults are about 4 mm in length.

Besides climatic factors which play an important role in natural control, there are many natural enemies of aphids, such as lady beetles, syrphid fly larvae, and aphidlion predators, and the parasite, *Aphidius testaceipes* (Cresson). Frequently there is no need for chemical control because of these natural control agents.

The commonly recommended ovicide is 2% oil emulsion applied alone or in combination with other chemicals in the dormant or delayed-dormant stage of plant development.

REFERENCES: *Ohio Agr. Exp. Sta. Bul.* 464, 1930; *Conn. Agr. Exp. Sta. Bul.* 552, 1952; *Calif. Agr. Exp. Sta. Leaflet* 76, 1966; *USDA Agr. Handbook* 290, 1965; *Ohio Sta. Univ. Ext. Bul.* 506, 1979.

WOOLLY APPLE APHID

Erisoma lanigerum (Hausmann), Family Aphididae

This woolly aphid occurs in practically all the apple-growing districts of the world. Besides apple, it attacks elm, mountain ash, and species of hawthorn. It feeds on plant sap from the roots as well as from the upper parts of the trees. Above ground it is found chiefly on the trunks, limbs, and twigs, in old pruning scars or wherever the bark is tender. This injury is not considered as serious as that produced by the root-feeding forms, which cause gall-like swellings. On young seedlings root injury may be quite severe. The insects are recognized by the white woolly covering which is predominantly at the posterior end of their blue-black bodies (Fig. 279).

The life history of this aphid is rather complicated. Eggs are laid, usually on the bark of elm trees in the fall; these eggs hatch in the spring, and wingless, parthenogenetic, ovoviviparous stem mothers establish colonies on the terminal leaves, which soon become curled and stunted from the feeding. By early summer a generation of winged forms appears; they fly to apple and other hosts

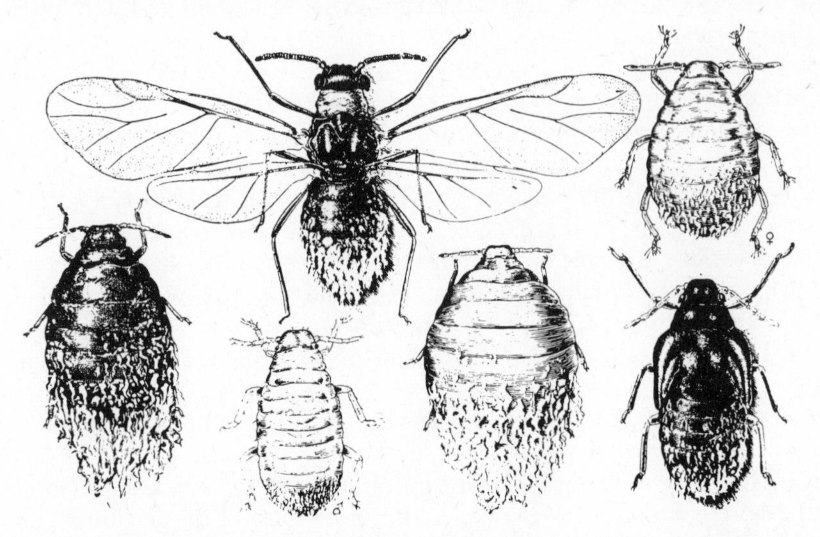

Figure 279. Various stages of the woolly apple aphid, *Eriosoma lanigerum* (Hausmann); much enlarged. (Baker, USDA)

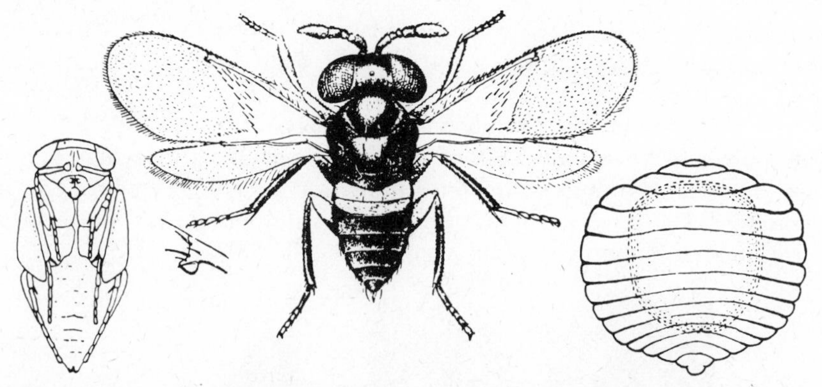

Figure 280. *Aphelinus mali* (Hald.), a very effective parasite of the woolly aphid; (left) pupa; (right) larva; greatly enlarged. (USDA)

and there establish new colonies. Repeated generations are produced, and some of the individuals crawl to the tree roots where they may continue to reproduce indefinitely. In the fall, winged individuals again develop; they fly back to elm where they give birth to the sexual forms which eventually mate and lay overwintering eggs. Not all the aphids return to elm, some wingless forms remaining on apple all winter, both above and below ground, thus maintaining a continuous infestation on this host.

The woolly pear aphid, *Erisoma pyricola* Baker and Davidson, is a related species similar in appearance and life cycle. Its alternate host is elm, and the control measures are the same as for woolly apple aphid.

Some variation exists in susceptibility of apple varieties to attack by woolly aphids, Northern Spy being resistant wherever grown. Syrphid fly larvae and lady beetles are considered important predators. A chalcid wasp, *Aphelinus mali* (Hald.) (Fig. 280), is an important parasite which has kept woolly aphids checked in all areas where it has been introduced. However, with the advent of new chemicals for orchard pest control, populations of this wasp have been reduced so much in some regions that woolly aphids are again a problem.

REFERENCES: *Va. Agr. Exp. Sta. Tech. Bul.* 57, 1935; *Maine Agr. Exp. Sta. Bul.* 256, 1916; *J. Econ. Ent.*, 43:463–465, 1950; 57:1009, 1964; *USDA Agr. Handbook* 290, 1965; *J. Hort. Sci.* 37:207–218, 1962.

CODLING MOTH

Laspeyresia pomonella (L.), Family Olethreutidae

The codling moth, long known as an apple pest, was introduced into this country from Europe by early settlers. It now occurs wherever apples are grown and is generally considered one of the most important pests of the crop. Pear, quince, English walnut, and occasionally other fruits are also attacked. Injury is caused by the larval stage, which tunnels into the fruits, usually to the core, lowering their commercial value and keeping quality. Blemish marks called "stings" also result when larvae chew into fruits that have been treated with slow-acting poison sprays, such as lead arsenate.

The moth has a wing expanse of nearly 18 mm; the fore wings are gray-brown, crossed with lines of lighter gray with deep golden or bronzed areas near the tips. The larva is white, often tinged with pink; it has a brown head, and is 12 mm in length when fully developed (Fig. 281).

Figure 281. The codling moth: *a*, egg; *b*, larva just emerging from the egg; *c*, larva in cocoon; *d*, pupa; and *e*, adult; enlarged; *f*, infested apples. (Slingerland)

Only fully grown larvae survive the winter, hibernating in silken cocoons in places of concealment on or near apple trees, or in and about packing sheds. Pupation takes place in the spring. Moths begin emerging about the time that apples are in bloom, live for 2 or 3 weeks, and deposit some 30 to 40 eggs per female. Most moths of the first generation are present in May and early June. They lay many of their eggs between sundown and 10 P.M.; few eggs are laid, however, if the temperature is below 60 F (16 C). The tiny, white, flattened eggs with crinkled edges are approximately the size of a common pinhead (Fig. 281). They are placed on leaves, twigs, and fruits. Usually the weather is cool when the first eggs are laid, and about 2 weeks are required for incubation. In hot weather the eggs hatch in 5 days. The young larvae bore into the fruits, often entering through the calyx, feed for nearly 3 weeks, then leave and spin cocoons in the places in which they normally hibernate. Approximately 2 weeks later the moths of the next generation begin to emerge, and the cycle is repeated. Sometimes the second-generation moths appear before those of the first cease their activity, resulting in overlapping of generations. In more northern areas some first-generation larvae do not pupate but remain in their cocoons until the following spring. Three nearly complete generations and a partial fourth are known to occur in the southernmost apple-growing areas.

Tremendous variation in codling moth populations in different orchards and orchard regions, and in the same orchard or region at different periods, indicates that the insect is much affected by natural controlling agencies. Climatic factors, especially temperature, are known to play a great part in determining abundance. Biological factors are also of some value, with the ichneumon, *Itoplectis conquisitor* (Say), the braconid larval parasites, *Macrocentrus ancylivorus* Rohwer, *M. delicatus* Cresson, *Ascogaster quadridentata* Wesm. (Fig. 282), and the egg parasite, *Trichogramma minutum* Riley, probably the most important.

Because of the necessity for controlling both insects and diseases on fruits

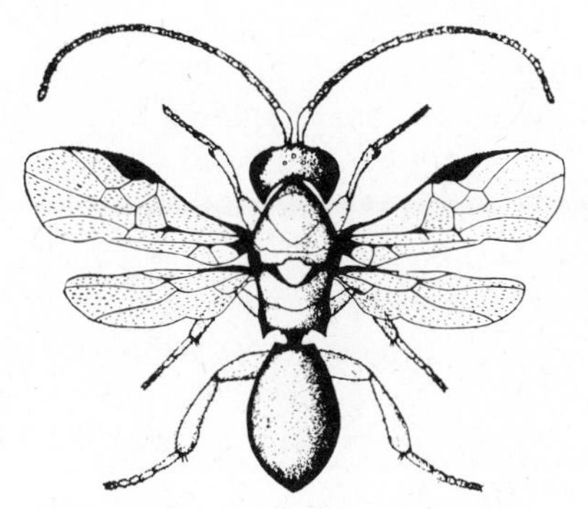

Figure 282. Ascogaster quadridentata Wesm., a braconid parasite of the codling moth, enlarged. (Cox, N. Y. Agr. Exp. Sta.)

most states and provinces have developed spray programs for their growers. These embody the best combination of compatible materials to bring about effective control for a particular region with the least damage to the natural enemies.

Although details of a spray program for codling moth control do vary widely, the main features can be given. The first spray is called the petal-fall and is applied when nearly all petals have fallen and before the calyx closes. It is followed by a variable number of cover sprays, usually 4 to 6 at approximately 2- to 3-week intervals. Proper timing of sprays is essential and may be determined locally by observations of moth activity by means of bait pails or pheromone traps. The total number of cover sprays depends on many factors. These are intensity of infestation, activity of natural enemies, seasonal variations, possibility of excess poisonous residues, and probable returns for the extra materials and effort expended. Regardless of the total cover sprays employed, it is important that a second-generation spray (sometimes designated as the fifth cover spray) be applied about 10 weeks after petal-fall, which would normally be in late July or early August for most fruit-growing areas.

REFERENCES: *Ohio Agr. Exp. Sta. Bul.* 583, 1937; *Res. Bul.* 930, 1963; *Conn. Agr. Exp. Sta. Bul.* 552, 1952; *USDA Cir.* 270, 1950; *Calif. Agr. Exp. Sta. Leaflet* 76, 1966; *Wash. Agr. Exp. Sta. Bul.* 340, 1936; *Ohio Sta. Univ. Ext. Bul.* 506, 1979.

APPLE MAGGOT

Rhagoletis pomonella (Walsh), Family Tephritidae

The apple maggot is a native insect widely distributed from North Dakota to Oklahoma eastward. It has been seriously injurious primarily in the northeastern states and Canada. Hawthorn, plum, pear, and cherries also serve as hosts. Crab apples are invariably infested by this pest. A related species, *R. mendax* Curran, is called the blueberry maggot, important as a pest of the blueberry crop. Of the orchard fruits, apples are the most seriously damaged.

The adult is a fly, a little more than 6 mm in length, dark brown, with light bands on the abdomen and both light and dark markings on the wings. Larvae are white tapered maggots slightly smaller than those of the house fly (Fig. 283).

Winter is passed as puparia in the soil. Flies begin emerging during the latter part of June, continuing for a month or more. However, some overwintering puparia remain inactive, the adults not appearing until the second summer. A week or more elapses after adult emergence before eggs are deposited. The tiny white eggs are inserted underneath the skin of the apple, the susceptible varieties being Cortland, Wealthy, Delicious, and others that mature early and have sweet or subacid characteristics. Hatching occurs within a few days and the larvae mine the flesh leaving irregular winding brown tunnels, which often cause premature dropping of the fruits. Infested fruits have very little market

Figure 283. The apple maggot, *Rhagoletis pomonella* (Walsh): adult, larva, and eggs; enlarged. (Porter, USDA)

value. Larval development requires from 2 weeks in drops of early maturing varieties to 3 or more months in hard winter apples. When fully developed they leave the fruits, crawl into the soil, and transform to puparia. One generation per year is typical; in some areas a partial second brood of adults appears late in the season. Most of these begin to emerge late in September and are of little importance since practically all the fruit is harvested except the hard winter varieties which are unattractive to the flies for oviposition.

Recorded parasites are *Opius melleus* Gahan, which attacks the larvae, and *Patasson conotracheli* (Gir.), which attacks the eggs. No predators of importance have been indicated.

Where economically feasible, the systematic destruction of infested dropped apples and the elimination of species of hawthorn in the vicinity of the orchard are considered valid practices for control. Apple maggots in fruits may be killed by cold-storaging the fruit at 32 F for a period of 40 days. Fruits on ornamental flowering crabs serve as sites for the development of high populations of this pest. Pesticides for killing adults before oviposition occurs are usually necessary in the third, fourth, fifth, and sixth cover sprays.

REFERENCES: *USDA Tech. Bul.* 66, 1928; *Cir.* 600, 1941; *J. Econ. Ent.*, 40:183–189, 1947; 44:147–153, 1951; 47:479:485, 1954; 57:163–164, 1964; *Conn. Agr. Exp. Sta. Bul.* 552, 1952; 604, 1957; *Ohio Sta. Univ. Ext. Bul.* 506, 1979.

PLUM CURCULIO

Conotrachelus nenuphar (Herbst), Family Curculionidae

Widely distributed east of the Rocky Mountains, this native American insect is an important pest of stone fruits. It also attacks apple, pear, quince, and related hosts. In apple, the worms reach their full growth only in fruits that fall prematurely.

Injury to all hosts results first from the spring feeding of the adults, then from the female egg punctures, next from the feeding of the larvae within the fruits, and finally from the early fall feeding of the beetles.

The adult is a hard-bodied snout beetle, nearly 6 mm in length, brown with faint gray markings, and has 4 humps on the elytra. The fully grown larva is legless, about 9 mm long, curved, and white with a brown head (Fig. 284).

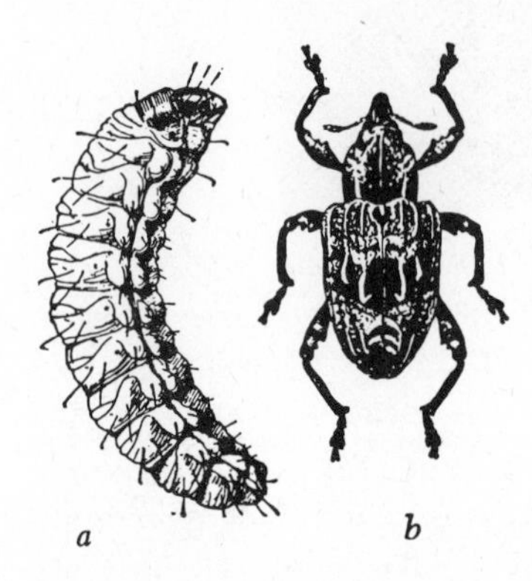

Figure 284. Plum curculio; *a,* larva; *b,* beetle; 5×. (USDA)

The beetles hibernate in protected places in or near the orchard and appear with the early foliage in the spring. They feed for a period of 5 or 6 weeks during which eggs are placed in fruits that have attained sufficient size. The white eggs are laid in cavities in the fruits made by the mouthparts; the fruits are also marked by crescent-shaped cuts just beneath (Fig. 285). Even when the larvae fail to develop, which often occurs in apples that remain on the tree, the oviposition scars persist and reduce the quality of the fruit. Both oviposition and early adult feeding punctures cause abnormal growth of fruits. Hatching requires about 5 days, and the larvae feed for 2 or 3 weeks in the fruits before

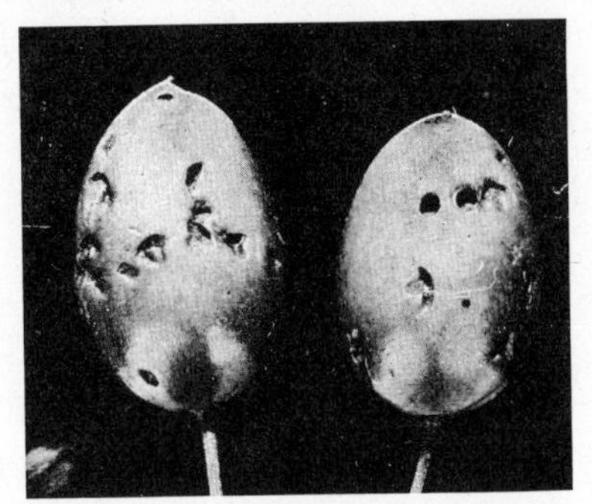

Figure 285. Plum curculio egg punctures in plums. (USDA)

they are fully developed. Mature larvae leave the fruit through a clean-cut exit hole, free of frass or webbing, drop to the ground, and pupate in the soil. The beetles emerge in a month or more and either feed on fruits for a period and then go into hibernation or lay eggs that develop into a second generation, the latter more common in the Georgia peach belt. Woodland areas and brushy fence rows adjacent to orchards serve as ideal hibernating quarters.

Natural control of the curculio results from winter mortality, attacks of birds and other predators, and from parasites. An ichneumon wasp, *Tersilochus conotracheli* (Riley), attacks the larvae in the fruit, as does the parasite, *Triaspis curculionis* (Fitch). An egg parasite, *Patasson conotracheli* (Girault), is of some importance, too. Both beetles and larvae are also attacked by a fungus, *Isaria anisopliae* Metch.

Mechanical control by jarring the sluggish beetles from the trees in the morning and capturing them on sheets was an early method of control which is occasionally employed even now in peach orchards in the South. Destruction of all infested fruits which fall to the ground is of value, especially in the 2-generation areas. Hogs have been put in orchards for this purpose. Cultivation of the soil containing the pupae destroys many of them. Sprays for this pest are usually applied at the petal-fall, first and second cover spray periods.

REFERENCES: *Conn. Agr. Exp. Sta. Bul.* 301, 1930; 552, 1952; 575, 1953; *Del. Agr. Exp. Sta. Bul.* 193, 1953; *Va. Agr. Exp. Sta. Bul.* 297, 1935; *N.Y. Agr. Exp. Sta. Bul.* 684, 1938; *USDA Tech. Bul.* 188, 1930; *Farmers' Bul.* 1861, 1954; *J. Econ. Ent.* 47:909–912, 1954; 51:131–133, 1958; 53:439–441, 1960; *Ohio Agr. Exp. Sta. Res. Bul.* 930, 1963; *Ohio Sta. Univ. Ext. Bul.* 506, 1979.

APPLE CURCULIO

Tachypterellus quadrigibbus (Say), Family Curculionidae

This native North American insect is found from Canada to Florida and westward to the Mississippi River. Common hosts are apple, pear, crab apple, and hawthorn. Knotty malformed fruits are caused by the feeding and egg punctures and represent a large proportion of the loss. Larval feeding takes place within the fruits. It is probable that, in orchards where the plum curculio does important damage, the work of the apple curculio often occurs and passes unnoticed. The injuries, except those resulting from egg punctures, are so similar that they are readily differentiated only by an experienced person.

The adult resembles the plum curculio, but is slightly smaller-bodied with a much longer snout which protrudes forward. Larvae also resemble those of the plum curculio but are thicker in the abdominal region.

Only 1 generation occurs each year. Eggs are laid during May and June in cavities hollowed out in the fruits by the long snouts of the adult females. Hatching occurs in a week, the larvae developing near the center of the fruits and in about 20 days transforming to pupae. New adults appear approximately 7 days later, do some feeding on maturing fruits, and soon enter hibernation in debris on the ground under the trees.

REFERENCES: *N.Y. Agr. Exp. Sta. Tech. Bul.* 240, 1936; *W. Va. Agr. Exp. Sta. Bul.* 126, 1910.

APPLE LEAFHOPPERS

ORDER HOMOPTERA, FAMILY CICADELLIDAE

Many species of leafhoppers attack apple. They are widely distributed throughout the apple-growing areas of the United States and Canada. Some have a much greater distribution, occurring on many other hosts. Damage is caused by removal of plant sap by the piercing-sucking mouthparts of both nymphs and adults. The resulting injury interferes with the normal photosynthetic processes of the plant, causing smaller, poorer quality fruits which are often speckled with excrement. Foliage often becomes whitened as the leafhopper population increases. Some comments on the life cycle, distribution, and habits of the more prevalent species follow.

White Apple Leafhopper, *Typhlocyba pomaria* McAtee, is widely distributed and has often been confused with the rose leafhopper. It is frequently the dominant species and is especially abundant in the central and eastern areas of the United States. Winter is passed in the egg stage underneath the bark of small branches. Hatching occurs near the pink stage, and the white adults appear in June and again in late August and September; there are 2 generations each year.

Rose Leafhopper, *Edwardsiana rosae* (L.), resembles *T. pomaria* and is also widely distributed (Fig. 286). It is the most important species in the Pacific Northwest. It also winters in the egg stage in the bark of trees, appearing as

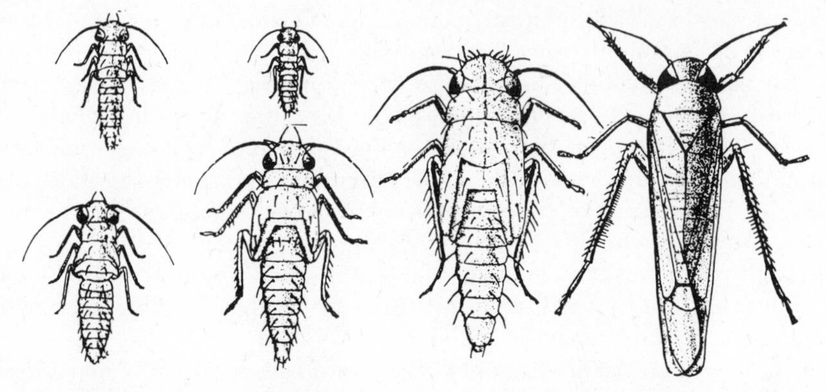

Figure 286. The rose leafhopper, *Edwardsiana rosae* (L.), 14×. (Childs, Ore. Agr. Exp. Sta.)

small raised blisters. Hatching occurs in April and May, and 3 to 6 weeks are required for nymphal development. First-generation adults appear in June, the second in late July and August. Adults may live for nearly 2 months. Besides apples, roses are seriously injured in some localities.

Apple Leafhopper, *Empoasca maligna* (Walsh), is a green species, slightly larger in size (Fig. 287). Its range extends from the north central states into the Pacific Northwest. Overwintering eggs are found in the bark of 2- or 3-year-old branches. These hatch in April or May, and the light green nymphs develop into adults 3 or more weeks later. Egg-laying continues until late July; the eggs remain in the bark until the following spring. This leafhopper has only 1 generation annually and does not become as numerous as the white species.

Potato Leafhopper, *Empoasca fabae* (Harris), has been discussed as a pest of potatoes, beans, and alfalfa (p. 259). This small green species occasionally

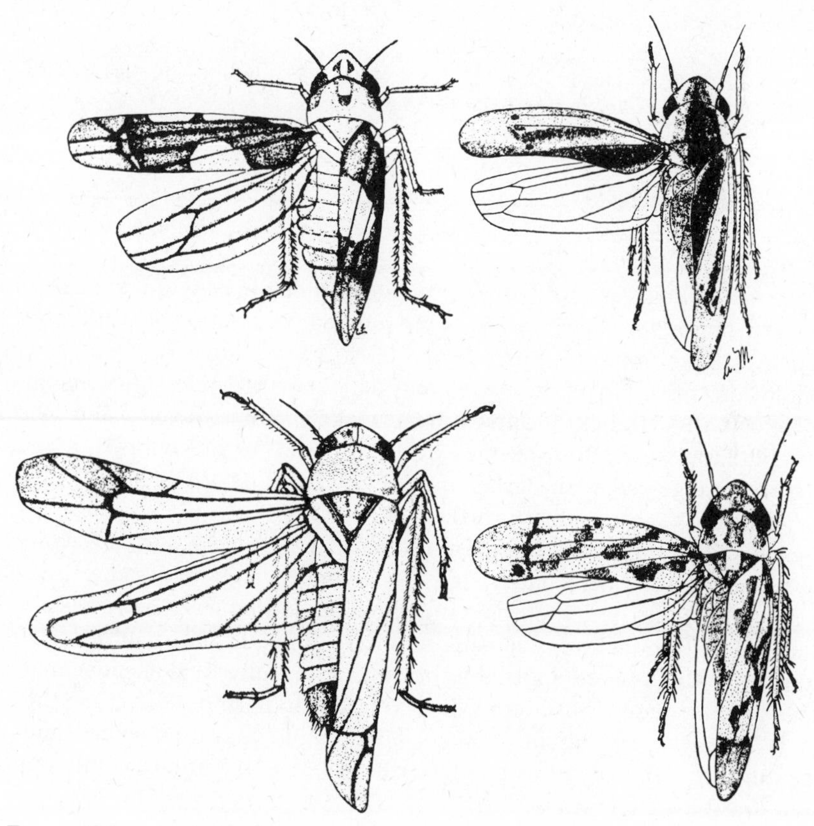

Figure 287. Apple leafhoppers: (L to R, above) *Erythroneura hartii* (Gillette) and *E. lawsoniana* Baker, (below) *Empoasca maligna* (Walsh) and *Erythroneura omani* Beamer. All enlarged. (Ackerman, USDA)

attacks apple, and is often a pest in nurseries. It overwinters only in the South, re-occupying its northern range by annual migration. The destructive range is almost entirely east of the Rocky Mountains. There may be 3 or more generations on apple, depending on the latitude.

Oblique-Striped Apple Leafhopper, *Erythroneura lawsoniana* Baker, is a common species found on apple. It is opaque white with a dorsal, scarlet oblique stripe (Fig. 287). Winter is passed in the adult stage, activity beginning early in the spring, with 2 or 3 generations developing annually. Other species sometimes abundant on apple and other hosts and having similar life cycles are *E. hartii* (Gillette), *E. omani* Beamer (Fig. 287), *E. dowelli* Beamer, and *E. magnacalx* Beamer.

Natural enemies include spiders, lacewings, *Orius tristicolor* (White), and other predators in the leaf bug family. Mymarid egg parasites (*Anagrus* spp.), and dryinid adult parasites in the genus *Aphelopus* are, at times, of considerable importance in keeping populations checked.

REFERENCES: *J. Econ. Ent.* 24:1214–1222, 1931; 47:361–362, 1954; *Conn. Agr. Exp. Sta. Bul.* 552, 1952; *Ohio Agr. Exp. Sta. Res. Bul.* 930, 1963.

APPLE RED BUGS

ORDER HEMIPTERA, FAMILY MIRIDAE

Two species of bugs similar in appearance and in life history do damage almost identical in character. They are the apple red bug, *Lygidea mendax* Reuter, and the dark or false apple red bug, *Heterocordylus malinus* Reuter. They have been injurious primarily in the north central and northeastern states and in Canada. Nearly all varieties of apples are attacked, Rome Beauty and Red Delicious perhaps least. Both species are bright red in the nymphal stage; the adults are orange-red with darker markings. Their length is about 6 mm. In size and form they resemble tarnished plant bugs.

Eggs of the red bugs are found on the bark, where they pass the winter. At apple-blooming time they hatch, and the young bugs puncture the fruits, just as they are forming, with their piercing-sucking mouthparts. Growth is arrested at these feeding punctures, causing stunted, dimpled, and malformed fruits, sometimes with russeted spots that become especially evident near maturity. Foliage feeding occurs but is considered of little importance. By June the adult stage is reached, and about 10 days later egg-laying commences and continues into July. Usually all adults have disappeared by August. Only one generation develops each season.

REFERENCES: *N.Y. Agr. Exp. Sta. Bul.* 716, 1946; *Ohio Agr. Exp. Sta. Res. Bul.* 930, 1963.

EYE-SPOTTED BUD MOTH

Spilonota ocellana (D. and S.), Family Olethreutidae

This insect, also called the bud moth, is said to have been imported from Europe over 100 years ago. It is now distributed in all the principal apple-producing sections of this continent, but it has been a serious pest only in the northeastern and northwestern states.

The moth is somewhat smaller than the codling moth, dark brown with a light-colored band which shows best when the wings are in a normal resting position. The fully grown larva is nearly 12 mm in length, brown with a shiny black head and thoracic shield, and a paler mid-dorsal stripe.

Eggs are laid during the midsummer period, singly or in small groups on either upper or lower leaf surfaces. Several days later hatching takes place and the larvae feed on the leaves, usually in shelters of leaves and silk. Cull apples result from occasional feeding on fruits that are in contact with the leaf shelters. Early in the fall the partly grown larvae form silken hibernating shelters on twigs at the base of spurs or in crevices in the bark of larger branches. Here they remain inactive until the following spring, when activity is resumed as the first buds are swelling. They often are found feeding inside the buds or in rolled or twisted leaves. In shelters formed by crumpled leaves and silk they transform to pupae in early summer, and new adults begin emerging by mid-June, thus completing the single-generation life cycle.

The cyclic nature of bud moth activity is caused by the 14 or more recorded parasites and predators and a nuclear polyhedrosis virus disease that play an important role in control, along with environmental forces, particularly low temperatures. Winter temperatures of −21 F (−29 C) usually eliminates this insect.

REFERENCES: *USDA Bul.* 1273, 1924; *Conn. Agr. Exp. Sta. Bul.* 552, 1952; *J. Econ. Ent.*, 55:930–934, 1962; *Ann. Rev. Ent.* 15:301, 1970.

FRUIT-TREE LEAF ROLLER

Archips argyrospilus (Walker), Family Tortricidae

This native leafroller occurs in the northern half of the United States, from coast to coast, and in Canada as well. Although it is commonly discussed as an apple pest, it injures all the orchard fruits and may require special control measures on any one of them. Early larval feeding on blossom buds may prevent setting of fruits. Serious damage results where the leaves are held against the fruits with silk, the larvae feeding within. Foliage injury is of less consequence although it may also be severe.

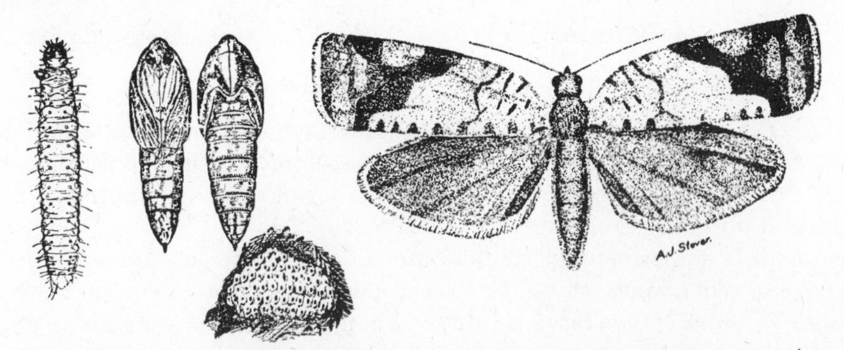

Figure 288. The fruit-tree leafroller, *Archips argyrospilus* (Walker), showing the egg mass, larva, pupa, and adult; 4×. (Childs, Ore. Agr. Exp. Sta.)

The adult moth is slightly larger than the codling moth, brown with variable lighter markings on the front wings (Fig. 288). The slender, pale green larva reaches a length of 18 mm; it has a black head and a black spot on the thorax just back of the head. Eggs are laid in gray compact oval masses on the bark of twigs and branches, each mass containing up to 100 or more. They are coated with a secretion from the moth which hardens and serves to protect the eggs.

The insect winters in the egg stage and hatching occurs when the buds begin to open. The larvae feed on the buds, blossoms, leaves, and fruits, becoming fully grown in June and transforming to pupae inside rolled or folded leaves. In about 2 weeks the moths begin emerging and shortly afterwards lay their eggs and die, only 1 generation developing annually.

Many parasites attack the various life stages of the leaf roller, but their combined effect is not always sufficient to keep it checked. Some of the more important parasites are *Itoplectis conquisitor* (Say), *I. 4-cingulatus* (Provancher), *Apanteles polychrosidis* Viereck, *Brachymeria ovata ovata* (Say), and *B. intermedia* (Nees).

REFERENCES: *USDA Cir.* 270, 1950; *J. Econ. Ent.* 61:348–352, 1968.

OBLIQUE-BANDED LEAFROLLER

Choristoneura rosaceana (Harris), Family Tortricidae

Of less importance as an apple pest than other species in the genus, this leafroller may be found attacking foliage in the spring, and fruit as well as foliage in the summer and fall. Widely distributed in southern Canada and the United States from Maine to California, it feeds on many plants and has at times been a pest in glasshouses, attacking roses. The young light green, black-headed larvae mine the leaves first, then feed inside rolled leaves tied with silk. Adults are brown, with 3 oblique darker bands on the front wings, and wingspread of nearly 2 cm. Winter is spent as partly grown larvae in tight-woven cases on woody hosts. The following spring they continue feeding and new adults emerge in June. One or 2 generations develop each season, depending on latitude. This species is attacked by many of the same parasites found on fruit-tree leafroller.

REFERENCE: *Ann. Ent. Soc. Amer.* 61:285–290, 1968.

RED-BANDED LEAFROLLER

Argyrotaenia velutinana (Walker), Family Tortricidae

The red-banded leafroller (Fig. 289) is a native insect that has become injurious in the section of the country north of the Ohio and east of the Mississippi Rivers. It is probably most abundant in the so-called Cumberland–Shenandoah apple region, but it occurs in other widely scattered localities. Host plants include apple, cherry, plum, peach, grape, several of the small fruits, vegetable crops, ornamentals, and weeds.

Larvae feed on foliage, often inside rolled or folded leaves held in place by silken threads. When leaves adjacent to fruits are tied with silk, the larvae cause many blemish marks by eating patches off the surface. Such feeding rarely extends to a depth of 6 mm, but it may be 25 mm or so across the area.

The adult is about the size of the codling moth, brown with a broad reddish band and other irregular light markings on the fore wings. The green larva is comparatively slender and about 16 mm in length.

Winter is passed as a pupa inside a silken cocoon under leaves or other objects on the ground. Moths begin emerging in the spring and deposit their eggs in flattened clusters usually on the bark of the trees. These hatch near the

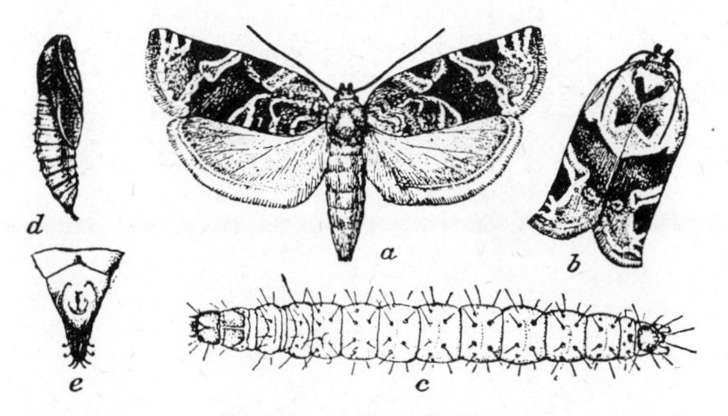

Figure 289. Red-banded leafroller, *Argyrotaenia velutinana* (Walker): *a,* female moth; *b,* moth with wings folded at rest; *c,* larva, dorsal view; *d,* pupa, lateral view; *e,* tip of abdomen of pupa, showing abdominal hooks; 2×. (USDA)

petal-fall period, and the larvae feed and develop, then pupate and emerge as adults in July. These soon lay eggs, and a second generation develops through late July and August. Three to 4 overlapping generations have been observed in southern areas.

Natural enemies include the egg parasite, *Trichogramma minutum* Riley, and 22 species of hymenopterous parasites of the larva and pupa.

REFERENCES: *Va. Agr. Exp. Sta. Bul.* 259, 1927; *Ohio Agr. Exp. Sta. Res. Bul.* 930, 1963; *J. Econ. Ent.*, 54:88–91, 1961; *Ohio Sta. Univ. Ext. Bul.* 506, 1979.

PEAR PSYLLA

Psylla pyricola Förster, Family Psyllidae

The pear psylla, of European origin, was first found in Connecticut in 1832 and has since spread throughout the area east of the Mississippi River, wherever pears are grown. It was discovered in the Pacific Northwest in 1939 and is now known to occur in most of the pear-growing regions in the West.

Injury is limited to pears. Honeydew excreted by the piercing-sucking nymphs and adults drops on the foliage and fruit and a sooty fungus develops in it. This results in black spots on the foliage and lower quality fruits. Severe infestations may cause leaf drop and prevent normal bud formation. This insect is the vector of a virus pathogen causing the disease known as "pear decline."

Adults are approximately 2 mm in length and resemble tiny red-brown cicadas, with 4 membranous wings (Fig. 290). These overwinter concealed in crevices on the trunks of trees and in shelter furnished by the ground cover

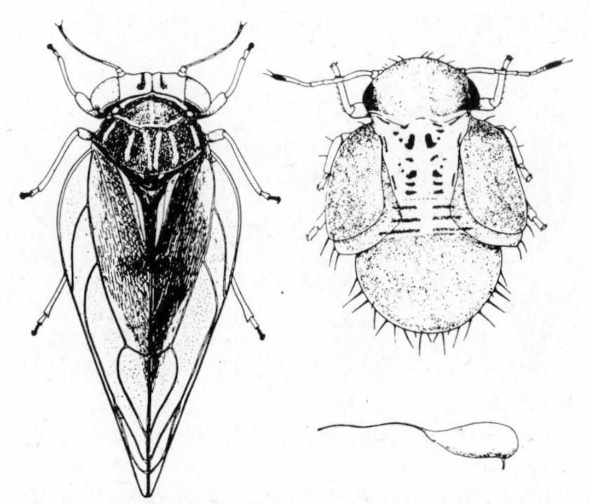

Figure 290. The pear psylla, *Psylla pyricola* Förster; adult, nymph, and egg; 25×. (Britton.)

or other material. Emergence takes place during the first warm days in the spring, and egg-laying begins soon afterwards. The elongate, white-to-yellow eggs are deposited in the crevices about the buds, and, after the foliage is out, on the leaves. Hatching takes place in 10 to 30 days. The flattened nymphs pass through 5 instars before reaching the adult stage. The nymphs are yellow at first, becoming green as they increase in size, with the final instar almost black. At least 4 generations may develop each year. Second-generation adults, and those produced later in the season, are smaller and lighter in color than those of the first generation. They were once considered a different species.

An application of oil emulsion spray in early spring while the trees are still dormant kills many adults and eggs. In some areas these dormant oils are applied in combination with ethion or other pesticides.

REFERENCES: *Conn. Agr. Exp. Sta. Cir.* 143, 1941; *USDA Cir.* 270, 1950; *Agr. Handbook* 290, 1965; *Calif. Agr.*, 17:14–15, 1963; *Calif. Agr. Exp. Sta. Leaflet* 71, 1966; *Can. J. Zool.* 41:953–961, 1963; *Ann. Rev. Ent.* 15:303–305, 1970; *Bul. Ent. Soc. Amer.* 21:247–249, 1975.

PEAR LEAF BLISTER MITE

Phytoptus pyri Pagenstecher, Family Eriophyidae

Introduced from Europe about 1870, this tiny mite is now generally distributed in North America. It attacks buds, foliage, and fruits of pear, apple, mountain ash, service berry, and cotoneaster. Its feeding causes the formation of galls or blisters on the leaves. These blisters are yellow to green at first, becoming red, and finally black on the leaves of pear (Fig. 291). Fruits may be russeted or deformed and will often crack open.

Figure 291. Damage to foliage by the pear leaf blister mite. (N.Y. Agr. Exp. Sta.)

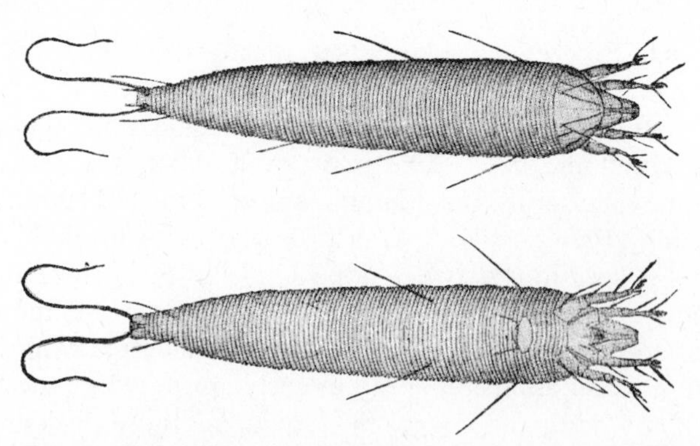

Figure 292. The pear leaf blister mite, dorsal and ventral views; 300×. (Nalepa.)

The white tapered four-legged mites are exceedingly small, less than 0.25 mm in length (Fig. 292). They winter beneath the bud scales, and become active and lay eggs when the buds begin to swell in warm spring weather. They produce their characteristic injury by feeding and burrowing into the leaf tissue. A succession of overlapping generations develops throughout the summer, with migration to other leaves and fruits occurring. With the approach of cold weather the mites return to the buds where they hibernate beneath the scales.

Dormant applications of lime-sulfur or oil emulsion sprays are often used to control this pest.

Other eriophyids, troublesome in some pear and apple growing regions, are the pear rust mite, *Epitrimerus pyri* (Nalepa), and the apple rust mite, *Aculus schlechtendali* (Nalepa).

REFERENCES: *N.Y. Agr. Exp. Sta. Cir.* 51, 1916; *USDA Cir.* 270, 1950; *Agr. Handbook* 330, 1967; *Calif. Agr.* 28(11): 15, 1974; *Ohio Sta. Univ. Ext. Bul.* 506, 1979.

PEAR THRIPS

Taeniothrips inconsequens (Uzel), Family Thripidae

The pear thrips were first noticed as pests in California in 1904, and since that time have become destructive in New York and adjacent states. In the western infestations they attack pear, plum, prune, cherry, apricot, apple, almond, and other plants, but in the East they have been primarily destructive to pear.

The dark brown adult is scarcely 1.5 mm in extreme length and of the form illustrated (Fig. 293). The nymphs are smaller in size, wingless, and somewhat lighter in color, often almost white. Both nymphs and adults feed by rasping the surface of plant tissues and then sucking out the juices.

Thrips spend several months in the soil as nymphs. Adults emerge in early spring and begin to feed, largely in the opening buds of the hosts. After feeding

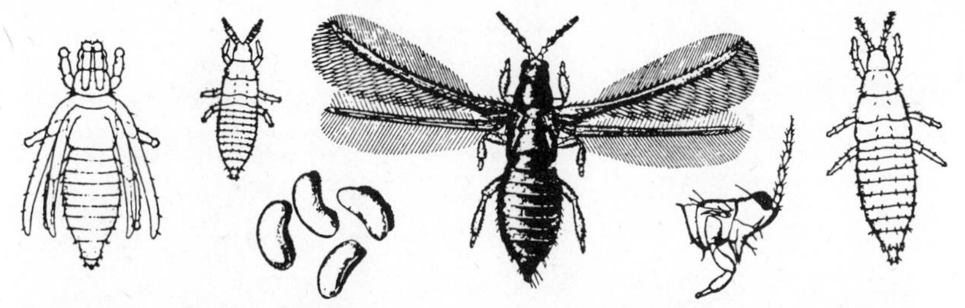

Figure 293. Life stages of pear thrips, *Taeniothrips inconsequens* (Uzel); 17×. (Foster and Jones, USDA)

for 3 weeks they begin laying tiny white eggs in the tissues of foliage and buds. Two weeks later the eggs hatch, and the young thrips feed and develop on foliage and fruits. After 3 weeks they drop to the ground, crawl into the soil to a depth of several inches and remain inactive, changing in late autumn to adults, which emerge the following spring. The most serious injury is caused by the adults feeding in early spring on the developing buds, causing deformed leaves and blossoms, and thus a reduction of the crop. Both nymphs and adults may cause young fruits to be scabbed, russeted, and deformed. Pear thrips are thus active on the trees for only about 2 months in the spring and are dormant in the ground the rest of the year. Bean thrips (p. 311) also attack pears in some regions.

Populations of this pest vary greatly from year to year.

REFERENCES: *USDA Cir.* 270, 1950; *Agr. Handbook* 290, 1965; 330, 1967.

PEAR SLUG OR SAWFLY

Caliroa cerasi (L.), Family Tenthredinidae

Of European origin, this widely distributed chewing insect (Fig. 294) skeletonizes the leaves of pear, cherry, plum, quince, and occasionally apple. Because in some areas it becomes quite abundant on cherry trees, it is also known as the cherry slug.

In their earlier instars the larvae are green-black, elongate, slim, and sluglike, with very little evidence of legs (Fig. 295). In the last instar they transform to typical sawfly larvae, nearly 12 mm in length, and resemble green-orange caterpillars. Their bodies are enlarged near the head and taper posteriorly. The adult is a glossy black, 4-winged sawfly, scarcely as large as a house fly.

The winter is passed in the soil inside a cocoon. Adults emerge in the spring after the trees are fully leaved; they deposit their eggs in slits in the leaves by means of a sawlike ovipositor. These hatch a week or more later. Larval development is completed in less than a month, and pupation takes place in cocoons in the soil. In late July adults of the second generation appear and lay

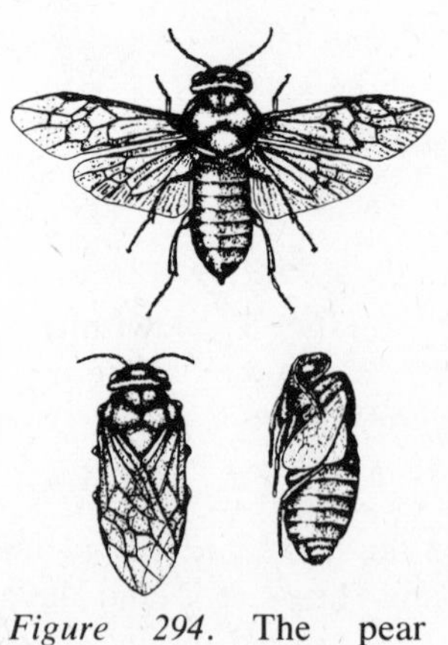

Figure 294. The pear sawfly, *Caliroa cerasi* (L.); adults; 4×. (Severin, S. D. Agr. Exp. Sta.)

Figure 295. Injury by pear sawfly larva; natural size. (USDA)

eggs. On hatching, the larvae develop to maturity, enter the soil, and spin cocoons in which they overwinter.

Where chemicals are applied in a spray program for controlling other pear insects, no special applications are needed for pear slug.

REFERENCE: *USDA Agr. Handbook* 330, 1967.

PEAR MIDGE

Contarinia pyrivora (Riley), Family Cecidomyiidae

The pear midge is an introduced insect, present in the northeastern states for many years. The adult is a very small mosquitolike fly. It emerges from the overwintering puparia in the soil and lays eggs in the swelling blossom buds of pear trees. The resulting creamy white-to-pale orange maggots feed inside the young fruits, causing them to be ill-shaped and to drop prematurely. In 5 weeks the maggots reach full development, drop to the ground, and enter the soil where transformation to puparia occurs; there is only 1 generation each year.

REFERENCE: *N.Y. Agr. Exp. Sta. Tech. Bul.* 247, 1937.

SINUATE PEAR TREE BORER

Agrilus sinuatus (Oliv.), Family Buprestidae

This insect occasionally attracts notice, but it is only a pest of importance in limited areas. Injury has been reported from Ohio, New York, New Jersey, and adjoining states. It is primarily a pest of pear, but mountain ash, cotoneaster, and hawthorn may also be attacked. Damage is caused by the larvae boring just beneath the bark, their slender sinuate galleries being found in both the trunk and branches.

The adult is a bronzy, slender, buprestid beetle, nearly 1 cm in length; the larva is elongate and slender, with the characteristic flattened head much reduced in width. Adults emerge in early summer and feed sparingly on the foliage before mating and laying their eggs. The larvae feed for 2 summers, hibernating each winter, then pupating the following spring and emerging as adults. There is 1 generation every 2 years.

Orchards receiving sprays containing lead arsenate or organophosphorus compounds are not troubled by this insect. Sprays should be timed to kill the emerging adults before they lay their eggs.

REFERENCE: *N.Y. Agr. Exp. Sta. Bul.* 648, 1934.

GREEN FRUITWORMS

ORDER LEPIDOPTERA, FAMILY NOCTUIDAE

Green fruitworms are the larval stages of moths and are so named because of their habit of eating deep holes into immature fruits of apple, pear, peach, cherry, almond, and apricot trees. Injury is most frequently noticed in southern Canada and northern United States orchards, but their range of distribution is wide and the host plants include shrubs, shade, and forest trees too.

Only 3 of 10 species associated with fruit trees are considered economically important pests in the United States. They are the white-striped fruitworm, *Lithophane antennata* (Walker), speckled green fruitworm, *Orthosia hibisci* (Guenée) (Fig. 296), and pyramidal fruitworm, *Amphipyra pyramidoides* Guenée. All are various shades of green and marked with white or yellow longitudinal stripes. Fully grown larvae are nearly 5 cm long.

Figure 296. The speckled green fruitworm, *Orthosia hibisci* (Guenée); 2×. (Courtesy of R. W. Rings and OARDC).

The adults resemble our common cutworm species. The overwintering stage may be the egg, pupa, or adult, depending on the species. Activity for all species begins in the spring with most larval damage occurring in May and June, first to the tender foliage or fruit blossoms and later to the fruits. One generation occurs per year for each species.

Orchards treated with chemicals for control of other pests are not troubled by fruitworms.

REFERENCES: *J. Econ. Ent.* 61:174–179, 1968; 62:1388–1393, 1969; 63:1562–1568, 1970; 66: 364–368, 1973; 68:178–180, 1975; *N.Y. Food and Life Sci. Bul.* 49:1–15, 1974.

MISCELLANEOUS INSECTS ATTACKING APPLE AND PEAR

Apple Flea Weevil, *Rhynchaenus pallicornis* (Say), attacks apple, hawthorn, crab apple, elm, hazelnut, and choke cherry. Ornamental crab apples frequently are damaged. This insect may be locally injurious in the area from the Mississippi River eastward and north of the Ohio River. The nearly black adults are snout beetles, almost 25 mm in length. They pass the winter under the duff on the ground. They become active and feed on newly opening buds and leaves in the spring. Soon eggs are laid on the leaves and hatch into legless white larvae. The larvae mine the leaves, especially at the edges, become fully grown in 2 or more weeks and pupate within the mines. By late May or early June new adults appear, which feed for nearly a month, causing damage similar to that of flea beetles. They then enter hibernation quarters and remain inactive until the following spring. Only 1 generation develops each year.

REFERENCE: *Ohio Agr. Exp. Sta. Bul.* 372, 1923.

Apple Seed Chalcid, *Torymus varians* (Walker), is an interesting pest species of the superfamily Chalcidoidea. The tiny wasps oviposit in apples and related fruits, causing a dimpled condition on the surface resembling plant bug injury. The larvae feed on the seeds. The range of the insect is northeastern United States extending into Canada. It has not been injurious in orchards receiving a regular spray program of present-day pesticides.

Lesser Appleworm, *Grapholitha prunivora* (Walsh), is the young of a congener of the codling moth, which it somewhat resembles. The larvae are smaller and more red than those of the codling moth. They attack apple, cherry, plum, and prune, and are found throughout the fruit-growing regions of North America. Early feeding by the worms on the foliage is relatively unimportant, the serious injury occurring late in the summer when the larvae mine the skin of the fruits and ruin their market value (Fig. 297). The injury resembles that done by the red-banded leafroller, except that the latter consumes the skin whereas the lesser appleworm feeds beneath it. There are 2 generations per year. Apples

Figure 297. Fruit injury caused by the lesser appleworm, *Grapholitha prunivora* (Walsh.) (Quaintance, USDA)

thoroughly sprayed for codling moth control do not suffer injury from this species. A major parasite is *Eurytoma tylodermatis* Ashmead.

REFERENCE: *N.Y. Agr. Exp. Sta. Bul.* 410, 1922.

Pistol Casebearer, *Coleophora malivorella* Riley, is considered a minor pest of apple. Occasionally, however, it becomes abundant and does extensive damage. Outbreaks have occurred from the Mississippi Valley eastward. The delicate gray moths have fringed wings with a spread of almost 12 mm (Fig. 298). The orange-yellow larvae spend almost their entire lives in gray-to-brown pistol-shaped cases, which are nearly 9 mm long when development is complete.

The winter is passed as partly grown larvae inside their cases attached to twigs or branches (Fig. 299). In early spring activity is resumed and they feed on the leaf buds, blossom buds, and foliage, the greatest damage being done to the buds. When fully grown the larvae again fasten their cases to twigs and

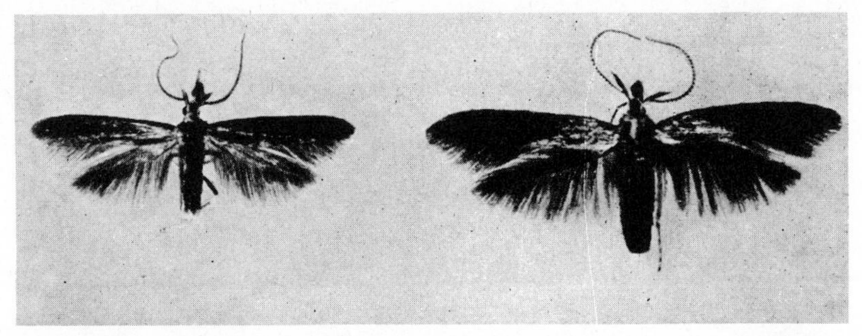

Figure 298. Adults of the pistol casebearer, *Coleophora malivorella* Riley; 4×. (Slingerland.)

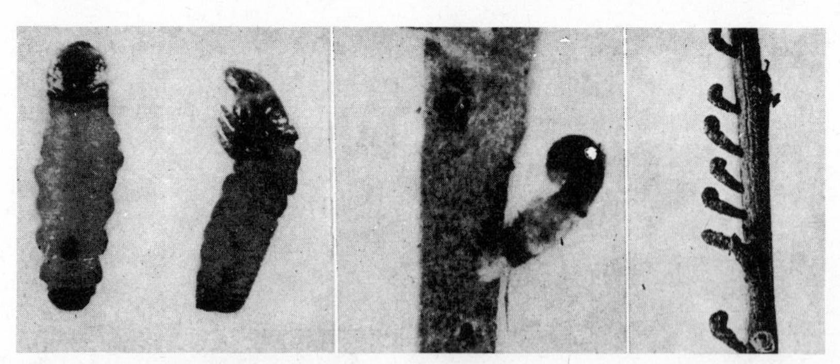

Figure 299. The pistol casebearer; young larvae, cocoon and injury to young twig, and group of cocoons; 4×. (Slingerland)

transform to pupae. The adult moths begin emerging 2 or 3 weeks later, usually late June, and deposit their eggs shortly thereafter. Eggs are laid on either upper or lower leaf surfaces, and hatching occurs 11 or more days later. The larvae leave the eggs at the point of attachment and burrow through the leaf, emerging on the other side a few days later with a tiny case already partly formed. Feeding continues throughout the summer with many entering hibernation by late August.

Numerous parasites and predators attack the pistol casebearer and undoubtedly are responsible for the normally low populations. Low temperatures during the brief egg-laying period also contribute to natural control.

REFERENCES: *W.Va. Agr. Exp. Sta. Bul.* 246, 1931; *Pa. Agr. Exp. Sta. Bul.* 406, 1941; *Ohio Agr. Exp. Sta. Res. Bul.* 930, 1963.

Cigar Casebearer, *Coleophora serratella* (L.), has much the same range and hosts as the pistol casebearer but has not attracted much attention in recent years. The life history and general appearance of the insect is also very similar, except that the larval case of this species is cigar-shaped. Damage to buds and leaves is caused by the chewing larvae, which feed while inside their cases. Another common name is birch casebearer.

Apple Leaf Skeletonizer, *Psorosina hammondi* (Riley), is a small pyralid moth with brown wings marked with white lines. The brown-green larvae feed on apple and pear leaves, sometimes skeletonizing large numbers and doing some damage to the trees. Where spraying is practiced, the insect is a rarity. Its range is eastern North America.

European Apple Sawfly, *Hoplocampa testudinea* (Klug), was discovered on Long Island in 1939 and has spread to many New England States and into Canada. It is a typical sawfly, 6 mm long, dark brown above and yellow-brown beneath. The larvae can be easily distinguished by their 7 pairs of prolegs. They tunnel under the skin of the fruit at first and later bore directly into the apple.

Winter is passed as mature larvae in the soil. Pupation takes place in the spring, and adults emerge when apple trees first come into bloom. Egg-laying occurs during full bloom, the eggs being inserted in the flesh of the calyx cup. The larvae leave the fruit by the middle of June, enter the soil, and remain inactive until the following spring. Commercial orchards receiving chemical sprays normally applied during the pink and petal-fall periods will not be troubled by this pest.

REFERENCE: *Conn. Agr. Exp. Sta. Bul.* 552, 1952.

Pear Plant Bug, *Lygocoris communis* (Knight), is a near relative of the tarnished plant bug but is slightly longer and darker, except at the tips of the wings. The bug deposits winter eggs in the bark of the host. Hatching occurs near blossoming time; the tiny nymphs feed mostly on the unfolding leaves, whereas the later instars feed very largely on the fruits. The sucking of sap causes many fruits to drop if the punctures are early and numerous. Drops of sap exude from the punctures (Fig. 300), and when these disappear the damage shows as black spots or points. As the pears grow these spots rupture, resulting in cracked corky areas and malformed fruits later in the season. The nymphs pass through 5 instars, becoming adults by late June. By late July all have laid eggs and died.

REFERENCE: *N.Y. Agr. Exp. Sta. Bul.* 368, 1913.

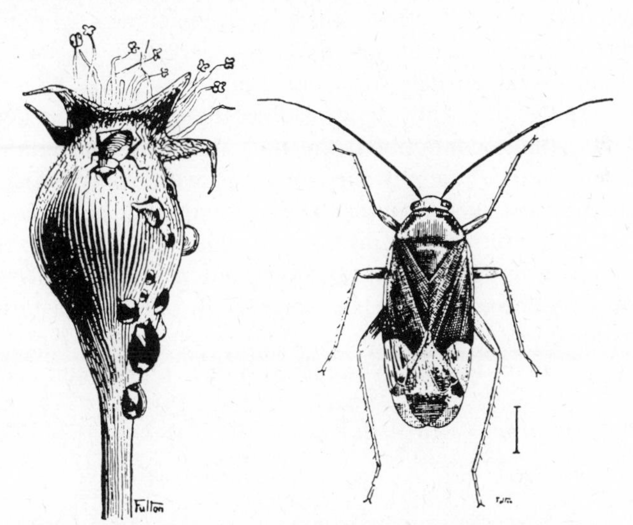

Figure 300. The pear plant bug, *Lygocoris communis* Knight; (right) adult bug; (left) nymph on young pear, which shows exudation from punctures later resulting in a deformed fruit; 5×. (Fulton, N.Y. Agr. Exp. Sta.)

Quince Curculio, *Conotrachelus crataegi* Walsh, resembles the plum curculio in appearance but differs somewhat in habits and life cycle. It is probably the most serious insect enemy of quinces, and it may also attack pears. Damage is caused by the legless white larvae feeding in the fruits and also by feeding punctures of the adults, resulting in malformations like those associated with the plum curculio. Larvae of this species winter in the soil, pupate in the spring, and transform to adults in early summer. Eggs are laid in adult feeding punctures in the fruits. On hatching the larvae spend nearly 3 months in the fruits before full growth is attained. Then they drop to the ground, burrow 2 or 3 inches below the surface, and remain there until the following spring. Only 1 generation occurs each season.

REFERENCES: *N.Y. Agr. Exp. Sta. Bul.* 148, 1898; *Conn. Agr. Exp. Sta. Bul.* 344, 1933.

Comstock Mealybug, *Pseudococcus comstocki* (Kuwana), has been a conspicuous pest of umbrella catalpa, holly, maple, boxwood, privet, mulberry, Japanese honeysuckle, and other hosts for some years. It was introduced into the United States sometime before 1918 and is now widely distributed. Since 1932 several serious infestations have been found in commercial apple orchards of the Cumberland–Shenandoah region and elsewhere. Injury results from removal of plant sap by the nymphs and adults, and from the fungus growth that develops on the copious supply of honeydew excreted by the insects. The honeydew and fungus cause much discoloration of fruits which cannot be removed even by the most effective fruit-washing machines. At picking time the insects themselves may remain on the fruits, especially in the calyces; it is also a problem to remove them from this area.

Winter is passed as tiny white eggs inside cottony masses, tucked in crevices of the bark. Hatching occurs about the time that leaves are unfolding in the spring. The flattened white powdery nymphs with spiny processes on the margins of their bodies migrate over the plant and feed, becoming fully grown in a month. In northern Virginia there are 3 generations a season.

Excellent control in apple orchards has resulted from liberating large numbers of parasites. Where this has been done with the platygasterid, *Allotropa burrelli* Muesebeck, and the encyrtids, *Clausenia purpurea* Ishii, and *Pseudaphycus malinus* Gahan, the mealybug is no longer a problem.

REFERENCES: *Va. Agr. Exp. Sta. Tech. Bul.* 29, 1925; *J. Econ. Ent.*, 44:123–124, 1951; *USDA Tech. Bul.* 1139, 1956.

19

PESTS OF STONE FRUITS

Other pests of stone fruits discussed in previous chapters are: Japanese beetle, leafrollers, leafhoppers, plum curculio, apple maggot, cicadas, stink bugs, spider mites, and coccoids. At times some of them are quite troublesome.

ORIENTAL FRUIT MOTH

Grapholitha molesta (Busck), Family Olethreutidae

The oriental fruit moth was introduced into this country around 1913 from Japan and was first discovered in Washington, D.C. in 1916, from which it has spread over most of the United States wherever peaches are grown. Moths were first caught in baits in California in 1942, in Idaho in 1944, and in Oregon and Washington in 1945. Peach and quince are the common hosts, but occasionally apple, pear, plum, and other fruit trees are attacked, especially if infested peach trees are adjacent.

The only injurious stage of the oriental fruit moth is the larvae, which tunnel into the tips of rapidly growing twigs (Fig. 301), thus preventing normal growth. New lateral shoots appear just below the point of attack giving the tree a bushy appearance. The larvae also enter the fruits and feed on the flesh, especially near the end of the growing season after twig growth becomes hardened. When fully grown they leave through holes chewed in the side of the fruit from which gummy exudates accumulate.

The oriental fruit moth is closely related to the codling moth; the adults are similar in shape but somewhat smaller, 6 mm in length, and of a dark gray coloration (Fig. 301). The larvae are white with a pink tinge, nearly 12 mm long when fully grown, the last abdominal segment bearing a dorsal, black, five-toothed, comblike structure which is a means of distinguishing it from codling moth larvae.

Figure 301. Oriental fruit moth larval damage: (left) characteristic wilted tip; (right) twig cut open showing larva in burrow. (Wood and Selkregg, USDA); (insert) adult moth; 6×. (Garman, Conn. Agr. Exp. Sta.)

Larvae hibernate in silken cocoons on the ground, in the tree, or on other objects near-by. Pupation takes place in the spring and moths begin emerging near peach tree blooming time. Eggs are laid on twigs and foliage, later on fruits. Early larvae tunnel into tips of tender green twigs, where they cause the characteristic wilting that indicates their presence. Full growth requires almost 2 weeks, after which they leave the twigs, spin silken cocoons on nearby objects, and transform to pupae. In 10 days adult emergence occurs. During the summer approximately 1 month is required to complete the life cycle. There are 4 or 5 generations each season in the latitude of New Jersey.

The rearing and liberation of numerous insect parasites has effected practical control in many localities. Important species are the egg parasite, *Tricho-*

gramma minutum Riley; the braconids, *Macrocentrus ancylivorus* Rohwer, *M. delicatus* Cresson, and *M. instabilis* Muesebeck; the chalcid, *Brachymeria ovata ovata* (Say); the eurytomid, *Eurytoma tylodermatis* Ashmead; and the ichneumons, *Pristomerus euryptychiae* Ashmead, *Cremastus forbesi* Weed, *C. epagoges* Cushman, *C. minor* Cushman, and *Glypta rufiscutellaris* Cresson.

Chemicals have given outstanding control of the oriental fruit moth. The commonly recommended materials are parathion, guthion, diazinon, imidan, and carbaryl. Since these chemicals also kill the parasites, Brunson and Allen have concluded, after a 5-year experiment, that 5 mass liberations, at 4-day intervals, of the parasite, *M. ancylivorus,* at the rate of 6 females per tree beginning about May 20, and 1 preharvest spray of parathion or guthion applied near the end of July, will give satisfactory control of the oriental fruit moth.

REFERENCES: *N.J. Agr. Exp. Sta. Bul.* 455, 1928; *Conn. Agr. Exp. Sta. Bul.* 313, 1930; 575, 1953; *J. Econ. Ent.,* 47:147–152, 1954; *USDA Tech. Bul.* 183, 1930; 1182, 1958; 1265, 1962; *Agr. Inf. Bul.* 182, 1958; 272, 1963; *Farmers' Bul.* 2205, 1964; *Calif. Agr. Exp. Sta. Leaflet* 78, 1966; *Ohio J. Sci.* 70:58–61, 1970; *Ohio Sta. Univ. Ext. Bul.* 506, 1979.

PEACH TWIG BORER

Anarsia lineatella Zeller, Family Gelechiidae

The twig borer occurs generally throughout the peach-growing areas of the United States. It has been only a minor pest in the eastern states but sometimes is a problem in the Pacific Coast states. Besides peach, it attacks plums, apricots, and almonds. The larvae injure the twigs and fruits in much the same manner as the oriental fruit moth.

The moth is very small, not exceeding 6 mm in length, with narrow ash gray wings fringed with hairs (Fig. 302). Larvae are small, red-brown caterpillars, with head and thoracic segments darker.

Larvae hibernate in silk-lined cavities at various locations on branches and twigs, sometimes at the base of new shoots, under loose bark, or in crotches of branches. When growth of the trees begins in the spring these larvae emerge

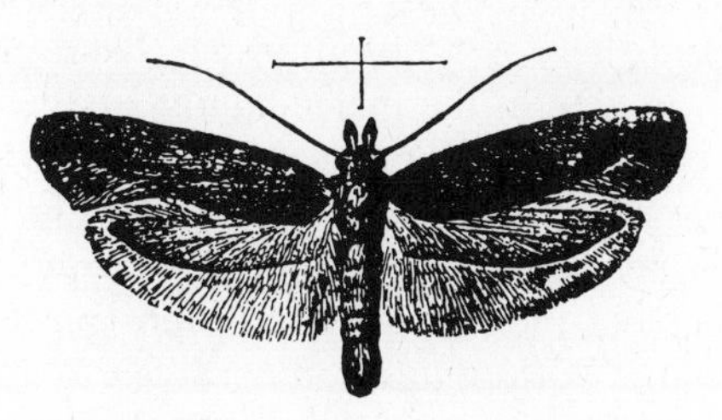

Figure 302. The peach twig borer, *Anarsia lineatella* Zeller; adult moth with wings spread; 4×. (USDA)

and bore into twigs and buds. They usually extend their tunnels only a short distance; this stops growth or kills the shoot. The feeding continues in other twigs until the larvae are fully grown, and then silken cocoons are made usually on the branches and trunk, followed by pupation. About 2 weeks later the adults emerge and begin laying eggs on leaves and fruits. The next generation of larvae feed on twigs and fruits; those which follow feed almost entirely in the fruits. There are 1 to 4 generations each season, depending on locality.

Natural enemies include the parasitic wasps, *Itoplectis 4-cingulatus* (Prov.) and *Paralitomastix pyralidis* (Ashmead); the predaceous thrips, *Leptothrips mali* Fitch; and the predaceous mite, *Pyemotes ventricosus* (Newport).

The twig borer is not a problem in orchards sprayed each year at the delayed dormant period with liquid lime-sulfur. A 3% oil emulsion has given almost as good control. Lime-sulfur is not recommended for apricot trees because of foliage injury. If no dormant sprays are applied, it is necessary to use parathion, basic lead arsenate, diazinon, guthion, endosulfan, or carbaryl. Basic lead arsenate is inferior to the other chemicals, but it has given what is considered satisfactory control in California.

REFERENCES: *J. Econ. Ent.*, 44:935–939, 1951; *USDA Agr. Inf. Bul.* 272, 1963; *Calif. Agr. Exp. Sta. Cir.* 449, 1955; *Leaflet* 78, 1966.

PEACH TREE BORER

Synanthedon exitiosa (Say), Family Aegeriidae or Sesiidae

The peach tree borer is a native American insect that attacks plum, prune, cherry, almond, apricot, and nectarine but is important mainly on peach. Injury is caused by the larvae boring just beneath the bark near the ground level, destroying the cambium and often girdling the trunk or roots. It is found throughout the United States wherever peaches are grown.

The adult is a clearwing moth, steel blue with yellow or orange markings; both pairs of wings of the male are very largely clear, and there are several narrow yellow bands on the abdomen; the female's fore wings are covered with metallic blue scales, and there is a broad orange band on the abdomen. The female is about 25 mm in length, the male slightly smaller. The moths are day fliers and may easily be mistaken for wasps. Fully grown larvae may exceed a length of 25 mm and are white with brown heads (Fig. 303).

The insects winter as larvae of all sizes in their burrows at the bases of trees. In the spring they complete their growth and then spin tough silken cocoons covered with their sawdust borings and soil particles. These are situated about an inch below the soil surface in an upright position near the base of the tree. Pupae may be found from early June to September, moths from late June to

Figure 303. The peach tree borer; (left), adult female and male moths, natural size; (right), larva, 3×. (Slingerland)

October; the latter are most prevalent in July, August, and early September, depending on the latitude. The brown eggs are usually laid on the tree trunks, sometimes on foliage of sprouts and even on other nearby objects. Incubation requires a little more than a week, and newly hatched larvae are present from midsummer to early fall. Young larvae make their way to the base of the tree, bore into and feed on the inner bark and cambium layer. Exudations of gum usually mixed with the borings of the larvae indicate the presence of the insect. There is only 1 generation each year.

Common parasites include the braconid, *Bracon sanninoideae* (Gahan); the bombyliid, *Villa lateralis* (Say), and the egg parasite, *Telenomus quaintancei* Girault. Predators include ants, chrysopid larvae, spiders, moles, birds, mice, and skunks.

Older control methods consisted of removing the borers by hand in fall or spring, fall treatment of the soil around each tree with paradichlorobenzene, and fall or spring treatment of the same area with ethylene dichloride emulsion. A new method of control is spraying the tree trunks with a pesticide to kill adults, eggs, and young larvae before they enter the bark. Sprays may be applied during 1 or more regular spray periods, thus requiring much less labor than hand-worming or the fumigation method. Pheromone baited traps are helpful in timing spray applications.

REFERENCES: *N.J. Agr. Exp. Sta. Bul.* 391, 1923; *J. Econ. Ent.*, 42:343–345, 1949; 45:611–615, 1952; 46:704–705, 1953; 47:359–360, 1954; 52:804–806, 1959; 56:463–465, 1963; *Conn. Agr. Exp. Sta. Bul.* 575, 1953; *USDA Tech. Bul.* 854, 1943; *Agr. Inf. Bul.* 272, 1963; *Bul. Ent. Soc. Amer.* 23:15–18, 1977; *Ohio Sta. Univ. Ext. Bul.* 506, 1979.

LESSER PEACH TREE BORER

Synanthedon pictipes (G. and R.), Family Aegeriidae or Sesiidae

The lesser peach tree borer is widely distributed in North America and is similar in many respects to the larger species but has some distinct differences in appearance and habits. Both males and females of the lesser borer have front wings devoid of scales, the color is metallic blue-black marked with yellow, and the female lacks the orange band (Fig. 304). Important over much of its range as a pest of peach, plum, and cherry, the lesser borer shows a preference for peach. A little less likely to be a primary pest, it tends to enter the wood through areas of the trunk and limbs of older trees which have been injured by implements, cankers, low temperatures, or sunscald. It works usually in the upper part of the trunk and in the scaffold branches. When found at the base of the tree, it is often in wounds made by the peach tree borer.

Early in the spring the overwintering larvae change to pupae and begin emerging as adults, usually a month earlier than the peach tree borer. Along the trunk and limbs of the tree are deposited the brown eggs which hatch into white larvae with brown heads. A second generation occurs in the South, but only one occurs in the latitude of the Dakotas.

This species can be controlled by thorough, properly timed application of parathion, endosulfan, or guthion sprays. Pheromone baited traps are useful in timing pesticide applications. Where these same pesticides are applied for the control of other peach pests, special sprays for the lesser peach tree borer may not be necessary.

REFERENCES: *Ohio Agr. Exp. Sta. Bul.* 307, 1917; *Res. Bul.* 768, 1956; *J. Econ. Ent.*, 45:611–615, 1952; *USDA Agr. Inf. Bul.* 272, 1963; *Bul. Ent. Soc. Amer.* 23:15–18, 1977; *Ohio Sta. Univ. Ext. Bul.* 506, 1979.

Figure 304. The lesser peach tree borer, 2×. (USDA)

PEACH BARK BEETLE

Phloeotribus liminaris (Harris), Family Scolytidae

The peach bark beetle is a native North American insect found generally east of the Mississippi River from Tennessee and Maryland northward. It attacks mainly peach, cherry, and plum, occasionally mountain ash, and resembles the shot-hole borer (Fig. 267) in both larval and adult stages (Fig. 305). A comparison of the figures shows that the shot-hole borer has the abdomen very sharply slanting from the tips of the elytra to near the base of the hind legs, whereas in the peach bark beetle there is no such declivity.

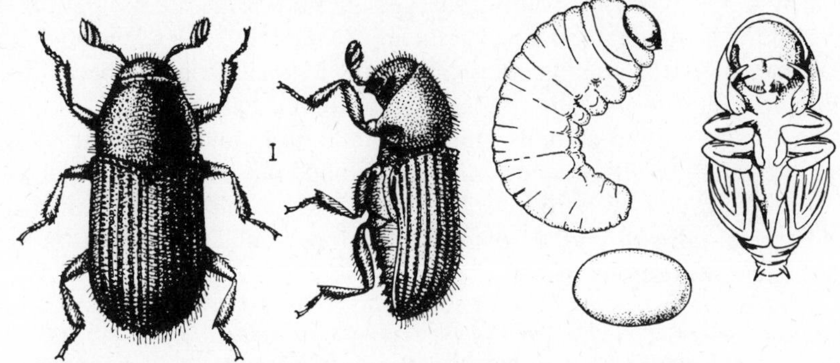

Figure 305. Life stages of the peach bark beetle; 16×. (USDA)

The peach bark beetle winters under the bark of the tree in the adult stage, and early the following summer emerges and lays its tiny white eggs. Two generations are produced each year. The presence of this species may be determined by the parent egg gallery, which almost always runs across the grain and is forked at one end, whereas the larval galleries run with the grain. The side branch of the fork enables the female to turn around within the egg gallery and it is occupied by the male at the time of mating. No male has been observed working. Sap flow from the plant parts under attack causes gummy exudates on the surface.

REFERENCES: *USDA Farmers' Bul.* 763, 1916; *Ohio Agr. Exp. Sta. Bul.* 264, 1913; 768, 1956.

TARNISHED PLANT BUG

Lygus lineolaris (P. de B.), Family Miridae

This bug commonly attacks apple, pear, peach, plum, and quince both in the nursery and in commercial orchards. Blossoms, buds, and developing fruits are most often attacked; peach trees also suffer much twig injury. Heavy feeding may cause fruits to become deformed. Peaches are especially susceptible, and the injury is sometimes described as "catfacing." Sunken areas more or less conical in shape with corky tissue at the bottom are indicative of early plant bug damage to fruits. Leguminous cover crops, or adjacent plantings of the same, favor high bug populations in the orchard.

Related species are *Lygus hesperus* Knight and *L. elisus* V. D. These occur in the western and southwestern states.

Adults are scarcely 6 mm in length. They are brown, tan, or green, always with darker markings (Fig. 156). Most specimens show the rusty appearance that is the basis for the common name. Nymphs are shades of green or yellow.

In its southern range, breeding may be continuous; northward, the adults hibernate. Activity begins in the spring, and injury to blossoms and buds is caused by these overwintering bugs. Eggs are laid in the stems, petioles, and midribs of the hosts. Hatching occurs in approximately 7 days, and the nymphs molt 5 times before becoming adults, requiring almost 30 days. There are 3 to 5 or more generations annually, depending on the latitude.

Clean culture around gardens, nurseries, and orchards, especially the elimination of a rank growth of weeds and leguminous plants, will do much to prevent injury. Spraying orchard trees with guthion, parathion, or imidan at shuck-split and petal-fall will control plant bugs. More applications may be needed where heavy infestations prevail.

REFERENCES: *Mo. Agr. Exp. Sta. Res. Bul.* 29, 1918; *J. Econ. Ent.*, 42:335–338, 1949; *USDA Agr. Inf. Bul.* 272, 1963; *Ohio Agr. Exp. Sta. Res. Bul.* 768, 1956; *Bul. Ent. Soc. Amer.* 21: 119–121, 1975; *Ohio Sta. Univ. Ext. Bul.* 506, 1979.

CHERRY FRUIT FLIES

ORDER DIPTERA, FAMILY TEPHRITIDAE

Three native species of fruit flies in the larval stage are important pests of wild and cultivated cherries. They are the cherry fruit fly, *Rhagoletis cingulata* (Loew); the black cherry fruit fly, *R. fausta* (Osten Sacken) (Figs. 306 and 307); and the western cherry fruit fly, *R. indifferens* Curran. The first 2 species are generally distributed in the cherry-growing areas of southern Canada and northern United States, from Nebraska to the Atlantic Coast, but *R. fausta* is found farther north and west. *R. indifferens* is found from the west coast eastward to Montana, or the range of its wild host *Prunus emarginata*. The quality and market value of the crop are greatly reduced by the maggots feeding in the flesh near the seed, often causing malformed fruits. The typical species attacks both sweet and sour cherries; the dark form shows preference for the sour varieties.

Adults of all species are a little more than half the size of the house fly and have wings marked with dark bands. The body color is dark with yellow markings, often distinctly darker in *R. fausta* which usually has more prominent dark bands on the wings. The larvae are white legless tapered maggots, nearly 7 mm long when fully grown.

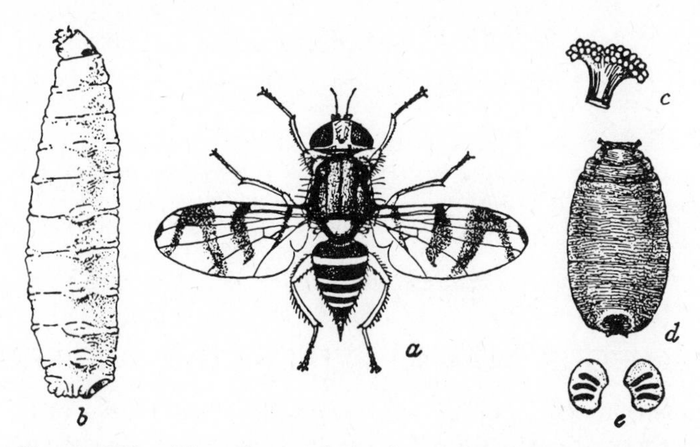

Figure 306. The cherry fruit fly, *Rhagoletis cingulata* (Loew): *a,* fly; *b,* maggot; *c,* anterior spiracles of maggot; *d,* puparium; *e,* posterior spiracular plates of puparium; 5×. (USDA)

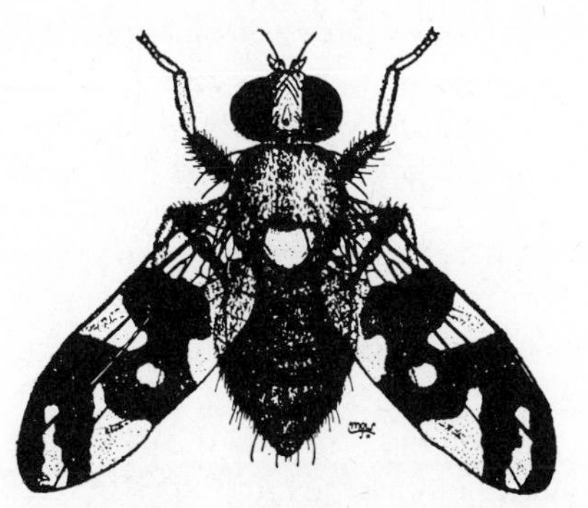

Figure 307. The black cherry fruit fly, *Rhagoletis fausta* (Osten Sacken); 6×. (Eichmann, Mont. Agr. Exp. Sta.)

Life histories and habits are also very similar for all species. Winter is passed as puparia in the soil. Adults emerge in late spring and lay their eggs in the fruits, one fly being capable of producing over 300. Larval development is completed within the cherries, after which they drop to the ground and change to puparia. Many larvae are likely to be in the fruits of early varieties at harvest time and may be distributed in shipment. Only 1 generation develops each season.

Both destruction of infested fruits and cultivation to destroy the puparia have been recommended as control practices. Pesticides are necessary for the production of a clean crop. Those commonly recommended are diazinon, parathion, guthion, or imidan. Proper timing is essential for successful control since the adults must be killed just as they are emerging and before they lay their eggs. Usually 2 to 4 applications at 7- to 14-day intervals are required, beginning when the first flies are trapped.

REFERENCES: *N.Y. Agr. Exp. Sta. Bul.* 325, 1912; *Mich. Agr. Exp. Sta. Bul.* 131, 1930; *Mont. Agr. Exp. Sta. Bul.* 313, 1936; *USDA Cir.* 270, 1950; *Agr. Handbook* 290, 1965; *Pa Agr. Exp. Sta. Bul.* 548, 1952; *Wash. Agr. Exp. Sta. Bul.* 470, 1945; *Tech. Bul.* 13, 1954; *J. Ent. Soc. B. C.* 68:29–32, 1971; *Ann. Rev. Ent.* 17:493–518, 1972; *Bul. Ent. Soc. Amer.* 19:103–104, 1973; 20:93–101, 1974; *Ohio Sta. Univ. Ext. Bul.* 506, 1979.

MISCELLANEOUS INSECTS ATTACKING STONE FRUITS

Cherry Fruitworm, *Grapholitha packardi* Zeller, is the larva of a very small mottled gray moth related to the oriental fruit moth. It has become quite injurious to cherries from Colorado northward and westward to British Columbia and Vancouver Island. It has been reported as being present in New Jersey, Iowa and Wisconsin. Its native host seems to have been wild cherry but it is known to attack blueberry.

Larvae hibernate in galleries in the bark or twigs and pupate in the spring. Moths begin emerging in early June and lay tiny flattened eggs on the fruits. A week later hatching begins, and the white larvae with black heads bore into the green cherries and feed about the pit. In almost 3 weeks the larvae mature to a length of 1 cm. There is only 1 complete generation each year, but a partial second may develop in some areas.

REFERENCE: *J. Econ. Ent.*, 45:800–805, 1952.

Cherry Fruit Sawfly, *Hoplocampa cookei* (Clarke), is a Pacific Coast pest of cherries and plums, sometimes peaches and apricots. The blackbodied adults with yellow appendages are only 3 mm in length. Larvae are yellow-white and about 6 mm in length when fully grown. Eggs are laid in the blossoms, and the young larvae enter the fruits and feed on the developing seeds. Some fruits wither, others drop. In less than a month development is completed and the larvae leave the fruits and enter the soil where pupation occurs. Adult emergence takes place the following spring; there is but 1 generation per year. Early sprays prevent injury, the applications being made before the blossoms open to kill the newly emerging adults before they lay eggs.

REFERENCES: *Calif. Cir.* 227, 1921; *Leaflet* 72, 1966.

Cherry and Hawthorn Leaf Miner, *Profenusa canadensis* (Marlatt), is an occasional pest of Morello cherries and often quite damaging to ornamental hawthorn in the north central and New England states. Adults are black-bodied sawflies, nearly 5 mm in length. The dorsal side of the prothorax is black in the males and orange in the females. In late April and early May they emerge from pupae in the soil and lay their eggs in the leaves of the hosts. The nearly white maggotlike larvae devour the tissue between the upper and lower leaf epidermis (Fig. 308), pass through 5 instars during a period of almost a month, then drop to the ground, enter the soil and remain in an earthern cell until late winter when pupation begins. Only 1 generation occurs each year.

Spraying ornamental hawthorn with emulsions of lindane or dimethoate just as soon as the leaves are fully expanded will control this pest. One thorough application should be sufficient. On Morello cherries the regular sprays of guthion or parathion should solve the problem.

REFERENCES: *N.Y. Agr. Exp. Sta. Bul.* 411, 1915; *J. Agr. Res.*, 5:519–529, 1915; *J. Econ. Ent.*, 43:694–696, 1950; 52:1218–1219, 1959.

Figure 308. The cherry and hawthorn leaf miner, *Profenusa canadensis* (Marlatt), Larvae can be seen in the mines; natural size. (Original))

Green Peach Aphid, *Myzus persicae* (Sulzer), is a vector of over 50 virus diseases. It is rather slender in form, light green or yellow, with indefinite darker stripes on the abdomen (Fig. 309). Its fruit tree hosts include all the stone fruits. The black shiny winter eggs are placed on these plants and, about the time peach trees bloom, hatching occurs, the nymphs developing into wingless females which give birth to succeeding generations. After 2 or 3 generations winged forms are produced that migrate to the various herbaceous alternate hosts, pepper, potato, sugar-beets, tobacco, spinach, lettuce, and other vegetables being quite important in this respect. Here repeated generations are produced, and with the approach of cold weather winged migrants again return to peach and other stone fruits and give birth to the sexual forms that eventually mate and lay the overwintering eggs. In the South all generations may be

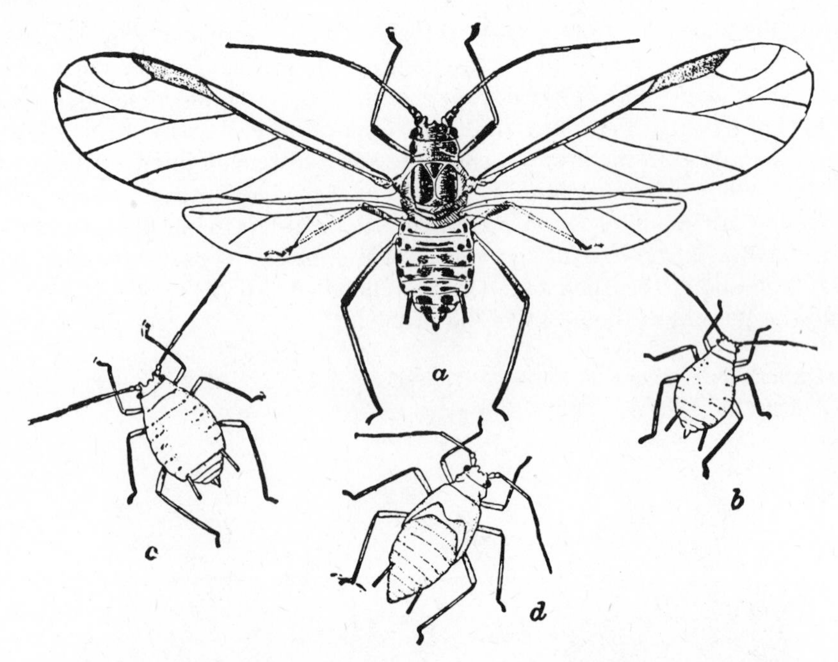

Figure 309. The green peach aphid, *Myzus persicae* (Sulzer); also known as the spinach aphid. *a,* winged adult; *b,* young nymph; *c,* older nymph; *d,* last stage of nymph; all greatly enlarged. (USDA)

parthenogenetic. Damage to the blossoms, leaves, and fruits of peach trees is rarely serious from these sap-feeding insects, but at times tobacco and garden crops are severely injured. Insects that reduce populations of this aphid are lady beetles, lacewing larvae, syrphid fly larvae, *Aphidius testaceipes* (Cresson) and *Aphelinus semiflavus* Howard. This aphid is a vector of the virus causing maize dwarf mosaic of corn, and yellows virus of sugarbeets.

REFERENCES: *Calif. Agr.* 17:10–11, 1963; *USDA Agr. Handbook* 290, 1965.

Black Peach Aphid, *Brachycaudus persicae* (Passerini), overwinters as black wingless forms feeding on the root sap of peach and sometimes plum. The life cycle of this aphid has not been thoroughly investigated, but it is known that some of the root forms migrate to new growth above ground in the spring and increase in numbers rapidly. Winged forms eventually develop that fly to other trees and possibly other hosts where new infestations are established. All aerial forms usually disappear by midsummer. It is not known whether sexual forms are ever produced. Young trees may be seriously injured by the forms on the roots, but older trees suffer little root injury and rarely much from the leaf-infesting forms. The root forms are attended by ants, but how important the ants are in maintaining the population has not been learned.

Mealy Plum Aphid, *Hyalopterus pruni* (Geoffroy), an importation from Europe, is rather light green but usually appears gray because of the blue-white powdery substance that covers the body (Fig. 310). It winters in the egg stage on plum and produces a few early generations in the spring on this host. In early summer winged forms develop and migrate to grasses, on which they feed and reproduce until fall; winged migrants then return to plum and give birth to the sexual forms that complete their development, mate, and deposit the overwinterings eggs. All generations are parthenogenetic ovoviviparous females, except the sexual forms that develop in the fall and lay eggs. Injury results from removal of plant sap, causing stunting of growth and smudging of fruits and foliage with honeydew. This aphid has been especially injurious in California where prunes and apricots are also attacked.

When aphid populations indicate the need for control measures apply a spray containing parathion, demeton, diazinon, malathion, or trithion.

REFERENCES: *Calif. Agr. Exp. Sta. Bul.* 606, 1937; *Leaflet* 77, 1966.

Rusty Plum Aphid, *Hysteroneura setariae* (Thomas), has also been called the southern plum aphid. Although it occurs in the North it is distinctly a southern pest. It attacks plum, cherry, occasionally peach and sometimes grain sorghums. It also transmits sugarcane mosaic virus. It is rusty in appearance, with white markings at the base of the antennae, the tibiae, and the tip of the abdomen, thus being readily distinguished from other aphids on the same hosts. The most important injury is done by those aphids that work in the blossoms and prevent the setting of fruits, although foliage injury may also be severe. The life history is similar to that of the mealy plum aphid. Eggs pass the winter on the woody hosts, and a few wingless generations of parthenogenetic ovoviviparous females are produced in the spring, followed by winged forms migrating to grasses, then a later migrating generation returning in late fall to the tree hosts, upon which the egg-laying sexual forms are produced. Controlling ants in sugarcane fields greatly reduces populations of this aphid on sugarcane.

REFERENCE: *J. Econ. Ent.*, 54:204, 1961.

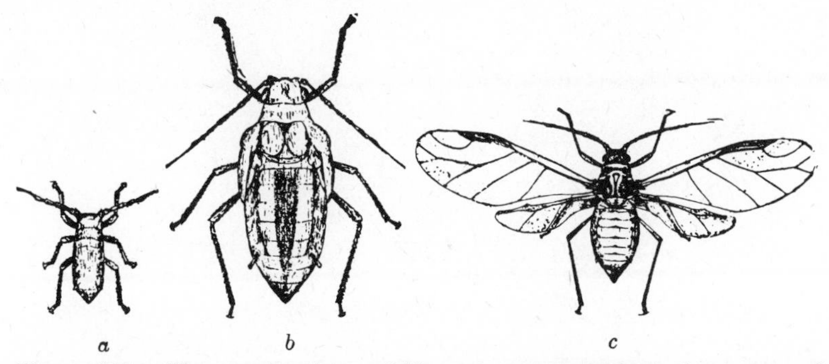

Figure 310. The mealy plum aphid: *a,* young nymph; *b,* last stage of nymph of winged form; *c,* winged female; all much enlarged. (Lowe.)

Black Cherry Aphid, *Myzus cerasi* (F.), is a European species that has become the most common member of the family attacking cherry in many parts of North America. Its shiny black coloration is sufficient for identification, as no similar insect attacks the same hosts (Fig. 311). Sweet cherries are the favored host; other varieties of cherries and stone fruits are rarely damaged. The winter eggs, which are tucked in among the buds, hatch about the time the buds are bursting. The nymphs suck plant sap and develop rapidly on the new growth, causing curling and distortion of the leaves. There are several generations of wingless parthenogenetic ovoviviparous females. Winged adults develop in midsummer, migrate to water cress, peppergrass, and other plants of the mustard family, and give birth to succeeding generations. In late autumn there again develop winged individuals that return to cherry and produce the wingless sexual forms that lay the overwintering eggs.

REFERENCES: *Calif. Agr. Exp. Sta. Leaflet* 72, 1966; *USDA Agr. Handbook* 290, 1965.

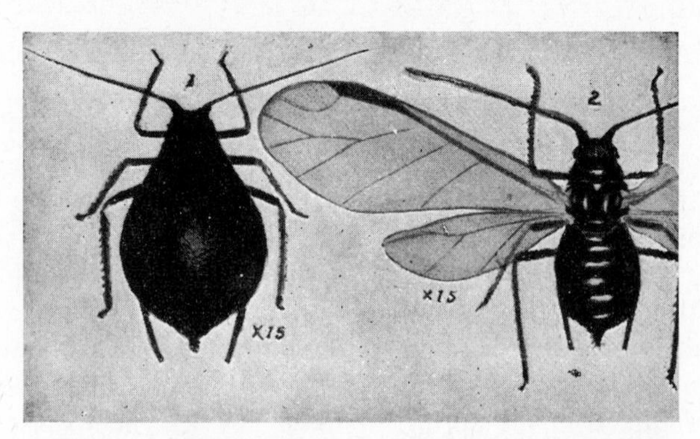

Figure 311. The black cherry aphid, *Myzus cerasi* (F.); 1, apterous female; 2, winged female; 15×. (Gillette and Taylor, Colo. Agr. Exp. Sta.)

20

PESTS OF GRAPES

In addition to the pests discussed in this chapter the following may also seriously damage grapes: climbing cutworms, Japanese beetles, spider mites, leafrollers, European fruit lecanium, and cicadas. These pests are covered in previous chapters.

GRAPE BERRY MOTH

Endopiza viteana Clemens, Family Olethreutidae

The berry moth is generally distributed east of the Rocky Mountains. It is much more troublesome in some sections than in others, being especially destructive in the large grape-growing areas east of the Mississippi River and north of the Ohio River. It is the only common insect that does extensive damage to grape berries; the grape curculio, which also feeds inside the fruit, is much less abundant and rarely a serious pest.

The brown moth, with an irregular pattern of darker and lighter markings, has a wingspread of slightly more than 12 mm. The larvae are green or gray-green, slightly hairy, and nearly 10 mm long (Fig. 312).

Moths begin to emerge from overwintering pupae when grape foliage is well unfolded, usually about the first of June, continuing over a period of several weeks. They lay their tiny scalelike eggs on the grape stems, blossom clusters, or berries. Newly hatched larvae feed on the blossoms and small berries, often leaving a silken thread wherever they crawl, resulting in webbed clusters. One larva may injure or destroy several grape berries. In 3 or 4 weeks they become fully grown and each makes a somewhat semicircular cut in the leaf, folds over this portion and ties it with silk, thus forming a cocoon inside which pupation takes place. Second-generation moths begin emerging 10 to 14 days later and once more lay their eggs, this time on the berries. Peak adult emergence usually

Figure 312. The grape berry moth; adults, larvae, infested grapes, and cocoons in leaf; all enlarged except bottom right. (Slingerland.)

occurs about mid-August. Larvae of this generation eat their way into the berries and feed on the pulp and seeds. There are normally 2 generations each year but a partial third may develop in more southern areas. Full-grown larvae of the last generation pupate inside the cocoons made in the leaves. Many cocoons become detached and fall to the ground; others remain and drop with the leaves in the autumn.

The only biological controlling factor that has been found to be of appreciable value is the egg parasite, *Trichogramma minutum* Riley.

When feasible, cultural practices aid greatly in reducing the overwintering population. Success of these practices depends on plowing or cultivating so as to bury the cocoons containing overwintering pupae, thereby preventing emergence of adults. A popular method is to throw the soil from the row centers into a low ridge under the grape trellis with a grape hoe, disk, or plow. Do this 30 to 45 days before harvest. Make the ridge flat and wide, letting it remain during the winter. Make the row centers almost level and seed them to a winter cover crop. In the spring, at least 15 days before grape bloom, pull the soil ridge with the cocoons on its surface from under the trellis into the row centers with a mechanical grape hoe. With a hand hoe pull out any islands of soil left around the posts and grapevines into the row centers. Then disk or plow the row centers to bury the cocoons. Use a straight disk followed by a cultipacker. Rains after this operation help to seal in the cocoons. This practice has reduced berry moth populations to a point where shortened spray schedules can be used.

It is impossible to make a grape spray schedule fit all situations and localities. In most vineyards with a berry moth problem at least 3 and more often 4, spray applications are advisable. Each grower should study his conditions and apply only necessary sprays. The first is made at postbloom, the second 7 to 10 days later, the third about the first of August, and the fourth in mid-August. Commonly recommended chemicals are: methoxychlor, carbaryl, diazinon, imidan, guthion, and parathion. A hooded spray boom is necessary for effective grape insect control.

REFERENCES: *Del. Agr. Exp. Sta. Bul.* 198, 1936; *J. Econ. Ent.*, 45:101–104, 1952; 46:77–84, 1953; *USDA Farmers' Bul.* 1893, 1972; *Agr. Inf. Bul.* 252, 1962; *Agr. Handbook* 290, 1965; *Ann. Ent. Soc. Amer.* 62:1374–1378, 1969; *Ohio Agr. Exp. Sta. Bul.* 1060, 1973; *Ohio Sta. Univ. Ext. Bul.* 506, 1979.

GRAPE LEAFHOPPERS

ORDER HOMOPTERA, FAMILY CICADELLIDAE

Throughout the United States and Canada where grapes are grown various species of leafhoppers will almost invariably be found sucking the juices from the lower surfaces of the leaves. This causes the foliage to become blotched with tiny white spots and under heavy infestation the leaves will turn yellow or brown, many falling from the vine. Since their feeding seriously interferes with the normal photosynthetic processes of the plant, the quantity and quality of the fruit are greatly reduced. Other hosts are Virginia creeper, apple, plum, cherry, currant, gooseberry, blackberry, raspberry, dahlia, and weeds.

Apparently the most common species of the eastern grape regions is *Erythroneura comes* (Say), followed by *E. tricincta* Fitch (Fig. 313). Other species of importance in some areas are *E. ziczac* Walsh, *E. vitifex* Fitch, *E. octonotata* Walsh, *E. dolosa* Beamer and Griffith, *E. vulnerata* Fitch, *E. vitis* (Harris), and *E. elegantula* Osborn. All species are scarcely more than 3 mm in length, pale yellow, with red, yellow, or black markings on the front wings.

There seems to be little important difference in the biology of the various species except as it concerns their relative abundance in different areas on different hosts. They pass the winter as adults in protected places, usually under plant remnants on the ground. With the first warm days of spring the leafhoppers become active and feed to some extent on any green plant before grape foliage appears. Eggs are laid in the leaf tissue, hatching taking place in about 2 weeks.

Figure 313. Grape leafhoppers: *a,* leaf showing appearance of injury; *b,* (right) *E. comes* (Say); (left) *E. tricincta* Fitch; *c,* a nymph; 9×. (USDA)

The pale wingless nymphs feed on the lower surface of the leaves, molting 5 times before transforming to adults. This developmental period requires 3 to 5 weeks depending on the temperature. There are 2 or 3 generations of these insects each season.

Some natural control of leafhoppers results from insect parasites and predators, and a fungus, *Entomophthora sphaerosperma* Fres. The mymarid egg parasite, *Anagrus epos* Girault, greatly reduces grape leafhopper populations, especially where vineyards are located near plantings of blackberry, according to California entomologists. They found this tiny wasp (0.3 mm long) producing 9 to 10 generations each season and surviving the winter in the eggs of the noneconomic leafhopper, *Dikrella cruentata* (Gillette), which occurs on native and cultivated species of blackberry. Presumably any host that harbors overwintering eggs of leafhoppers could also serve in this capacity because *Anagrus epos* has been found parasitizing eggs of other species of leafhoppers. This idea should be considered in any integrated control program.

Chemical control is accomplished by applying at the prebloom or postbloom periods, and at other times if deemed necessary, one of the following pesticides: methoxychlor, carbaryl, diazinon, or guthion. Treat only if on the average 10 nymphs per leaf are found for the first generation and 5 nymphs per leaf for the second generation; use the higher figure if *Anagrus* spp. is present. Frequent use of chemicals induces mite problems on grape.

REFERENCES: *Del. Agr. Exp. Sta. Bul.* 198, 1936; *USDA Farmers' Bul.* 1893, 1972; *Can. Dept. Agr. Pub.* 35, 1946; *J. Econ. Ent.* 40:195–198, 487–495, 1947; *Calif. Agr. Exp. Sta. Leaflet* 79, 1966; *Calif. Agr.* 19:3, 1965; *Ohio Sta. Univ. Ext. Bul.* 506, 1979.

GRAPE ROOTWORM

Fidia viticida Walsh, Family Chrysomelidae

Widely distributed from the Mississippi Valley eastward, this native insect is considered an important pest of grapes and related host species. It thrives in neglected vineyards. The larvae devour small roots and eat pits in the outer portion of larger roots, causing a general unthrifty condition of the plant and reduction in yield. When the larvae are abundant vines may be killed in three or more years. One of the best diagnostic characteristics of the presence of these insects is the chainlike feeding marks produced by the adult beetles (Fig. 314). They devour narrow strips of upper leaf epidermis, causing the lower surface to die. Beetles also feed on the surface of the green grape berries.

The adult beetle is hairy, chestnut-brown, and almost 6 mm long. The white larva is curved and hairy with a brown head. Adults make their appearance in vineyards about the end of the blooming period. After feeding on the foliage for a week or more they begin depositing clusters of creamy white elongated eggs on the canes, most often under loose bark. These hatch in a week or 2, the larvae dropping to the ground and entering the soil where they feed until the approach of cold weather. Winter is passed in the larval stage in the soil at

Figure 314. Leaf damage by adults of the grape rootworm; natural size. (Original)

depths varying from a few inches to nearly 2 feet. They resume feeding and complete their development the following spring, pupation occurring in earthen cells usually near the surface of the soil and not far from the base of the vine. Approximately 2 weeks are spent in the pupal stage. One generation may develop each year, but many larvae require a second year to complete their development.

Control measures are directed toward eliminating the adults soon after emergence and before they deposit their eggs.

A related species is the western grape rootworm, *Bromius obscurus* (L.), which is well known in Europe and ranges in North America from Alaska through the Rocky Mountain states. The appearance, life cycle, and habits are similar to those of the eastern species, and control is accomplished in the same way.

REFERENCES: *USDA Farmers' Bul.* 1220, 1926; 1893, 1972; *Ark. Agr. Exp. Sta. Bul.* 426, 1942; *OARDC Bul.* 1060, 1973.

ROSE CHAFER

Macrodactylus subspinosus (Fabr.), Family Scarabaeidae

The name of this insect usually associates it with the rose; in discussions it is often included with the grape pests. Actually it is a general feeder and it is somewhat misleading to call it a pest of any one crop. Besides grape and rose, it also attacks many tree fruits, raspberry, blackberry, strawberry, peony, iris, dahlia, hollyhock, and other garden plants. The beetle is native to this country and occurs generally west to the Rocky Mountains. More abundant in areas having light sandy soil, the beetles feed on flowers, leaves, and fruits, and the larvae feed in the soil on roots of various grasses and weeds.

The larvae resemble small white grubs. The rastrel pattern is a way to differentiate them (Figs. 80, 315). They pass the winter deep in the soil and migrate upward in the spring, completing their development and transforming to pupae in early May. Light tan long-legged beetles nearly 12 mm in length begin emerging in late May and early June (Fig. 315). Some years they are extremely abundant and cause serious damage. Feeding, mating, and egg-laying occur into early July, the eggs being deposited singly a few inches below the surface of the soil. Hatching takes place in 2 or more weeks, and the grubs feed until cold weather, becoming nearly fully grown. One generation occurs each year.

On grape, petal-fall sprays for berry moth kill rose chafers. However, control is seldom needed on grape or other hosts.

REFERENCES: *USDA Farmers' Bul.* 1893, 1972; *OARDC Bul.* 1060, 1973.

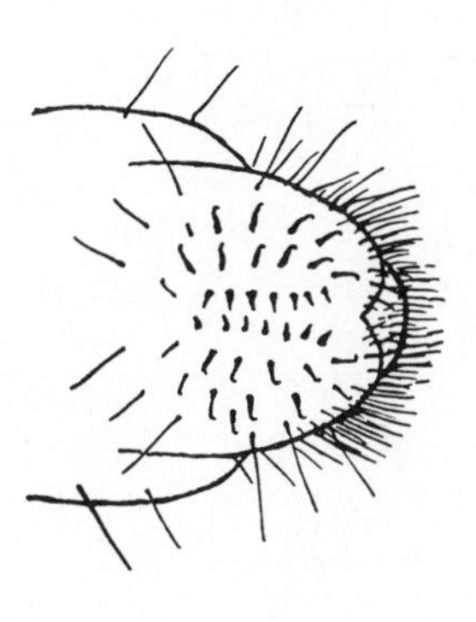

Figure 315. The rose chafer; 3×. (left); rastrel pattern of larva (right).

GRAPE FLEA BEETLE

Altica chalybea Illiger, Family Chrysomelidae

This flea beetle is a native species occurring generally throughout the Mississippi Valley and eastward. In addition to cultivated and wild grapes, it is reported as feeding on Virginia creeper, plum, apple, pear, quince, beech, and elm. The insect thrives best in neglected vineyards. Damage is caused by the adults eating the buds and unfolding leaves of grape, and the larvae skeletonizing the foliage in a manner resembling adult rootworm feeding.

Adults of this robust, metallic, blue-green beetle, almost 5 mm in length, emerge from hibernation and soon afterward deposit their light yellow eggs largely in cracks in the bark, at the base of buds, between bud scales, or on the leaves. These eggs hatch in a few days and the brown larvae, marked with black spots, feed for a 3- to 4-week period, then drop to the ground and pupate in the soil, emerging as adults a week or two later. The new adults feed the remainder of the summer and then go into hibernation in the fall, 1 generation developing each season.

Growers following a spray schedule for controlling berry moth will not be troubled with grape flea beetle.

REFERENCES: *USDA Farmers' Bul.* 1893, 1972; *OARDC Bul.* 1060, 1973.

GRAPE PHYLLOXERA

Daktulosphaira vitifoliae (Fitch), Family Phylloxeridae

The phylloxera is an insect native to eastern United States. It is established in Europe where it has been one of the most important enemies of European grapes; it has also been carried to California where it attacks European varieties of grapes. It is of relatively little importance on cultivated grapes of eastern United States. Wild grapes are often infested.

The biology of this insect is very complicated, and anyone interested in the details should consult the references cited. Both winged and wingless individuals develop, reproduction is sexual and parthenogenetic, both oviparous and ovoviviparous females occurring. One form lives on the foliage, causing numerous knots or gall-like growths to develop. This form is most abundant in eastern United States and is found primarily on wild grape species. Other forms are found on the roots (Fig. 316) where the most serious damage is done, galls being formed and the nutritional processes so seriously affected by these piercing-sucking insects that the vines are stunted, rendered unproductive, or die. This form is not common in the eastern states; it is the only important one in the Pacific Coast states.

Growers of American varieties of grapes, with a few exceptions, will not find phylloxera of sufficient importance as a pest to require control measures. Those

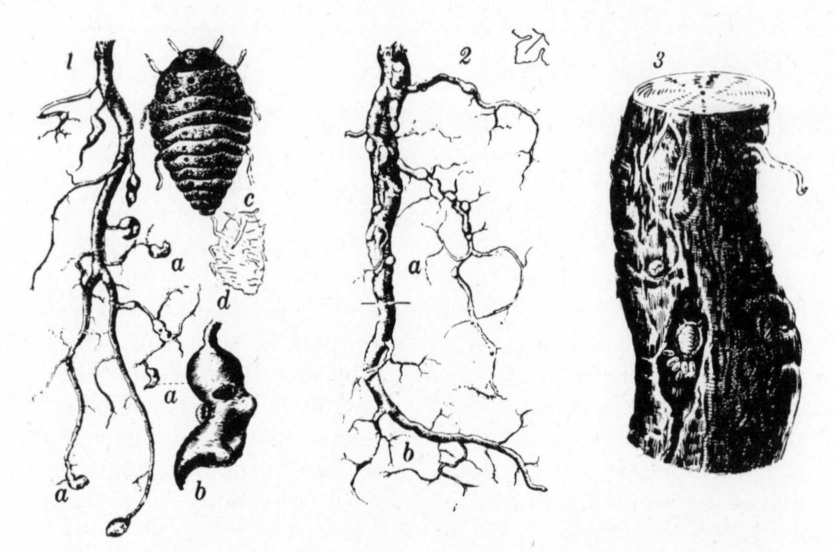

Figure 316. The grape phylloxera; 1, *a,* root galls or nodosities; *b,* same, enlarged; *c,* adult; *d,* molted exoskeleton of same; 2, *a,* infested portion of small root; *b,* normal root; 3, large root showing insects and eggs *in situ*. (Davidson and Nougaret, USDA)

planting European varieties should select vines grafted on resistant rootstocks of eastern United States. This practically eliminates any injury by this insect.

REFERENCES: *USDA Dept. Bul.* 903, 1921; *Tech. Bul.* 20, 1928; *Farmers' Bul.* 1893, 1972.

GRAPE ROOT BORER

Vitacea polistiformis (Harris), Family Aegeriidae or Sesiidae

This native insect attacks wild and cultivated grapes and is found in the eastern and central states as well as in the Pacific Northwest. It is seldom abundant in any part of its range and injuries resulting from it are rarely noticed and usually local in distribution. The larvae bore into the roots, often girdling them, which reduces or destroys the productivity of the vine. An unthrifty condition of the plant is often the first symptom of their presence.

The adult is a moth related to the peach tree borer. It is wasplike in appearance, dark lustrous brown with orange bands on the abdomen, and an over-all length of about 25 mm. The white larvae with brown heads are 25 mm or more in length, and their form and the nature of their damage are illustrated in Fig. 317.

Eggs are laid in late summer on foliage and on canes of grape or other plants in the vineyard, a single moth laying as many as 500. These hatch in 2 weeks, the larvae burrowing into the soil and tunnelling the roots for almost 2 years, then leaving the roots and pupating in cocoons near the soil surface in June. About 1 month later the new adults emerge and fly about several days before egg-laying begins.

Figure 317. The grape root borer, *Vitacea polistiformis* (Harris); larvae in roots and adults on foliage; natural size. (W. Va. Agr. Exp. Sta.)

A high degree of egg parasitism by a proctotrupid wasp has been reported. Cultivation to destroy the pupae and extra fertilization are mechanical and cultural control methods. The usual spray program for the important grape insects evidently keeps the root borer controlled since reports of damage are not common.

REFERENCES: *W.Va. Agr. Exp. Sta. Bul.* 110, 1907; *USDA Farmers' Bul.* 1220, 1926; *Proc. N.C. Branch Ent. Soc. Amer.*, 18:86, 1963; *OARDC Bul.* 1060, 1973.

GRAPE MEALYBUG

Pseudococcus maritimus (Ehrhorn), Family Pseudococcidae

Widely distributed in the United States, the grape mealybug is a pest which has been recorded on a remarkably long list of hosts. Besides pears and grape vineyards, it has also become troublesome in nurseries and on ornamental plantings, particularly *Taxus*. Damage is caused by removal of plant sap by the nymphs and adults, and from the honeydew voided by them which spots the leaves and fruits and serves as a medium for development of sooty fungus.

Their appearance is typical of all mealybugs. Winter is passed in the first instar within the shelter of the white cottony ovisac produced by the female. In April and May they become active, migrate over the twigs, and soon begin feeding. The adult stage is reached by late June, and eggs are deposited inside the white cottony ovisac. Several days later newly hatched crawlers appear and begin feeding, many getting into the grape clusters (Fig. 318). Adults again appear in late August, and egg-laying continues until cold weather, the hatched eggs forming the overwintering population. Eggs that fail to hatch before cold weather in the autumn are not viable the following spring.

Figure 318. The grape mealybug, *Pseudococcus maritimus* (Ehrhorn); natural size. (USDA)

Mealybug eggs are frequently attacked by a hemerobiid predator and an encyrtid parasite, which often keep the insect checked. All life stages are preyed upon by lady beetles.

Several organophosphorus pesticides will control the mealybug.

REFERENCES: *J. Econ. Ent.*, 42:41–44, 1949; 45:340–341, 1952; 55:849–850, 1962; 57:1–3, 372–374, 1964; *USDA Farmers' Bul.* 1893, 1972; *Calif. Leaflet* 79, 1966; *OARDC Bul.* 1060, 1973.

GRAPE CURCULIO

Craponius inaequalis (Say), Family Curculionidae

The grape curculio occasionally is found feeding on grape berries, and its damage may be mistaken for that of the berry moth. It occurs from New England to Florida and west to the Mississippi Valley. The adult is a snout beetle, slightly more than 2.5 mm in length and of a black shade (Fig. 319). They winter in sheltered situations and become active about the time that Concord grapes bloom. They spend almost 2 weeks feeding before egg-laying begins. Their feeding marks are on the upper surface of the leaves and appear as short, somewhat curved lines, usually in groups. Eggs are inserted into cavities formed by the mouthparts under the skin of the grape berries. The larvae develop inside the berries, feeding on the flesh and seeds. The legless condition of the white larvae distinguishes them from the caterpillars of berry moths. In 3 weeks the larvae are fully grown, after which they drop to the ground and pupate. New adults emerge a few weeks later, feed on grape foliage the rest of the summer, and go into hibernation with the approach of cold weather. The sprays applied for grape berry moth control this insect.

REFERENCE: *USDA Farmers' Bul.* 1893, 1972.

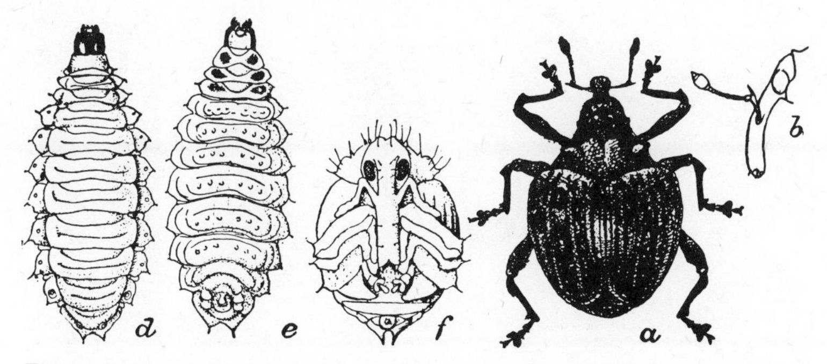

Figure 319. The grape curculio, *Craponius inaequalis* (Say): *a,* beetle; *b,* head of same from side; *d,* larva from above; *e,* same from below; *f,* pupa; 12×. (USDA)

GRAPE CANE GALL MAKER

Ampeloglypter sesostris (LeC.), Family Curculionidae

This small red-brown snout beetle, 3 mm in length, causes red gall-like swellings to form in grape canes in the region where the female deposits her eggs. With her chewing mouthparts the female punctures a grape shoot, usually just above one of the lower joints; in this puncture an egg is placed, and additional punctures are made above the first in which no eggs are deposited. On hatching the larva feeds in the pith, becoming mature, pupating, and emerging as an adult by late summer. The beetles hibernate and lay their eggs the following spring and early summer. The only control measure likely to be necessary is the destruction of infested canes as soon as the swellings have developed enough to indicate the presence of the insect. Vineyards following a regular spray program are not troubled by this insect but proper timing is essential for control.

REFERENCES: *W.Va. Agr. Exp. Sta. Bul.* 119, 1909; *OARDC Bul.* 1060, 1973.

GRAPE CANE GIRDLER

Ampeloglypter ater LeC., Family Curculionidae

Like its congener, the grape cane girdler is a minor pest of grape. The two insects resemble each other in their life cycle as well as in size and form, but the girdler is black instead of rufous. The normal host is Virginia creeper, but it readily attacks grape. It is found primarily in central and eastern United States.

Eggs are laid in late spring in cane punctures made by the mouthparts. After an egg has been laid, the female continues to make a series of punctures until the cane is encircled, but an egg is placed only in the first puncture (Fig. 320).

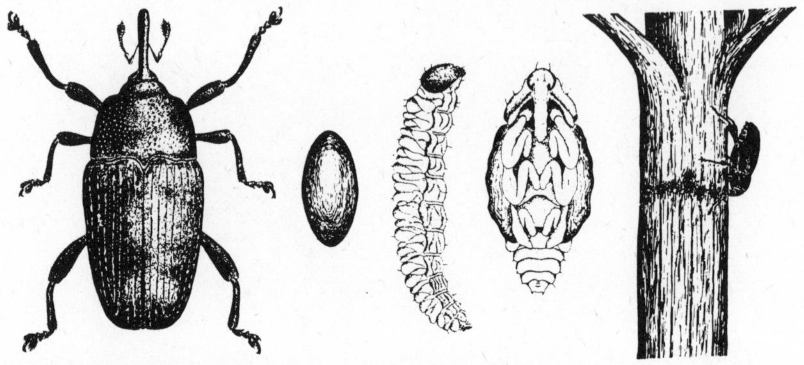

Figure 320. The grape cane girdler, *Ampeloglypter ater* LeC. Left to right: adult, egg, larva, pupa, damage; 6×. (Brooks)

A similar girdle is made at a point higher on the cane, causing the tip of the cane to break. The grub feeds in the pith of the cane, and both injured portions may break off. After larval development is completed, pupation occurs, adults appearing again in late summer; these hibernate during the winter. Destruction of the injured canes a few inches below the girdled areas usually controls this insect. It is never a problem in a vineyard following a spray program for more important pests.

REFERENCES: *W.Va. Agr. Exp. Sta. Bul.* 119, 1909; *OARDC Bul.* 1060, 1973.

MINOR GRAPE INSECTS

Grape Leaf Folder, *Desmia funeralis* (Hübner), is widely distributed from coast to coast. It is the larva of a dark brown moth spotted with white (Fig. 321). Full-grown caterpillars are 2 cm long, light green with faint dark markings. They feed on the tissue inside folded portions of the leaves and rarely become numerous in sprayed vineyards. The winter is passed in the pupal stage in the folded and fallen leaves; the moths, emerging in the spring shortly after grape foliage appears, lay their eggs on the leaves. Newly hatched larvae soon begin feeding and folding the leaves. There are 2 generations each season with a third in more southern areas. If control measures are necessary, carbaryl, parathion, permethrin, and *Bacillus thuringiensis* have been used with success.

REFERENCES: *J. Econ. Ent.*, 41:441–442, 1948; *Calif. Agr. Exp. Sta. Leaflet* 79, 1966; *Calif. Agr.* 23(4):4–5, 1969; 26(7):5, 1972.

Figure 321. The grape leaf folder, *Desmia funeralis* (Hübner); *a,* male; *b,* female; *c,* larva; *d,* head and thoracic segments of larva, enlarged; *e,* pupa; *f,* tip of pupa, enlarged; *g,* grape leaf folded by larva. (USDA)

Grape Leaf Skeletonizer, *Harrisina americana* (Guér.), occurs rather widely in the eastern half of the United States. It is rarely noticed except on wild grape and on vines grown in home gardens. Its yellow larvae attract attention by their habit of feeding in groups. Often several worms feed side by side in a row extending across a leaf. Young larvae consume only the upper surface, but the last instars may eat the entire leaf. When fully grown they are slightly more than 1 cm long. The adult is a small, smoky black, narrow-winged moth. Winter is passed as pupae in cocoons on leaves and in debris on the ground. Adults appear late in the spring and lay their lemon-yellow eggs in clusters on the lower leaf surfaces. There are 2 and a partial third generation each year. Another species, the western grape leaf skeletonizer, *H. brillians* Barnes and McDunnough, occasionally develops to pest proportions in the West. It has habits and a life cycle like its congener. Vineyards receiving the regular sprays for important grape pests are not troubled with skeletonizers.

Spotted Pelidnota, *Pelidnota punctata* (L.), is a pale yellow beetle spotted with black, of the form of May or June beetles but considerably larger. These beetles occur generally west to the Rocky Mountains. They attack grapes in midsummer, feeding on foliage and fruits. They are never a problem in sprayed vineyards, but unsprayed isolated vines are occasionally fed upon.

Grapevine Aphid, *Aphis illinoisensis* Shimer, is reported from most of the eastern half of the United States and attacks both wild and cultivated grapes. It is seldom a problem in sprayed vineyards. The winter host is viburnum, and on plants of this group the winter eggs are laid. Hatching begins in early spring and, after a few wingless parthenogenetic generations, winged individuals are produced that migrate to grape, where colonies of this dark brown aphid develop on the young shoots and leaves. When very abundant they may infest the fruit clusters causing some of the grapes to drop. In the fall winged individuals again develop that return to viburnum and produce the egg-laying females. Natural enemies of aphids such as lady beetles, aphidlions, syrphid fly larvae, and parasites usually keep this insect controlled.

REFERENCES: *USDA Farmers' Bul.* 1220, 1926; *OARDC Bul.* 1060, 1973.

Grape Colaspis, *Colaspis brunnea* (F.), feeds on a wide variety of hosts in both the larval and the adult stages. Adults feed on the foliage of soybeans, strawberry, grape, corn, beets, potato, melons, roses, okra, and many leguminous plants. In addition to grapes and strawberries, the larvae feed on the roots of various clovers, soybeans, rice, and grasses. The larvae are sometimes called clover white grubs because of their resemblance to these insects. Adults are light brown and about 5 mm in length.

Partly grown larvae pass the winter in the soil, complete their development in the spring, pupate, and emerge as adults in June. Eggs are laid during the summer and fall, the newly hatched larvae feeding until cold weather. Only 1 generation develops each season.

Newly set strawberry plants may be attacked by the grape colaspis if the planting is made after a spring-plowed leguminous crop. Injury seldom occurs if crop rotation is practiced.

REFERENCE: *Ark. Agr. Exp. Sta. Bul.* 624, 1960.

21

PESTS OF SMALL FRUITS

Pests of strawberries covered in previous chapters are: cyclamen mites, spider mites; cutworms, white grubs, leafhoppers, spittlebugs, plant bugs, stink bugs, sap beetles, snails, and slugs. The cane fruits may also be damaged by Japanese beetle, cicadas, and scale insects.

STRAWBERRY WEEVIL

Anthonomus signatus Say, Family Curculionidae

This little weevil is found from Canada and the Atlantic Coast to Texas, and north through the Mississippi Valley. It is apparently a native insect that feeds on wild strawberry, dewberry, brambles, and redbud, in addition to cultivated strawberries.

The adult is a brown snout beetle with black patches on the wings, and is scarcely 3 mm in length (Fig. 322). Larvae are correspondingly small white thick-bodied curved grubs. Hibernating beetles emerge in early spring, feeding first on whatever food plant is available. When strawberry blossom buds are formed the beetles lay eggs in the feeding punctures that they make in this part of the plant. Then they move down a short distance and partly cut through the stem, causing the bud to wilt, fall over at a sharp angle, or drop to the ground. Because of this habit the insect is sometimes known as the "clipper." In the buds the larvae complete their development before mid-summer, change to pupae, and emerge as adults. After a short feeding period they go into hibernation; there is 1 generation each year. Pistillate varieties of strawberries are relatively immune from attack since only varieties with staminate flowers seem to furnish the proper food or developmental conditions.

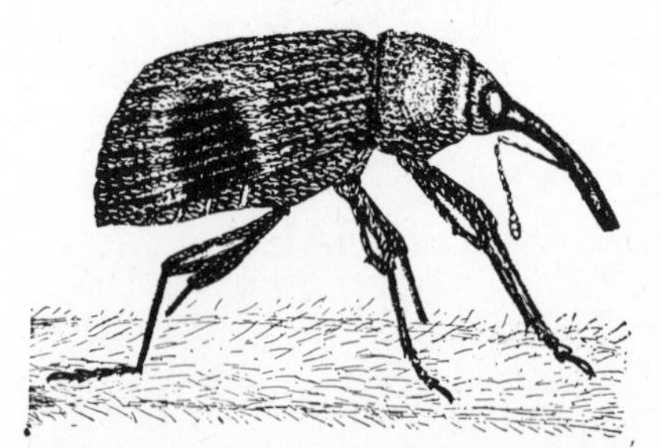

Figure 322. The strawberry weevil, *Anthonomus signatus* Say; 12×. (Baerg, Ark. Agr. Exp. Sta.)

If chemical control measures are applied in the spring at the first sign of adult weevil activity, very little crop damage accrues and no dangerous residues on the berries result.

REFERENCES: *Ark. Agr. Exp. Sta. Bul.* 185, 1923; *Purdue Agr. Ext. Leaflet* 344, 1952; *USDA Farmers' Bul.* 2184, 1973; *Ohio Res. Bul.* 987, 1966; *Ohio Sta. Univ. Ext. Bul.* 506, 1979.

STRAWBERRY LEAFROLLER

Ancylis comptana fragariae (W and R.), Family Olethreutidae

The leafroller, or leaf folder, is present in Canada and a large part of the United States from the Mississippi Valley eastward. Damage results from the larvae feeding within the folded, rolled, or webbed leaves causing them to turn brown and die.

The adult is a rusty brown moth with markings of light yellow and a wing expanse of 12 mm or more (Fig. 323). The larvae are pale green to brown and exceed 12 mm in length when fully grown.

Both larvae and pupae hibernate, the pupae in folded leaves and the larvae sometimes in the same situation and sometimes in other shelters, which they construct in the litter near the plants. Adults emerge in April and early May, and lay tiny eggs on the foliage. Hatching occurs in a week or more and the larvae complete their development, pupate, and emerge as adults in 40 to 50 days. There are 2 generations and a partial third each season.

Several parasites attack this insect; some of importance are *Macrocentrus ancylivorus* Rohwer, and *Spilochalcis albifrons* (Walsh). At times, pesticides may be necessary to suppress the leafroller.

REFERENCES: *Ohio Agr. Exp. Sta. Bul.* 651, 1944; *Purdue Agr. Ext. Leaflet* 344, 1952; *USDA Farmers' Bul.* 2184, 1973; *Ohio Sta. Univ. Ext. Bul.* 506, 1979.

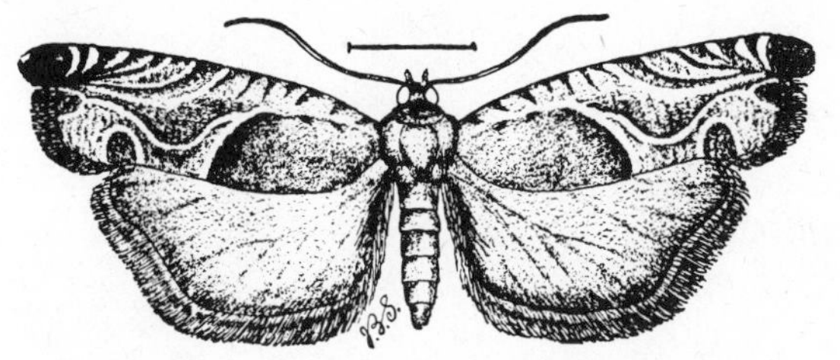

Figure 323. The strawberry leafroller moth; 6×. (J. B. Smith.)

ROOT WEEVILS

ORDER COLEOPTERA, FAMILY CURCULIONIDAE

These weevils are widely distributed in northern United States and southern Canada. In the Pacific Northwest infestations may become numerous and correspondingly injurious. Besides strawberries, they also attack raspberries, loganberries, blueberries, grapes, azalea, Taxus, hemlock, rhododendron, primrose, red clover, grasses, and many other nursery and flower garden plants. Damage is caused by the larvae devouring the roots and the adults feeding on the leaves. Adults are sometimes a nuisance in and around homes.

There are 4 species that cause damage. They are the strawberry root weevil, *Otiorhynchus ovatus* (L.); the black vine weevil, *O. sulcatus* (F.); the rough strawberry root weevil, *O. rugosostriatus* (Goeze); and the clay-colored root weevil, *O. singularis* (L.) (Figs. 324, 325, and 326).

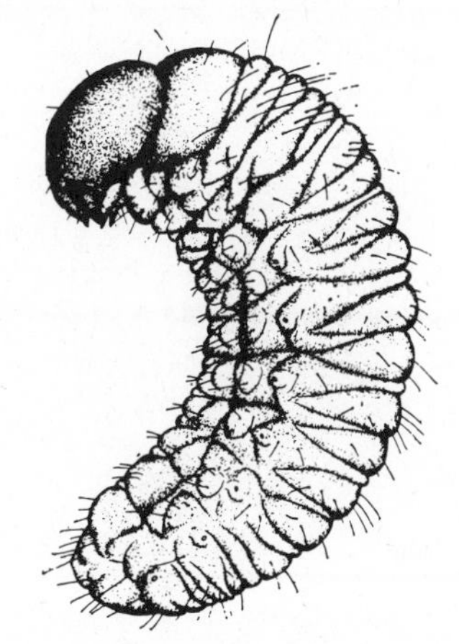

Figure 324. The strawberry root weevil: (left) adult, 2.5×; (right) larva, 5×. (Pettit and Downes.)

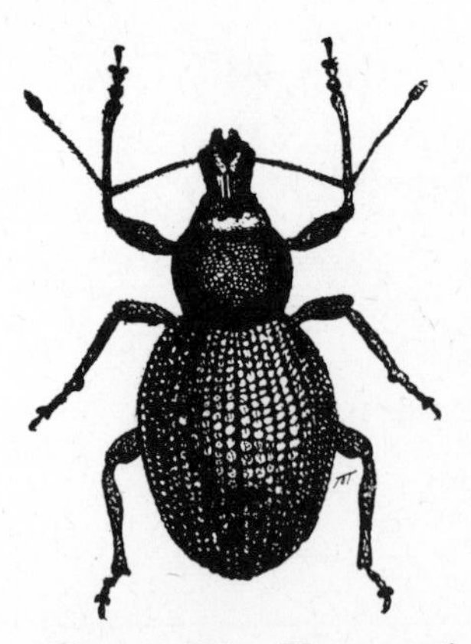

Figure 325. The rough strawberry root weevil; 5×. (Utah Agr. Exp. Sta.)

Figure 326. The black vine weevil; 3.5×. (USDA)

All these chewing weevils are snout beetles, with hard-shelled bodies and rows of small round pits on the elytra. They are light brown to black and vary in length from 5 to 10 mm, the largest species being the black vine weevil. The flightless adults are females that reproduce parthenogenetically. The white larvae tinged with pink are curved legless grubs with brown heads; they are nearly 1 cm in length (Fig. 324).

The life cycles of all 4 species are very similar. Most of them spend the winter as nearly grown larvae in the soil among the roots of the host plants. In the spring they change to pupae and begin emerging as adults about June. In 2 or more weeks they begin depositing eggs near the crowns of the plants, each female laying many eggs, over 1600 being recorded for the black vine weevil. Hatching occurs in about 10 days and the tiny larvae burrow into the soil and feed on the roots. A small percentage of the root weevils overwinter as pupae or adults in the soil. These adults appear in May and soon begin egg-laying. There is usually only 1 generation per year.

Damage from root weevils can be greatly reduced where new plantings are placed in uninfested soil. In crop rotations, avoid planting a susceptible host after grass or clover sod. Nursery stock, especially yew, is often treated with chemicals. Where feasible, surface soil treatment with chemicals just as adult emergence begins is effective, since migration from the soil is entirely by crawling.

REFERENCES: *USDA Tech. Bul.* 325, 1932; *Farmers' Bul.* 2184, 1973; *Conn. Agr. Exp. Sta. Cir.* 174, 1950; 211, 1960; *Utah Ext. Cir.* 186, 1953; *J. Econ. Ent.* 46:234–237, 1953; 48:207–208, 1955; *Can. Dept. Agr. Pub.* 78, 1953; 990, 1956; *Ohio Res. Bul.* 987, 1966.

STRAWBERRY CROWN BORER

Tyloderma fragariae (Riley), Family Curculionidae

The crown borer is considered a native American pest. It is found rather generally distributed through the eastern half of the country. Damage is caused by the larvae boring into the crown of the plant and feeding on the interior; sometimes the crown is hollowed out so completely that growth is checked or the plant killed. Some injury to the crowns also results from the feeding cavities made by adults in which they deposit their eggs. Adults also eat small holes in the leaves.

The flightless adults are dark brown snout beetles having three darker spots on each wing cover and a total length of about 4 mm. They hibernate during the winter, becoming active about the time strawberries blossom. Egg-laying soon begins and continues into August. The white curved legless larvae with dark heads feed in the crowns, becoming fully grown, pupating, and emerging as adults the same summer. After feeding for a period they go into hibernation in any convenient shelter. One generation develops each year.

Rotation of the crop effects some control since the beetles migrate only by crawling. Avoid placing new plantings closer than 300 yards to old plantings or border wastelands containing wild strawberries or cinquefoil. Deep plowing and compacting the soil after the second crop is harvested will destroy many borers. Care should be taken, in the establishment of new plants, to secure uninfested ones. All these cultural measures will control this insect.

REFERENCES: *Ohio Agr. Exp. Sta. Bul.* 651, 1944; *Purdue Agr. Ext. Leaflet* 344, 1952; *USDA Farmers' Bul.* 2184, 1973.

STRAWBERRY ROOTWORM

Paria fragariae Wilcox, Family Chrysomelidae

Widely distributed in Canada and the United States, this insect attacks strawberries, raspberries, blackberries, roses, and other plants. Adults feed primarily at night on foliage, and the larvae feed on the roots and crowns of strawberries and related hosts.

The beetles are glistening dark brown with black markings, and are slightly more than 3 mm in length (Fig. 327). The larvae have the appearance of small white grubs, being curved and white, with 3 pairs of true legs which help distinguish them from crown borer larvae. Adults hibernate under plant remnants in the field, become active in the spring, feed and lay eggs on the older leaves near the ground. Oviposition may continue until mid-July. Hatching occurs in a week and the larvae, after passing through 4 instars, pupate and transform to adults. The entire life cycle is completed in about 2 months, with 1 generation in the North, 2 in California, and 1 and a partial second in North Carolina.

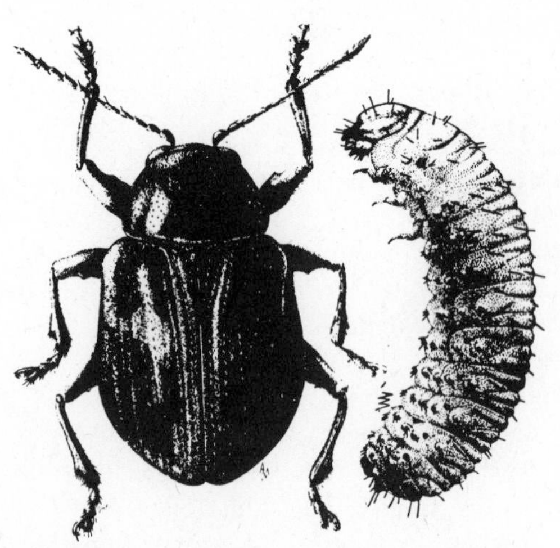

Figure 327. The strawberry rootworm, *Paria fragariae* Wilcox, adult and larva; 13×. (USDA)

Cultural controls consist of plowing infested fields in early July while the beetles are in the larval stage, and setting new plantings some distance from infested fields or from adjacent favorable hibernating quarters.

REFERENCES: *Ohio Agr. Exp. Sta. Bul.* 651, 1944; *Purdue Agr. Ext. Leaflet* 344, 1952; *J. Econ. Ent.*, 46:1101–1102, 1953; *Can. Dept. Agr. Bul.* 990, 1956; *USDA Agr. Handbook* 290, 1965.

STRAWBERRY ROOT APHID

Aphis forbesi Weed, Family Aphididae

This aphid is a pest of strawberries primarily in the eastern half of the United States. Both nymphs and adults suck the sap from the roots, and their presence can be detected by the conspicuous abundance of ants.

Winter is passed as shiny black eggs on the stems and foliage of strawberries. In early spring they hatch into blue-green female aphids which feed on new strawberry leaves. Ants, especially the cornfield ant, soon carry them to the roots where repeated generations are produced throughout the summer. These individuals are all parthenogenetic wingless females that give birth to their young. Winged females (Fig. 328) develop in October and produce sexual individuals that mate and lay the overwintering eggs. The root forms may persist through the winter as more or less active individuals, especially in mild climates.

Usually no aphids are found in plantings that have been treated with a pesti-

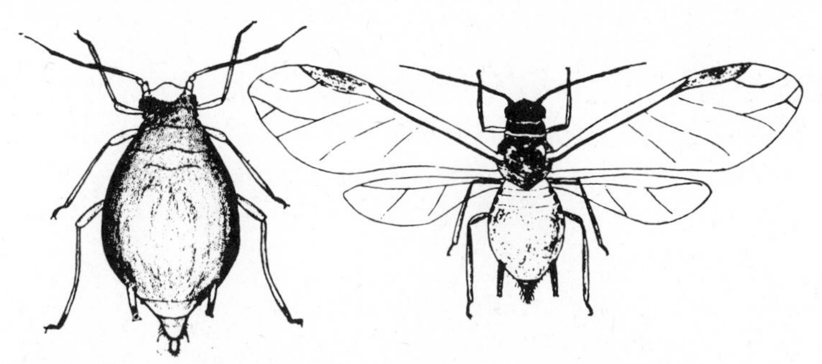

Figure 328. The strawberry root aphid, early season forms; 10×. (Marcovitch, Tenn. Agr. Exp. Sta.)

cide for the control of root weevils or ants. Dipping plants in demeton before setting in the field is a recommended procedure.

REFERENCES: *Purdue Agr. Ext. Leaflet* 344, 1952; *USDA Farmers' Bul.* 2184, 1973; *Agr. Handbook* 290, 1965; *Calif. Agr. Exp. Sta. Leaflet* 81, 1966.

STRAWBERRY APHID

Chaetosiphon fragaefolii (Cockerell), Family Aphididae

Widely distributed in North America, this small pale yellow aphid is often a problem for growers of strawberry plants. It does not attack the roots but confines its feeding to the foliage, principally the undersides of the leaves. Besides removal of plant sap the foliage is often smutted from fungus growth which develops on the honeydew excretion of the aphids. This species also transmits a virus causing the disease known as "yellows." Recommended pesticides are organophosphorus compounds.

REFERENCES: *USDA Farmers' Bul.* 2184, 1973; *Agr. Handbook* 290, 1965.

RASPBERRY CROWN BORER

Pennitsetia marginata (Harris), Family Aegeriidae or Sesiidae

The adults are clear-winged moths having some resemblance to wasps. They are black, with yellow bands on the abdomen, and bands and stripes on the thorax (Fig. 329). Female moths are larger than the males; the former are more than 25 mm in body length. The larvae reach a length of over 25 mm and are thick-bodied, soft, and white. Eggs are laid on the foliage of raspberry, blackberry, and boysenberry in late summer. On hatching, the larvae crawl down to the base of the canes, excavate small cavities, and remain there in winter. Dur-

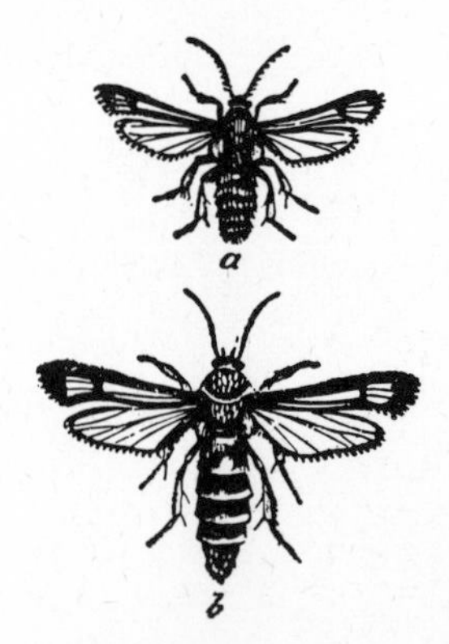

Figure 329. The raspberry crown borer, *a,* male; *b,* female; about natural size. (Summers.)

ing the following spring and summer they continue their growth, feeding in the crowns and roots. They make large galleries which may be extensive enough to ruin the plants but are smaller than the ones that follow the next summer when the entire crown may be hollowed out. Early in the second summer growth is completed and pupation takes place in the canes, the moths emerging by midsummer and laying their eggs, thus completing the life cycle which requires 2 years.

Removal and burning of all wilting canes in June and July is of help in control. Drenching the lower cane and crown areas with diazinon or guthion in fall or spring is an approved procedure.

REFERENCES: *Can. Dept. Agr. Pub.* 880, 1952; *Ohio Sta. Univ. Ext. Bul.* 506, 1979.

RASPBERRY CANE BORER

Oberea bimaculata (Olivier), Family Cerambycidae

If the tips of young shoots of raspberry, blackberry, and sometimes roses are found wilted and dying in early summer, it is a sign that they have been girdled by the cane borer. This insect is generally distributed from Kansas eastward, and is often quite destructive in the province of Quebec.

The adult is a slender black beetle about 12 mm long, with prominent antennae and usually 2 black dots on the yellow prothorax. They appear in June and may be present until late August. After laying an egg, the female girdles the cane about 6 mm above and again 6 mm below the egg puncture, causing the tip of the shoot to wilt and die. On hatching, the larvae bore downward in the cane, passing the winter not far below the point of girdling. The next season they continue boring until the crown is reached; there the second winter is passed at or below ground level. The following spring full growth is attained when the larvae are nearly 2 cm in length. Pupation follows, and new adults begin emerging in June. Two years are required to complete the life cycle.

The best control practice is the destruction of the canes that show characteristic injury. If pruning is done within a few days after the wilted tips appear, only an inch or so more than the wilted part need be removed.

REFERENCES: *Can. Dept. Agr. Pub.* 880, 1952; *Ohio Sta. Univ. Ext. Bul.* 506, 1979.

RED-NECKED CANE BORER

Agrilus ruficollis (Fabr.), Family Buprestidae

Although widely distributed in eastern United States and Canada, this insect is seldom of much importance as a pest. The adults feed on foliage, and the larvae produce gall-like enlargements on the cane from their spiral tunnelling within. Affected canes are weakened and may die or produce no fruit. Raspberry, blackberry, and dewberry are the common hosts.

This bronzed, blue-black beetle, with a coppery red prothorax, is slightly more than 7 mm long. The white larva attains a length of more than 12 mm. Adults are present from May to August, feeding on foliage and ovipositing on canes. The young larvae make spiral burrows just beneath the bark, and when they become larger they also bore in the pith. Winter is passed in the larval stage, with pupation occurring in the spring and new adults appearing in late May. One generation develops each year.

Destruction of the infested canes is the simplest remedy, but to be effective this must embrace not only the cultivated plants but also wild hosts in the vicinity.

REFERENCES: *USDA Farmers' Bul.* 1286, 1922; *Can. Dept. Agr. Pub.* 880, 1952; *Ohio Sta. Univ. Ext. Bul.* 506, 1979.

RASPBERRY CANE MAGGOT

Pegomya rubivora (Coq.), Family Anthomyiidae

The raspberry cane maggot is prevalent over the entire northern half of the United States and southern part of Canada. It seldom causes serious injury except in British Columbia. Wilted tips of the canes are sometimes cut off almost as cleanly as if the cut were made with a knife; canes of both raspberry and blackberry also show gall-like swellings. Both these symptoms are caused by the cane maggot, an insect which resembles the cabbage maggot. The larvae are white, tapered, and scarcely exceed 8 mm in length. The flies are not unlike the house fly but are only about ½ as large and are much more pointed near the tip of the abdomen.

Winter is passed as puparia in the infested canes; the adults emerge in the spring and lay their eggs on the new shoots of brambles. Developing larvae feed in the layer of wood outside the pith, thus girdling the canes. In raspberries the injured tips usually break off; in blackberries this usually does not happen, since the larvae are unable to complete their development in such plants. The larvae tunnel downward in the canes and when fully grown transform to puparia, which is the hibernating stage.

Since the presence of the insects is indicated by the wilted tips, it is an easy operation to cut off the infested portions a few inches below the girdle and destroy them by burning.

BRAMBLE LEAFHOPPER

Ribautiana tenerrima (H.-S.), Family Cicadellidae

This is a major pest of cane fruits in Europe as well as in parts of the United States and Canada. By sucking the sap from the leaves it prevents the fruits from developing and ripening, thus reducing the yield. The injury has the appearance of tiny white spots or mottled areas on the leaves.

Adults are about 3 mm long, pale yellow-green, wedge-shaped, with red eyes. The nymphs are similar in shape but are wingless and pale white. The winter is passed in the egg stage under the bark of the canes. Hatching takes place usually the first 2 weeks of May, and the nymphs become fully grown in 3 or 4 weeks. Adults are abundant in late June and July, and they lay eggs which result in a second generation. In southern areas there may also be a third generation.

Since this leafhopper survives the winter as eggs, it may serve as an overwintering host for the important leafhopper egg parasite, *Anagrus epos* Girault. If so, chemical control measures should be avoided wherever feasible.

REFERENCE: *Can. Dept. Agr. Pub.* 116, 1950.

RASPBERRY FRUITWORMS

ORDER COLEOPTERA, FAMILY BYTURIDAE

These important pests of red raspberries and loganberries are widely distributed in northern United States and southern Canada. Damage to the plants results from both adult and larval stages. In the spring the adult beetles feed on the young unfolding leaves, buds, and blossom clusters, causing a marked reduction in fruit set. The larvae penetrate the flower buds and developing fruits, often causing them to drop or decay before harvest. Wormy fruits are unmarketable.

The common species are the western raspberry fruitworm, *Byturus bakeri* Barber, and the eastern raspberry fruitworm, *Byturus rubi* Barber.

The elongate oval light brown beetles (Fig. 330), with a slightly pilose sur-

Figure 330. A raspberry fruitworm; adult (14×) and feeding larvae. (Walden, Conn. Agr. Exp. Sta.)

face, are about 3 mm in length. They emerge from the soil in April and May, begin feeding and laying eggs on the blossom clusters, and later on the developing berries. On hatching, the light yellow larvae feed in the blossoms and young fruits, becoming almost 8 mm in length when fully developed. Mature larvae leave the plant, burrow into the soil, and transform into pupae. The pupae change to adults during the summer, remaining in this stage in the soil until the following spring.

Satisfactory control results from the application of diazinon; the first should be made when the blossom buds appear and the second just before blooming.

REFERENCES: *Conn. Agr. Exp. Sta. Bul.* 251, 1923; *USDA Misc. Pub.* 468, 1942; *J. Econ. Ent.*, 41:436–440, 1948; *Can. Dept. Agr. Pub.* 880, 1952; *Ohio Sta. Univ. Ext. Bul.* 506, 1979.

RASPBERRY SAWFLY

Monophadnoides geniculatus (Htg.), Family Tenthredinidae

This is one of the common pests of raspberries in the north central states eastward to New England and in Canada. The adult is a small thick-bodied black sawfly, little more than 6 mm in length, which lays its eggs in the leaf tissue of the host in May and June. Larvae are light green and marked by conspicuous bristles, which arise from small swellings on the body (Figs. 331 and 332). These larvae may be as much as 12 mm in length. In the course of their development they consume the leaf tissue, and a heavy infestation may result in loss of the crop berries. The larvae complete their feeding in less than 2 weeks and then estivate in cocoons which they construct in the ground. They remain in these cocoons through the winter and pupate in early spring.

REFERENCES: *Can. Dept. Agr. Pub.* 880, 1952; *Calif. Agr. Exp. Sta. Leaflet* 75, 1966.

Figure 331. Larva of the raspberry sawfly; 2×. (USDA)

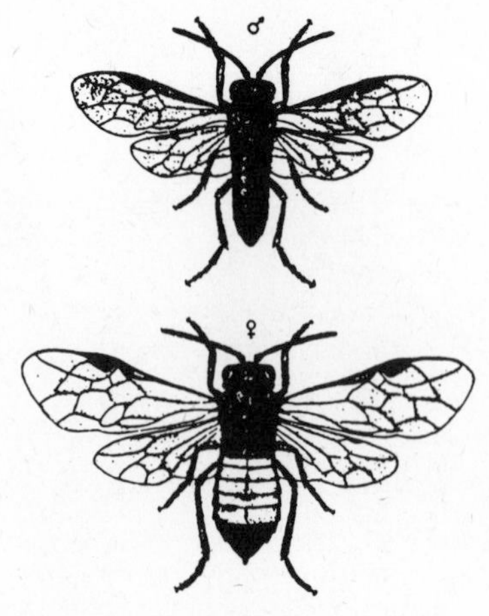

Figure 332. Both sexes of the raspberry sawfly, *Monophadnoides geniculatus* (Htg.); 3×. (Lowe, N.Y. Agr. Exp. Sta.)

TREE CRICKETS

ORDER ORTHOPTERA, FAMILY GRYLLIDAE

Tree crickets are native insects, species of which are to be found in most regions of southern Canada and in the United States. They are minor pests of orchard trees and bush fruits which are damaged primarily by oviposition. Later instar nymphs may feed on the foliage, and the adults feed on both foliage and ripening fruits. Additional injury may result from disease organisms gaining entrance through the egg punctures, and from the splitting of canes by the freezing of moisture which has gained access. Common species are the prairie tree cricket, *Oecanthus argentinus* Saussure; black-horned tree cricket, *O. nigricornis* Walker; four-spotted tree cricket, *O. quadripunctatus* Beutenmüller; and snowy tree cricket, *O. fultoni* Walker.

The various species are similar in appearance and development (Fig. 333). Eggs are laid in late summer or early fall, and the winter is passed in this stage (Fig. 334). These eggs hatch rather late in the spring, and the young crickets feed as predators for most of their lives, sometimes eating a small amount of vegetation when they are nearly grown. Adults appear in late summer, thus completing the cycle. Only the males are capable of producing the familiar tree cricket song. The prairie tree cricket oviposits just beneath the bark, primarily in fruit trees or other solid wood plants. The other species lay their eggs in plants with a pithy center, such as brambles and elder. Raspberry canes sometimes are severaly damaged. The eggs are placed in elongated rows, each con-

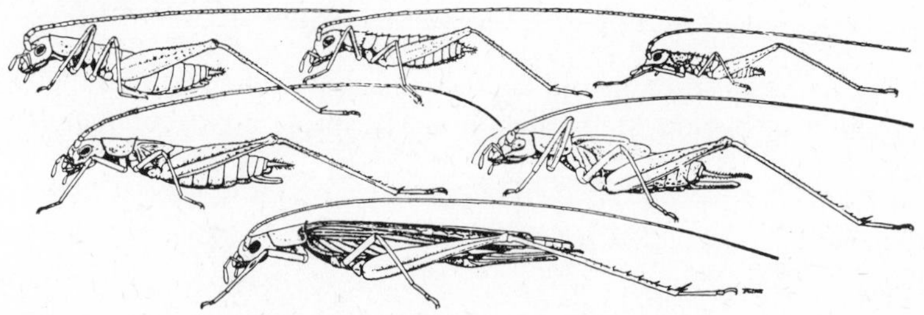

Figure 333. Nymphal instars and adult of the tree cricket; 2×. (Parrot and Fulton, N.Y. Agr. Exp. Sta.)

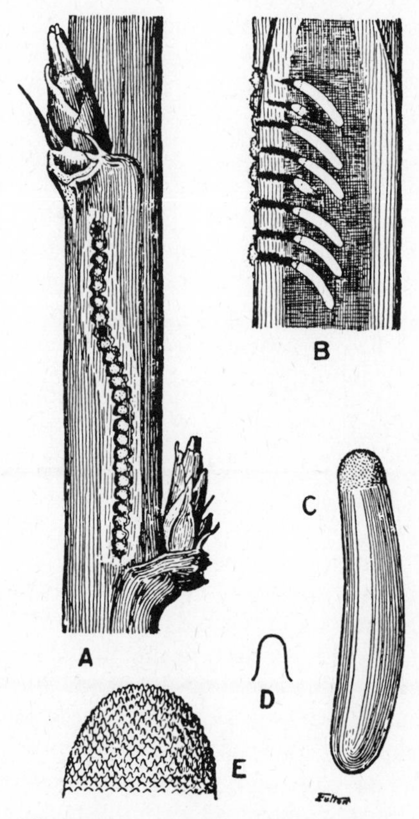

Figure 334. Raspberry stem injured by the tree cricket; *A,* wound made by egg punctures; *B,* longitudinal section through same, showing eggs; *C,* egg; *D, E,* cap of egg; all enlarged. (Fulton.)

taining as many as 80. Pruning old fruiting canes after harvest and the removal and destruction of egg-infested canes in early spring are recommended practices.

REFERENCES: *N.Y. Agr. Exp. Sta. Bul.* 388, 1914; *Can. Dept. Agr. Pub.* 880, 1952; *Ann. Ent. Soc. Amer.*, 56:772–789, 1963.

ROSE SCALE

Aulacaspis rosae (Bouché), Family Diaspididae

Frequently rose bushes, raspberry and blackberry canes become encrusted with these scale insects. They not only spoil the appearance of the plants but their sucking of plant sap greatly reduces plant vigor. Female scales are almost circular, nearly 2 mm in diameter, dirty white, with an orange-yellow dot in the center. The male scales are much smaller, long and narrow, and snow white. These scales are most abundant in humid situations where the amount of sunlight is reduced.

Winter is passed as pink eggs under the female scale coverings in some areas; in other regions full-grown females survive. The eggs begin hatching in late May or early June and after the tiny crawlers molt the first time the scale covering begins to form. By July they complete their development, and a second generation develops in August and September.

Remove and burn all heavily infested canes that can be spared. Dormant spraying with liquid lime-sulfur gives effective control.

REDBERRY MITE

Acalitus essigi (Hassan), Family Eriophyidae

This mite causes a condition, primarily in Himalaya and Mammoth blackberries, known as "redberry disease." It is present when some drupelets on a berry remain red, hard, and sour after the fruit is ripe. The condition is caused by a microscopic mite (Fig. 335) related to the blister mite of pear; it is apparently an imported species, being known in England. This mite has been injurious in the Pacific Coast areas since 1921. The mites spend the winter in the buds and attack the drupelets soon after the flowers open in the spring. Many overlapping generations develop each summer. Predaceous mites are of some value in checking this pest. Control is accomplished by spraying with lime-sulfur in autumn, after removal of old canes, or in spring at the delayed dormant period.

REFERENCES: *Wash. Agr. Exp. Sta. Bul.* 279, 1933; 155, 1938; *Tech. Bul.* 6, 1952; *Calif. Ext. Cir.* 87, 1944; *Agr. Exp. Sta. Leaflet* 75, 1966; *Can. Dept. Agr. Bul.* 880, 1955.

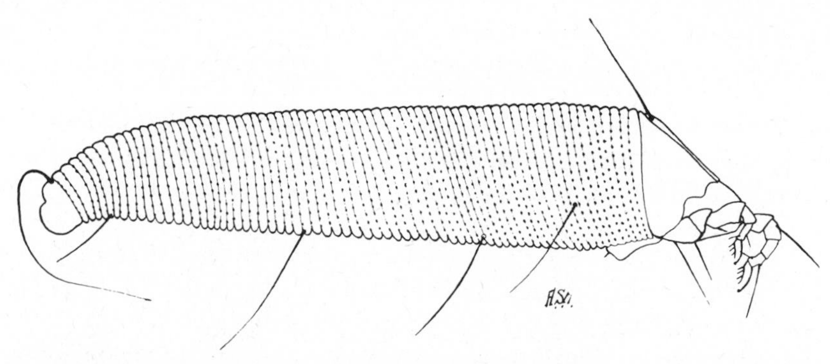

Figure 335. The redberry mite which causes "redberry disease," primarily in Himalayan and Mammoth blackberries. It is so small it cannot be seen without magnification. Figure greatly enlarged. (Drawing by A. S. Hassan.)

CURRANT FRUIT FLY

Epochra canadensis (Loew), Family Tephritidae

Found throughout the northern and western states and adjacent areas of Canada, this insect attacks the fruits of currants and gooseberries in the larval or maggot stage. The adult is 8 mm in length, yellow-bodied with darker shadings and conspicuously banded wings (Fig. 336).

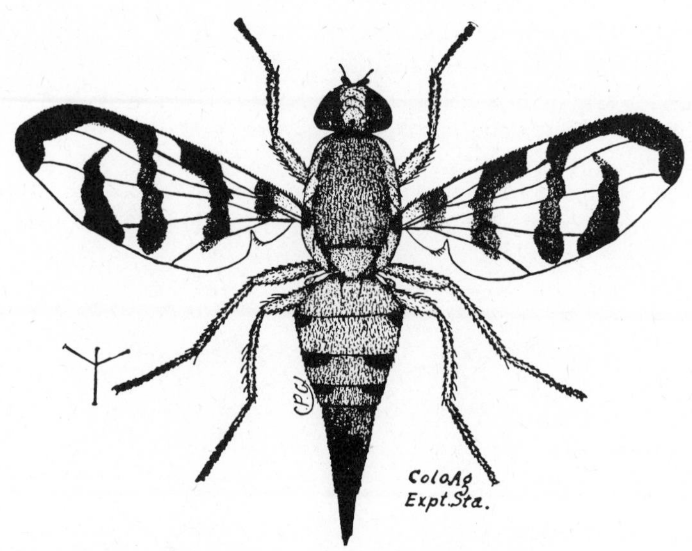

Figure 336. The currant fruit fly, *Epochra canadensis* (Loew); 7×. (Gillette, Colo. Agr. Exp. Sta.)

The insect hibernates in the pupal stage in the soil. Adults appear in the spring and lay their eggs under the skin of well-formed berries in which the larvae feed. Their presence is indicated by a dark spot on the berry which may be surrounded by red coloration. Usually the infested berries fall off before the larvae are fully developed and their feeding continues on the ground for a few days before they enter the soil and change to puparia. This stage survives the winter.

Widely advocated control measures are destruction of all infested berries before the larvae leave them and cultivation of the soil containing the puparia. Early-maturing varieties will escape much of the injury done by the fruit fly. Properly timed sprays or dusts control this insect. Applications should be made just as the first adults appear, to kill them before egg-laying begins. This is the period when 80% of the blossoms have withered and fallen; a repeat application is usually needed 10 days later.

REFERENCES: *Can. Dept. Agr. Pub.* 46, 1951; *Wash. Agr. Exp. Bul.* 155, 1938; *Maine Agr. Exp. Sta. Rept.*, 1896.

IMPORTED CURRANTWORM

Nematus ribesii (Scop.), Family Tenthredinidae

The most conspicuous and by far the best known of the currant pests is this sawfly larva, which appears in great numbers and devours the foliage before ripening of the fruits. Gooseberries are also a favored host. This insect was imported from Europe before the Civil War and has become generally distributed in southern Canada and the United States.

The adult sawflies, nearly the size of a house fly with a dark head and thorax and red-yellow abdomen (Fig. 337), emerge about the time the foliage of the host plants appears and, when the leaves are well grown, lay rows of watery-white eggs along the veins. From these hatch the green spotted caterpillarlike larvae which consume the foliage. In the last instar these larvae are light green

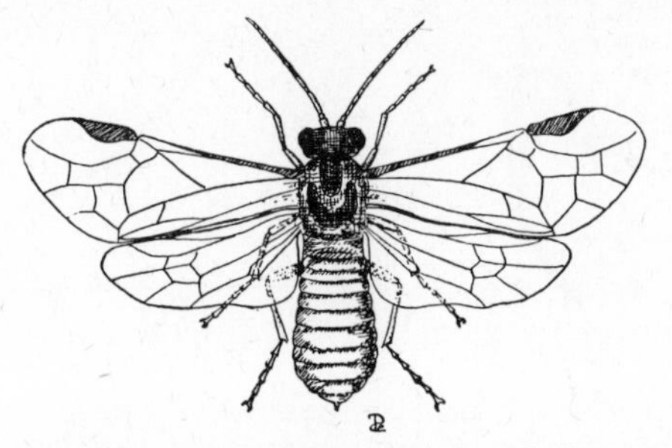

Figure 337. Adult female of the imported currantworm; 3×. (After Herrick)

in color. Pupation takes place in the litter on the ground and adults emerge soon afterwards. A second generation appears about July; possibly because of natural enemies this generation is usually so small as to escape notice. The pupae from the second generation hibernate.

Since the worms do their damage and attract notice just about picking time, the pesticides chosen must not leave a residue poisonous to man. Those commonly recommended are rotenone, pyrethrum, carbaryl, or malathion.

REFERENCE: *Can. Ent.* 52:106, 1920.

FOUR-LINED PLANT BUG

Poecilocapsus lineatus (Fabr.), Family Miridae

This bug is a general feeder, attacking weeds, legumes, fruits of many sorts, and ornamental and vegetable garden plants, but it is usually listed as a pest of currants (Fig. 338). Both nymphs and adults suck plant sap from the leaves causing distortion, curling, and browning on some hosts; on others spotting of the leaf surface is the characteristic damage.

The adult is 6 mm in length, yellow-green with 4 dark stripes on the back. Eggs are placed in the stems of both woody and herbaceous plants, as shown in Fig. 338, and the insect passes the winter in this stage. The orange nymphs appear in the spring and feed on the leaves of the plants in which the eggs were deposited. This causes an early concentration on such hosts as currants, the

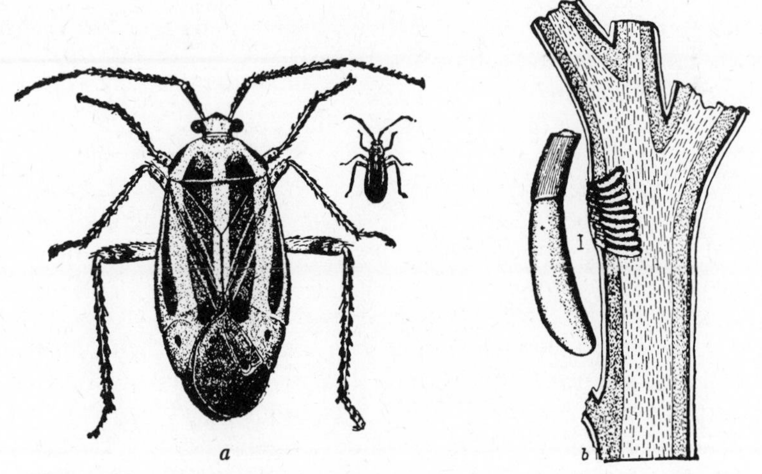

Figure 338. The four-lined plant bug, *Poecilocapsus lineatus* (Fabr.): *a*, adult (6×); *b*, cross-section of stem showing eggs in position and a single egg greatly enlarged. (Slingerland.)

stems of which are attractive to egg-laying females. Most of the damage is done by these early individuals; later the bugs disperse and the damage is little noticed. There is 1 generation each year.

REFERENCES: *Cornell Bul.* 58, 1893; *Mo. Agr. Exp. Sta. Bul.* 342, 1934.

CURRANT APHID

Cryptomyzus ribis (L.), Family Aphididae

Early-season foliage of currants, gooseberries, and snowball is often distorted due to the extraction of plant sap from the undersurface by this aphid. Other species than the one named are found from time to time on these hosts.

Winter is passed as tiny black eggs on the currant stems. When leaves appear in the spring these eggs hatch into green wingless female aphids which, when fully grown, give birth to succeeding generations. By early summer winged forms are produced that fly to weeds or other hosts and continue their reproduction. In the fall winged migrants are again produced that return to currants and give birth to males and females which, on reaching maturity, mate, the females depositing the overwintering eggs.

REFERENCE: *N.Y. Agr. Exp. Sta. Bul.* 517, 1924.

CURRANT BORER

Synanthedon tipuliformis (Clerck), Family Aegeriidae or Sesiidae

This well-known European insect is widely established in North America and tunnels the canes of currants and gooseberries in the larval stage. Red currants are the most susceptible host.

The adult is a relative of the borers attacking peaches. It is 13 mm long with wings almost devoid of scales and a spread not exceeding 25 mm, the general appearance being somewhat wasplike (Fig. 339). The moths appear in late May or early June and deposit their eggs on the canes. Hatching takes place normally in 10 or more days when the pale yellow larvae bore into the canes and tunnel in the pith and wood. This often causes yellowing of the foliage and sometimes death of the canes. Decreased vigor of surviving canes also reduces the yield. Larvae feed all summer and pass the winter in their tunnels. When fully grown they are nearly 15 mm long. In early May they pupate, and adults begin emerging about 2 weeks later, 1 generation developing each year.

Destruction of the infested canes provides a simple remedy, but this is often not practical in commercial plantings because of the lack of sufficient labor competent to do the work and the expense involved to make the practice effective. A high degree of control can be effected by 1 application of a spray con-

Figure 339. The currant borer, adult; 2×. (Beutenmüller.)

taining parathion 10 to 14 days after the first moths emerge. Laboratory experiments reveal that parathion is highly toxic to the eggs, suggesting that thorough coverage of the canes with the spray is needed for maximum effectiveness. Guthion and diazinon are other control chemicals.

REFERENCES: *J. Econ. Ent.,* 46:394–400, 1953; 57:123–130, 1964; *USDA Farmers' Bul.* 1398, 1933.

GOOSEBERRY FRUITWORM

Zophodia convolutella (Hübner), Family Pyralidae

Widely distributed from Maine to Oregon in northern United States and southern Canada, this insect attacks both gooseberries and currants. Damage results when the larvae bore into the fruits, which they may completely hollow out. One larva may be responsible for the destruction of several berries during the course of its feeding.

The adult is a moth with a wing expanse of nearly 2.5 cm. The wings are ash-colored with dark markings. Larvae at maturity are green with a yellow tinge and a pink cast; along the sides are darker lines or stripes. The winter is spent in the ground or under litter in the pupal stage. Moths emerge and lay their eggs in the flowers of gooseberries. On hatching, the larvae bore into the developing fruits. Rotenone is a commonly recommended pesticide and 2 applications are usually required.

REFERENCE: *N.Y. Agr. Exp. Sta. Bul.* 423, 1916.

22

PESTS OF CITRUS

Like all other crops citrus is attacked by many pests. Only the important ones are discussed in this chapter. For further information, consult Ebeling* and the publications of state experiment stations where citrus is grown. Many of these insects are also pests of plants in glasshouses.

ARMORED SCALES

PURPLE SCALE

Lepidosaphes beckii (Newman), Family Diaspididae

Worldwide in distribution purple scale is one of the oystershell scales, so-called because its shape somewhat resembles the shell of the oyster. This species is purplish brown and attains 2 to 3 mm in length (Fig. 340); it is slightly narrower in form than the average of the better known oystershell scale of northern fruits. This scale attacks citrus in Florida and in California, especially in the coastal areas.

Infestations of these insects suck sap from the foliage and branches, then spread to the fruits where contamination and disfiguration reduce their market value.

Where there is any hibernation period for the scale it is passed in the egg stage, the eggs being produced and sheltered under the armor or scale covering of the female insect. Eggs hatch in early summer; there are 3 or more generations each year with a tendency toward some overlapping.

Natural enemies include the lady beetles, *Scymnus marginicollis* Mann., *Lindorus lophanthae* (Blaisd.), and *Chilocorus stigma* (Say), along with the

* Walter Ebeling, *Subtropical Fruit Pests,* Univ. Calif. Press, 1959.

Figure 340. Purple scale, *Lepidosaphes beckii* (Newman), on grapefruit; 3×.

predatory thrips, *Aleurodothrips fasciapennis* Franklin, a lacewing, *Chrysopa lateralis* (Guerin), and various species of predaceous mites. Parasites include *Prospaltella aurantii* (Howard) (Fig. 236), *Aspidiotiphagus lounsburyi* (Berlese and Paoli), *A. citrinus citrinus* (Crawford), and *Aphytis lepidosaphes* Compere. Since the latter species was introduced into Florida and other areas, purple scale has been reduced to the status of a minor citrus pest.

Chemical control has been mainly thorough coverage with summer to fall sprays of 2% light medium or medium grade oil emulsions. Combination sprays of these oils plus ethion, diazinon, parathion, malathion, or guthion have also been employed successfully during the postbloom period. These are all considered emergency treatments to kill off scale populations. Oil sprays improperly used may impair the quality and appearance of the fruit as well as increase the susceptibility of trees to cold injury.

REFERENCES: *J. Econ. Ent.*, 43:305–309, 1950; *Proc. Fla. State Hort. Soc.*, 64:66–71, 1951; *Fla. Agr. Exp. Sta. Bul.* 479, 1951; *USDA* E-870, 1953; *Agr. Handbook* 290, 1965; *Calif. Citrus Exp. Sta. Pest Cont. Guide*, 1977; *Fla. Citrus Com. Spray Prog.* 1977.

CALIFORNIA RED SCALE

Aonidiella aurantii (Maskell), Family Diaspididae

Red scale, as this species is most often called, is one of the larger armored scales. The females are sometimes 3 mm in diameter, the average size being closer to 2 mm; males are still smaller and are oval in shape. The circular-shaped females are red, the color showing through the thin scale covering.

Introduced into this country from Australia, the red scale now occurs in California and the Gulf states, and in glasshouses over the country. There are numerous host plants and it is injurious in California to several deciduous fruits as well as being one of the most destructive citrus pests of the region.

Females of this scale give birth to young. There are 3 or 4 overlapping generations each year. The scale feeds on all parts of the plant, but it is most numerous on foliage, fruits, and younger branches. Lemons are said to be the favorite food plant but all citrus may be badly injured. The salivary secretions injected while feeding seem to produce a toxic effect on the plants attacked.

The lady beetles, *Chilocorus stigma* (Say), *Lindorus lophanthae* (Blaisdell), and *Microweisia coccidivora* (Ashmead) are important in natural control of this insect in the United States. Three introduced eulophid parasites, *Aphytis lingnanensis* Compere, *A melinus* DeBach, and *Comperiella bifasciata* Howard, complemented by the encyrtid, *Prospaltella perniciosi* Tower, are controlling red scale populations in California and elsewhere.

REFERENCES: *J. Econ. Ent.*, 43:610–614, 1950; 47:100–102, 1954; 57:322–324, 1964; *Calif. Agr.* 16(12):2–3, 1962; 27(11):3–7, 1973; *Calif. Citrus Exp. Sta. Pest Cont. Guide*, 1977.

YELLOW SCALE

Aonidiella citrina (Coq.), Family Diaspididae

This scale has been listed as a variety of California red scale and as a separate species. It differs from the red scale in that it is less red and has a more transparent scale covering. Its distribution and habits also vary somewhat from those of the red scale. Yellow scale is most abundant in the interior districts of California and tends to feed on foliage and fruits rather than on twigs and branches. It also occurs on citrus in Florida, but populations are not high. All citrus may be badly injured but oranges are said to be a favored host. The life cycle is the same as for California red scale.

Natural enemies include the parasites, *Aphytis citrinus* Compere, *Comperiella bifasciata* Howard, *Aspidiotiphagus citrinus* (Craw.), and *Prospaltella aurantii* (Howard); they usually keep the scale under control.

FLORIDA RED SCALE

Chrysomphalus aonidum (L.), Family Diaspididae

This species has been one of the two most destructive scale insects in Florida. Where established and favored by a succession of good seasons it may develop into a very serious pest of citrus; in other localities it may be absent or rare. Since it tends to feed on fruit and foliage rather than on wood, it is greatly reduced by freezing weather, which results in defoliation of citrus. This accounts, in part at least, for its periodical scarcity. In California the species has never become established except in glasshouses. The insect is similar in form to the California red scale, but its scale is dark red or brown and the insect beneath is yellow rather than red. The introduction of the parasite, *Aphytis hóloxanthus* DeBach, into Florida has reduced populations of this scale to the point that it is no longer a problem.

OTHER ARMORED SCALES

ORDER HOMOPTERA, FAMILY DIASPIDIDAE

Chaff Scale, *Parlatoria pergandii* Comstock, found in many parts of the world, is primarily a pest of citrus in the Gulf Coast states of North America. It is occasionally a pest of ornamentals both inside and outside glasshouses. Adult females are dark purple and have a gray-brown scale covering, nearly circular in shape. The male scale covering is much smaller, narrow in form, and white. This is an egg-laying species and 4 generations per year may develop. A recorded parasite is *Aphytis hispanicus* (Mercet). Other close relatives of this scale are parlatoria date scale, *P. blanchardi* (Targ.-Tozz.), and olive scale, *P. oleae* (Colvée).

Dictyospermum Scale, *Chrysomphalus dictyospermi* (Morgan), is a species with wide distribution in the world. It attacks oranges, lemons, grapefruits, tangerines, avocados, and a number of other hosts both inside and outside glasshouses. The scale covering varies from yellow-brown to deep brown for the immature insects and shades of gray to dark brown for the adults. All stages are yellow beneath the scale. Female scale coverings are circular; those of the male are smaller and more oval-shaped. This oviparous species produces 3 to 6 generations each year, and is controlled in the same manner indicated for other armored scales.

Glover Scale, *Lepidosaphes gloverii* (Packard), occurs in Florida and in a small area in California. It may develop on citrus along with purple scale, which it resembles except for its narrower form. The eggs are placed in two rows beneath the scale covering, whereas in the purple scale they are irregularly placed. This scale is sometimes called "long scale."

Greedy Scale, *Hemiberlesia rapax* (Comstock), is light gray resembling oleander scale (p. 341), but the scale covering of the female is strongly convex and the yellow-brown nipple is off to one side. Primarily a glasshouse pest in many parts of the United States, it is also found outdoors in California and other tropical regions. It occurs on a wide variety of woody ornamental plants, and occasionally on citrus. Control is the same as for other armored scales.

Citrus Snow Scale, *Unaspis citri* (Comstock), world-wide in distribution, is considered a severe problem in the northeast portion of the Florida citrus belt. It is very destructive, often killing large branches and sometimes young trees.

The female scale covering is oystershell-shaped but is broader at the base than the purple scale. The males are small, slender, ridged, and white, the basis for the common name. Control is accomplished by thorough coverage with pesticides recommended for other armored scales. In Florida oil sprays have been the least effective.

Populations of many of these scale insects are kept at sub-economic levels by the following parasites: *Aphytis maculicornis* (Masi), *A. proclia* (Walker), *A mytilaspidis* (LeBaron), *A. chrysomphali* (Mercet), *A coheni* DeBach, *A. melinus* DeBach, *Prospaltella aurantii* (Howard), *P. elongata* Dozier, *Coccophagoides utilis* Doutt, *Aspidiotiphagus citrinus citrinus* (Craw.), and *A. lycimnia* (Walker).

UNARMORED SCALES

BLACK SCALE

Saissetia oleae (Olivier), Family Coccidae

An unarmored species of tropical origin, black scale is one of the most widespread and destructive citrus scales in California, where now it is also a major olive pest. In Florida and other Gulf states it is becoming more of a problem, especially on grapefruit. Black scale may become a serious glasshouse pest too.

It is a large black or dark brown species, often reaching 6 mm in length; the average size, however, is somewhat smaller. Females are globular with characteristic H-shaped ridges on the dorsal side and many tiny rounded pale areas thus giving it a speckled appearance (Fig. 341). Male scales are smaller and much more flattened than the females. Except in some coastal areas of California, the black scale has only 1 generation annually. The white-to-red orange eggs are deposited under the shell-like dorsal wall of the female scale. The shrivelled remnants of the undersurface of her body can be found inside, near the peak of the dorsal wall. Most eggs are laid in May and June but the period may extend from April to September. Egg-laying occurs again in late fall in the 2-generation area. Hatching may occur in 16 days or be extended over a period

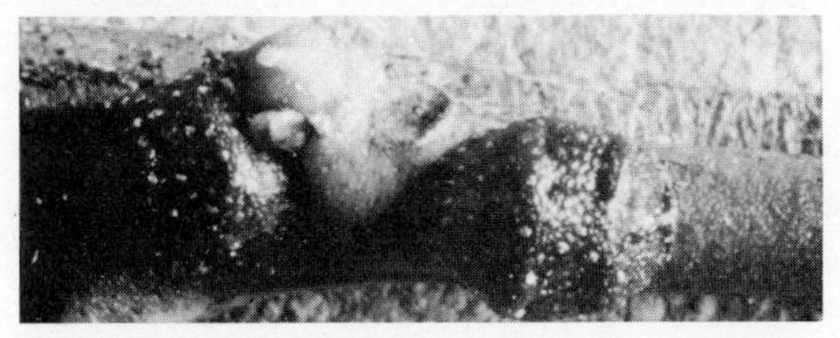

Figure 341. Female black scales, *Saissetia oleae* (Olivier); 4×. (Courtesy of A. S. Deal, UCR)

of 6 weeks in cold weather. The young scales migrate over the plant before settling on the foliage and green twigs to feed. This is the important period for dissemination. Most of the scales migrate to the twigs and branches after the second molt.

As in other scales, injury consists of removal of great quantities of plant sap which may result in stunting, defoliation, or death of branches. In addition, a copious supply of honeydew is excreted which serves as a medium for the development of sooty fungus. This blemishes the fruits and may affect the photosynthetic processes of the plant when it covers the leaf surfaces.

There are numerous parasites and predators of black scale. Some important parasites are *Scutellista cyanea* Mots., *Aphycus helvolus* Compere, *A. stanleyi* (Compere), *A. lounsburyi* Howard, *A. luteolus* Timberlake, and numerous species of *Coccophagus*. Predators include aphidlions, syrphid fly larvae, and lady beetles.

Pesticides are sometimes necessary and are more effective if applied immediately after egg-hatching is completed. Commonly recommended materials are light to medium summer oil alone or in combination with organophosphorus or carbamate chemicals.

REFERENCES: *USDA Agr. Handbook* 290, 1965; *Calif. Citrus Exp. Sta. Pest Cont. Guide*, 1977; *Calif. Agr.* 30(11):12–13, 1976.

CITRICOLA SCALE

Coccus pseudomagnoliarum (Kuwana), Family Coccidae

This unarmored scale is supposed to have been imported from Japan sometime during the present century. It exists only in the dry interior valleys of Arizona and California and attacks the wood and foliage of all kinds of citrus; it has also been found on a few other plants growing near citrus trees. Considerable quantities of honeydew are produced, disfiguring the fruits of citrus especially when sooty fungus develops on it.

The scale is quite flat and transparent while on summer foliage, becoming gray with approaching maturity. Egg-laying takes place in the spring and early summer and extends over a period of nearly 4 months. The first molt occurs a month after the eggs hatch, the second a month later. Migration to the twigs begins in November and continues throughout the winter and spring. Maturity is reached in late April or May. Only 1 generation develops each year.

Important parasites are *Aphycus luteolus* Timb., *A. stanleyi* Compere, *A. helvolus* Compere, and species of *Coccophagus*.

This scale is controlled in the same manner as that outlined for black scale.

COTTONY-CUSHION SCALE

Icerya purchasi Maskell, Family Margarodidae

The cottony-cushion scale is commonly regarded as a pest of citrus, but it may infest several kinds of fruit and shade trees, ornamentals, as well as other plants. It was introduced into California in 1868, on acacia, and subsequently into Florida where the infestation has never become a problem. Damage from this insect is similar to that done by all soft or unarmored scales.

The scale itself is rather large, red-brown, and covered with white or pale yellow, waxy filaments. White cottony egg masses with a fluted appearance are commonly seen and easily recognized so that other descriptive features of the scale are unimportant (Fig. 342). These masses are nearly 12 mm in length and may contain up to 1000 bright red eggs. Hatching occurs in a period of a few days to nearly 2 months depending on the temperature; the red nymphs begin to feed, soon becoming covered with waxy cottony filaments and powder. The small delicate 2-winged flylike male scales are rarely seen.

Within 20 years after its introduction this scale was seriously threatening the citrus industry of California and no means of control was available. A search was made in Australia for the natural enemies that held it in check there. The vedalia lady beetle, *Rodolia cardinalis* (Muls.), and a parasitic fly, *Cryptochaetum iceryae* (Williston), were found to be effective natural enemies. Within 18 months after their introduction into California citrus groves the infestation of cottony-cushion scale was reduced to the level of an occasional pest of minor importance. They have also been established in Florida and in other areas where needed. This is one of the classic examples of biological control of a destructive insect species.

REFERENCE: Huffaker, C. B., ed. *Biological Control,* Plenum Press, pp. 167–168, 1971.

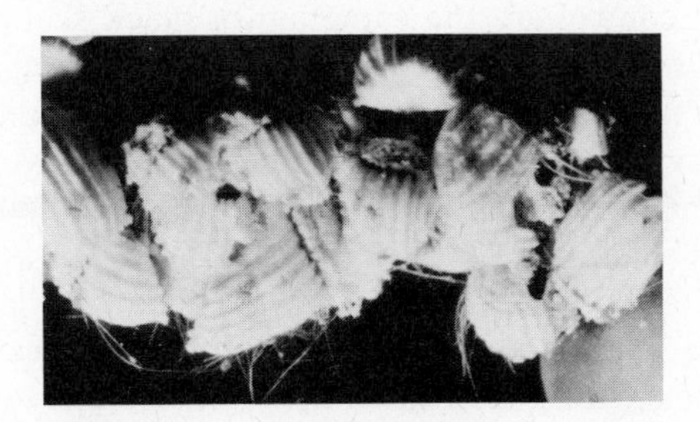

Figure 342. Cottony-cushion scale, *Icerya purchasi* Maskell; natural size. (Courtesy of R. N. Jefferson, UCR)

BROWN SOFT SCALE

Coccus hesperidum L., Family Coccidae

Commonly known as soft brown scale, this unarmored species occurs outdoors throughout the world in tropical and subtropical areas, and is also known universally as a glasshouse pest. Like most of the scales that attack citrus, it has a wide range of host plants. On citrus it is often attended by ants, and control of the ants helps reduce scale populations.

The adult female is brown, oval, flattened, and may be nearly 4 mm in length. Females lay only a few eggs at a time which hatch quickly and are therefore rarely seen. Males are rarely observed. There are several overlapping generations each year in glasshouses and 3 to 5 in outdoor subtropical areas. Like all unarmored or soft scales this species does not lose the legs and antennae at the first molt as do the armored scales.

This scale is not a problem where its many hymenopterous parasites are present. Williams and Kosztarab* list 46 parasitic species.

The lady beetle, *Chilocorus cacti* (L.), is an important predator.

REFERENCE: *Conn. Agr. Exp. Sta. Bul.* 578, 1954.

OTHER UNARMORED SCALES

ORDER HOMOPTERA, FAMILY COCCIDAE

Hemispherical Scale, *Saissetia coffeae* (Walker), is a common glasshouse species found living in the open in the warmer regions, occasionally attacking citrus. It is also a pest of many ornamental plants, ferns being a common host. Adult females are strongly convex, smooth shiny brown, and nearly 3 mm in diameter. There are 2 generations each year; males are not common. The life cycle, damage, and control are the same as for black scale.

Nigra Scale, *Parasaissetia nigra* (Nietner), is found in glasshouses and outdoors in tropical areas. Besides citrus it attacks other fruit crops, as well as ornamental plants, ivy, holly, and Japanese aralia being favored. Adult females are elliptical, flattened, and shiny black. The life cycle and control measures are the same as for black scale.

Tessellated Scale, *Eucalymnatus tessellatus* (Signoret), is found almost entirely in glasshouses on various tropical plants. Females are elliptical and flattened, green-black with a network of pale lines. They are oviparous, but the nymphs emerge from the eggs immediately after they are deposited. Control is the same as for other soft scales.

* *Res. Bul.* 74, VPI and Sta. Univ., 1972.

CITRUS MEALYBUG

Planococcus citri (Risso), Family Pseudococcidae

Of the several species of mealybugs found on citrus in Florida, this one is considered to be most important. It is found out-of-doors in the South and in glasshouses all over the country. It attacks primarily citrus and ornamental glasshouse plants but may occasionally occur on other hosts. Injury resulting from the feeding by these piercing-sucking insects consists primarily of reduced plant vigor because of sap loss, and the sooty fungus growth that develops on the honeydew excretions which contaminate fruits and disfigure foliage.

Adult females are oval flattened light brown insects with short spines projecting from the body margin and posterior end of the abdomen. Large specimens are 3 to 4 mm long. Their bodies are covered with a waxy or mealy white powder from which they take their common name (Fig. 343). Breeding is continuous; yellow eggs are produced in a cottony mass of waxy filaments which covers most of the female at the time of egg-laying. Each female may deposit over 500 eggs during her life span. Hatching takes place in a week or more depending on the temperature, and the tiny yellow nymphs, resembling crawler scales, migrate over the plant and soon begin sucking sap. There are 2 or 3 overlapping generations per year outdoors and sometimes more in glasshouses. Immature males and females are similar but after 4 weeks of nymphal development those to become males form a mass of silken filaments in which transformation to a gnatlike two-winged adult occurs nearly 2 weeks later. Upon mating the males soon die. There is little winter survival of mealybugs outside glasshouses in the North.

There is a high degree of natural control of mealybugs. Important predators include the brown lacewing, *Sympherobius angustatus* Banks, several species of lady beetles, of which *Cryptolaemus montrouzieri* (Muls.) is considered important, some chrysopids, syrphid fly larvae, and larvae of other true flies.

Figure 343. Citrus mealybug, *Planococcus citri* (Risso); 4×. (Courtesy of R. E. Treece, OARDC)

Important parasite species are the encyrtid, *Leptomastidea abnormis* (Gir.), and the platygasterid, *Allotropa citri* Muesebeck.

Control of Argentine and other species of attendant ants contributes to mealybug control.

This mealybug is normally controlled by oil emulsions alone or in combination with organophosphorus pesticides.

REFERENCE: *Calif. Agr. Exp. Sta. Bul.* 713, 1949.

OTHER MEALYBUGS

ORDER HOMOPTERA, FAMILY PSEUDOCOCCIDAE

Citrophilus Mealybug, *Pseudococcus calceolariae* (Maskell), was found in California in 1913. It may be distinguished from *P. citri* by its anal filaments of which a pair are nearly half as long as the body, and by 4 rows of thinly waxed depressions down the back, the middle pair being most conspicuous. There is no marked difference in the appearance or biology of this species and *P. citri*. A large number of host plants are attacked. Control out-of-doors has been secured largely by the encouragement of parasite and predator populations. In addition to the natural enemies listed for *P. citri,* 2 parasites from Australia have now reduced the populations of this mealybug to a level approaching complete control. These chalcid parasites are *Coccophagus gurneyi* Compere and *Tetracnemus pretiosus* Timberlake.

Long-Tailed Mealybug, *Pseudococcus longispinus* (Targ.-Toz.), occurs on a wide range of hosts both inside and outside glasshouses. This species is slightly smaller than *P. citri,* the body is thinly covered with white wax, and a broad faint dark stripe is evident on the back. Filaments along the sides are about ½ the body width, and the terminal pair is as long as, or longer than, the body (Fig. 344). No eggs are laid, the females giving birth to their young. In addition to the natural enemies given for the other species of mealybugs, this species is effectively checked by *Anarhopus sydneyensis* Timberlake, an encyrtid parasite imported from Australia.

REFERENCES: *Calif. Agr. Exp. Sta. Bul.* 258, 1915; *USDA Ent. Bul.* 1040, 1922; DeBach, P., *Biological Control by Nat. Enemies,* Cambridge Univ. Press, 135–139, 1974.

CITRUS WHITEFLY

Dialeurodes citri (Ashmead), Family Aleyrodidae

Although the infestation in California was reported eradicated in 1942, it was again found in 1967; this insect is still an important citrus pest in Florida and occasionally becomes destructive in other Gulf Coast states. Injury results

Figure 344. Long-tailed mealybug, *Pseudococcus longispinus* (Targ.-Toz.); 9×. (Courtesy of R. N. Jefferson, UCR)

directly from loss of plant sap, and indirectly from the sooty fungus growth that develops on the copious honeydew excretions of the whiteflies. The fungus often covers both foliage and fruits, retarding growth and reducing the market value of the fruits.

The lemon-yellow eggs are elliptical, 0.25 mm in length, and attached to the leaf by a short stalk. Hatching takes place in a week or more depending on the temperature, and the tiny pale green nymphs migrate over the plant, insert their mouthparts and suck sap. After the first molt the legs and antennae are shed, and the nymphs resemble young soft scales. The last instar is thick-bodied and is sometimes called a pupa. Nymphal development is completed in 3 to 4 weeks, and the four-winged adults, scarcely 1.6 mm in length, begin emerging. The wings, legs, and antennae are milk white, the body is pale yellow, and the eyes black. When at rest the wings are held rooflike over the body (Fig. 345). Development from egg to adult requires nearly 50 or more days depending on the temperature. There are usually 3 generations per year in Florida.

Natural control results from lacewing and lady beetle predators, and the introduced parasite, *Prospaltella lahorensis* Howard, plus the presence of several species of entomophagous fungi. Some important species are the red fungus, *Aschersonia aleyrodis* Webber, and the brown fungus, *Aegerita webberi* Fawcett. These fungi are now generally present in all citrus groves and will increase in numbers when the proper environmental conditions prevail.

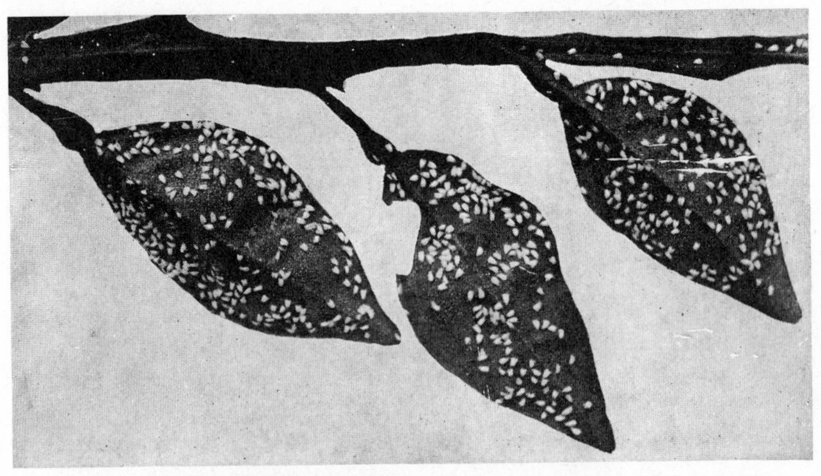

Figure 345. Citrus whitefly, *Dialeurodes citri* (Ashmead); adults on foliage; natural size. (Morrill and Back, USDA)

OTHER WHITEFLIES

ORDER HOMOPTERA, FAMILY ALEYRODIDAE

Cloudy-Winged Whitefly, *Dialeurodes citrifolii* (Morgan), resembles the citrus whitefly but lays black eggs and has a darkened or cloudy area on each wing; also, events in the life cycle are about 2 weeks later. This species has nearly the same distribution and is usually of lesser importance; in some areas of Florida, however, it appears to be on the increase. The yellow fungus, *Aschersonia goldiana* (S. and E.), attacks only this species and, along with *Aegerita webberi* Fawcett, brings about some natural control.

Woolly Whitefly, *Aleurothrixus floccosus* (Maskell), was first observed as a Florida citrus pest in 1909. It has never become important, apparently because its natural enemies have held it in check. The common name refers to the woolly waxy filaments that cover the last nymphal instar or pupa. The eggs are brown and slightly curved, resembling a tiny wiener. Adults are sluggish and have more yellow coloration in the body and wings than the cloudy-winged species. In California this pest is being controlled by the introduced parasites, *Cales noacki* DeSantis, and *Amitus spiniferus* (Brethes).

Citrus Blackfly, *Aleurocanthus woglumi* Ashby, occurs in India, the Philippine Islands and Ceylon, and in 1935 was found in Mexico and the West Indies where it is considered to be a serious pest of citrus and other fruits. Infestations in the United States have been found in Texas and Florida where eradication programs have been initiated. The adults have a slate-blue appearance. They lay their tiny eggs largely in a spiral path on the undersides of the leaves. The eggs are creamy white at first, becoming almost black before hatching. They are attached to the leaf by a short stalk. There are 3 to 6 overlapping generations per year. Parasites introduced from India have shown promise as biological control agents. The most promising species is *Amitus hesperidum* Silvestri, complemented by *Prospaltella clypealis* Silvestri, and *P. opulenta* Silvestri.

Avocado Whitefly, *Trialeurodes floridensis* (Quaintance), is occasionally found on citrus and commonly attacks avocados and guavas in Florida. It resembles the citrus and greenhouse whiteflies in appearance and life cycle. Control measures are the same as those for the citrus whitefly.

REFERENCES: *Fla. Agr. Exp. Sta. Bul.* 123, 1914; *Bul.* 126, 1915; *Bul.* 148, 1918; *Calif. Agr.* 30(5):4–6, 1976.

CITRUS THRIPS

Scirtothrips citri (Moulton), Family Thripidae

Citrus thrips are prevalent in most citrus regions of California, Arizona, and Texas. Oranges, lemons, nectarines, and grapefruits are commonly attacked, and there are alternate hosts. Damage is caused by both nymphs and adults rasping the surface of foliage and fruits and sucking the cell contents. This stunts growth and produces a scabby or scurfy appearance of fruits. Buds are often killed; new leaves are dwarfed, distorted, and of a characteristic gray appearance.

Citrus thrips overwinter as tiny bean-shaped eggs deposited during the autumn in the tissues of leaves and stems. Hatching occurs generally by early March and the pale wingless nymphs with red eyes begin feeding. As they increase in size they become yellow to orange, the eyes dull red. After the second instar, growth is completed; the nymphs drop to the ground and transform to a prepupal or third instar and later a pupal or fourth instar before emerging as adults. Pale orange-yellow adults, scarcely 2 mm long, have 4 wings fringed with hairs which at rest are folded down the back (Fig. 346). Females are larger-bodied than the males and can reproduce without mating, but then the progeny

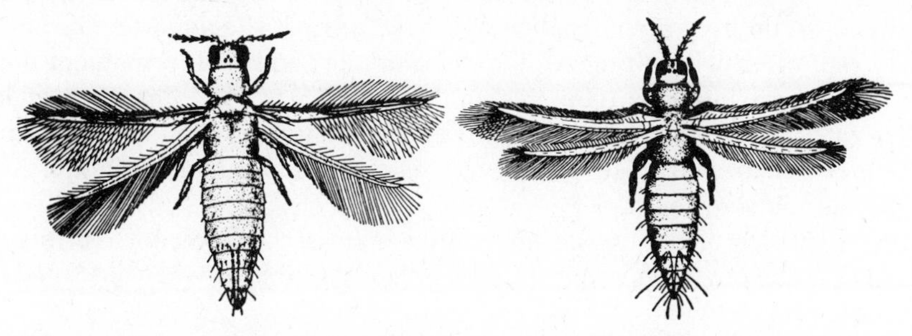

Figure 346. Citrus thrips, *Scirtothrips citri* (Moulton), and flower thrips, *Frankliniella tritici* (Fitch) (L. to R.); 15×. (Courtesy of USDA and Fla. Agr. Exp. Sta.)

are all males. Development from egg to adult requires a month in cool weather and about 2 weeks in hot weather. There are probably 10 or 12 generations per year in warmer localities.

Other species that may attack citrus are greenhouse thrips, *Heliothrips haemorrhoidalis* (Bouché), western flower thrips, *Frankliniella occidentalis* (Pergande), and flower thrips, *Frankliniella tritici* (Fitch) (Fig. 346). They are all similar to citrus thrips in appearance and life cycle, but they are more often present in the blossoms and this feeding has been considered less important.

Natural control by predatory thrips, lady beetles, mites, and spiders has been observed. Pesticide applications are sometimes necessary. Use only those least toxic to natural enemy populations. Tartar emetic plus sugar is a recommended mixture.

REFERENCES: *USDA Cir.* 708, 1944; *J. Econ. Ent.,* 45:578–593, 1952; *Calif. Citrus Exp. Sta. Pest Cont. Guide,* 1977.

CITRUS RED MITE

Panonychus citri (McGregor), Family Tetranychidae

This world-wide citrus pest is known as the purple mite in Florida and as the citrus red spider in California. Although found in many of the Gulf Coast states, the citrus red mite has been considered a major pest only in Florida and California. Both nymphs and adults extract the sap from foliage, fruits, and tender branches with their piercing-sucking mouthparts, producing tiny gray or silvery spots on the leaves and fruits. When leaf damage is severe, the normal photosynthetic processes of the plant are greatly inhibited, resulting in leaf drop, decreased plant vigor, and smaller, poorer quality fruits.

This mite closely resembles the European red mite, which is a major pest of deciduous trees in the North. The bright red eggs are spherical but slightly flattened, with a dorsal hairlike projection. In warm weather, hatching occurs in a week or more and the tiny six-legged nymphs (often called larvae) migrate over the plant and begin feeding. After the first molt all the succeeding stages, called protonymphs, deutonymphs, and adults, have 8 legs. The adult female is dark velvety red, often with a tinge of purple, globular, and with prominent dorsal white bristles arising from tubercles on the body (Fig. 347). The adult male is smaller with a more pointed abdomen. Development from egg to adult requires less than 3 weeks, and 12 to 15 generations may occur annually. Unfertilized eggs develop into males.

Various species of phytoseiid mites are important as predators of this and other mite species. *Amblyseius hibisci* (Chant) is one of the very effective predators.

REFERENCES: *J. Econ. Ent.,* 46:1014, 1953; 47:356–357, 1954; 54:55–60, 1961; *USDA Agr. Handbook* 290, 1965; *Calif. Agr. Ext. Cir.* 87, 1944.

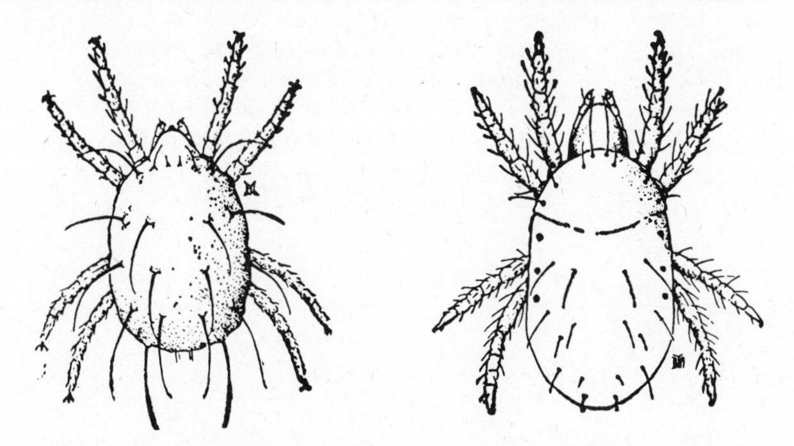

Figure 347. Citrus mites: citrus red mite, *Panonychus citri* (McGregor), (left); six-spotted mite, *Eotetranychus sexmaculatus* (Riley), (right); 40×. (Quayle, Calif. Agr. Exp. Sta.)

CITRUS RUST MITE

Phyllocoptruta oleivora (Ashmead), Family Eriophyidae

This mite is one of the most common and serious pests of citrus in Florida, Texas, and other Gulf Coast states as well as in some citrus districts in California. Mites feed on the sap of leaves, twigs, and fruits of all kinds of citrus; lemon, lime, grapefruit, and orange are most severely damaged. Leaves lose their glossy appearance, becoming bronzed and stunted under heavy infestations, often dropping prematurely. New twig growth is stunted and discolored. Damaged fruits become light brown to black in the areas where the mites feed; this discoloration is called russeting and losses occur every year owing to reduced size and poor appearance of the fruits.

Citrus rust mites are very small, averaging 0.1 mm in length. They are lemon-yellow, elongate, tapering posteriorly, with 2 pairs of legs (Fig. 348). Only females have been found. Pale yellow spherical eggs are commonly deposited on fruits and foliage. Hatching takes place in 2 or more days and the young develop to maturity in 7 to 10 days in summer. Many overlapping generations occur throughout the year but the mites are less abundant in winter and in July and early August.

Excellent control can be obtained using sulfur as a dust or spray. Dusts are more easily applied but require more applications because they are quickly washed away by rains. Applications should be made as needed. Follow the specific recommendations for your citrus area. Weekly examination of the leaves and fruits with a hand magnifier to determine whether mites are present is of

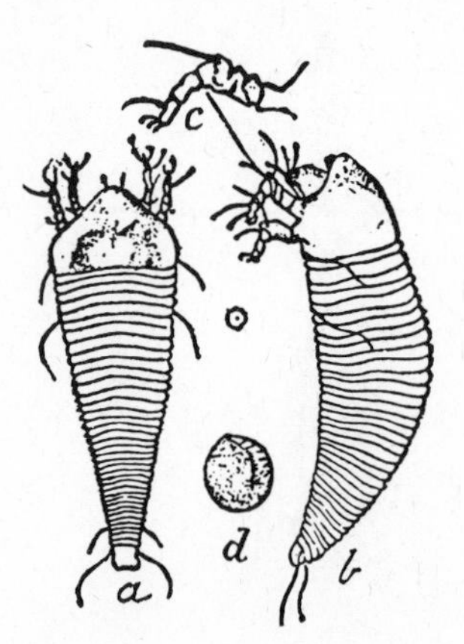

Figure 348. The citrus rust mite, *Phyllocoptruta oleivora* (Ashmead): *a,* dorsal view; *b,* lateral view enlarged, the dot in circle indicating natural size; *c,* leg; *d,* egg with embryo just about ready to hatch, more enlarged. (Hubbard, USDA)

help in deciding need for treatment. When 15 to 20% of the leaves or fruits are infested and 1 mite per magnifier lens field is found, pesticides are usually necessary.

REFERENCES: *USDA Tech. Bul.* 176, 1930; *Farmers' Bul.* 2012, 1950; *Agr. Handbook* 290, 1965; *Calif. Citrus Exp. Sta. Pest Cont. Guide,* 1977.

OTHER MITE SPECIES

Six-Spotted Spider Mite, *Eotetranychus sexmaculatus* (Riley), is occasionally a pest in many citrus-growing areas. It resembles the two-spotted spider mite in general appearance, life cycle, and habits. However, the eggs bear a filamentous stalk like the citrus red mite, and the females have six pigmented dorsal spots which suggest the common name. These spots are not always conspicuous and they may sometimes be absent (Fig. 347).

Citrus Flat Mite, *Brevipalpus lewisi* McGregor, is present as a pest in a few counties in California and Arizona. It is said to attack only the fruits and only when the color begins to appear in them. Their feeding causes oranges to become russeted and lemons to appear silvery. The egg has a central filamentous stalk and the adult is amber with some spots on the lateral margins of the body. Webbing is produced by this mite.

Texas Citrus Mite, *Eutetranychus banksi* (McGregor), at times is an important species in the Texas citrus belt and in Florida. It forms little webbing, feeds on the upper leaf surface, and is more rapid in its movements than the citrus red mite which it closely resembles in appearance and life cycle.

Citrus Bud Mite, *Eriophyes sheldoni* Ewing, is a world-wide pest, but in the United States it is found principally along the coastal areas of California, Florida, and the Hawaiian Islands, where it attacks all kinds of citrus and is most damaging to lemons. It feeds primarily in the buds which may be killed or may cause the growth of twigs, leaves, blossoms, and fruits to be stunted and malformed. This eriophyid mite resembles the citrus rust mite in appearance and life history. Development from egg to adult is completed in 10 or more days depending on the temperature.

All these mites have many predators.

REFERENCES: *J. Econ. Ent.*, 44:823–832, 1951; 45:271–273, 1952; *USDA Agr. Handbook* 290, 1965; *Calif. Citrus Exp. Sta. Pest Cont. Guide*, 1977.

SPIREA APHID

Aphis citricola Van der Goot, Family Aphididae

Often called the citrus aphid because it is the most important species attacking that crop in both California and Florida, this aphid is widely distributed elsewhere in the United States and was named for its common host, spirea. In addition to citrus and spirea it feeds on apple, pear, quince, and haw. This species is closely related to the apple aphid, *A. pomi,* with which it has often been confused. Injury consists of removal of plant sap by the piercing-sucking nymphs and adults which results in curled, stunted, or distorted foliage and fruits. Also, under heavy infestation the large amounts of excreted honeydew serve as a medium for sooty fungus growth that smudges the fruits and leaves. Newly planted trees are said to suffer the greatest injury.

In the more southern citrus areas, this typically green aphid produces only parthenogenetic ovoviviparous generations throughout the year; from approximately Gainesville, Florida, northward sexual forms are produced in the fall which deposit the overwintering eggs on spirea. In the South the life cycle may be completed in 6 days. Since it probably does not average more than 10 to 12 days, many generations are produced each year. Heavy populations may be observed on the tender growing tips of spirea in the spring. Ants are often associated with this aphid.

Natural control results from numerous species of lady beetles, aphidlions, syrphid fly larvae, and parasites, especially *Aphidius testaceipes* (Cresson).

Other species of aphids that may attack citrus are potato aphid, *Macrosiphum euphorbiae* (Thomas); black citrus aphid, *Toxoptera aurantii* (Fonscolombe); and cotton aphid, *Aphis gossypii* Glover.

REFERENCE: *Fla. Agr. Exp. Sta. Bul.* 203:431–476, 1929.

MEDITERRANEAN FRUIT FLY

Ceratitis capitata (Wiedemann), Family Tephritidae

A native of Africa the Mediterranean fruit fly has long been known as one of the most destructive insects attacking citrus and many other fruits in subtropical areas of the world except North America. Quarantines have been maintained to keep it out of the United States. In spite of these precautions, the fly was discovered in Florida in 1929. Prompt action was taken by state and federal authorities, and through their combined efforts the fly was exterminated in less than a year. This fly was again discovered in the vicinity of Miami, Florida, in 1956; this and subsequent infestations have also been eradicated.

The fruit fly is somewhat smaller than the house fly, yellow-and-black-bodied, with transparent wings having darker-banded areas (Fig. 349). Some females lay 600 eggs during their life span. The eggs are laid under the skin of the fruits and the larvae feed inside. Larvae are nearly white tapered maggots almost 1 cm long when fully grown. After about 10 days larval development is completed; they leave the fruits, drop to the ground, enter the soil, and change to brown puparia. Development from egg to adult may occur in less than 3 weeks, but it may require 3 months or more under unfavorable conditions. There are usually several generations per year depending on the locality. Winter is passed as puparia in the soil, or as adults in cooler regions, or breeding may be continuous in warmer regions.

No control measures are needed in the United States owing to complete eradication of the infestation. In other countries promising results have been obtained with phosphate-type pesticides applied to kill the adults before they lay eggs. Residual sprays, bait sprays, and soil poisons are of value in control and eradication programs.

REFERENCES: *J. Econ. Ent.*, 45:274–279, 1952; 54:30–35, 1961; 61:438–443, 1968.

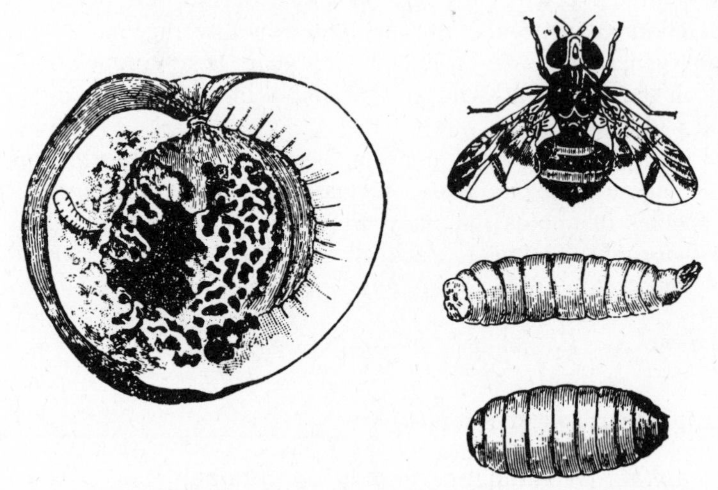

Figure 349. Life stages of the Mediterranean fruit fly: adult, larva, puparium, and larva working in a peach; peach about natural size; others 3×. (Compere.)

MEXICAN FRUIT FLY

Anastrepha ludens (Loew), Family Tephritidae

This fruit fly has long been known as the orange maggot, but it attacks other citrus as well. Native to Mexico, it has also been found in the Rio Grande Valley in Texas. Quarantines are maintained to prevent its spread from these areas. Authorities of Mexico cooperate in the enforcement of the necessary quarantines and control programs. Efforts are made to exterminate the fruit fly wherever it has become established.

The yellow-brown adult is larger than the house fly. The wings are transparent except where mottled with brown areas. Females have an elongated tip on the abdomen which encloses the ovipositor (Fig. 350). The green eggs are deposited just beneath the skin of the fruit; the larvae are creamy white legless maggots with a tapered body. Damage is caused by the maggot working in the fruit. There are 4 to 6 generations a year.

All fruits shipped from the infested area are fumigated. The continuance of these efforts in extermination and the maintenance of quarantine regulations are therefore of vital interest to both producers and consumers of citrus fruits.

REFERENCES: *USDA PA*–265, 1964; *Misc. Pub.* 531, 1944; *Tech. Bul.* 1330, 1965; *J. Invert. Path.* 8(4):542–543, 1966; *Ann. Ent. Soc. Amer.* 59:298–300, 1966; *J. Econ. Ent.* 59:1400–1402, 1966; 60:992–994, 1759–1760, 1967; 62:53–56, 511–512, 1255–1257, 1969; *Bul. Ent. Soc. Amer.* 16:186–193, 1970.

ORANGE CATERPILLARS

ORDER LEPIDOPTERA

Foliage and fruits of citrus as well as other fruits are sometimes attacked by various species of caterpillars.

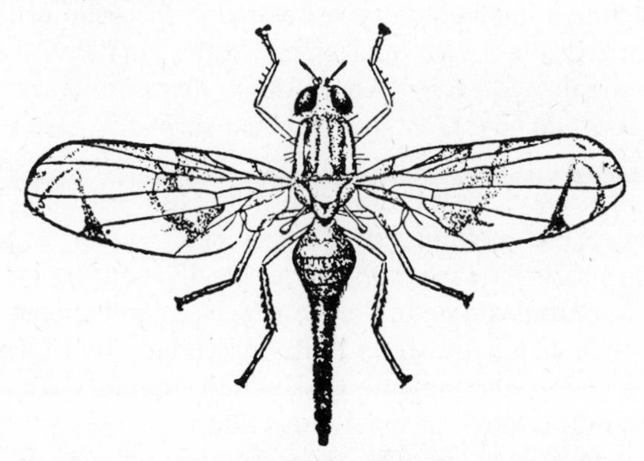

Figure 350. The Mexican fruit fly, *Anastrepha ludens* (Loew); 5×. (Riley.)

The "orange dogs" are larvae of swallowtail butterflies. *Papilio xuthus* L. is the common species in California, whereas the giant swallowtail, *P. cresphontes* Cramer, is more prevalent in Florida and the Gulf States.

One of the pests of deciduous fruit trees has become of some importance as a pest of citrus trees in California. This is the fruit-tree leaf roller, *Archips argyrospilus* (Walker) (see p. 403).

Orangeworms of several species have been recorded in California. The larvae of one of these, the orange tortrix, *Argyrotaenia citrana* (Fernald), feed on tender foliage and also enter fruits. Resembling the orange tortrix is another species called the garden tortrix, *Clepsis peritana* (Clemens). The pink scavenger caterpillar, *Pyroderces rileyi* (Walsingham), feeds primarily on fruits that are injured by other insects, but it occasionally attacks sound fruits. Other occasional pests are the navel orangeworm, *Amyelois transitella* (Walker), often a serious pest of walnuts and almonds; the omnivorous leafroller, *Platynota stultana* Walsingham (see p. 314); the western tussock moth, *Orgyia vetusta* (Boisduval); and various species of cutworms.

Control of these pests is usually accomplished by the regular sprays for more important citrus pests. However, special treatments are sometimes required. *Bacillus thuriengiensis* is used for navel orangeworm. Early harvest of walnuts is a means of avoiding navel orangeworm damage.

REFERENCES: *Hilgardia,* 31:129–171, 1961; *Calif. Agr. Exp. Sta. Bul.* 764, 1958; *Calif. Agr.* 18:10–12, 1964; 29(9):3, 1975; *Calif. Citrus Exp. Sta. Pest Cont. Guide,* 1977.

ARGENTINE ANT

Iridomyrmex humilis (Mayr), Family Formicidae

The Argentine ant is a South American insect introduced in New Orleans about 1890. Since that time it has become established in the southern states of North America, particularly the lower Mississippi Valley and in California. It is apparently unable to survive except where the winters are warm.

This ant is important in citrus groves because of its relations with aphids, scales, and mealybugs. It is attracted to these insects because of the honeydew they produce and it may carry them from place to place thus furthering their spread. It readily seeks colonies of *Coccus hesperidum* L.; apparently the honeydew of this species is especially attractive. The ant transports rusty plum aphids to sugarcane, thus aiding in the transmission and spread of mosaic. It is also detrimental because it destroys both eggs and active forms of predators and parasites. Observations indicate that, where the ants are absent from the citrus groves, natural control of mealybugs and other pests is greater.

The worker ants in this species are exceedingly small, usually less than 2 mm long; the queens may attain a length of 6 mm (Fig. 351). Their color is dark

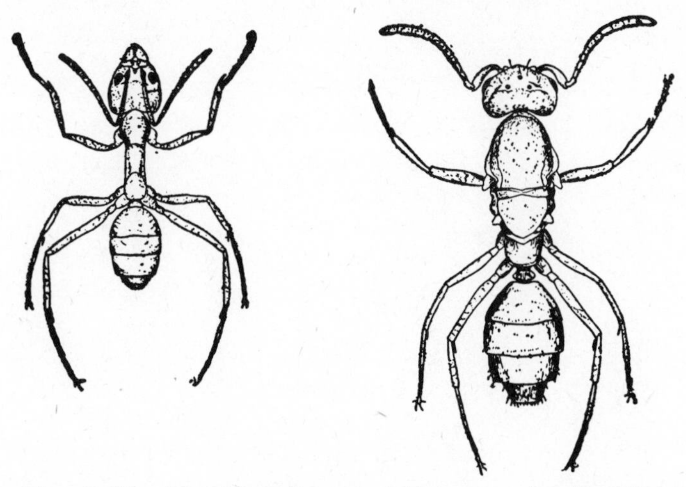

Figure 351. The Argentine ant, *Iridomyrmex humilis* (Mayr): (left) worker; 15×. (right) wingless queen; 7×. (Newell.)

brown and the antennae are elbowed, the basal portion being much longer in the worker caste of these ants than in other species. Nests are made in the soil of orchards, fields, and gardens.

Effective control can be accomplished by treating the nests with approved pesticides. Sprays, dusts, or granular formulations may be applied to the soil over large areas. In citrus groves, direct the pesticide to the region under the tree, particularly next to the trunk. Avoid getting the toxicants on fruits.

REFERENCES: *J. Econ. Ent.*, 47:591–593, 1954; *Ann. Ent. Soc. Amer.* 49:441–447, 1956; 63: 1238–1242, 1970.

23

PESTS OF STORED PRODUCTS AND HOUSEHOLD GOODS

Insects destroy at least 5% of the world production of all cereal grains after they are harvested and while they are in storage on the farm, in elevators, or in warehouses. These losses consist of lowered weight and food value, insect adulteration, heating of grains with resultant mold and spoilage, and low germination of seeds. The actual amount of grain loss annually has been estimated at 300 million bushels.

Processed and packaged foods are also subject to attack and, unless frequent examinations are made and control measures initiated, serious damage may result in processing plants, wholesale warehouses, retail stores, and homes. Annual losses to agricultural commodities in storage is estimated to be $1 billion.

Loss of clothing, rugs, furniture, and other household furnishings, because of clothes moths, carpet beetles, and similar pests has been estimated at $300 million to $800 million annually.

Only the most important pests can be discussed in this chapter, and, since all insects of stored products are controlled in much the same way, these measures are discussed after all the insects are described.

Since references include information about the insects as well as control, they are listed after the discussion on control.

GRANARY, RICE, AND MAIZE WEEVILS

ORDER COLEOPTERA, FAMILY CURCULIONIDAE

The granary weevil, *Sitophilus granarius* (L.), and the rice weevil, *Sitophilus oryzae* (L.), differ from each other only slightly. The rice weevil can fly, has round pits on the thorax, and is nearly black except for two faintly light spots on each elytron (Fig. 352). The granary weevil cannot fly, has oval pits on the

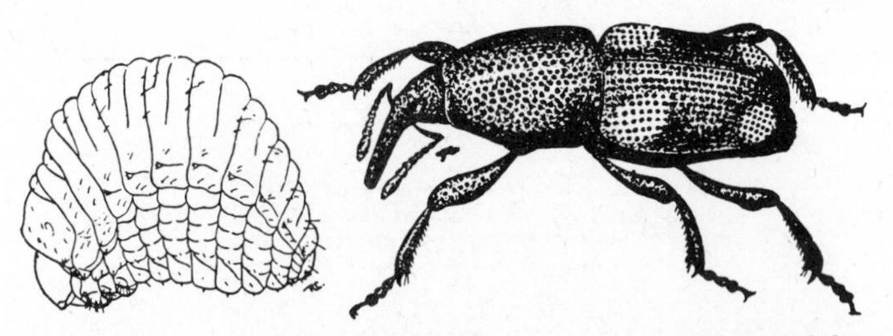

Figure 352. Larva and adult of the rice weevil, *Sitophilus oryzae* (L.); 10×. (USDA)

thorax, and is uniformly dark brown or black. The maize weevil, *S. zeamais* Motschulsky, is darker and larger than its congeners. Another close relative is the broadnosed grain weevil, *Caulophilus oryzae* (Gyllenhal), found in abundance in Florida and a few other southeastern adjoining states. The length of each beetle is approximately 3 to 4 mm, and all have chewing mouthparts at the end of their snout or prolonged head. All species may be found infesting stored grains all over the world. Grains on farms, in transit, or in elevators are subject to attack. In the South, the rice weevil also attacks grain in the field.

Each female may live several months and deposit 200 to 400 eggs during that period. Before oviposition she bores a small hole into the grain with her mouthparts, deposits the egg in this cavity, and covers it with a gelatinous fluid. After hatching, the small white fleshy legless larvae devour the inside portion of the grains and, when fully developed, transform to pupae, and emerge as adults. Although the life cycle may be completed in 4 weeks, this period is greatly prolonged by cool weather. The entire larval and pupal periods are spent inside whole grains. These insects are favored by grains with a high moisture content.

CONFUSED AND RED FLOUR BEETLES

ORDER COLEOPTERA, FAMILY TENEBRIONIDAE

The confused flour beetle, *Tribolium confusum* duVal, and the red flour beetle, *Tribolium castaneum* (Herbst), are found throughout the world attacking primarily milled-grain products. In whole grains they feed only on grain dust and broken kernels. The confused flour beetle is undoubtedly the most abundant and injurious insect pest of flour mills in the United States. In temperate regions the confused species predominates; the red species is more subtropical.

The two species are very similar (Fig. 353); both are about 4 mm long and red-brown. The head and dorsal sides of the thorax are densely covered with minute punctures. The easiest distinguishing character is that segments of the antennae of the confused flour beetle increase in size gradually from base to tip, whereas in the red flour beetle the last few segments are abruptly much larger than preceding ones.

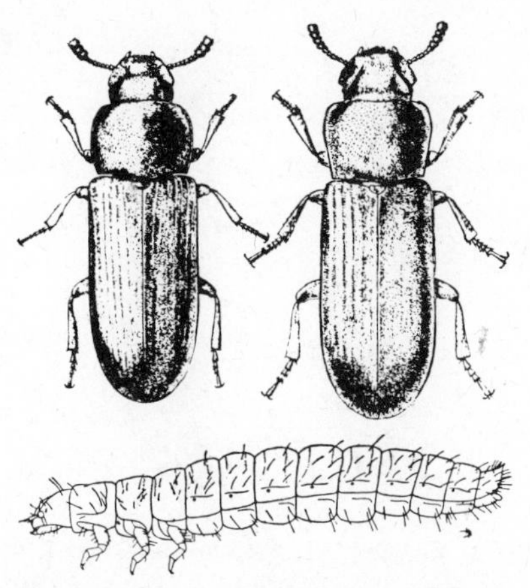

Figure 353. The red flour beetle, *Tribolium castaneum* (Herbst) (left), and the confused flour beetle, *T. confusum* du Val (right). Below, larva of *T. castaneum;* 8×. (USDA)

Another congener is the black flour beetle, *T. audax* Halstead, found principally in flour mills of the northern and western states of North America.

The average life span of the beetles is around 1 year. In flour or other foods each female lays 400 to 500 white eggs which hatch in a week or more into slender cylindrical white larvae tinged with yellow. When fully grown they transform to naked pupae, and a week later adult emergence takes place. The period from egg to adult requires about 4 weeks in warm habitats. Development is retarded by cool temperatures and unfavorable food.

SAW-TOOTHED GRAIN BEETLE

Oryzaephilus surinamensis (L.), Family Cucujidae

The saw-toothed grain beetle is the most prevalent of the beetles in stored foods. It is a cosmopolitan species, and its origin is not definitely established.

This red-brown beetle is one of the smallest of the grain pests, being not more than 3 mm long, and of slender, flattened form (Fig. 354). Edges of the prothorax are distinctly serrated. The slender larva is white with some brown on the head and margins of the segments; it has well-developed legs.

Breeding is continuous when temperatures permit, and all life stages may be found together in infestations, both larvae and adults feeding on the products. The adults live, on the average, from 6 to 10 months; some individuals may live as long as 3 years. Each female deposits 40 to 280 white eggs, which hatch in a few days. When larval development is completed they pupate within deli-

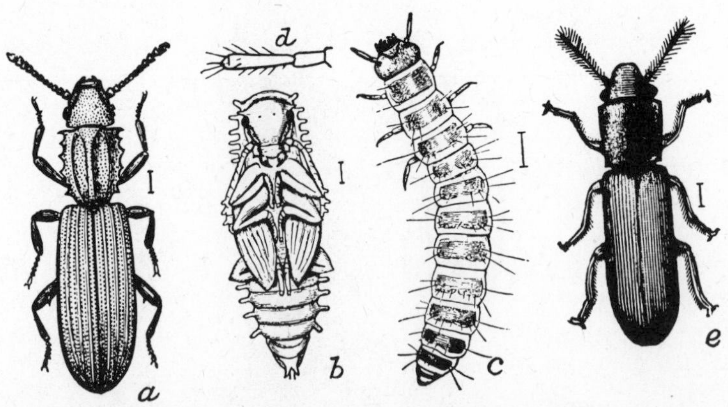

Figure 354. Grain beetles. (Left) saw-toothed grain beetle, *Oryzaephilus surinamensis* (L.): *a,* adult; *b,* pupa; *c,* larva; *d,* antenna; 13×; (right) *e,* square-necked grain beetle, *Cathartus quadricollis* (Guér.); 12×. (USDA)

cate cocoon-like coverings made of food particles held together with a sticky secretion. After a week the adults emerge, completing the cycle, which may take 3 to 4 weeks at summer temperatures.

Its congener, the merchant grain beetle, *O. mercator* (Fauvel), is similar in appearance, habits, and distribution.

SQUARE-NECKED GRAIN BEETLE

Cathartus quadricollis (Guér.), Family Cucujidae

Similar to the saw-toothed species, the square-necked grain beetle is a trifle larger, darker in color, and lacks the serrations on the prothorax (Fig. 354). This beetle is chiefly abundant in the South where it is found in great numbers infesting seed pods of various plants. Stored corn as well as exposed ears in the field are attacked. The life cycle is very similar to that of the saw-toothed grain beetle.

ANGOUMOIS GRAIN MOTH

Sitotroga cerealella (Olivier), Family Gelechiidae

The most common of the moths infesting whole grains is the angoumois grain moth. World-wide in distribution, the larvae attack grains both in storage and in the field, and are especially destructive in southern North America.

The larval stage overwinters in the North; in the South breeding is continuous throughout the year, retarded when the temperature is low and accelerated when it is high. Heated buildings such as mills and warehouses provide ideal conditions for development in areas of cold weather.

The small buff moths with narrow wings fringed with hairs have a wingspread of 16 mm (Fig. 355). They emerge in the spring and either lay their white eggs on stored grains or fly to fields and deposit them on developing heads of small grains. The average number deposited by each female is 40, but some are known to lay almost 400. The eggs change to red as they age. Upon hatching the tiny white larvae chew into the grains and devour the inside portion. Before pupation each larva prepares an exit through the seed coat by cutting the surface approximately ¾ the circumference of the hole, leaving a weakly fastened flap which the adult moth pushes out of the way to effect emergence. Damaged ears of corn have the appearance shown in Fig. 356. Development from egg to adult may be completed in 5 weeks in warm areas; it is prolonged by lower temperatures.

INDIAN MEAL MOTH

Plodia interpunctella (Hübner), Family Pyralidae

A native of Europe but now found world-wide, this insect is considered the most troublesome of the grain-infesting moths. Damage is caused by the larvae spinning silken threads as they feed and crawl, thus webbing the particles of food together. Besides infesting all cereal products and whole grains (Fig. 357), this species also feeds on a wide variety of foods, such as dried fruits, nuts, dog biscuits, dried milk, and seeds.

The moth has a wing expanse of nearly 18 mm, the apical portion of the fore wings being red-brown or coppery and the basal portion gray (Fig. 358). Each female deposits 100 to 300 gray-white eggs, singly or in groups, on food materials. When fully grown the brown-headed larvae are nearly 12 mm long, white-bodied tinged with pink. They spin silken cocoons, in which transformation to pupae occurs, and later emerge as adults. The entire life cycle requires 4 to 6 weeks during the summer.

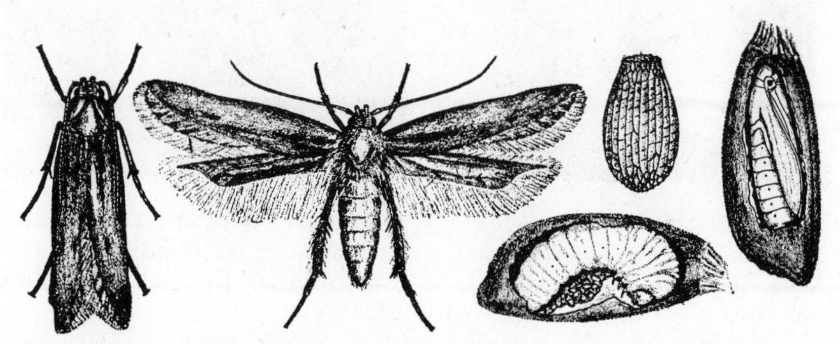

Figure 355. The angoumois grain moth, *Sitotroga cerealella* (Olivier); 4×. (King, Pa. Dept. Agr.)

Figure 356. Larval damage to popcorn by the angoumois grain moth. (Original.)

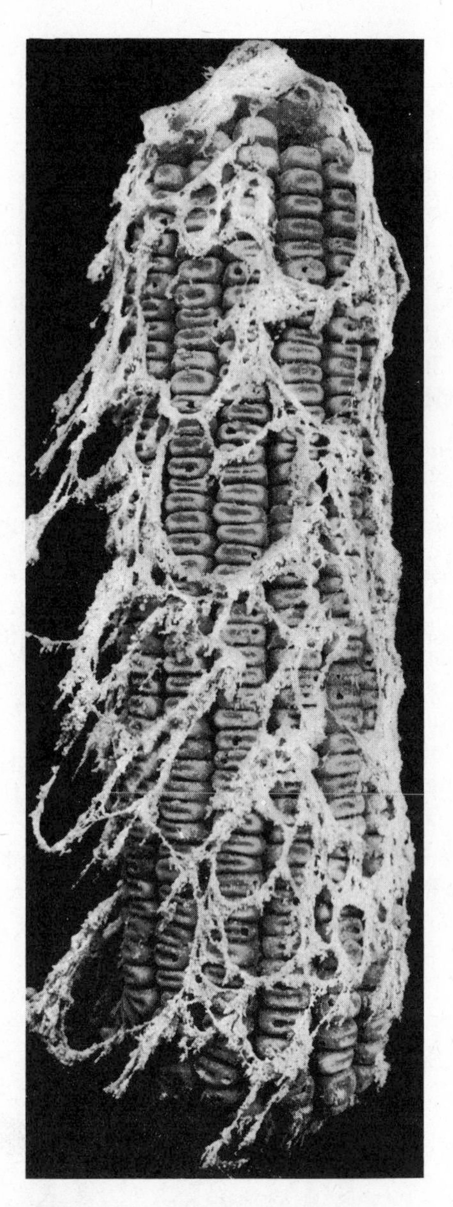

Figure 357. Work of the Indian meal moth showing the characteristic webbing by the larvae. (Back and Cotton, USDA)

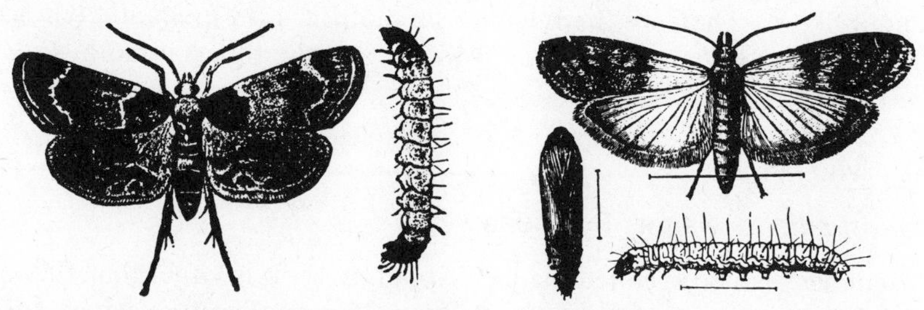

Figure 358. (Left) the meal moth, *Pyralis farinalis* (L.); 2×; and (right) Indian meal moth, *Plodia interpunctella* (Hübner); 2½×. (USDA)

MEDITERRANEAN FLOUR MOTH

Anagasta kuehniella (Zeller), Family Pyralidae

The Mediterranean flour moth, now found world-wide, is a native of Europe and was discovered in California in 1892. It is considered most troublesome in flour mills. Slightly larger than the angoumois grain moth, this insect has a wing expanse of nearly 2.5 cm and is gray with darker markings on the fore wings (Fig. 359). The larvae spin silken threads wherever they crawl, webbing and matting together particles of food on which they feed. Although milled cereal grain products are most often infested, whole grains may also be damaged. In flour mills the machinery may become so clogged with matted flour that operations are halted. Infestations in homes usually can be traced to purchased infested cereal products.

The females lay tiny white eggs in flour, meal, or grains, which hatch in a few days into white larvae with brown heads. These feed, becoming pink-tinged as development proceeds, finally spinning silken cocoons in which pupation takes place. The adults emerge soon afterwards, completing the cycle which normally requires 8 or 9 weeks.

Three other members of this family occasionally are found in dried fruits, nuts, grains, cereal products, and tobacco but are not considered major pests in

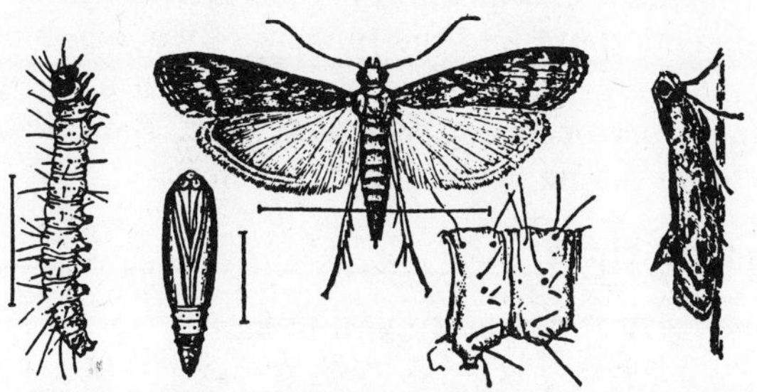

Figure 359. Mediterranean flour moth, *Anagasta küehniella* (Zeller); 2×. (USDA)

the United States. They are the raisin or fig moth, *Cadra figulilella* (Gregson), the almond moth, *Ephestia cautella* (Walker), and the tobacco moth, *Ephestia elutella* (Hübner).

MEAL MOTH

Pyralis farinalis L., Family Pyralidae

This insect has also been called the meal snout moth. It is larger than the Indian meal moth, having a wing expanse of about 25 mm. The fore wings are red-brown, each having 2 wavy transverse white bands (Fig. 358). Widely distributed, although less abundant, the meal moth is a general feeder in the larval stage, attacking milled grains but more likely to be abundant in damp, spoiled grains or grain products in poor condition. The bags of infested sacked feeds or grains may be cut by the larva, allowing the materials to sift out. Female moths deposit 200 to 400 eggs which hatch into white larvae with dark heads. They construct tubes of silk mixed with particles of food in which they feed from the openings at the end. When fully developed the larvae spin silken cocoons in which pupation occurs, the adults emerging several days later. The developmental period requires 6 to 8 weeks in warm weather.

CADELLE

Tenebroides mauritanicus (L.), Family Trogositidae

This cosmopolitan insect, thought to be a native of America, is a shiny black elongate beetle measuring 10 mm (Fig. 360). The fleshy white elongated larva with a black head and thoracic shield, and 2 dark horny projections at the posterior end, is easily recognized. Both adults and larvae feed on cereal products and whole grains. They often devour only the germ, leaving the endosperm portion of grains. In mills they frequently damage the silk bolting cloth, and are known as "bolting cloth beetles."

Each female is capable of depositing almost 1000 eggs during her lifetime, which may extend over a period of 1 or 2 years. These white eggs are laid in masses in the food material and hatch in a week during warm weather. The larvae feed and frequently bore into the timbers of wooden bins. When fully developed they migrate to secluded places where transformation to pupae takes place, with adult emergence occurring soon afterward. The entire life cycle can be completed in 70 days during hot summer weather.

LESSER GRAIN BORER

Rhyzopertha dominica (F.), Family Bostrichidae

Although one of the smallest beetles attacking grain in North America, the lesser grain borer is one of the most destructive. Both adults and larvae cause

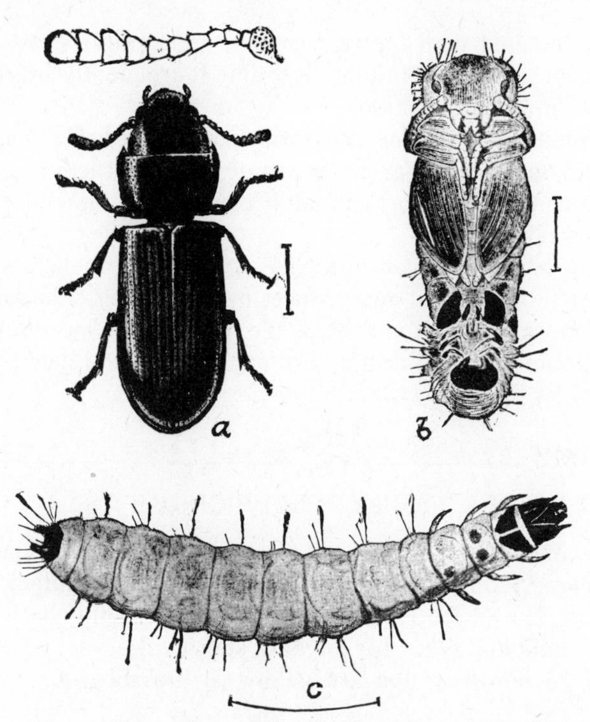

Figure 360. The cadelle, *Tenebroides mauritanicus* (L.): *a,* adult beetle with enlarged antenna; *b,* pupa; *c,* larva; 5×. (USDA)

serious damage to a wide variety of grains. This has been noted especially in the South, but infested samples have been found in all large grain centers. Frequently tunnelling of wood in storage bins results from their feeding, weakening the structures.

The beetle is easily recognized by its slender cylindrical form, shiny brown or black color, roughened surface of fore wings, and small size, less than 3 mm long. Its head is hidden under the pronotum. Each female lays 300 to 500 eggs, dropping them singly or in groups in loose grain. Hatching takes place in a few days and the white larvae feed on the flour produced by the feeding beetles, or they bore directly into grains, feeding until growth is complete. Pupation follows and in a short time the adults eat their way out. Development from egg to adult requires approximately 1 month at summer temperatures.

FLAT GRAIN BEETLE

Cryptolestes pusillus (Schönh.), Family Cucujidae

The flat grain beetle is about 2 mm long, flattened, oblong, red-brown, with antennae almost ⅔ as long as the body. It is world-wide in distribution and is often found associated with the rice weevil, apparently because the adult is unable to survive in sound, uninjured grain. It frequently infests grains and cereal products in poor condition.

The tiny white eggs are dropped in farinaceous material. In whole grains the larvae often devour only the germ. When development is complete they form cocoons, followed by pupation and adult emergence, the entire life cycle requiring 5 to 9 weeks.

The rusty grain beetle, *Cryptolestes ferrugineus* (Stephens), is similar to the flat grain beetle in habits and appearance; it differs in that the antennae of the males are not more than half as long as the body. Being more resistant to cold weather, it is more common in the northern states than related species.

MEALWORMS

ORDER COLEOPTERA, FAMILY TENEBRIONIDAE

The yellow mealworm, *Tenebrio molitor* L. (Fig. 361), and the dark mealworm, *Tenebrio obscurus* F., are the larvae of the largest beetles attacking grains and cereal products. The adults are nearly 2 cm in length, shiny dark brown for the yellow species, and dull black for the dark species. Larvae, when fully grown, are well over 25 mm long and are yellow or gray-brown, according to the species.

Both are seldom found in cereal products in homes but are often numerous in neglected grains and milled products that accumulate in dark corners, under sacks, in bins, or in places where livestock feeds are stored. Moist areas are their natural habitat.

The larval stage overwinters and adults appear in June, the females laying almost 500 bean-shaped white eggs. Hatching occurs in about 2 weeks, with larval development continuing throughout the summer followed by hibernation during the winter and transformation to pupae in late spring. There is only 1 generation each year.

Sanitation is often the simplest way to eliminate these insects.

CIGARETTE BEETLE

Lasioderma serricorne (F.), Family Anobiidae

The cigarette beetle is found in temperate, subtropical, and tropical areas attacking tobacco, seeds, spices, pepper, drugs, and occasionally grain and cereal products. It is a small robust oval light brown beetle, with the head bent down sharply, giving a humped appearance when viewed from the side (Fig. 362). Although variable in size, it is usually about 2.5 mm in length. The antennae are serrate.

The adults live 2 to 4 weeks, and during this time the females lay almost 100 white eggs which hatch into whitish, curved, hairy larvae. Pupation and adult emergence completes the life cycle, which requires about 6 weeks during the summer.

Figure 361. The yellow mealworm, *Tenebrio molitor* L.; adult beetle, pupae, and larvae, with wheat grains to indicate size. (Back and Cotton, USDA)

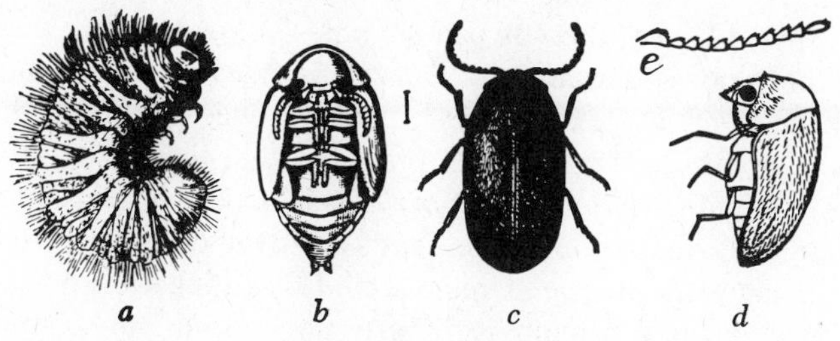

Figure 362. The cigarette beetle, *Lasioderma serricorne* (F.), a pest of miscellaneous household materials: *a,* larva; *b,* pupa; *c,* adult; *d,* side view of adult; *e,* antenna; all greatly enlarged. (USDA)

DRUG-STORE BEETLE

Stegobium paniceum (L.), Family Anobiidae

Very similar in appearance and life cycle to the cigarette beetle with which it is closely allied, the drug-store beetle differs primarily in being more elongate, having distinctly striated wing covers, 3 slightly enlarged segments at the tip of the antennae, and a less hairy larval stage (Fig. 363). It is a very general feeder, attacking many drugs, pepper, spices, seeds, and processed foods of all kinds.

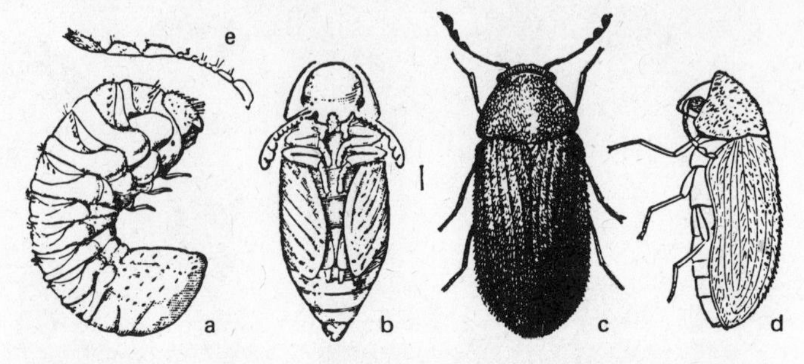

Figure 363. Drug-store beetle, *Stegobium paniceum* (L.): *a,* larva; *b,* pupa; *c,* adult, dorsal view; *d,* adult, side view; *e,* antenna; 12×. (USDA)

PEA AND BEAN WEEVILS

ORDER COLEOPTERA, FAMILY BRUCHIDAE

Some species of beetles that are field pests of beans, peas, and cowpeas also attack the dried seeds of these crops in storage, often causing serious damage. An important species in this group is the bean weevil, *Acanthoscelides obtectus* (Say), which is widely distributed in North America and causes damage as shown in Fig. 364. The broadbean weevil, *Bruchus rufimanus* Boheman, is found on the West Coast; the cowpea weevil, *Callosobruchus maculatus* (F.), and the southern cowpea weevil, *C. chinensis* (L.), are more abundant in the tropics. All these species except the broadbean weevil can produce repeated generations in dried seeds in storage, provided that the temperature is favorable. The pea weevil, *Bruchus pisorum* (L.), is a pest of peas in the field and may complete its development in the seeds after the crop is harvested but does not attack dry peas in storage. Unless killed by fumigation or by other treatment, pea weevils may remain in hibernation within the container of seed throughout the winter. There is only 1 generation a year. More information on this group of insects is given under pests of leguminous crops.

Figure 364. Bean weevils and damaged beans; slightly enlarged. (Original)

CONTROL OF INSECT PESTS OF STORED GRAINS AND OTHER FOOD PRODUCTS

A control program for this group of pests is essentially the same for the home, farm, mill, elevator, or warehouse, the main difference being in the scope of operation and the equipment used. With this in mind, the following are recommended procedures: provide clean storage, practice sanitation, use residual sprays, apply seed protectants, inspect frequently, fumigate, supercool and superheat. All these measures are not necessary in every situation.

Proper storage means clean, tight, dry bins or other easily accessible facilities, which not only afford protection from insects, rodents, birds, and poultry, and can be easily inspected or fumigated, but also provide reasonable safety from fire and wind damage. Clean, insect-free grains stored with a moisture content of 10% or less are seldom subject to attack. It is well to keep in mind that the factors favorable for preserving the keeping quality of grain and cereal products are generally unfavorable for the development of stored-grain insects.

Sanitation measures consist of thoroughly cleaning out old grain remnants from granaries, bins, or other places where these accumulations may serve as breeding areas. All infested materials should be fumigated, fed to livestock, or destroyed.

The wall and floor surfaces of thoroughly cleaned storage facilities should then be sprayed with residual pesticides of either 2% malathion, 2.5% methoxychlor, or 0.5% pyrethrins or allethrin in combination with synergists at the rate of 2 gallons per 1000 square feet of surface. This should be done at least 3 weeks before grain or cereal products are stored.

Newly harvested grain can be protected from infestation through the first season of storage by adding a commercially prepared powder (Pyrenone) composed of 0.08% pyrethrins, 1.1% piperonyl butoxide, and 98.82% wheat flour. This is distributed throughout the grain as it is put into bins at the rate of 75 pounds per 1000 bushels. A 1% premium grade malathion-wheat flour dust at the rate of 60 pounds per 1000 bushels is also approved as a grain protectant. Malathion may be sprayed on the grains by mixing 1 pint of 57% emulsifiable concentrate in 2 to 5 gallons of water and applying this to 1000 bushels. Apply to grain stream going into storage.

Inert dusts of hydrated lime, diatomaceous earth, magnesium oxide, silica aerogel, rock phosphates, or aluminum oxide protect seeds having a moisture content of 12% or less when applied at the rate of 1 ounce per bushel. Beans or other seeds so treated must be thoroughly washed before they are used as food. This control method is little practiced. Seeds for planting can be protected by mixing 1 ounce of either 3% methoxychlor dust or 1% lindane dust per bushel. These chemicals are effective for long periods regardless of the moisture content of the seed. They are poisonous and the seed should be so labelled and never used as food for any animals.

Surface activity of Indian meal moth larvae can be prevented in stored shelled corn if the top layer is sprayed in June and August with white mineral oil at the rate of 2 quarts per 100 square feet of surface grain. In some regions Indian meal moths are becoming resistant to malathion. Where this happens they are being controlled with dichlorvos plastic strips that give off a lethal fumigant. To be effective the treated area must be enclosed, nonventilated, and the strips hung above the grain in the spring before moths begin emerging. When strips become dust covered or spent, replace them.

Storing insect-free grains or grain products in fabric or multiwall paper bags impregnated with pyrethrins plus synergists has afforded considerable protection against penetration by insects and may be of practical value for food shipments and storage.

Inspections should be made every 30 days of all grains in storage in regions 2, 3, and 4 (Fig. 365) to determine the need for control measures. The presence of adult grain insects indicates the need for immediate fumigation. These inspections can be less frequent in region 2 because lower temperatures prevail which retard development. Where grain is held over from 1 year to the next, it should be examined carefully and fumigated at the first sign of infestation.

Fumigate all old grain which cannot be removed from storage before new grain is binned. In grain elevators wheat that is to be stored for over 1 month should be fumigated within a week after it is received. Unprotected farm-stored grains should be fumigated within 6 weeks after harvest.

The accompanying table suggests fumigants and dosages for each. The first four materials listed can be used in any fumigation but are more often recommended for farm grain storage. They are liquids whose vapors are heavier than air and therefore should be applied at the top of the grain or other materials to

Fumigant	Dosage per 1000 Bushels	
	Wooden Bins	Concrete or Metal Bins
Carbon tetrachloride, 100%	6 gal.	3 gal.
Carbon tetrachloride-carbon disulfide, 80%–20%	4 gal.	2 gal.
Ethylene dichloride-carbon tetrachloride, 75%–25%	8 gal.	4 gal.
Carbon tetrachloride-ethylene dichloride-ethylene dibromide, 60%–35%–5%	4 gal.	2 gal.
Calcium cyanide	20 lb.	10 lb.
Chloropicrin	4 lb.	2 lb.

be treated. Chloropicrin and calcium cyanide are generally used in grain elevators and should be applied continuously to the grain stream as the bins are filled. Chloropicrin on grains with a moisture content of 12% or higher results in lowered germination. Methyl bromide alone, or in combination (4 to 1) with chloropicrin or ethylene dibromide, has met with success in fumigating storage warehouses and controlling surface infestations of Indian meal moths in grain

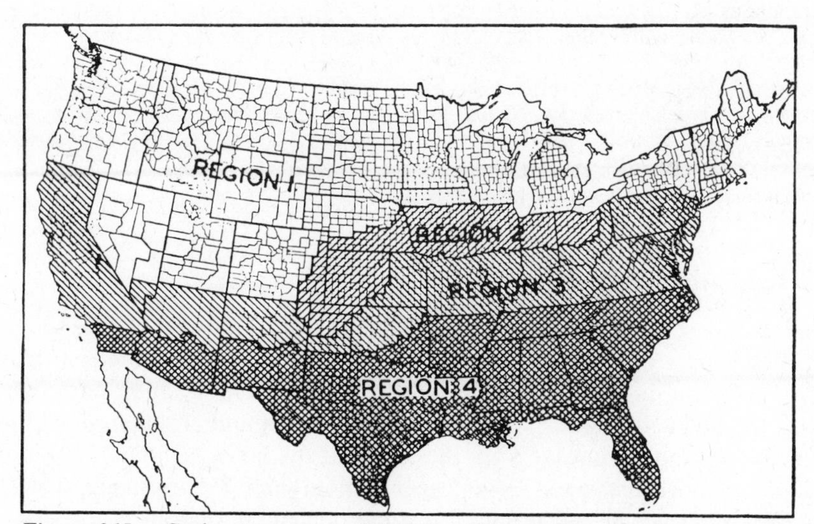

Figure 365. Grain storage conditions in different regions: region 1, relatively safe; region 2, loss may occur in some years; region 3, hazardous every year—inspection and fumigation recommended; region 4, farm storage unsafe and not recommended. (Cotton, USDA)

elevators. These materials are generally applied at the rate of 1 to 1.5 pounds per 1000 cubic feet of empty storage space. Tablets of aluminum phosphide (Phostoxin), on exposure to air, generate phosphine gas, which has proved effective in killing stored grain pests. Slight modification of some of the basic fumigant mixtures in the table are also available; some of these contain a small percentage of sulfur dioxide or other irritating chemicals as warning gases.

All fumigants are poisonous and care should be exercised in their use. For commercial products follow the directions of the manufacturer.

Most insects are not active at temperatures below 50 F (10 C). If infested grain or other food products are cooled artificially or naturally, owing to the prevailing winter temperatures of a given region, development is retarded. Most stored-product pests are killed by an exposure of 7 days to temperatures of 15–20 F (−9 to −6 C).

Superheating is also employed as a control measure at times. An exposure of 10 minutes to a temperature of 140 F (60 C) is fatal to all stored-product pests. Because of the insulating effect of these products considerable time is required to allow for heat penetration in large quantities of grain. Special grain driers are available which force hot or warm air through layers of grain. Tests have shown that the germination of wheat, rye, oats, and buckwheat apparently is not impaired by artificial drying with heated air at 120, 140, and 160 F (49, 60, and 71 C).

REFERENCES: R. T. Cotton, *Insect Pests of Stored Grain and Grain Products,* Burgess Publishing Co., Minneapolis, Minn., 1963; *Nebr. Agr. Exp. Sta. Cir.* 62, 1940; *Mont. Agr. Exp. Sta. Bul.* 297, 1935; *Ill. Ext. Cir.* 512, 1941; *Minn. Agr. Exp. Sta. Bul.* 340, 1947; *Ky. Agr. Exp. Sta. Bul.* 571, 1951; *USDA Farmers' Bul.* 1880, 1941; 1906, 1942; 1260, 1953; *Cir.* 462, 1938; 720, 1945; E–783, 1949; *Leaflet* 195, 1940; 235, 1943; 553, 1971; *Yearbook of Agriculture,* pp. 629–639, 1952; *Marketing Res. Rpt.* 780, 1967; *Minn. Agr. Exp. Sta. Bul.* 425, 1954; *Can. Dept. Agr. Pub.* 1131, 1961; *Proc. Ent. Soc. Wash.* 64:43–50, 1962; *Kans. Agr. Exp. Sta. Bul.* 416, 1960; *Bul. Ent. Soc. Amer.* 21:165–168, 1975.

MISCELLANEOUS PESTS OF FOODS

Cheese Skipper, *Piophila casei* (L.), is the larva of a shiny black fly, 4 to 5 mm long. It is a pest of wide distribution in this country and in other parts of the world. The larva is a white slender-bodied maggot, about 8 mm long, tapering toward the head. It attacks cheese of various kinds and cured smoked meats. The name "skipper" comes from the fact that the larva is capable of jumping or skipping short distances. Several generations may be produced during the year. It is reported to cause mild intestinal myiasis in man.

Control is accomplished by providing insect-free storage with 30-mesh screening on all doors and windows. The paraffin-coated cloth covering on cheese should be carefully applied to prevent possible places of entry. Infestations in storage can be eliminated by fumigation, raising the temperature to

125 F (52 C) or lowering the temperature to 40 F (4.5 C). Proper disposal of infested foods is also recommended.

Pomace or Vinegar Flies, *Drosophila melanogaster* Meigen and other congeners are often extremely numerous and annoying especially during the fruit-canning season and particularly at processing plants. These tiny flies, 3 mm long, are attracted to well-ripened and fermenting fruit produce. The life cycle may be completed in about 10 days at temperatures of 77 F (25 C), resulting in many overlapping generations.

Any practice in harvesting and marketing that prevents damage to produce and accelerates the time between picking and ultimate consumption or processing will help prevent infestation by this pest. Prompt disposal of fermenting fruits and vegetables will greatly reduce fly populations and should be practiced any place such produce exists.

Growing tomato varieties resistant to cracking is a way to avoid this pest.

Field grown tomatoes have been protected by applications of diazinon, naled, or malathion, at 4 to 5 day intervals. To determine the need for treatment place several intentionally split ripe tomatoes under the vines at various places in the field in late afternoon; if eggs are seen in the slits the next morning, following examination with a magnifying glass, it is time to begin insecticide applications.

Good control is achieved without residue problems by treating harvested fruits and vegetables with a freshly mixed dust containing 1% piperonyl butoxide and 0.1% pyrethrins immediately following picking and preceding processing.

Outside walls of canning plants and inside walls, ceilings, and other areas of structures used for holding unprocessed produce may be sprayed with diazinon, naled, or ronnel. Avoid contaminating processing equipment or produce with these chemicals. Culls and refuse at processing plants may be treated with diazinon, malathion, or ronnel.

REFERENCES: *USDA Farmers' Bul.* 2189, 1962; *Ohio Sta. Univ. Ext. Bul.* 459, 1977.

Larder Beetle, *Dermestes lardarius* L., and the black larder beetle, *Dermestes ater* DeGeer, are occasional pests of cured meats, cheese, and other food products of animal origin. Primarily scavengers, these insects are often found in dried remains of dead animals, including insects. Cultures of both species are often maintained to clean skeletons for museums. Both species are approximately 6 mm in length, robust, and black; the larder beetle has a dull yellow band across the base of the elytra. The larvae are also similar, brown and hairy, like most dermestids. Another closely related species, the hide beetle, *D. maculatus* DeGeer, is reported to be a carrier of salmonella. Ordinary sanitation prevents serious infestation. Providing insect-free storage by applying EPA approved residual sprays is recommended.

Flour, Grain, or Cheese Mites in the family Acaridae may become abundant in stored foods, as their common names imply. Some mites are associated with flour or grains and the condition known as "baker's" or "grocer's itch" is caused by certain species. One important species in flour, grain, and cereal products is the grain mite, *Acarus siro* L. Other species are the mold mite, *Tyrophagus putrescentiae* (Schrank), and the cheese mite, *Tyrolichus casei* Oudemans, which are pests of mushrooms, cheese, as well as cereals. Fumigating with 1 pound of methyl bromide per 1000 cubic feet has given good control. Direct contact sprays containing synergized pyrethrins has also given satisfactory control. Dipping waxed cheese blocks in a 4% solution of sodium o-phenylphenate mixed with 5% gelatin kills resting stages and eggs. Wiping lightly infested cheese blocks with mineral oil or cottonseed oil also helps eliminate mites.

REFERENCES: *J. Econ. Ent., 46:844*–849, 1953; 48:754–755, 1955; 52:237–240, 514–518, 1959; *OARDC Res. Bul.* 977, 1965.

Khapra Beetle, *Trogoderma granarium* Everts, is known to have been present in the San Joaquin valley in 1946. It has spread through southern California into Arizona, New Mexico, Texas, and Mexico. Occasionally it is found in widely scattered areas in shipments of infested products. It is native to India. Grain and cereal products are more seriously damaged, but there seems to be little in the way of dried vegetable and animal products that it will not attack. Development is more rapid in milled cereal products than in whole grains. Larvae of this dermestid are able to develop in foods with a moisture content of less than 2%. The larvae are very resistant to starvation; they may live for months or years without food. Adults are red-brown to black and less than 3 mm long. As many as 12 generations develop each year, the number depending on the temperature and food supply. Eradication is hampered owing to its habit of crawling into spaces of infested structures difficult to reach with residual sprays and toxic concentrations of fumigants. Control recommendations are the same as those for other grain insects. Infested areas are under federal quarantine and an eradication program is in effect.

REFERENCES: *Proc. N.C. States Branch, Ent. Soc. Amer.,* 10:70–71, 1955; *J. Econ. Ent.,* 48:332–333, 1955; 52:312–319, 1959; 57:305–314, 1964; *USDA* PA-436, 1961.

HOUSE ANTS

ORDER HYMENOPTERA, FAMILY FORMICIDAE

Several species of ants invade houses. They are important largely because of the food they contaminate and the annoyance caused by their presence. A few large species occasionally come into houses, but it is mainly the smaller ones that cause the most trouble. Ants are predatory and do reduce pest populations of insects. Natural control of termites is primarily from various species of predatory ants.

Ants are social insects that live in colonies or nests, in which remain the egg-laying queens, the larvae, pupae, and many young ants. The workers, all sterile females, care for the colony and forage for food. Usually in the spring or early summer ant colonies produce winged males and females, which have the potentialities for starting new colonies; many, however, are unsuccessful. After mating the males or kings die (see Fig. 366).

True ants are constricted where the abdomen joins the thorax. They have elbowed antennae and 2 pairs of wings differing in size (first pair larger) and held at an angle above the body.

The more common species are: crazy ant, *Paratrechina longicornis* (Latreille); Pharoah ant, *Monomorium pharaonis* (L.); little black ant, *M. minimum* (Buckley); pavement ant, *Tetramorium caespitum* (L.); thief ant, *Solenopsis molesta* (Say); black imported fire ant, *S. richteri* Forel; fire ant, *S. geminata* (F.); red imported fire ant, *S. invicta* Buren; southern fire ant, *S. xyloni* McCook; small yellow ant, *Acanthomyops claviger* (Roger); large yellow ant, *A. interjectus* (Mayr); Argentine ant, *Iridomyrmex humilis* (Mayr); black car-

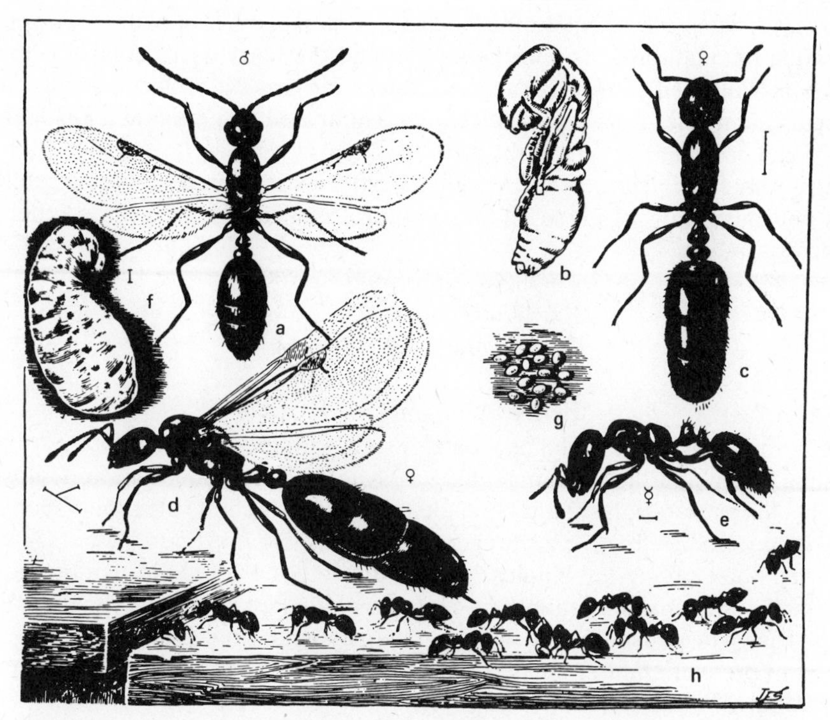

Figure 366. The little black ant, *Monomorium minimum* (Buckley): *a,* male; *b,* pupa; *c,* female; *d,* winged female; *e,* worker; *f,* larva; *g,* eggs; *h,* group of workers in line of march; *a* to *g* much enlarged; *h,* 5×. (USDA)

penter ant, *Camponotus pennsylvanicus* DeGeer; red carpenter ant. *C. ferrugineus* (Fabr.); and Florida carpenter ant, *C. abdominalis floridanus* (Buckley).

Pharoah ants are about 2 mm long and red-yellow, commonly nesting in the walls of heated buildings and feeding on a wide variety of foods. The thief ant is very similar in appearance but smaller in size; it nests in the soil and is troublesome only during warm seasons. The little black ant is about 3 mm long, nearly black, and nests both in the soil and in houses. The pavement ant is black, much larger, and nests in the soil, often under stones and other objects. Argentine ants are discussed on p. 486. In the South they are serious household pests. Carpenter ants are black and approach or exceed 13 mm in length, commonly infesting stumps, logs, dead branches of trees, and timber in houses and other buildings. The imported fire ant builds mounds in the field and does damage to agricultural crops. In addition, it has a severe sting that makes it annoying to persons in the infested area, which is primarily southern United States.

Ants in the home can be eliminated by following sanitary practices and by treating the trail of the workers to the point of entrance and then the nest itself, if it can be reached, with locally approved pesticides. Some effective chemicals are chlorpyrifos, chlordane, diazinon, malathion, propoxur, and ronnel. Treating foundation walls and the soil around the house, particularly near the kitchen, will often eliminate the ant problem and make it unnecessary to apply a pesticide inside. Carpenter ants can be eliminated by applying a pesticide to the trees or timber in which they are working.

In ant-infested places, where residual pesticides are not desirable, the following poisoned bait has given effective control of Pharoah as well as other ant species:

Thallium sulfate	5.5 g.
Extracted honey	3 oz.
Granulated sugar	1 lb.
Water	1 pt.
Benzoate of soda	5 g.

Mix the sugar, honey, and benzoate of soda with the water and heat to almost boiling. Then add the thallium sulfate, stir until all is in solution, and allow to cool. This preparation should be made in a chemical hood since the vapors are *poisonous*. To be effective for longer periods, the bait must not dry out. A cardboard pillbox having 1 or 2 holes 3 mm in diameter punched through the side of the lid, and the inside bottom part coated with melted paraffin, will serve as an inexpensive dispenser for the baits and help avoid drying. Care should be exercised to prevent children or pets from getting into poisoned baits. Store properly labeled, unused bait in a cool place.

REFERENCES: *USDA Leaflet* 147, 1937; PA-368, 1962; *Home and Garden Bul.* 96, 1976; 28, 1975; *Tech. Bul.* 1326, 1965; *J. Econ. Ent.*, 43:565, 1950; 54:45–47, 1961; 57:331–333, 1964.

SUBTERRANEAN TERMITES

ORDER ISOPTERA, FAMILY RHINOTERMITIDAE

Termites, often called white ants, are neither true ants nor are they always white. The termite abdomen is broad at the juncture with the thorax. The majority of the individuals in a termite colony are white, but the winged swarm stages are black. Termites have moniliform antennae and 2 pairs of wings alike in size and appearance held flat over the body.

There are many kinds of termites in the world, some of which live entirely in dry or damp wood. These species are in the genera *Kalotermes* and *Incisitermes* and are found principally in the southern coastal states of the United States. Subterranean termites are by far the most destructive species in North America and commonly infest wood in the soil or wood that can be reached from the soil by means of covered runways. The eastern subterranean termite, *Reticulitermes flavipes* (Kollar) is generally distributed throughout eastern United States and adjoining Canadian provinces; the western subterranean termite, *R. hesperus* Banks, occurs widely on the Pacific Coast and eastward to Nevada and Idaho; the arid land subterranean termite, *R. tibialis* Banks, is found in abundance primarily west of the Mississippi River. Other species found in the southeastern states are *R. hageni* Banks and *R. virginicus* Banks. They occur primarily in the region southward from Philadelphia, Pennsylvania, to Oklahoma. The Formosan subterranean termite, *Coptotermes formosanus* Shiraki, is now found in southern and southeastern United States in addition to Formosa, Hawaii, Guam, Midway, and other Pacific Islands.

The principal food of termites is cellulose obtained from wood and other plant tissues. Serious damage results to wooden buildings (Fig. 367), fence posts, telephone poles, paper, fiber board, and fabrics derived from cotton and

Figure 367. Damage to a wooden structure by the subterranean termite. (Mich. Agr. Exp. Sta.)

other plants. Termites occasionally injure living plants, but this is of minor importance. They are able to obtain nourishment from a cellulose diet because of the presence in their digestive tracts of certain flagellated protozoa and other microorganisms, which possess enzymes capable of converting cellulose into starches and sugars.

Termites are social insects and produce different forms or castes. These include winged, dark-colored males and females which develop into kings and queens (Fig. 368), short-winged and wingless supplementary reproductives of both sexes, wingless workers of both sexes, and wingless forms with large brown heads and jaws, called soldiers. The winged forms are produced by a colony only once a year, usually in the spring, and these are capable of establishing new colonies. All wingless forms are white, soft-bodied, and scarcely 6 mm long (Fig. 369). After a colony is started the workers have the duties of feeding the queen and young, making runways, and searching for food. The soldiers guard the colony from attacks of other insects, principally ants. Should the queen be killed, the supplementary reproductives can reproduce their own kind, as well as workers and soldiers. This results in a colony that never swarms.

Subterranean termites are most numerous in moist, warm soil containing an

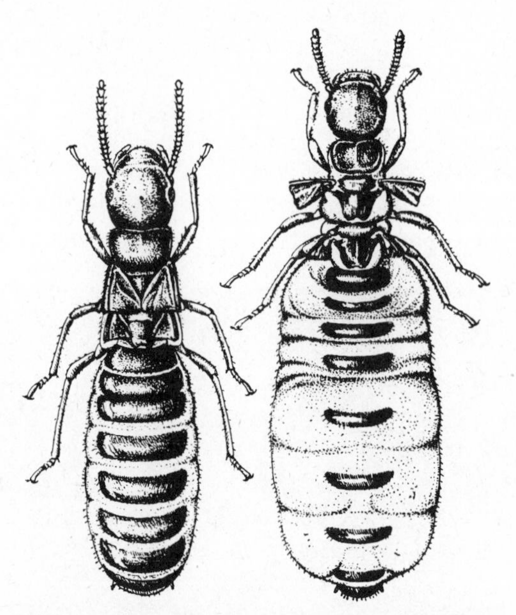

Figure 368. The subterranean termite, *Reticulitermes flavipes* (Kollar); king on left, enlarged; queen on right, about natural size. The bases of their shed wings are still evident. (USDA)

Figure 369. Termite workers, and a soldier at top right; 3×. (Original)

abundance of wood or other cellulose material. They may invade buildings by constructing runways composed of soil, chewed wood, excrement, and saliva, over the surfaces of foundation walls or through cracks in the foundation or concrete floors. Damage to wood is often not evident from outward appearance because the termites leave the thin outer layer of wood intact and eat only the central portions. An infestation can often be discovered by jabbing at suspected damaged timbers with a sharp tool such as a pocket knife or screw driver. The emergence of the winged reproductive termites from a building or the soil nearby is often the first indication of the presence of a termite colony.

When planning new structures it is important to consider means of preventing termite damage by proper design and construction. The following are suggested preventive measures: avoid leaving pieces of wood forms or scraps of lumber adjacent to the foundation or in fill material under porches, terraces, or steps; place the building on a solid, carefully built foundation; allow clearance of at least 18 inches beneath all wood substructures; provide for thorough drainage of the soil beneath the building and around the foundation; provide for ventilation under buildings without basements; avoid contact between the woodwork of the building and the soil; install wooden basement partitions, posts, and stair carriages after the concrete floor is poured; properly install metal shields of copper or zinc on top of the foundation.

Slow-growing heartwood of some species of trees is quite resistant to termites, but not immune. The following kinds of lumber are the most resistant to termite attack: foundation grade California redwood, southern tidewater red cypress, pitchy southern pine, and heartwood of eastern red cedar.

Natural control of termites is primarily from various species of predatory ants.

Prevention of damage to railroad ties, telephone poles, fence posts, or bridge timbers is accomplished by pressure-impregnating these items with coal-tar creosote, zinc chloride, chromated zinc chloride, zinc arsenite, or pentachlorophenol. This is done by commercial firms with special equipment and trained personnel.

Chemical control in buildings is accomplished by treating the soil around the foundation or other places of possible termite entry with chlordane or heptachlor. These chemicals applied as emulsions give long-lasting protection because they possess both contact and fumigant action, do not have objectionable odors, and are not phytotoxic at the required concentrations. An accepted procedure is to prepare a 1% chlordane emulsion and apply this to the soil at the rate of 2 gallons per lineal 5 feet for the first foot at ground level and 1 gallon per lineal 5 feet from grade level to footing. In some situations it will be necessary to trench along the entire perimeter of the building and gradually add the chemical as the trench is filled. For homes with basements the dosage is 4 gallons per lineal 5 feet. Considerable saving of time results if a soil auger can be used to make holes near the foundation about 18 inches apart, in which the emulsion is poured to saturate the soil and form a chemical barrier to termite entry. Slab floors and porches should be drilled and treated at the rate of 1 gallon per 10 square feet of area. Treating the foundation area of new homes and other buildings at the time of construction is good insurance against termite attacks. If heptachlor is used in place of chlordane, prepare it at a concentration of 0.5%. Dry wood termites are controlled by fumigation with HCN or methyl bromide.

If the termite problem is especially difficult, consult your extension entomologist or a reliable pest-control operator.

REFERENCES: *USDA Home & Garden Bul.* 64, 1975; *Farmers' Bul.* 2018, 1958; *J. Econ. Ent.*, 45:235–237, 1952; 46:527–528, 1953; *Fla. Ent. Bul.* 157, 1954; *Conn. Agr. Exp. Sta. Cir.* 218, 1961; *Bul.* 695, 1968

POWDER POST BEETLES

ORDER COLEOPTERA

These destroyers of seasoned wood rate next to the termites in the amount of damage they do. Most of this damage is caused by the larvae eating and tunnelling through timber in lumberyards, homes, other buildings, and manufactured wood products, such as furniture and handles of tools. The interior portion of the wood is reduced to a fine powder, and the surface becomes perforated with small holes (Fig. 370). Several species in each of the following families are usually the most destructive: Anobiidae, Lyctidae, and Bostrichidae. Where feasible, sanitation measures are of value in controlling these

Figure 370. Damage to wood by powder post beetles. Note adult emergence holes. (Original)

pests. Furniture and other wood products can be fumigated, methyl bromide or HCN being the common fumigants. Where facilities are available, heat treatment at 125 F (52 C) or higher will kill all stages if this temperature is maintained for an hour or more throughout the infested materials. More permanent protection is provided by treating wood with oil base solutions containing either diazinon, or pentachlorophenol.

REFERENCES: *USDA Leaflet* 558, 1972; 501, 1972; *Farmers' Bul.* 2104, 1972; *Tech. Bul.* 1157, 1957; *Home and Garden Bul.* 96, 1976.

COCKROACHES

ORDER ORTHOPTERA, FAMILIES BLATTIDAE AND BLATTELLIDAE

Many species of cockroaches occur throughout the world, some of which infest the household and are frequently found in restaurants, hotels, hospitals, grocery stores, slaughterhouses, offices, and libraries. The damage they do may be relatively slight, but their presence is so objectionable that they are considered among the worst of domestic pests.

Some of the more common species are the American cockroach, *Periplaneta americana* (L.) (Fig. 371), German cockroach, *Blattella germanica* (L.), oriental cockroach, *Blatta orientalis* L., Australian cockroach, *Periplaneta australasiae* (F.) (Fig. 372), and brown-banded cockroach, *Supella longipalpa* (Fabr.).

Eggs of roaches are produced in capsules or öothecae, which may be carried for several days by the female, protruding from the tip of her abdomen. Some species drop the capsules in places that they frequent; others, like the German

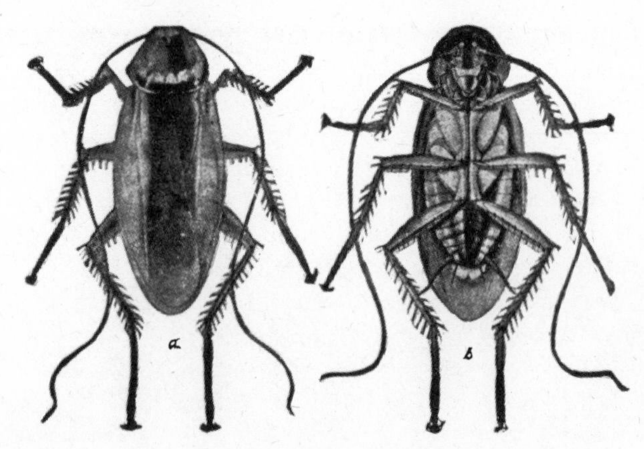

Figure 371. The American cockroach, *Periplaneta americana* (L.): *a*, view from above; *b*, from beneath; slightly reduced. (USDA)

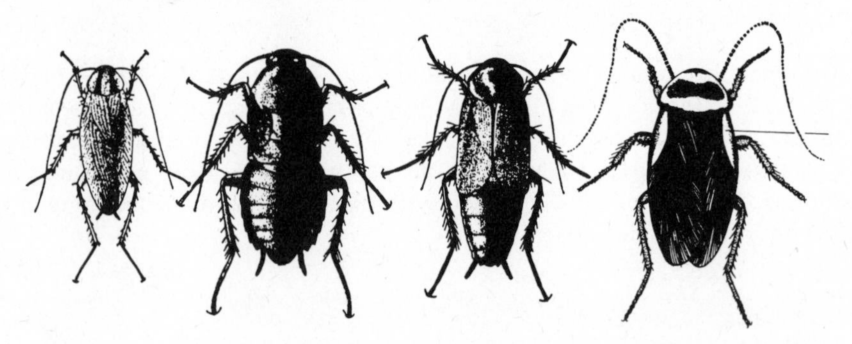

Figure 372. The German cockroach, *Blattella germanica* (L.); oriental cockroach, *Blatta orientalis* L., female and male; and the Australian cockroach, *Periplaneta australasiae* (Fabr.); from left to right. German enlarged; others reduced. (Conn. Agr. Exp. Sta.)

roach, carry them until hatching takes place (Fig. 11). Hatching of the egg capsules that are dropped may take place in 2 or more weeks. Egg capsules of the German roach contain 25 to 30 eggs, those of the American and oriental 12 to 13, and those of the brown-banded about 18. Young nymphs develop rather slowly and are found in the same places as adults. The German and brown-banded roaches have 2 or 3 generations per year, whereas the larger species require about a year or more to mature.

Adults of the German and brown-banded roaches are about 12 mm long and light brown, the German having 2 dark stripes on the pronotum and the brown-banded having 2 light yellow cross-bands on the wings. The female brown-banded has darker brown wings with the same markings. These 2 species are common in dwellings and restaurants. Adults of the American roach are red-brown and 4 to 5 cm long; they frequent slaughterhouses, hotels, and bakeries. Oriental roach adults are almost black and about 35 mm long. The females are practically wingless, and the males have wings that do not reach the tip of the abdomen. Roaches of this species thrive best in damp places and are sometimes called "waterbugs." The Australian roach resembles the American species but is slightly smaller and has a bright yellow stripe on the outer edge of the basal half of the fore wings. It may be found in houses but is more common in glass-houses.

All roaches live concealed during the day, and scurry about and feed at night. They are omnivorous and foods of all kinds may be contaminated with their excretions and offensive odor. Because of their flattened bodies they can hide in crevices behind baseboards, kitchen cupboards, under sinks and drain-boards, cracks in the plaster, or any darkened area.

At times 2 tropicopolitan evaniid wasps are found parasitizing cockroach eggs; they are *Evania appendigaster* (L.), and *Prosevania punctata* (Brullé). They cannot be depended upon to control a cockroach infestation.

Control may be accomplished by treating the areas where they hide during the day or the surfaces on which they are apt to crawl at night with malathion, ronnel, diazinon, bendiocarb, chlorpyrifos, or propoxur. Eliminating feeding and breeding areas should not be overlooked in the control program.

REFERENCES: *USDA Leaflet* 430, 1978; *Home and Garden Bul.* 96, 1976; *Can. Dept. Agr. Pub.* 109, 1953; *J. Econ. Ent.*, 57:327–328, 1964; *Ann. Ent. Soc. Amer.* 47:575–592, 1954; *Conn. Agr. Exp. Sta. Cir.* 224, 1970; *Bul.* 717, 1971.

SILVERFISH AND THE FIREBRAT

ORDER THYSANURA, FAMILY LEPISMATIDAE

Both of these insects are known as silverfish. They are similar in that both are slender, wingless, scale-covered insects, measuring slightly more than 10 mm in length, with 3 long slender tail-like appendages at the tip of the abdomen and 2 long slender antennae on the head. They differ in size and coloration, the silverfish, *Lepisma saccharina* L., being smaller and uniform silver or pearl gray; the firebrat, *Thermobia domestica* (Packard), larger with dusky markings on the body (Fig. 373).

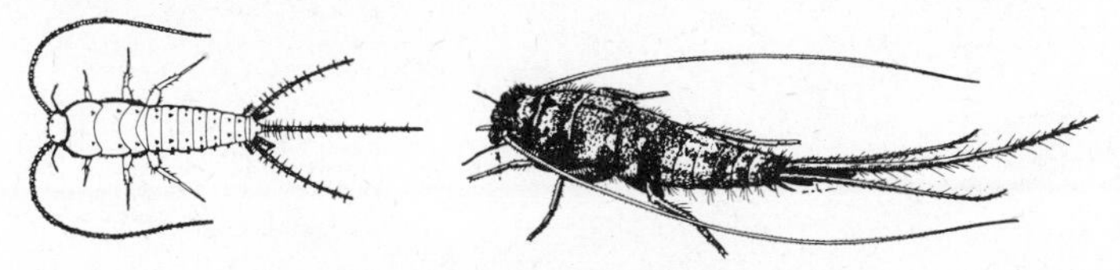

Figure 373. Silverfish (left); firebrat (right); about 2×. (Courtesy of U.S. Pub. Health Serv. and Ill. Nat. Hist. Surv.)

Silverfish thrive best in damp warm basements but may cause damage in any part of a house or public building. The firebrat is more likely to be found in very warm places. Old-fashioned bake ovens often became infested and this gave rise to the name "firebrat." With their chewing mouthparts both insects eat the sizing on paper, bookbindings, wallpaper or insulation materials. Starched clothing and curtains, or rayon, may be seriously damaged, and foods contaminated.

All life stages may be found throughout the year. The number of generations produced varies with prevailing temperature and humidity. Silverfish are said to mature in 7 to 24 months, firebrats in 3 to 24 months.

Control of these insects is accomplished by treating the places they frequent with malathion, chlorpyrifos, diazinon, dichlorvos, propoxur, or ronnel.

REFERENCES: *USDA Leaflet* 412, 1971; *Home and Garden Bul.* 96, 1976.

CLOTHES MOTHS

ORDER LEPIDOPTERA, FAMILY TINEIDAE

Quite small in size, less than 5 mm long, clothes moths cause damage by the larvae feeding on articles containing wool, mohair, feathers, fur, hair, casein, fish meal, or other products of animal origin. They are world-wide in distribution.

The most common species, the webbing clothes moth, *Tineola bisselliella* (Hummel), is straw-colored without markings on the wings, and a trifle smaller than the other species. The tiny white larvae with brown heads make webs or tubes on the surface, where they feed somewhat concealed (Fig. 374). The casemaking clothes moth, *Tinea pellionella* (L.), has gray-yellow wings with faint dark markings. Its larvae form cases of silk and chewed fibers, which are carried about wherever they crawl and in which pupation later takes place. The carpet moth, *Trichophaga tapetzella* (L.), is larger than either of the other species and has wings very dark at the base with light tips (Fig. 375). This insect is smaller than the Indian meal moths, which often occur in houses and are

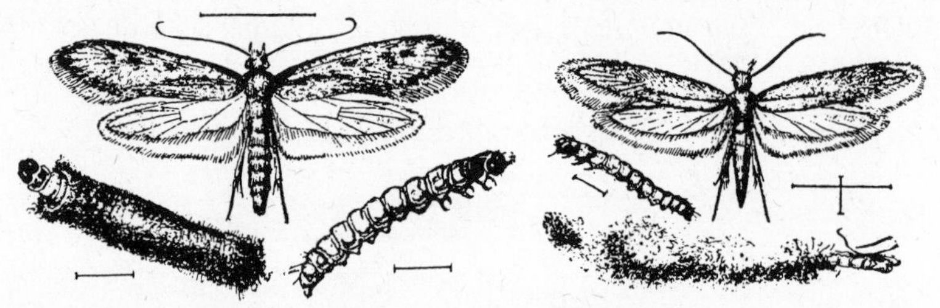

Figure 374. The casemaking clothes moth, *Tinea pellionella* (L.), (left), and webbing clothes moth, *Tineola bisselliella* (Hummel) (right). Actual size indicated by lines adjacent to each figure. (Riley)

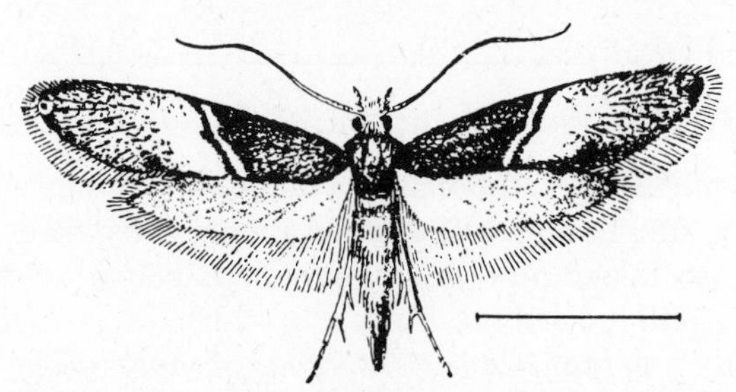

Figure 375. The carpet moth, *Trichophaga tapetzella* (L.); 3×. (Back, USDA)

mistaken for clothes moths. Larvae of carpet moths feed on coarse fabrics, making webs and constructing rather prominent cocoons in which they pupate.

Female moths lay 100 to 300 white eggs which hatch normally in 5 or more days. Larval development requires about 6 weeks, and the pupal period approximately a week. The length of the life cycle is quite variable because of uneven temperatures encountered, but at 80 F (27 C) the period from egg to adult requires about 50 days. At lower temperatures this period is greatly extended.

Careful sanitation, washing, dry cleaning, vacuum cleaning, brushing of articles subject to attack, and their storage in mothproof containers prevent damage. Winter clothing from the dry cleaners can be protected during the summer months when moths are most active by placing a small amount of paradichlorobenzene crystals in the storage container or closet. Clothing not moth-free should be placed in a closet set aside for storage and fumigated with 1 pound of PDB crystals to each 100 cubic feet of space, sealing the door with masking tape.

Heavily infested homes can be freed of clothes moths by fumigating the entire building with hydrogen cyanide, or fumigating clothes closets with a 3-to-1 mixture of ethylene dichloride and carbon tetrachloride. The modern method is the repeated use of aerosols of synergized pyrethrins in clothes closets, or treating infested articles or storage-closet walls and baseboards with residual sprays or dusts of methoxychlor, ronnel, or diazinon. Spraying until the surface is moist is sufficient. Sodium fluosilicate solutions also give good protection but only when applied directly to woolen articles. These solutions can be purchased in stores under various trade names. Whatever treatment is used, it is important to include all articles that may become infested.

Cedar-lined closets or chests high in cedar-oil content protect garments if free of all life stages of the insects when placed in such storage. Placing furs or woolen clothing in cold storage at temperatures of 42 F (5.5 C) or lower, or temperatures of 110 F (43 C) or higher will destroy these insects.

REFERENCES: *Cornell Univ. Memoir* 262, 1944; *USDA Farmers' Bul.* 1655, 1931; *Leaflet* 145, 1938; E–858, 1953; *Home and Garden Bul.* 113, 1971; 96, 1976.

CARPET BEETLES

ORDER COLEOPTERA, FAMILY DERMESTIDAE

Carpet beetles can cause great damage to home furnishings and clothing containing wool, hair, fur, feathers, and other animal substances. They can also subsist on dead insects and food products such as cereals. Four species commonly found in dwellings are the black carpet beetle, *Attagenus megatoma* (Fabr.), the common carpet beetle, *Anthrenus scrophulariae* (L.), the varied carpet beetle, *A. verbasci* (L.), and the furniture carpet beetle, *A. flavipes* LeConte. All species are widely distributed in the world.

None of these insects in any stage of growth has a body length greater than 6 mm, except very large specimens of black carpet beetle larvae, which may be almost 12 mm long. The adults are beetles that are broad or elongate oval. They feign death when disturbed.

The adult black carpet beetle (Fig. 376) is uniformly black with brown legs. The other species have a black-to-brown body color, which is concealed by a dense covering of small scales that form patterns helpful in differentiating the species. Adults of the carpet beetle (Fig. 376) have a dull red band extending down the center of the back. The adults of the furniture carpet beetle are mottled with patches of white, yellow, and black, and are light underneath. The varied carpet beetle adults are the smallest species and resemble furniture carpet beetles but have scales of less brilliance.

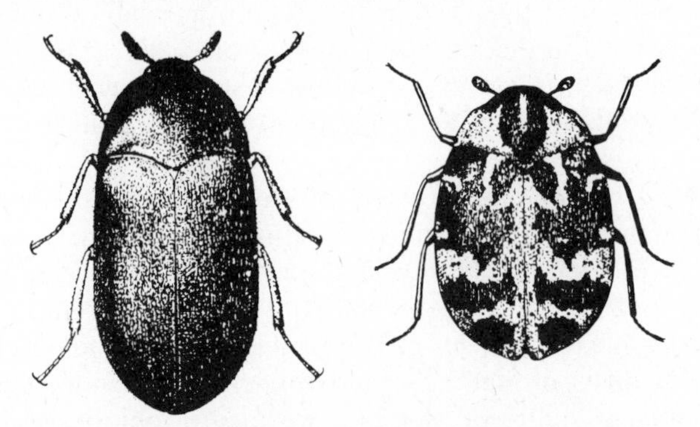

Figure 376. The black carpet beetle, *Attagenus megatoma* (Fabr.) (8×) and carpet beetle, *Anthrenus scrophulariae* (L.) (10×). (Conn. Agr. Exp. Sta.)

Larvae of all species, except the black carpet beetle, are more or less oval in shape, the bodies covered with black, brown, or tawny hairs and three tufts of bristles on each side of the posterior end (Fig. 377). The larvae of black carpet beetles are elongated, golden to chocolate brown, with a tuft of long hairs at the end of the body (Fig. 378).

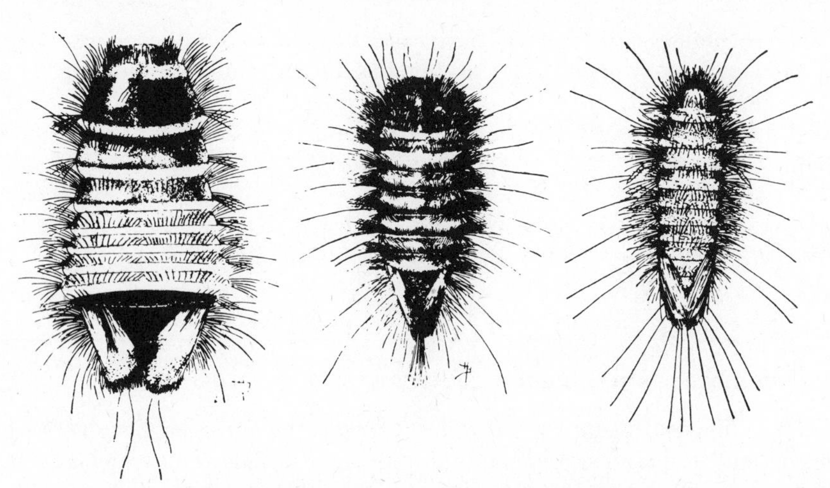

Figure 377. Larvae: of the varied carpet beetle (left), furniture carpet beetle (center), and carpet beetle (right); 6×. (Back, USDA, and Turner and Walden, Conn. Agr. Exp. Sta.)

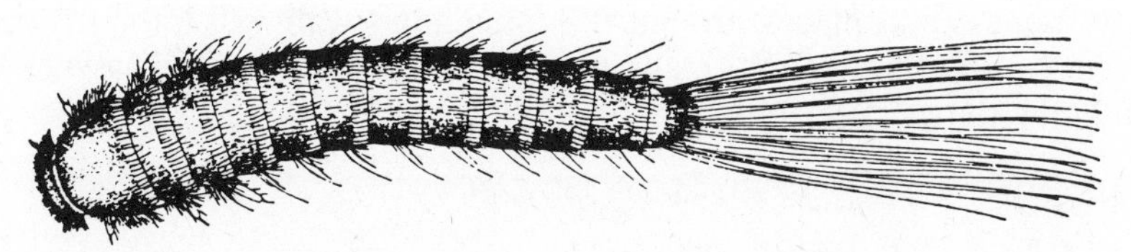

Figure 378. Larva of the black carpet beetle; 5×. (Turner and Walden, Conn. Agr. Exp. Sta.)

Adult beetles fly readily, are attracted to light, and are often found crawling on curtains and windows. They react favorably to sunlight, are readily seen out-of-doors late in the spring feeding on pollen of flowers, and in city areas undoubtedly fly from house to house on warm days. Each female beetle deposits approximately 100 white eggs which hatch normally in 8 to 15 days. Larval growth requires the most time and varies considerably with the temperature and food supply. They molt 6 to 10 or more times, and the old exoskeletons are often seen about infested articles. The pupal stage may last almost 2 weeks. The period from egg to adult is quite varied for each species, requiring 274 to 600 days for the black carpet beetle, 251 to 657 days for the varied carpet beetles, 126 to 422 days for the furniture carpet beetle, and 78 to 439 days for the common carpet beetle. Usually only 1 or 2 generations occur each year.

Methods of control suggested for clothes moths are generally effective against carpet beetles. However, carpet beetles are more difficult to kill and the residual sprays of diazinon, methoxychlor, malathion, or ronnel are strongly recommended, especially for use on and under the edges of rugs, under and on top of rug pads, and around baseboards and floor moldings. Tank-type vacuum cleaners are also suggested for reaching the covers and crevices around floors, baseboards, and moldings to eliminate sources of food and developing life stages.

REFERENCES: *USDA Leaflet* 150, 1938; *Yearbook of Agriculture*, p. 474, 1952; *Home and Garden Bul.* 96, 1976; *Cornell Univ. Memoir* 240, 1941; *Conn. Agr. Exp. Sta. Cir.* 224, 1970.

HOUSE CRICKET

Acheta domesticus (L.), Family Gryllidae

Sometimes house crickets enter dwellings and can be annoying because of their presence and frequent habit of eating holes in clothing and furnishings. This species (Fig. 379) is amber or light brown and active chiefly at night. The overwintering eggs hatch in late spring and adults appear in late summer, with only 1 generation occurring per year. House crickets can be eliminated with diazinon, chlorpyrifos, propoxur, or ronnel. Treating window screens, basement window frames, and places around the foundation wall will kill the crickets before they gain entrance. Normally no control measures are necessary. Most people enjoy the cricket chirping.

REFERENCE: *USDA Home and Garden Bul.* 96, 1976.

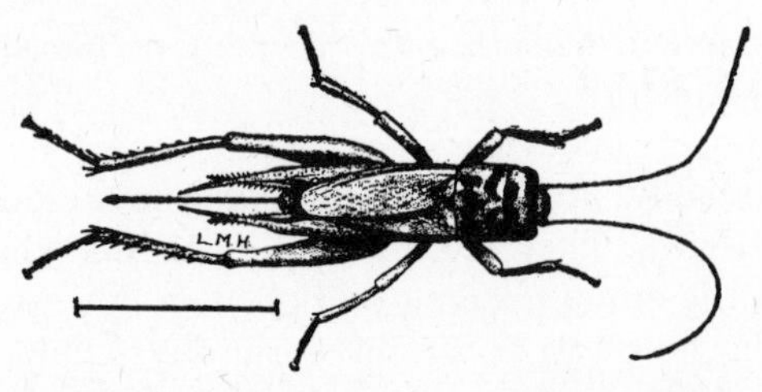

Figure 379. Female house cricket, *Acheta domesticus* (L.); slightly enlarged. (From Lugger)

24

PESTS OF DOMESTIC ANIMALS AND HUMANS

It is estimated that in the United States livestock pests alone cause annual losses amounting to 900 million dollars. Wasted food, damaged hides, and lowered production of milk, meat, eggs, and wool comprise these losses. All animals lose energy and weight fighting off attacks of blood-sucking arthropods. In addition, some species cause great annoyance simply by egg-laying. Others carry and transmit organisms causing disease, some of which are fatal; others result in reduced performance of the animal.

It is difficult to make an accurate estimate of the monetary loss suffered from insects directly detrimental to man, but time lost from work, loss in business at resort and vacation areas, cost of screening homes and buildings, lowered efficiency, medical expense, and similar items would amount to a considerable sum.

With the proper use of many of the more recently developed pesticides it is now possible to greatly reduce these losses. Although much still remains to be learned, progress in combating this group of pests has been rapid in recent years.

LICE ON LIVESTOCK

Lice on domesticated animals are world-wide in distribution. They are of two types, the biting or chewing lice and the sucking lice; on livestock there may be several species of each. Both are wingless, quite small in size, usually 2 to 4 mm long, with well-developed claws on their legs. Biting lice (order Mallophaga) generally have broader heads, and their mandibles are used in feeding on hair, scales of skin, and blood which oozes after the skin has been gnawed. Scabs that form where the skin is broken either by the mouthparts or claws may also be eaten. Sucking lice (order Anoplura) have narrow heads, puncture the skin, and suck up the blood directly in the liquid state. Irritation from these parasites causes the animals to rub and scratch, resulting in partly denuded areas of skin, lowered vitality, and loss in production of meat or milk. Some species of lice are also involved in the transmission of organisms causing disease.

Lice in both groups pass their entire life cycle on the host animals. The eggs are usually attached to hair or feathers, and the young resemble the adults except for their smaller size. Development from egg to adult varies from 2 to 4 weeks depending on the species. Repeat generations occur throughout the year. Spread from one animal to another is through contact. Horses, cattle, sheep, and goats are attacked by both kinds of lice. Only sucking lice attack hogs and man, and only biting lice infest birds, including poultry.

The biting lice on livestock are placed in the family Trichodectidae and the sucking lice in the families Haematopinidae and Linognathidae.

The horse biting louse, *Bovicola equi* (Denny) (Fig. 380), is 2 mm long and red-yellow with faint darker bands. It may become numerous on neglected animals but is rare on horses that receive proper care. A less common species is *Trichodectes pilosus* Giebel.

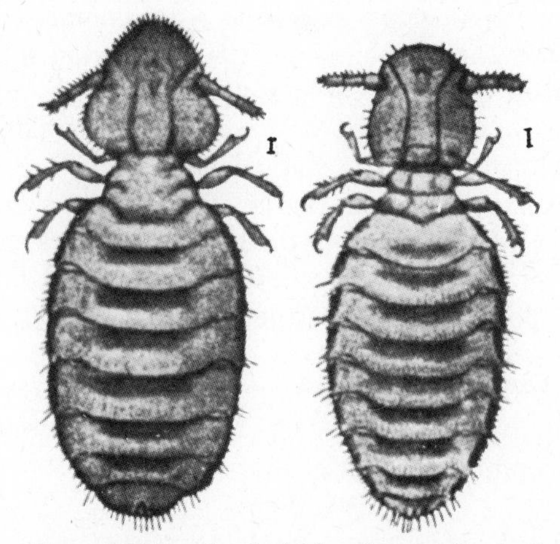

Figure 380. Cattle biting louse, *Bovicola bovis* (L.) (left), and horse biting louse, *Bovicola equi* (Denny) (right); 26×. (Osborn, USDA)

The horse sucking louse, *Haematopinus asini* (L.) (Fig. 381), is considerably larger than the biting species and is gray-brown to almost lead color. It is more likely to infest poorly kept horses on the range than those used regularly, and tends to become more numerous and to injure the animals more than do other species.

The cattle biting louse, *Bovicola bovis* (L.), is probably the most common species attacking cattle, although not so injurious as some of the others. It is

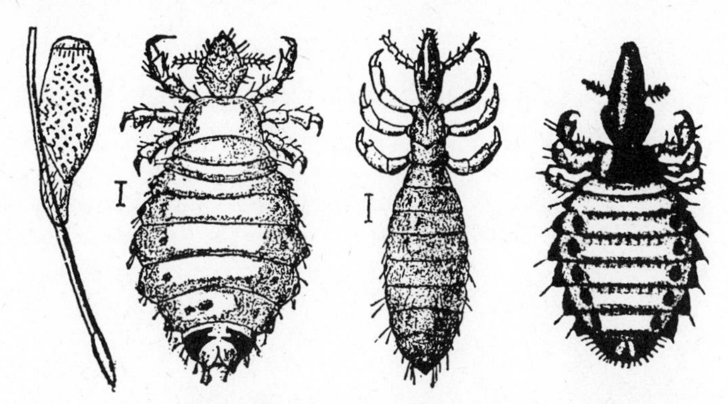

Figure 381. The short-nosed cattle louse, *Haematopinus eurysternus* (Nitzsch), and its egg; long-nosed cattle louse, *Linognathus vituli* (L.); and horse sucking louse, *Haematopinus asini* (L.), named from left to right; 15×. (Osborn.)

distinguished by its red-brown color, which is the basis for the other common name, "little red cattle louse." A small species, 2 mm maximum length, it is considerably flattened with darker pigmented areas on the dorsal side of the abdomen (Fig. 380).

The short-nosed cattle louse, *Haematopinus eurysternus* (Nitzsch) (Fig. 381), is slate gray, larger than the biting louse, with a short, narrow, more pointed head. It is more commonly found on mature animals and is generally more resistant to the usual control measures.

The long-nosed cattle louse, *Linognathus vituli* (L.) (Fig. 381), is blue-black, smaller than the other cattle sucking louse, and much more slender with a longer, narrower head. It is generally more abundant on young stock and dairy cattle. Another sucking louse, *Solenopotes capillatus* Enderlein, is about half as large as the short-nosed species and is quite blue in color, suggesting the common name "little blue cattle louse." The cattle tail louse, *Haematopinus quadripertusus* Fahrenholz, is a sucking louse sometimes found on cattle in the South.

The hog louse, *Haematopinus suis* (L.) (Fig. 382), is the largest of the sucking lice; it is 3 to 4 mm in length, with a rather broad body and prominent head and proboscis. It is the only louse found on hogs and often becomes very abundant.

The sheep biting louse, *Bovicola ovis* (Schrank), is the smallest and the most important species attacking sheep. Other species of lesser importance are the sucking foot louse, *Linognathus pedalis* (Osborn), and the sucking body louse, *Haematopinus ovillus* (Neuman) (Fig. 383).

The angora goat biting louse, *Bovicola limbatus* (Gervais), the goat biting louse, *B. caprae* (Gurlt), and the goat sucking lice, *Linognathus stenopsis* (Burm.), and *L. africanus* Kellogg and Paine, are species found on goats.

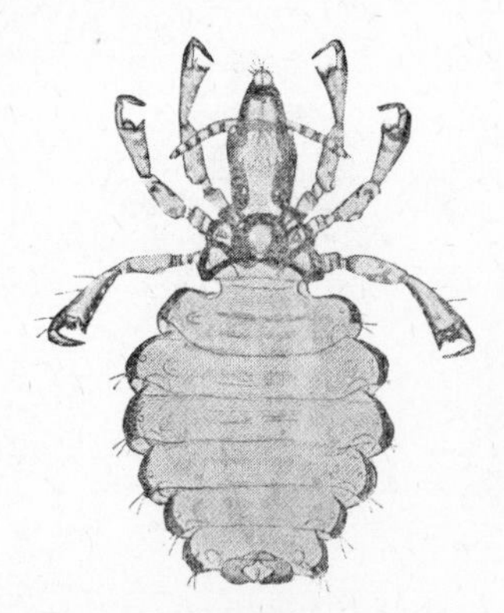

Figure 382. Hog louse, *Haematopinus suis* (L.), female; 10×. (Osborn, USDA)

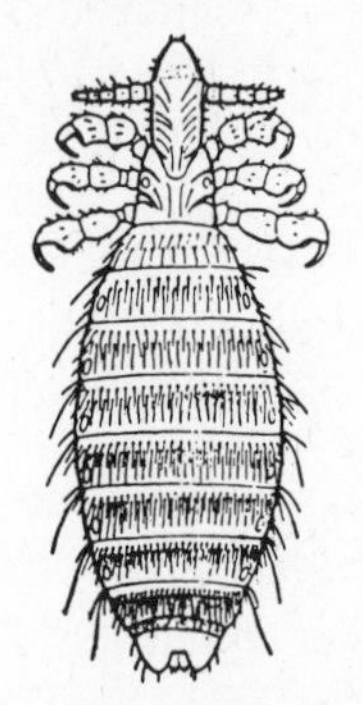

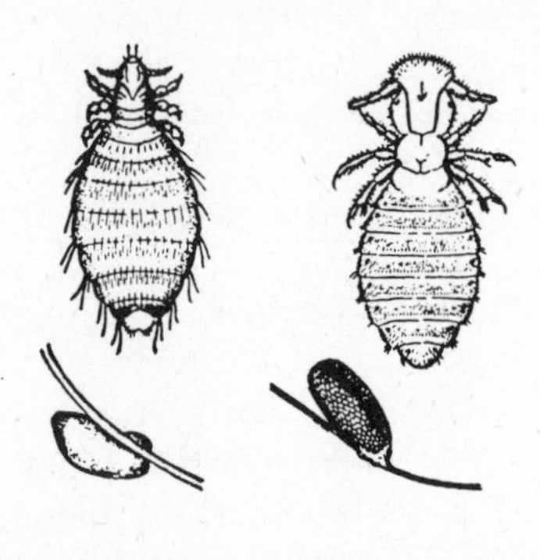

Figure 383. The sucking body louse of sheep, *Haematopinus ovillus* (Neuman); 12×; sheep sucking foot louse, *Linognathus pedalis* (Osborn) 7×; and sheep biting louse, *Bovicola ovis* (L.) 8×; named from left to right. Eggs of the last two species, greatly enlarged, are illustrated. (Mote.)

Common species on dogs are the dog biting louse, *Trichodectes canis* (DeGeer), and the dog sucking louse, *Linognathus setosus* (Olfers).

If all animals were treated 2 or 3 times at 2-week intervals, lice could be eradicated. Materials recommended as powders or sprays on various animals, dairy cattle in production, or beef animals about to be slaughtered are rotenone, synergized pyrethrins, coumaphos, crotoxyphos plus dichlorvos, methoxychlor, ronnel, or tetrachlorvinphos. Nonlactating dairy cattle and beef animals not ready for slaughter may be treated with any of the previously named materials or one of the following as pour-ons applied directly to the backline: crufomate, famphur, fenthion, ronnel, or trichlorfon. Granules of 5% ronnel applied at the rate of ½ pound per 100 square feet of bedding area have proved effective in hog louse control. The ease with which this treatment is made makes it popular. Dust bags or properly operated back-rubbers set up in beef cattle feed lot areas will reduce louse populations. Commercial or homemade devices that enable hogs to oil themselves by rubbing contribute to the control of hog lice. Dipping is also sometimes practiced.

REFERENCES: *Cornell Bul.* 832, 1946; *USDA Bur. Ent. Cir.* E-762, 1951; *Yearbook of Agriculture,* pp. 662–666, 1952; *Farmers' Bul.* 909, 1953; *Leaflet* 456, 1969; *J. Econ. Ent.,* 52:980–981, 1959; 54:821, 1961; 57:42–44, 1964; *Can. Dept. Agr. Pub.* 1006, 1957; *Ohio Sta. Univ. Ext. Bul.* 473, 1979.

HORSE AND DEER FLIES

ORDER DIPTERA, FAMILY TABANIDAE

The familiar blood-sucking flies of this family constitute one of the major problems of the livestock industry in this country, as well as in other parts of the world. Horse and deer flies feed on many warm-blooded animals but cause the greatest annoyance and injury to cattle, horses, and deer. It is estimated that under high fly populations the average daily loss of blood is 100 ml per animal. They also frequently attack man along summer beaches. Besides the lowering of vitality from loss of blood and pain from their bites, these flies are capable of carrying and transmitting the organisms causing such diseases as anaplasmosis, anthrax, surra, and tularemia. Many species have been described, and the dominant forms in one region may not be important in other areas. Some common destructive species are the black horse fly, *Tabanus atratus* Fabricius, the striped horse fly, *T. lineola* Fabricius, *T. sulcifrons* Macquart, *T. quinquevittatus* Wiedemann, *T. nigrovittatus* Macquart, *T. pumilus* Macquart, *T. punctifer* Osten Sacken, *Hybomitra lasiophthalma* (Macquart), and the deer flies: *Chrysops atlanticus* Pechuman, *C. frigidus* Osten Sacken, *C. macquarti* Philip, *C. niger* Macquart, *C. univittatus* Macquart, *C. parvulus* Daecke, and *C. vittatus* Wiedemann.

Horse flies are considerably larger than deer flies, heavybodied and from 1 cm to over 2.5 cm long (Fig. 384). The smaller species are black, brown, or gray and often have brilliant green eyes, sometimes crossed by red golden bands which disappear when the fly is dead. The largest species are brown to black and often slightly striped. Horse flies may be present from June to the middle of September.

Deer flies are slightly larger than house flies, mostly yellow or black with darker stripes on the abdomen and dark markings on the wings. The season for these flies is short, usually 3 or 4 weeks in June or July.

Both horse and deer flies breed in water or in wet soil. Dark-colored masses of eggs are deposited on low-growing plants or other objects near the edge of ponds, lakes, or streams. Hatching occurs in a week or more, and the larvae burrow in the mud, feeding on other insects, tiny crustacea, and aquatic or semiaquatic organisms. The fully grown gray larvae of the larger species may

Figure 384. Adult and larva of the black horse fly, *Tabanus atratus* Fabricius; 2×. (Ky. Agr. Exp. Sta.)

reach a length of almost 5 cm and have darker-colored bands (Fig. 384). Development from egg to adult may require only a few months, or 2 years, depending on the species.

No satisfactory methods have been developed for control of horse flies and deer flies. It is impractical in most regions to eliminate the breeding places. Impoundment of marsh areas coupled with trapping reduces some deer fly populations. Prevention of attack by providing daytime shelters for animals and allowing them to pasture at night is of value since the flies do not bite at night. Sprays containing 0.1% pyrethrins and 1.0% piperonyl butoxide give repellency protection of one to three days. Chemicals applied for controlling other species of flies attacking livestock offer some relief from horse and deer fly annoyance.

REFERENCES: *Ark. Bul.* 332, 1936; *USDA Misc. Pub.* 305, 1938; *Yearbook of Agriculture*, pp. 659–660, 1952; *Agr. Handbook* 290, 1965; *Tech. Bul.* 1295, 1964; *Ann. Ent. Soc. Amer.* 62:1429–1433, 1969; 63:27–31, 1970.

STABLE FLY

Stomoxys calcitrans (L.), Family Muscidae

Found throughout the United States and most of the world, stable flies commonly attack mules, horses, cattle, hogs, dogs, cats, sheep, goats, and man.

In some parts of the country, stable flies are known as dog flies because they often attack dogs viciously, especially around the ears. The adults feed only on blood, and the irritation from their puncturing of the skin along with loss of blood results in nervous animals. Constantly fighting the flies is also fatiguing. During severe outbreaks beef animals make very little weight gains, and milk production of dairy cows may be reduced 10 to 50%.

The adult is almost identical to the house fly, but it can be readily distinguished from the latter by the piercing-sucking beak which projects forward (Fig. 385). The house fly mouthparts are sucking or sponging and project downward from the head.

Breeding takes place in wet straw, manure, vegetable and fruit refuse, peanut litter, and marine grass windrows along the coast. The tiny, cream white eggs are laid in these moist habitats; hatching occurs in 2 to 3 days. The larval period lasts 12 to 30 days or longer, depending on the prevailing weather. Development from egg to adult requires from 3 to 6 weeks or more. In northern climates the overwintering stage is larvae or pupae; in southern areas breeding is continuous.

Control may be accomplished through one, or a combination, of the following methods: destruction of breeding places; application of residual pesticides to buildings, sheds, corrals, and other places where the flies rest; and application of pesticides to the animals or breeding areas.

Where manure or other breeding material cannot be spread often on the fields, treating with emulsions of diazinon, dimethoate, tetrachlorvinphos, dichlorvos, ronnel, or malathion will kill the larvae and prevent reproduction until the chemicals are dissipated. Do not use diazinon in poultry houses or dairy barns. Residual wall sprays containing dimethoate, tetrachlorvinphos, ronnel, malathion, fenthion, dichlorvos, or methoxychlor may be applied inside dairy barns, but only synergized pyrethrins or dichlorvos resin strips are approved for use inside milk processing rooms. Remove animals before spraying

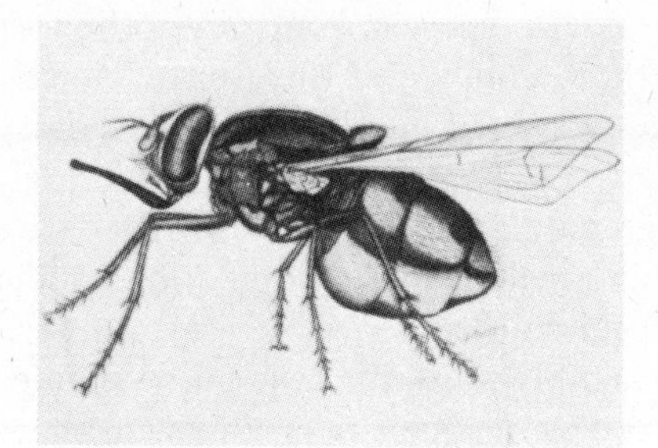

Figure 385. The stable fly, *Stomoxys calcitrans* (L.), adult female; 5×. (USDA)

and avoid contamination of feed troughs, water fountains, and milking equipment. Spray until the surfaces are wet to the point of run-off.

Space sprays, aerosols, or barn-fogging devices containing synergized pyrethrins or allethrin are useful in temporarily reducing fly populations in dairy barns.

Stable flies stay on animals only long enough to get a blood meal; therefore spraying the animals is less effective than spraying the premises. It is, however, a desirable additional control measure during severe outbreaks. Dairy cows are the animals most often treated but beef cattle, horses, and mules should not be neglected if high populations of flies prevail. Oil-base mist-type sprays containing crotoxyphos, dichlovos, or synergized pyrethrins or allethrin are recommended. Treat only the outer hair coat of the animal. Treadle sprayers are useful for such treatments.

REFERENCES: *USDA Leaflet* 338, 1953; *Agr. Handbook* 290, 1965; *Conn. Agr. Exp. Sta. Bul.* 650, 1962.

The false stable fly, *Muscina stabulans* (Fallen), is often mistaken for the stable fly. It resembles both the stable fly and the house fly but lacks the piercing-sucking mouthparts of the stable fly and is a trifle larger than the house fly. Its abdominal region is gray rather than yellow. At times this insect makes up a high percentage of the fly population in a given area, but it is usually less important than either of the species it so closely resembles. The false stable fly is also similar to the stable fly in breeding habits and is controlled in the same way.

HORN FLY

Haematobia irritans (L.), Family Muscidae

Introduced into North America from Europe about 1890, horn flies have since spread over the entire country. The blood-sucking adults feed chiefly on cattle but may attack sheep, goats, horses, hogs, and other animals. Most of their adult life is spent on the animals and their bites cause extreme annoyance, resulting in 10 to 20% reduction of milk production in dairy cows and reduction of weight gains of beef cattle by as much as ½ pound per day. On cattle, they have the habit of congregating about the ears, the base of the horns, on the flanks, withers, back, and belly. Higher populations occur on black cattle and lower populations on Brahman cattle.

The horn fly resembles the house fly; it is more slender and only about half as large (Fig. 386). Its proboscis or beak is similar to that of the stable fly. The eggs, laid only on fresh animal feces, hatch in a few days, and the larvae feed and develop, changing to puparia in the soil or feces, with adults emerging soon afterwards. About 2 weeks are required for development from egg to adult.

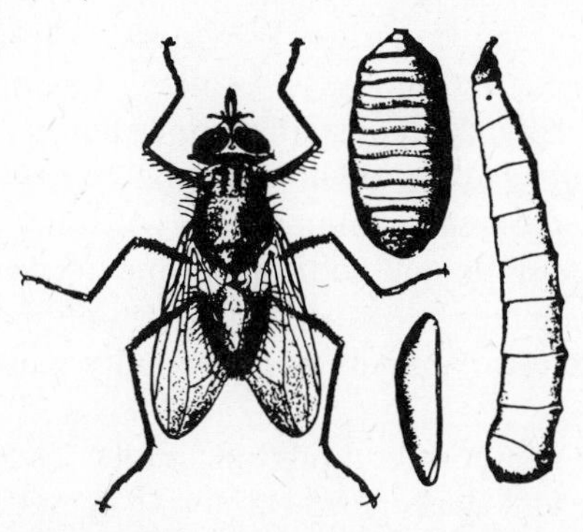

Figure 386. Egg, larva, puparium, and adult of the horn fly; 5×. (USDA)

The most effective way to rid animals of horn flies is to apply a pesticide. Approved chemicals for dairy cows in production are sprays containing crotoxyphos, dichlorvos, or synergized pyrethrins. Daily applications are usually necessary; treadle sprayers are helpful.

Good control has resulted by applying powders of methoxychlor or malathion to dairy or beef animals every 2 or 3 weeks. The recommended dosage is 1 heaping tablespoonful (10 grams) of 50% methoxychlor WP or 3 tablespoonfuls (1½ oz.) of 5% malathion powder, applied to the back of each animal and rubbed into the hair.

Commercially prepared or homemade dust bags or backrubbers are useful; they enable the animals to treat themselves. Approved chemicals for them are coumaphos, crotoxyphos, dichlorvos, methoxychlor, and stirofos. A homemade backrubber consists of a cable (chain, wire rope, or several strands of heavy wire), suspended from two 4-foot fence posts set 16 feet apart, so as to sag 18 inches from the ground. Or the cable may be fastened to the top of a 4-foot post and anchored at ground level 15 to 20 feet away. The cable is carefully wrapped with burlap bags which overlap one another and are tied securely with twine. The burlap is soaked with 1 gallon of the appropriate chemical; this is replenished at 1- to 2-week intervals.

REFERENCES: *USDA Leaflet* 291, 1950; 388, 1965; *Misc. Pub.* 1278, 1974; *J. Econ. Ent.*, 45:121–122; 329–334, 1952; 57:371–372, 1964.

FACE FLY

Musca autumnalis DeGeer, Family Muscidae

This native European insect was first found in North America at Middleton, Nova Scotia, Canada, in 1952 and has since spread throughout most of the

United States and adjoining Canadian Provinces. The fly habitually clusters around the eyes, nostrils, and mouths of cattle, sucking up the secretions in those areas as well as salivary deposits left on other parts of the body caused by animals licking themselves or others. During epidemic outbreaks over 100 flies have been observed on each animal. This constant annoyance is reflected in lowered production of milk and additional time necessary to finish beef animals for slaughter. These flies may also be carriers of pink eye and eye worms. Occasionally other domestic animals in the vicinity may be disturbed by face flies.

The insects pass the winter in buildings as adults. They resemble house flies but are slightly larger and darker. Warm spring temperatures activate those flies that survive the winter; they seek fresh cattle dung where feeding and oviposition occur. The tiny pale yellow eggs, which have a gray-black stalk projecting at one end, are laid in groups of 5 to 8, each female depositing approximately 230 in her life span. Hatching occurs in 1 day. The white maggots feed in the dung and pass through 3 instars before changing to white puparia in the soil. In another 10 days new adults begin to emerge, starting the cycle anew. This is continued with repeated generations throughout the summer. Although flies overwinter in buildings they avoid entering them in the summer months.

Natural enemies include a braconid, *Aphaereta pallipes* (Say) and a figitid, *Xyalophora* spp., both of which parasitize puparia; an anthomyiid predator, *Scatophaga stercoaria* (L.); and several species of rove beetles.

Fly sprays, dust bags, and backrubbers as outlined for controlling stable and horn flies give some relief from face fly attacks.

REFERENCES: *Can. Entom.* 85:422–423, 1953; 92:360–365, 1960; 101:561–576, 1969; *J. Econ. Ent.*, 53:450–451, 1960; 54:1147–1151, 1961; 57:631–636, 1964; 62:255–256, 1969; 65:1636–1638, 1972; *Proc. N.C. Branch Ent. Soc. Amer.* 17:135–136, 1962; *Ann. Ent. Soc. Amer.* 57:563–569, 1964.

HORSE BOT FLIES

ORDER DIPTERA, FAMILY GASTEROPHILIDAE

Three species of bot flies universally attack horses, mules, and donkeys. They are the common horse bot fly, *Gasterophilus intestinalis* (DeGeer), the nose bot fly, *G. haemorrhoidalis* (L.), and the throat bot fly, *G. nasalis* (L.).

Horses become nervous and excited during the period adults are flying and ovipositing, and seem to instinctively fear them. Adult flies are present when the animals jerk their heads, stamp their feet, stand in pastures with their throats over each other's backs, rub their noses and lips on objects, or run as if being chased. The larvae or bots infest the alimentary tract causing diges-

Figure 387. The horse bot fly; 3×. (Froggatt.)

tive disturbance, irritation to the lining, and obstruction of the normal passage of food materials.

Adult flies of all three species are much alike in appearance and resemble honey bees in size and coloration, but have only 1 pair of wings and a curved-under abdomen (Fig. 387). In the northern states they appear early in June and are present until the first killing frost in the fall. However, each individual may live a period of only a few days to possibly as long as 3 weeks. Egg-laying starts soon after the adults emerge and continues throughout the summer, some eggs remaining viable for almost 3 months. Females can lay nearly 500 eggs during their life span. Larvae (Figs. 388 and 389) occur in the stomach throughout the year but are most abundant during late fall and winter. Only 1 generation develops each year.

Eggs of the common bot fly (Fig. 390) are yellow and deposited principally on the fore legs but are also found on the mane, belly, and hind legs. After an

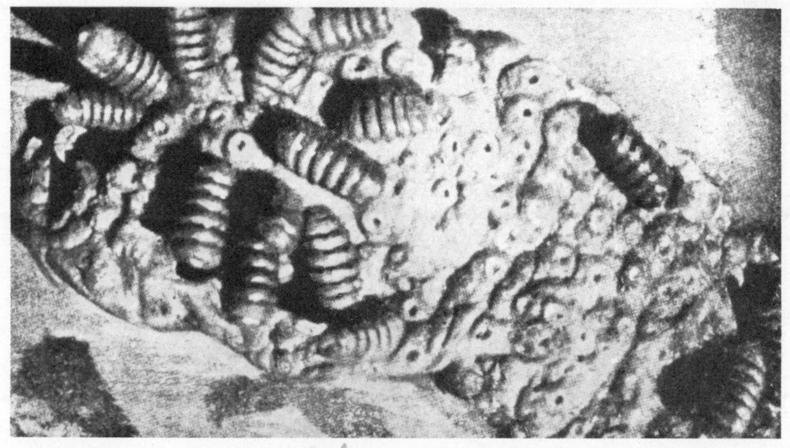

Figure 388. Larvae of bot flies attached to the wall of the stomach of a horse; slightly enlarged. (Osborn, USDA)

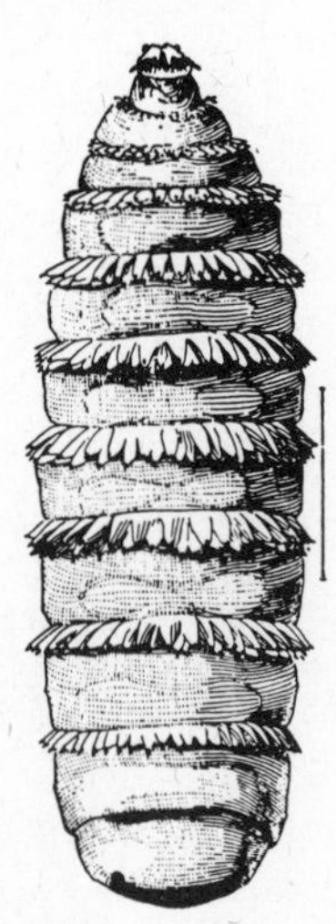

Figure 389. Larva of the horse bot fly; 4.5×. (Froggatt.)

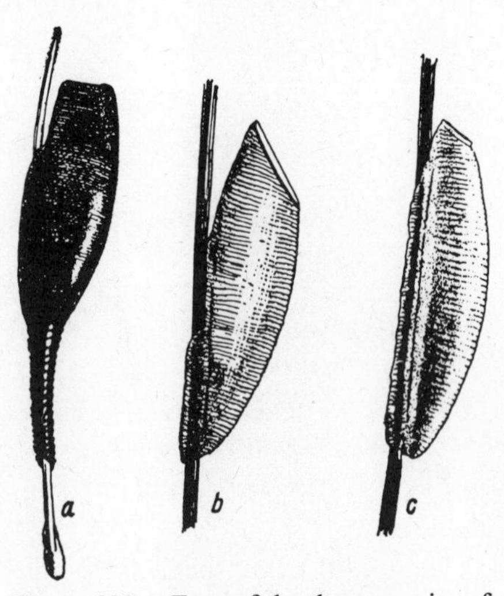

Figure 390. Eggs of the three species of bot flies attacking horses: *a, G. haemorrhoidalis* (L.); *b, intestinalis* DeGeer; *c, G. nasalis* (L.); 35×. (Dove, USDA)

incubation period of approximately a week or more they hatch immediately upon exposure to the warmth and moisture of the horse's lips in nibbling or scratching the legs. In this way the larvae get into the mouth where they burrow into and along the upper surface of the tongue. After 3 or 4 weeks they leave the tongue and pass into the stomach, where they complete their growth in 9 or 10 months while attached to the lining (Fig. 388), then move out of the body with the feces. Pupation takes place in the soil; the period lasts about a month or more.

Throat bot fly eggs are yellow (Fig. 390) and are laid on the hairs of the chin, jaws, and throat. These hatch in 5 days and the tiny larvae crawl to the mouth where they are found principally around the gums of the molar teeth. After 3 or 4 weeks they pass through the stomach into the duodenum where they remain attached until the following spring, then move out of the body with the feces, and pupate in the soil. The pupal period lasts about 44 days.

Black eggs of the nose bot fly (Fig. 390) are laid on the short hairs along the edges of the lips; hatching takes place in about 3 days. The tiny larvae burrow into the lining of the lips and mouth for a few weeks causing great irritation, then move to the stomach where they remain attached until spring. This species has the habit of attaching itself to the walls of the rectum and anus before leaving the body with the feces. Pupation occurs in the soil, the period varying from 21 to 68 days.

Horses may suffer from bot infestation during all periods of the year because many of the grown larvae do not pass from the intestinal tract before the new ones begin to appear.

Heavy frosts in autumn kill the adults; after a period of 3 or 4 weeks, the young throat and nose bots leave the tissues of the mouth, and no unhatched eggs of these species remain. However, many eggs of the common bot fly remain unhatched but viable. These eggs can be removed by clipping. A faster method of eliminating them is to vigorously apply warm water (115–120 F; 46–49 C) with a rag or sponge. Hatching takes place instantly, and the tiny larvae die from exposure. Research indicates that it is not necessary to add 2% phenol or coal-tar creosote to the water, as was done formerly. Warming the eggs with water should be done 2 weeks after frost has killed all adults. Four weeks later, or about December in northern regions, there should be community-wide internal treatment of all animals. Farther south bot fly activity is extended, and the most practical procedure is to treat in September and again in January. Commonly used chemicals are carbon disulfide, piperazine, trichlorfon, and dichlorvos. Follow directions and precautions on the package. Community-wide control programs are worthwhile efforts.

Keeping animals in the stable during the day and pasturing them at night will avoid infestation. Mechanical protectors offer some relief from nose bot fly annoyance. Sponging the nose and throat area or spraying with repellent-type fly sprays containing pyrethrins and piperonyl butoxide is of some value in control of nose and throat bot flies.

REFERENCES: *USDA Leaflet* 450, 1976; *Can. Dept. Agr. Pub.* 604, 1938.

CATTLE GRUBS OR OX WARBLES

ORDER DIPTERA, FAMILY OESTRIDAE

One of the most important insect problems with which the owner of cattle has to deal is the control of cattle grubs or heel flies. The losses to the livestock industry have been estimated at $160 million annually. These losses occur in several ways. Cattle instinctively fear the adult flies during egg-laying. They run wildly, sometimes injuring themselves in attempting to get away. The result is lower milk flow, poor weight gains, or even loss in body weight. Skin penetration causes irritation to the animal and damage to the flesh, which must be trimmed out in slaughtered animals. Perforation of the skin on the back means a reduction in the value of the hides as leather.

Two species are important, the common cattle grub, *Hypoderma lineatum* (de Villers), which is found throughout the United States, in Canada and Mexico, and the northern cattle grub, *H. bovis* (L.), which is found in southern Canada and the northern three-fourths of the United States. The northern cattle grub is a trifle larger in the adult stage and differs in some details of habits and

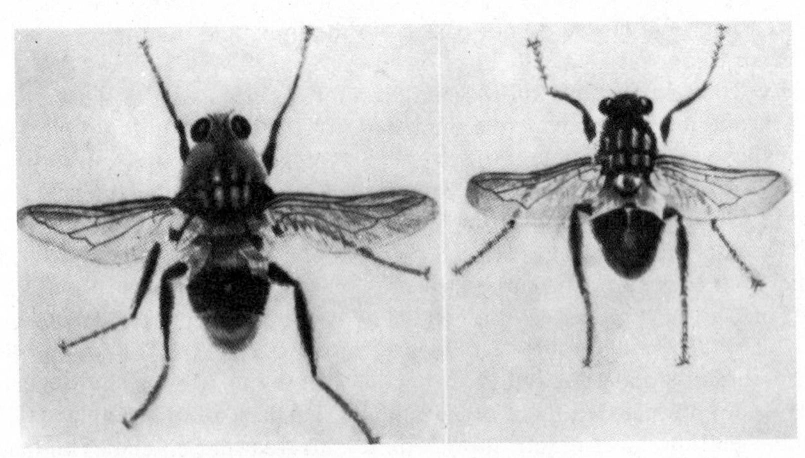

Figure 391. Adults of northern cattle grub, *Hypoderma bovis* (L.), and common cattle grub, *H. lineatum* (de Villers); (L. to R.) 2×. (Mote, Ohio Agr. Exp. Sta.)

developmental history, but both species are essentially the same in the damage they do.

Adults of the grubs are hairy, yellow and black flies about the size of the honey bee (Fig. 391). The more common species fastens its yellow white eggs to the hairs of the legs, usually on the heels, arranged in rows of several on a hair. The northern species attaches only one egg to a hair. These eggs are not often seen because they are attached close to the skin and sometimes hidden by adjacent hairs. Egg-laying ordinarily takes place only on sunny days; each female is capable of depositing about 500 eggs during a life span of a week or more.

Hatching occurs in a few days, and the spiny larvae (Fig. 392) penetrate the skin and migrate upward between the muscles to the abdominal and chest cavities. During the next 6 to 8 months they burrow about over the surface of the

Figure 392. Larvae of the cattle grub or ox warble; slightly enlarged. (Bishopp et al., USDA)

Figure 393. A cattle grub larva in position under the skin of the animal. (USDA)

paunch, intestines, spleen, and other organs. Larvae of the common species are especially numerous between the muscular and mucous layers of the gullet. In these situations the larvae are slender and range in length from 0.5 to 25 mm. In the fall, winter, and spring they migrate through the muscle tissues of the back, sometimes in the spinal canal (northern species), eventually reaching the permanent locations under the skin on the back of the animal (Fig. 393). Here the larva cuts a small hole in the skin, which it enlarges as growth proceeds until finally, after a period of 35 to 90 days, full development is reached; it squeezes through the opening, falls to the ground, burrows in, and pupates. The northern cattle grub develops in the same way but requires a period of 50 to 100 days to complete its growth and consequently appears later as lumps under the skin. The pupal period lasts 18 to 77 days for the common species and 15 to 25 days for the northern species. Adults are usually present from May to July and begin to lay eggs about an hour after emerging from the pupae. There is only 1 generation per year for both species (Fig. 394).

Control of grubs consists of treating the backs of all infested animals on a community-wide basis during the period the grubs are in that region (and before they emerge) with derris or cubé powder as a dust, wash, spray, or ointment. Three or 4 treatments are recommended at intervals of 3 or 4 weeks, starting around 3 weeks after the first holes are detected on the back. Dusts should contain 1½% rotenone. Apply the dust from a can with a perforated top, and rub it in thoroughly. A wash consisting of 12 ounces of derris or cubé powder mixed in a gallon of water containing 2 ounces of neutral soap powder, can be applied from a quart jar with a perforated lid and worked into the grub holes with a stiff bristled brush. An ointment consisting of 1 part of derris or cubé powder and 10 parts of petrolatum has given excellent results if applied directly in each grub hole. High-pressure sprays applied directly to the backs of the animals is a more rapid method of treating large numbers. Add 8 pounds of derris or cubé powder (containing 5% rotenone) to each 100 gallons of water. With the sprayer adjusted to deliver 5 gallons per minute, one person can treat the back of an animal every 10 seconds.

Rotenone is the only pesticide recommended for grub control in lactating dairy cows in some countries. However, beef and nonlactating animals can also be treated with the new systemic pesticides, coumaphos, ronnel, crufom-

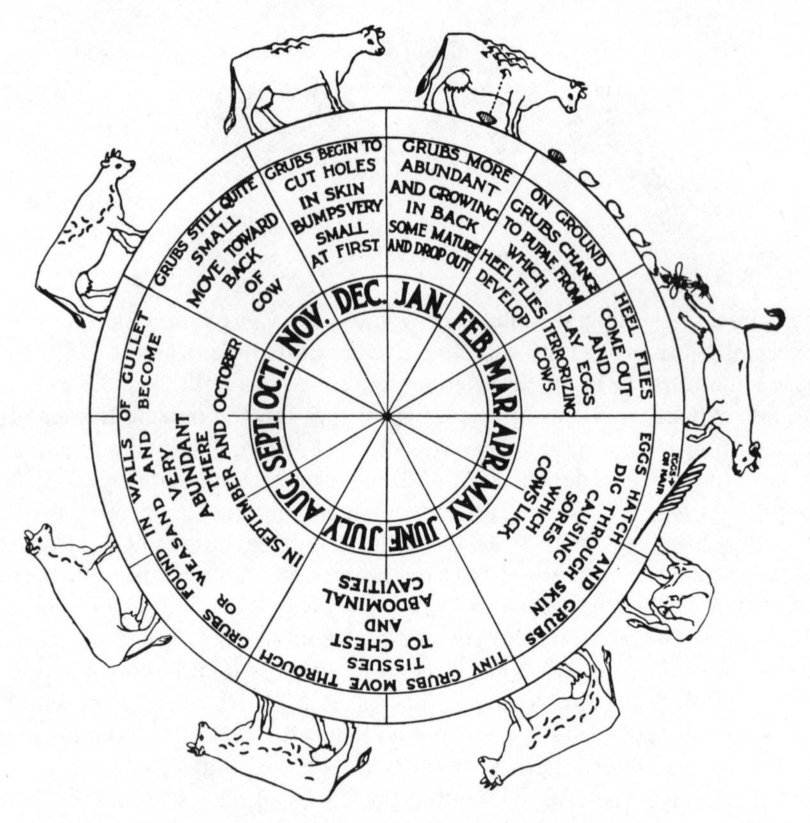

Figure 394. Diagram of the life cycle of the common cattle grub in southern United States. (USDA)

ate, famphur, fenthion, or trichlorfon. Eradication of grubs has resulted in some countries where these chemicals were widely used.

Infestation may be avoided by keeping cattle in the stable during the day. Hand extraction with forceps or squeezing has proved effective, especially for animals with elastic skin such as the Jerseys and Guernseys. Holsteins have thick, tough hides, and it is difficult to remove the grubs by squeezing. Care must be exercised to avoid crushing the grubs within the skin because their body contents may cause anaphylactic shock. Some cattlemen claim that they have lighter infestations when all cattle are sprayed with residual-type fly sprays, especially on the legs and body, during the period that adults are present and eggs are hatching.

REFERENCES: *J. Econ. Ent.*, 41:783–787, 1948; *USDA Leaflet* 527, 1972; *Ann. Rep. Ent. Soc. Ont.*, 76–80, 1950; *Ohio Ext. Bul.* 473, 1979; *Ann. Ent. Soc. Amer.* 63:1465–66, 1970.

SCREWWORM

Cochliomyia hominivorax (Coq.), Family Calliphoridae

Screwworms have been known in Texas since about 1842 and are still found in that state; also in New Mexico, Arizona, Central and South America, and nearby islands. All other infestations in the United States have been eradicated by releasing laboratory reared sterile flies in those areas. In past years, during the summer, this pest has been found as far north as South Dakota, Kentucky, Virginia, and New Jersey. This is due primarily to shipments of infested animals into these areas. Such infestations are eliminated with the coming of cold winter weather.

Any warm-blooded animal is subject to screwworm attack. Infestations have been found on all kinds of wild and domestic animals, as well as on man. Before an animal can become infested a break must occur in its body surface. Any cut or scratch, dehorning or castration operation, fly or tick bites, the navel opening of new-born animals, or a diseased condition of the skin may attract the adult flies. They are blue-green with three dark stripes on the thorax (Fig. 395) and, in size, are almost identical to the common blow flies.

Each female can lay 3000 eggs during her life span, depositing them in masses of 200 to 400 at 4-day intervals. They are securely cemented together and placed along the edges of the wounds. Hatching occurs within 12 hours; the white tapered maggots feed only on living flesh within the wound, becoming fully grown in 5 days. They crowd together with their heads pointed inward. Infested wounds develop a bloody discharge and a disagreeable odor which attract more egg-laying females; maggots increase rapidly, and the animal dies if nothing is done to check them.

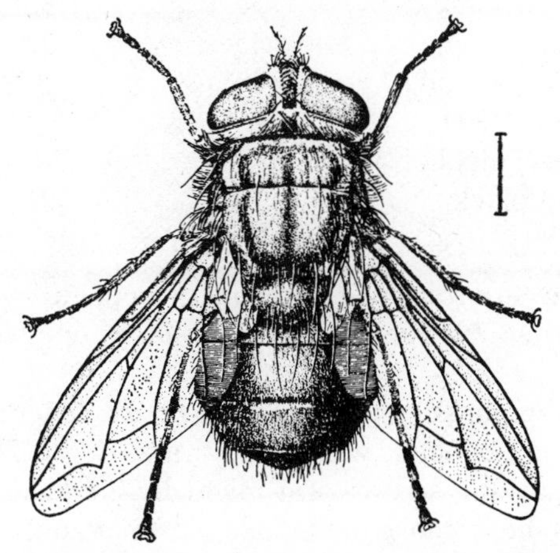

Figure 395. The screwworm fly, *Cochliomyia hominovorax* (Coq.); 5×. (USDA)

Mature maggots leave the wound, fall to the ground, enter the soil, and change to puparia. Adult flies normally begin emerging in one or 2 weeks, but this may be prolonged to almost 2 months by cool weather. A few days after emerging the flies mate once and the females begin laying eggs on wounded animals. There may be 6 to 10 generations per year depending on the latitude. If average daily temperatures lower than 54 F (12 C) prevail for 2 months or more, the pupae die in the soil.

Injury from this insect may be lessened by measures which help to prevent the wounds through which entrance is gained. For example, there should be careful livestock handling, proper management so that the young are born out of fly season, rounding off sharp horns of animals, controlling ticks and flies, prompt treatment of the dam and the navel of new-born animals, limiting branding, castration, lamb docking, and dehorning to the winter period. During the fly season all livestock in a community should be examined carefully twice each week. Any wound found, whether infested or not, must be promptly treated with a good screwworm remedy. Those recommended are EQ 335 or smears 62 and 82, preference being in the order named. EQ 335 contains (in percentages by weight) lindane, 3; pine oil, 35; mineral oil, 40 to 44; emulsifier, 8 to 12; silica aerogel, 8 to 12. This mixture does not stain and is not highly volatile. It is best applied to the entire wound area, both inside and out, with a 1-inch paint brush. Treatments should be made at 7-day intervals until the wound is healed.

Smears 62 and 82 are recommended if EQ 335 is not available. They contain the following ingredients in per cent by weight:

	Smear 62	**Smear 82**
Diphenylamine	35	35
Benzol	35	32
Turkey red oil	10	—
Triton X-300	—	2
n-Butyl alcohol	—	10
Lamp black	20	21

Additional remedies are ronnel or coumaphos applied once or twice a week until the wound is healed. Neither chemical should be applied to lactating dairy cows.

The eradication program is being continued by the federal government, and if it is done extensively enough, with Central and South American countries cooperating, eventually the screwworm might be eliminated. Success of the program is thought to be because the female flies mate only once (a laboratory observation). Releasing large numbers of gamma radiated puparia reared in the laboratory results in sterile females and males. Sterile females lay infertile eggs as well as fertile females, in the area, that mate with infertile males. So far eradication is not as easy as originally thought.

REFERENCES: *Fla. Bul.* 86, 1936; *USDA* E-813, 1951; *Agr. Handbook* 290, 1965; *Yearbook of Agriculture,* pp. 666–672, 1952; *J. Econ. Ent.,* 46:648–656, 1953; 53:1110–1116, 1960; 55:826–827, 1962; 57:324–325, 1964; *Bul. Ent. Soc. Amer.* 21:23–26, 1975.

BLOW FLIES

ORDER DIPTERA, FAMILY CALLIPHORIDAE

Some blow fly species may infest man, cattle, sheep, dogs, and other animals causing myiasis. Larvae of such flies are very similar to the screwworm larvae except that they feed on dead as well as living animals. Common species are the black blow fly, *Phormia regina* (Meigen), the secondary screwworm, *Cochliomyia macellaria* (F.), and the greenbottle fly, *Phaenicia sericata* (Meigen).

Sheep are most often attacked; the insects are then called fleece worms or wool maggots. The flies lay eggs in the wool, soiled by urine, feces, or warm spring rains, or combinations of these, and the maggots feed in these regions, frequently invading healthy tissues, which become inflamed and often infected with other organisms that result in blood poisoning and death.

Practices designed to prevent attacks of blow flies are essentially the same as those employed in screwworm control. EQ 335 is recommended for infested animals when diluted 1 part with 9 parts of water. Saturate the affected area thoroughly with the emulsion. Dioxathion and diazinon are recommended for fleeceworm control in addition to the other remedies given under screwworms.

REFERENCE: *VPI and State Univ. Res. Bul.* 123, 1977.

SHEEP BOT FLY

Oestrus ovis L., Family Oestridae

In the larval stage this insect is also called the sheep grub or head maggot. The adult is a yellow, hairy, beelike fly (Fig. 396), nearly 8 mm long. It is active most of the year in southern latitudes but only for 3 or 4 months in its northern range. During this period the eggs hatch within the body of the female, and she in turn deposits the larvae in the nostrils of sheep and goats. The maggots make their way through the nasal passages and invade the frontal sinuses, where the pressure produced as they increase in size may affect the semicircular canals, thus causing staggers or gid. It is said that they may penetrate the thin septa and reach the brain; death of the animal then results. Full-grown maggots often exceed 25 mm in length and are heavy-bodied (Fig. 397). Larval

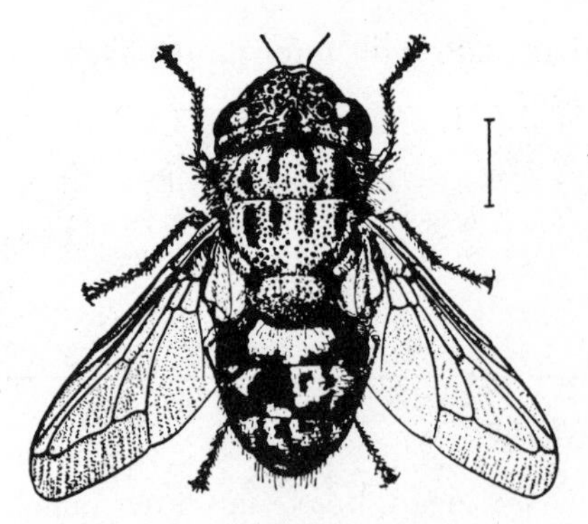

Figure 396. Adult female of the sheep grub, *Oestrus ovis* L.; 5×. (USDA)

Figure 397. Larvae of the sheep grub, *Oestrus ovis* L.; 0.5×. (Courtesy of F. R. Koutz.)

development may take from 4 to 10 months, after which they leave the host, change to puparia in the soil, with adult emergence 4 to 6 weeks later.

The flies are active only in the daytime. Their presence can be detected by the sheep crowding together with their heads held low; this is accompanied by stamping of the feet and shaking of the head. Maggot-infested sheep often sneeze and have a mucous discharge from the nostrils.

Darkened shelters for the sheep during the months when the flies are active will afford protection from infestation. Treating the nostrils of the sheep with pine tar is of value in deterring the flies from larvipositing. Placing salt in holes bored into a log and smearing the edges of the holes with pine tar enables the sheep to renew the repellent themselves. Once the grubs have become established in the animal, injection of 2 fluid ounces of 3% saponified cresol into each nostril has given good control. It is suggested that this treatment be made in late autumn by a veterinarian.

Crufomate, coumaphos, dichlorvos, dimethoate, and ronnel show promise of controlling this pest more efficiently.

REFERENCES: *J. Am. Vet. Med. Assoc.* 97:565–570, 1940; *Vet. Med.* 54:377–383, 1959; *64th Proc. U.S. Livestock Sanitary Assoc.* 178–186, 1960; *J. Econ. Ent.*, 56:530–531, 1963.

SHEEP KED

Melophagus ovinus (L.), Family Hippoboscidae

The sheep ked, also called sheep "tick," is not a true tick but a wingless, red-brown fly (Fig. 398). It is found throughout the year in the adult stage, feeding on the blood of sheep and goats with its piercing-sucking mouthparts. In addition to loss of blood their feeding causes a sheepskin leather defect called "cockle" which lowers the value of the hides. The larval stages develop within

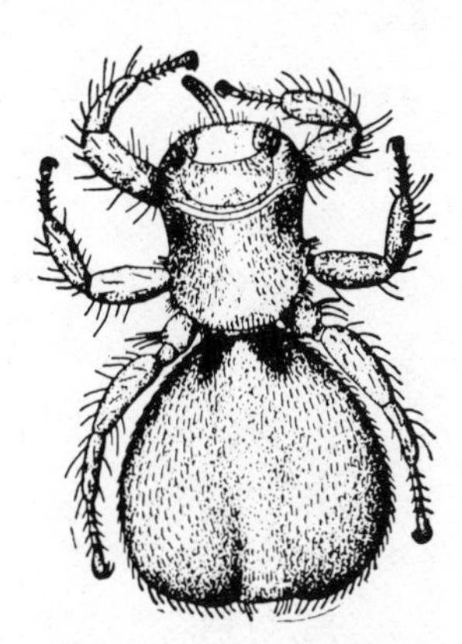

Figure 398. The sheep ked *Melophagus ovinus* (L.); 7×. (Ky. Agr. Exp. Sta.)

the uterus of adult female keds until fully grown; they are then deposited on the wool of the sheep and in a few hours transform to puparia, which are light in color at first but change to dark brown in several hours. These remain glued to the wool, and after a period of 3 weeks during summer weather and a longer period in cold weather the adults emerge, mate, and begin the cycle anew. Each female may give birth to 10 to 15 larvae during her life span of 100 or more days. There are several generations per year and populations are higher in winter.

These insects are often more numerous on lambs and it is important that ewes be freed of them before lambing time. Control is accomplished by dipping, spraying, or dusting. Excellent control has resulted from one treatment in a dip, consisting of 8 ounces of derris or cubé powder (containing 5% rotenone) in 100 gallons of water, to which has been added 2 tablespoonfuls of household detergent. This formula can be sprayed with equally good results. A dust containing 1% rotenone also controls keds. Other approved control chemicals are dioxathion, coumaphos, malathion, ronnel, methoxychlor, and toxaphene. Sheep should be dipped as soon as the shearing cuts have had an opportunity to heal. If keds are discovered too late in the season for dipping or spraying, the small flock owner will find it profitable to give individual dust treatments. Once the flock is ked free keep it so by treating all new additions to it because spread of the pest is by contact with infested animals.

REFERENCES: *Cornell Univ. Bul.* 844, 1948; *USDA Farmers' Bul.* 2057, 1953; *Agr. Handbook* 290, 1965; *Wyom. Agr. Exp. Sta. Bul.* 327, 1953.

SCAB AND MANGE MITES

ORDER ACARI

Scab and mange of domestic animals are caused by various parasitic mites. Different varieties of the scab mite, *Psoroptes equi* (Raspail), and *P. ovis*

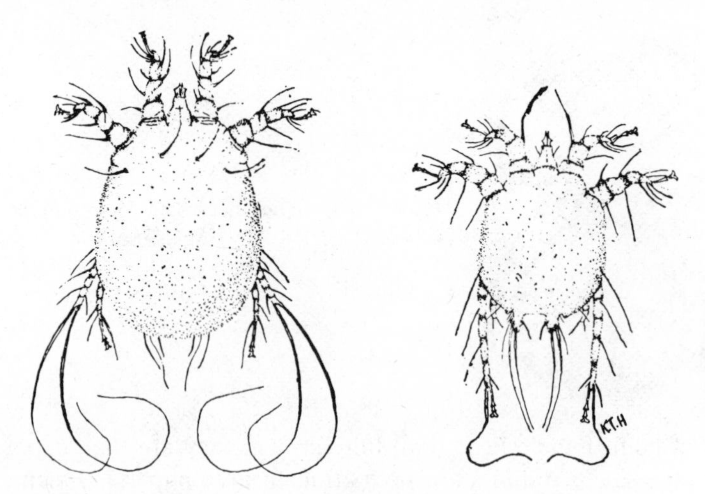

Figure 399. The sheep scab mite, *Psoroptes ovis* (Hering); female (left), male (right); 80×. (Ky. Agr. Exp. Sta.)

(Hering), affect cattle, sheep, goats, horses, and rabbits (Figs. 399 and 400). The mites live on the surface of the skin on any part of the body thickly covered with hair, feeding upon serum and lymph, and migrating to the periphery of the lesions formed as these become larger and encrusted with scabs. This form of mange itches severely and causes animals to scratch and rub against objects until the hair or wool comes out. The skin becomes thickened, wrinkled, and is frequently stained with blood. Transmission is usually by direct contact with infested animals or with objects against which they have rubbed.

Various species of *Chorioptes* scab mites attack the same hosts, have similar habits, and produce lesions very much like those described for *Psoroptes,* but the infestations are more localized, rarely involving the entire body of the host. For example, *C. bovis* (Gerlach) is commonly known as tail mange of cattle.

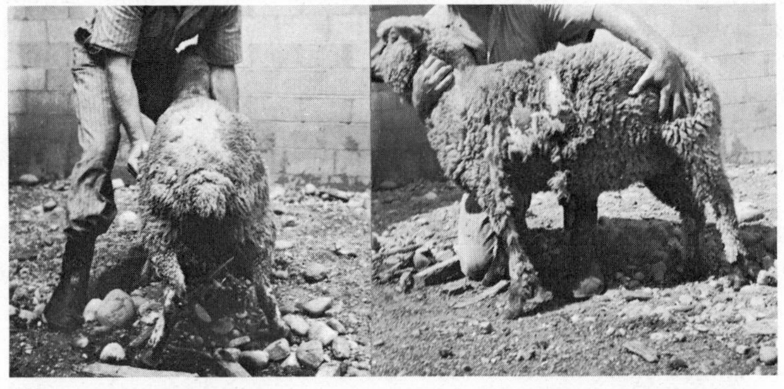

Figure 400. Sheep affected by scab mites. (Courtesy of F. R. Koutz.)

Mange mites, *Sarcoptes scabiei* (DeGeer) and *S. bovis* Robin, are also known as itch mites (Fig. 401). Several varieties attack cattle, horses, hogs, sheep, goats, dogs, rabbits, and man. The mites burrow into the tender skin areas of the host where the hair is sparse and often continue to spread until large portions of the body are affected. Nodules usually appear over and around the burrows; these burst and ooze serum, which hardens to form scabs. Intense itching causes the animals to rub and scratch, leaving open sores which frequently are invaded by bacteria. The skin becomes wrinkled and thickened as the infestation spreads. Transmission is usually by direct contact with mangy animals or with objects against which affected hosts have rubbed. Repeat overlapping generations occur throughout the year at intervals of 2 or 3 weeks.

Follicle mites of the genus *Demodex* attack domestic animals and man, causing manifestations of disease similar to those already described, by burrowing into hair follicles and oil glands. The nodules formed vary in size from a pinhead to as large as a marble and are filled with thick pus resulting from secondary bacterial infection. Demodectic mange is of minor importance in cattle, sheep, goats, and swine in the United States; it is, however, a serious pest of dogs, causing what is known as red mange. This mange is transmitted by direct contact; it does not itch severely, frequently not at all. Common species are: cattle follicle mite, *Demodex bovis* Stiles; dog follicle mite, *D. canis* Leydig; sheep follicle mite, *D. ovis* Railliet; goat follicle mite, *D. caprae* Railliet; hog follicle mite, *D. phylloides* Csokor; horse follicle mite, *D. equi* Railliet; cat follicle mite, *D. cati* Mégnin; and follicle mite, *D. folliculorum* (Simon).

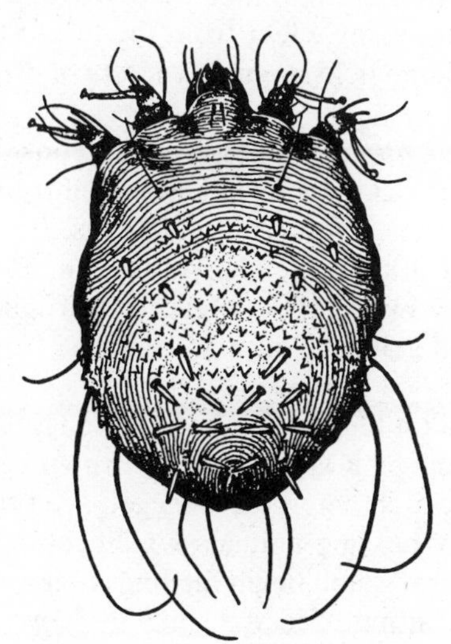

Figure 401. Sarcoptic mange or itch mite; 100×. (Imes, USDA)

Isolation of infested animals and treatment by spraying or dipping is a recommended procedure. Sheep scab mite has now been eradicated from the United States by supervised state-wide dipping of all infested sheep in a 0.5% concentration of a special livestock formulation of toxaphene. In addition to toxaphene, infested animals may also be treated with 0.06% lindane and proprietary nicotine or lime-sulfur dips. Commercial lime-sulfur diluted 1 to 15 with water has been used successfully as a dip to control mange mites. Dipping livestock is sometimes done on a community-wide basis and supervised by the county agent, extension entomologist, or state veterinarian. Control of demodectic mange on small animals has been successful using a 20% benzyl benzoate ointment or a 1% rotenone solution in oil. With these materials repeat applications to the affected areas are necessary. Crotoxyphos and dichlorvos are also effective chemicals.

REFERENCES: *USDA Misc. Pub.* 606, 1946; *Leaflet* 438, 1958; *S.D. Agr. Exp. Sta. Tech. Bul.* 10, 1952; *Bul. Ent. Soc. Amer.* 24; 401–406, 1978.

TICKS

ORDER ACARI, FAMILY IXODIDAE

Several species of hard-backed ticks attack both wild and domesticated animals, and occasionally man. All are ectoparasites which attach themselves to the host and suck blood, causing irritation, annoyance, and loss in vigor. This results in lower production of milk, meat, leather, and fiber. In addition to the painful wounds, the paralyzing effect from the bites, known as tick paralysis, is often more serious. Some ticks are also vectors of organisms causing dreaded diseases.

Cattle Tick, *Boöphilus annulatus* (Say), once considered to be the worst of the ticks that affect livestock, has been eliminated from the United States, except possibly in the extreme southern tip of Texas. It is responsible for spreading from affected to healthy cattle the protozoan, *Babesia bigemina,* which produces the disease known as Texas fever or piroplasmosis. Both the cattle tick and the disease still exist in South and Central America and in other parts of the tropics.

When the fertilized adult female tick becomes engorged with blood she drops to the ground and produces a large mass of brown eggs, as many as 3000 or more (Fig. 402), from which hatch the 6-legged seed ticks. These crawl up on vegetation, cling to any passing animals which rub against them, insert their piercing-sucking mouthparts and begin feeding. In a week or more they molt to the 8-legged nymph, continue feeding, and in 7 or more days transform to adults. These, after engorging, drop to the ground, the females laying their eggs. There may be 3 or more generations per year. If seed ticks fail to attach to an animal within 3 to 8 months, they die. If the female which has produced the

Figure 402. The cattle tick, *Boöphilus annulatus* (Say); adult female with egg mass; 3×. (USDA)

eggs was from a diseased animal, the young ticks will also carry the organisms causing the disease. Although cattle reared in infested areas develop immunity to the disease, those from outside such regions are fatally affected. A congener is *B. microplus* (Canestrini).

American Dog Tick, *Dermacentor variabilis* (Say), is common in North, Central, and South America; it is especially numerous in areas covered with grass or underbrush. A common pest of dogs and other wild or domesticated animals, it is also known to attack man. Aside from the annoyance and irritation from their blood-sucking habits, these ticks transmit the rickettsial organisms causing Rocky Mountain spotted fever in man and rodents. They are also capable of transmitting the organisms of tularemia and possibly relapsing fever.

In northern latitudes this tick overwinters in all stages except the egg; unengorged adults (Fig. 403) live more than 3 years under favorable conditions of moderate temperature and considerable moisture. Since these ideal conditions do not always prevail many ticks live only a few weeks to 3 or 4 months. Adults are more prevalent in the spring, and after engorgement and mating the females deposit their brown eggs in large masses on the ground over a period of 2 to 4 weeks. During this time as many as 6000 eggs have been recorded from

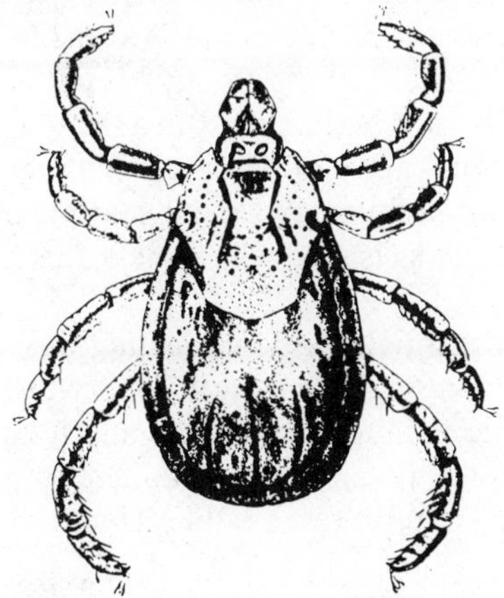

Figure 403. An unengorged female American dog tick, *Dermacentor variabilis* (Say); 5×. (USDA)

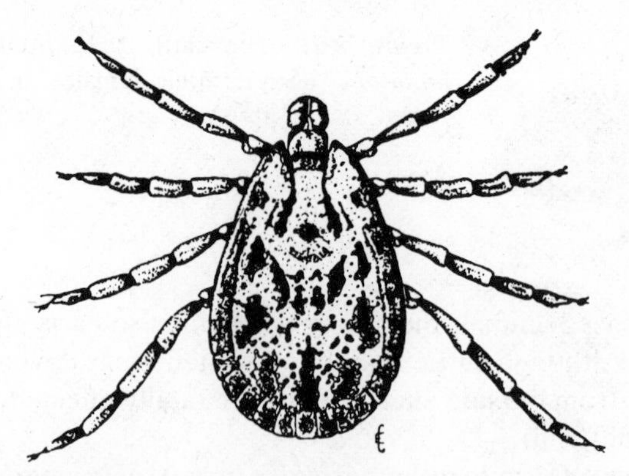

Figure 404. Male Rocky Mountain wood tick, *Dermacentor andersoni* Stiles; 5×. (Mont. Agr. Exp. Sta.)

a single female. Hatching occurs in 30 or more days, and the 6-legged stage crawls up on surrounding vegetation and clings to any passing animals which rub against them. The mouthparts are inserted and blood is sucked until they are fully engorged. This requires from 3 to 12 days, after which the ticks drop off and molt to the 8-legged nymph. After feeding on another host they again drop off and molt into the adult stage. The cycle may be completed during a period of a few weeks to several months or a year or more.

Rocky Moutain Wood Tick, *Dermacentor andersoni* Stiles (Fig. 404), is a blood-sucking pest of both wild and domesticated animals, and frequently of man. Besides irritation from the bites, this species transmits the rickettsial organisms causing spotted fever and the bacteria causing tularemia.

This tick is found in brushy areas of the Rocky Mountain states, from British Columbia to Arizona and New Mexico, and in isolated spots of the Pacific Coast states. The Bitter Root Valley of Montana has been the focal area endemic to the highly fatal spotted fever, but the disease is now present throughout the United States. Tularemia is likewise widely distributed in the United States.

The life cycle of this tick is much the same as that described for the American dog tick, except that only nymphs and adults overwinter successfully. Only adults feed on man and large animals; the larvae and nymphs feed on smaller mammals. Immunity to spotted fever can be induced by means of the vaccine injections developed by the United States Public Health Service with seasonal "booster shots."

Winter Tick, *Dermacentor albipictus* (Pack.), is common in Canada, the Southwest, Midwest, and north central states. It is found most often on horses, moose, deer, and may attack cattle. The life cycle is similar to those species previously described.

Pacific Coast Tick, *Dermacentor occidentalis* Marx, is more common in the Pacific States and attacks both wild and domesticated animals. Its life cycle is very similar to its congeners.

Brown Dog Tick, *Rhipicephalus sanguineus* (Latr.), is a cosmopolitan species in North America and a vector of the protozoans causing canine piroplasmosis and anaplasmosis of cattle. Although dogs seem to be the major host, this tick attaches to other animals and occasionally to man. It frequently invades homes and becomes a nuisance. A generation may be completed in less than 2 months in heated buildings or during summer months if there is an abundance of food.

Gulf Coast Tick, *Amblyomma maculatum* Koch, occurs along the Gulf of Mexico. It attaches to all farm animals but is more common on cattle, especially about the ears. In midsummer as many as 100 may be found on a single animal; in addition to inflammation, their bites serve as places of entry for screwworms. The hife history is the same as for the cattle tick.

Lone Star Tick, *Amblyomma americanum* (L.), is most abundant in the southwestern states and is recognized by the white spot on the back of the adult. It attaches to any part of the body of man, all kinds of livestock, and many wild animals, especially deer.

Other troublesome ticks in the same genus are the Cayenne tick, *A. cajennense* (Fabr.), and the gopher tortoise tick, *A. tuberculatum* Marx.

Tick Control. Climatic factors are probably the most important in natural control of ticks. Some species studied showed egg hatching was much better when relative humidity was within the range of 65 and 95%. Low woody deciduous forest type habitats are favorable to high tick populations.

Elimination of cattle ticks from most of the United States was achieved on a state-wide basis by dipping all animals in arsenical materials every 2 weeks from March to November, and placing them in pastures known to be tick-free. Recently developed chemicals are much more effective and are of great help in destroying all kinds of ticks. Approved chemicals for tick control on lactating dairy animals are crotoxyphos, dichlorvos, synergized pyrethrins, and rotenone. Additional chemicals approved for application to beef cattle and other animals are coumaphos, carbaryl, trichlorfon, famphur, fenthion, chlorpyrifos, malathion, and propoxur. In some places area treatment to eliminate ticks is desirable; recommended chemicals are chlorpyrifos, diazinon, malathion, and propoxur. Do not allow animals to graze in such treated areas.

Ticks may invade homes, the brown dog tick being most common. They may be eliminated by sprays of diazinon, ronnel, malathion, lindane, or chlorpyrifos.

Persons working in tick-infested areas may find repellents such as diethyl toluamide, dimethyl phthalate, dibutyl adipate, benzyl benzoate, or N-butyl acetanilide of protection value. Should ticks successfully attach to man or pets they can be stimulated to disengage their mouthparts by applying chloroform, turpentine, alcohol, or a hot needle to their body. After removal, sterilize the wound with tincture of iodine, merthiolate, or mercresin.

REFERENCES: *USDA Cir.* 478, 1938; *Leaflet* 387, 1963; *Yearbook of Agriculture*, pp. 662–663, 1952; E–762, 1951; *Farmers' Bul.* 1057, 1932; *Home and Gard. Bul.* 96, 1976; *J. Econ. Ent.*, 43:698–701, 1950; 44:1025–1026, 1951; 45:889–890, 1952; 57:340–346, 1964; *Ann. Ent. Soc. Amer.* 62:235–238, 285–287, 628–640, 1969; 63:128–133, 1970.

EAR TICK

Otobius megnini (Dugês), Family Argasidae

This soft-backed tick is found primarily in the semiarid sections of southwestern United States, the Rocky Mountain and Pacific Coastal states, and in Mexico and South America.

It attaches itself deep in the ears of cattle, sheep, goats, dogs, horses, and many wild animals. Its blood-sucking habit causes irritation and annoyance, and the bites occasionally become infected with pus-forming bacteria, giving rise to a condition known as "ear canker." Animals heavily infested scratch their ears, producing wounds which attract screwworm flies.

The life cycle is similar to that of other ticks. The young are covered with spines, giving rise to the common name "spinose ear tick." The spines do not occur on the adult and to our knowledge only the larval and nymphal stages are parasitic. Unless destroyed or dislodged they remain in the ear from 1 to 7 months or until fully grown and engorged with blood.

Control is accomplished by injecting ½ ounce of 1% lindane-pine oil mixture into each ear with a rubber tipped oil can, or gently dusting deeply inside each ear with 0.5% coumaphos, 5% ronnel, or 5% malathion. Squeeze bottle dusters are available as applicators.

REFERENCES: *USDA Farmers' Bul.* 980, 1953; *Agr. Handbook* 290, 1965; *Manual of Livestock Ticks*, 1965; *Can. Dept. Agr. Pub.* 930, 1956.

POULTRY LICE

ORDER MALLOPHAGA

Lice constitute one of the important problems of the poultry raiser. At least a dozen species attack chickens and others occur on turkeys, ducks, geese, guineas, and pigeons.

Infested individuals appear droopy and pale of comb and wattles; they lack appetite, and older birds fail to lay eggs. This is caused by the annoyance and irritation from the claws and bites of the lice, whose food consists of skin scales, scabs, blood, or blood clots which form after they bite the skin with their chewing mouthparts.

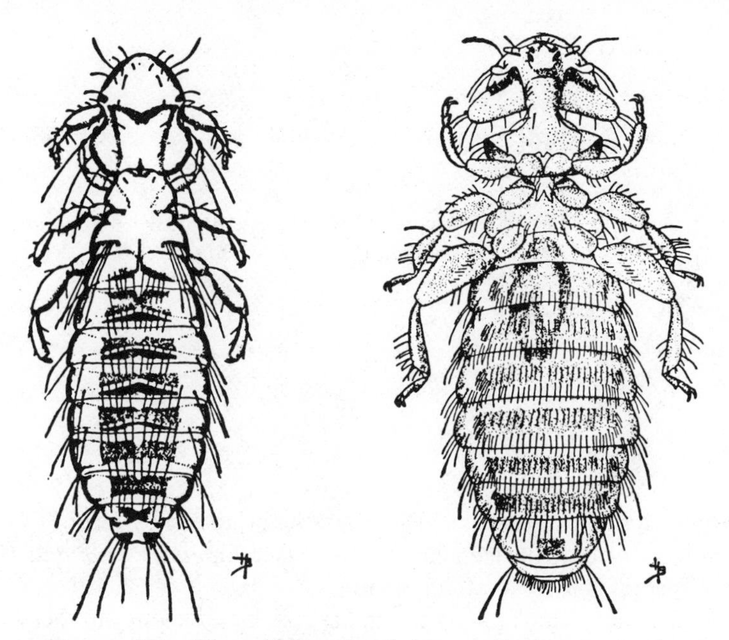

Figure 405. The chicken head louse, *Cuclotogaster heterographus* (Nitzsch) (left), and chicken body louse, *Menacanthus stramineus* (Nitzsch) (right); 25×. (USDA)

Poultry lice live on the host throughout the year but are more abundant during the summer. Their eggs adhere to the feathers; the wingless nymphs, hatching from them, molt several times and become adults in a few weeks, with many overlapping generations occurring. Adults are 2 to 4 mm long depending on the species.

The chicken head louse (Fig. 405), *Cuclotogaster heterographus* (Nitzsch), is most injurious to young chicks. It occurs on the head at the base of the feathers or down, and passes readily from one chicken to another through contact.

The chicken body louse (Fig. 405), *Menacanthus stramineus* (Nitzsch), is most injurious to grown fowls but often affects young chicks. It is found directly on the skin of any part of the body but especially below the vent.

The shaft louse (Fig. 406), *Menopon gallinae* (L.), is much smaller than the body louse, which it otherwise resembles. It tends to cling to the feathers rather than to the skin of the bird but undoubtedly causes irritation. It is often the most common of the lice on chickens. Other chicken lice are the large chicken louse, *Goniodes gigas* (Tasch.), brown chicken louse, *G. dissimilis* Denny, wing louse, *Lipeurus caponis* (L.), and fluff louse, *Goniocotes gallinae* (DeGeer) (Fig. 407). More species in the genus *Goniocotes* may infest chickens.

Besides the chicken body louse and the shaft louse, there are two species

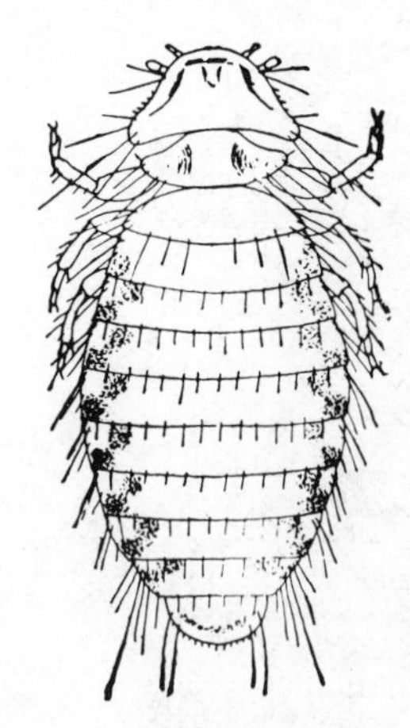

Figure 406. The shaft louse, *Menopon gallinae* (L.); 25×. (USDA)

that are quite specific to turkeys. These are the large turkey louse, *Chelopistes meleagridis* (L.), and the slender guinea louse, *Lipeurus numidae* (Denny). The latter species may also attack peafowl.

Lice on geese and ducks are seldom present in sufficient numbers to cause noticeable annoyance. The common species are the goose body louse, *Trinoton anserinum* (F.), and the large duck louse, *Trinoton querquedulae* (L.). Young

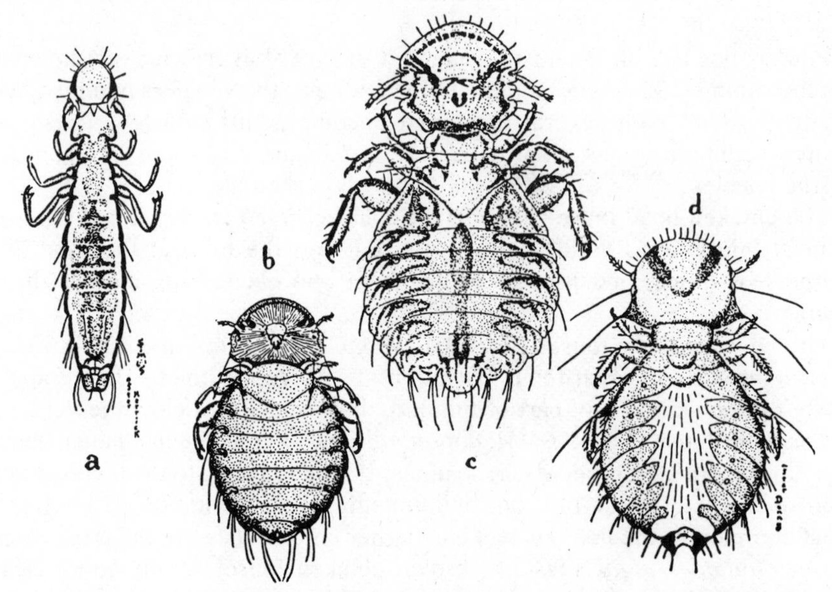

Figure 407. *a,* wing louse, *Lipeurus caponis* (L.); *b,* fluff louse, *Goniocotes gallinae* (DeGeer); *c,* large chicken louse, *Goniodes gigas* (Tasch.); *d,* brown chicken louse, *G. dissimilis* Denny; 15×. (Mote, Ohio Agr. Exp. Sta.)

ducks or geese may become infested with chicken head lice if closely associated with chickens.

Lice of the genera *Menacanthus, Menopon,* and *Trinoton* belong to the family Menoponidae; those in the genera *Chelopistes, Cuclotogaster, Goniocotes, Goniodes,* and *Lipeurus* belong to the family Philopteridae.

Poultry lice may be controlled by applying undiluted 40% nicotine sulfate to the top surface of the perches by spraying or with a paint brush just before the fowls go to roost. The fumes generated by the body heat of the birds kill the lice for about a 3-day period. Head lice are least affected. Since the louse eggs are not killed, repeat the treatment frequently. Roost paints containing 3% malathion have also proved to be effective louse killers. Dusting the birds, roosts, litter, and nests with 1% rotenone or 4% malathion at the rate of 1 pound per 100 birds gives good louse control. Other approved dusts for treating the birds are 0.5% coumaphos and 5% carbaryl; coumaphos may also be used to treat the litter. The carbaryl treatment may be repeated in 4 weeks if needed, but not more often. Providing dust-bath boxes containing 1% rotenone, 4% malathion, 5% carbaryl, or 0.5% coumaphos allows the birds to treat themselves. Spray formulations of the pesticides are sometimes substituted for the dusts. Do not contaminate feed or water with these chemicals.

REFERENCES: *USDA Farmers' Bul.* 1652, 1953; *Leaflet* 474, 1966; *J. Econ. Ent.*, 48:141–146, 1955; 54:1114–1117, 1961; *Ann. Ent. Soc. Amer.* 58:802–805, 1965.

MISCELLANEOUS POULTRY PESTS

Chicken Mite, *Dermanyssus gallinae* (DeGeer), sucks blood from fowls at night and remains secluded during the day in and about the nests and perches. Rarely do any occur on the fowls during the day. When numerous, these pests greatly reduce the vigor of a flock which is reflected in low weight gains and in low egg production. The red and gray mites are just large enough to be seen without a magnifying glass (0.8 mm).

The mite is inactive during cold weather but continues to breed the year around where warm weather prevails. It is always most abundant in summer. The white eggs are laid in the same place where hiding occurs during the day, and a generation may occur as often as every 7 to 10 days (Fig. 408).

Effective control is accomplished by thoroughly cleaning the poultry house and then treating all probable hiding places with one of the chemicals recommended for poultry louse control, with the exception of rotenone which is ineffective. Spray formulations give more satisfactory control of this pest. Anthracene oil diluted with an equal quantity of kerosene and applied to the perches, perch supports, and nests is an older recommended treatment which has merit.

Northern Fowl Mite, *Ornithonyssus sylviarum* (C. and F.), also called the feather mite, has been quite injurious in northern areas of the United States.

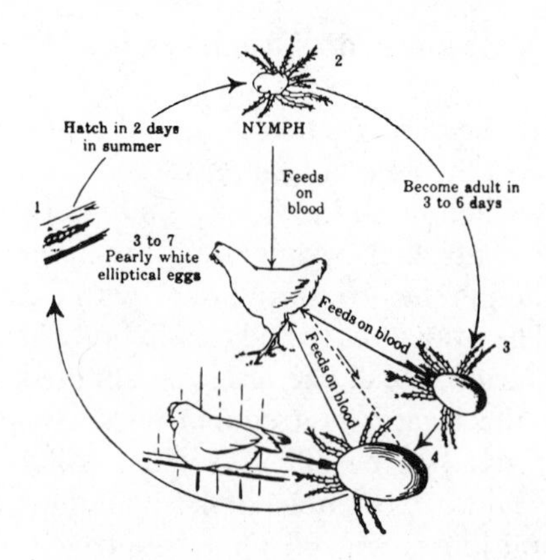

Figure 408. Life cycle of the chicken mite, *Dermanyssus gallinae* (DeGeer); 12×, except hens. (USDA)

This mite closely resembles the chicken mite but differs in that it breeds among the feathers and spends its entire life cycle on the host. The presence of the mites is often first noticed on the hens' eggs, and examination of the chickens will reveal groups of mites, their eggs and excrement among the feathers, especially about the tail. Their presence causes annoyance, loss of blood, and lack of thrift in the birds attacked.

This pest may be introduced into poultry flocks by infested English sparrows, pigeons, and mice. Destroying their nests in the vicinity of poultry houses and eradicating mice are recommended procedures.

Control measures for the northern fowl mite are the same as given for poultry lice (p. 551), with the exception of rotenone. Treatment of the birds, litter, and nests is necessary. Dusting sulfur has given excellent control when applied to the litter at a dosage of 4 pounds per 100 square feet. Dust baths of sulfur (in addition to those suggested for poultry lice) are recommended. All roosters should be given individual treatment because they do not take dust baths.

Tropical Fowl Mite, *Ornithonyssus bursa* (Berlese), is another dermanyssid mite that resembles the northern fowl mite in appearance and habits but does not spend its entire life on the bird. Control measures are the same as for northern fowl mite.

Depluming Mite, *Knemidokoptes gallinae* (Raill.), is an itch mite that lives at the base of the feathers of fowls and pigeons (Fig. 409). It produces intense itching causing the fowls to pull out their feathers. If the stumps are examined soon after breaking of the quill, mites will be found surrounded by skin scales and often encrusted. For complete control treat the entire flock with dusting sulfur, as suggested for northern fowl mite, or dip in a mixture consisting of 4

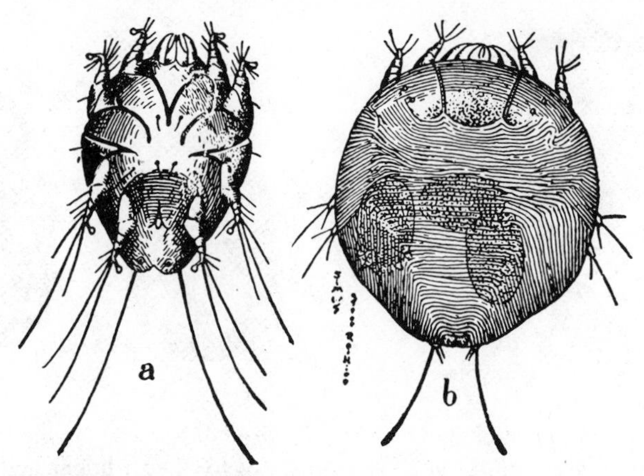

Figure 409. The depluming mite: *a,* adult male (114×); *b,* adult female (108×). (Mote, Ohio Agr. Exp. Sta.)

ounces of sulfur and 4 ounces powdered soap in 2 gallons of water. Dipping should be done on warm sunny days.

Scaly Leg Mite, *Knemidokoptes mutans* (R. and L.), attacks both wild and domesticated fowls and is the cause of a condition sometimes described as leg mange. The minute mites (Fig. 410) burrowing under the scales cause irritation, resulting in raised scales, exudation of blood and lymph, and the formation of a powdery substance. In severe cases the joints become so inflamed that the birds are lame and scarcely able to walk. This mite occasionally attacks the comb and wattles. Isolate the affected birds, and dip the feet and legs into crude oil, kerosene, or another light oil. Usually one application is sufficient to bring about control. To prevent spread of the mites, treatment of the perches with oil is also recommended.

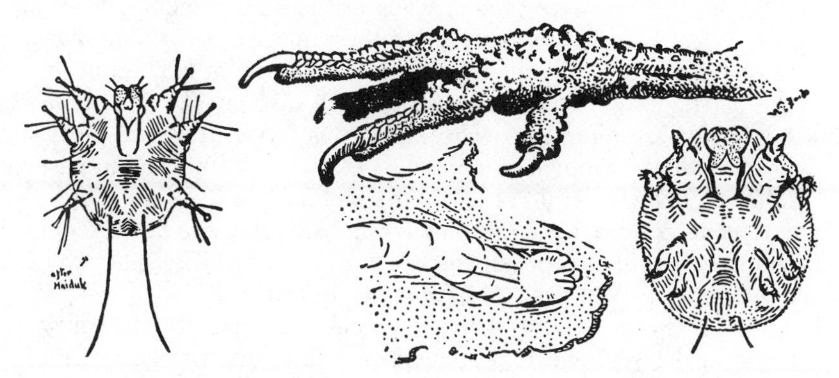

Figure 410. Scaly leg mite: male (left), female (right), infested leg of fowl, above, and mite in burrow in skin, below; 100×, except chicken leg. (Mote, Ohio Agr. Exp. Sta.)

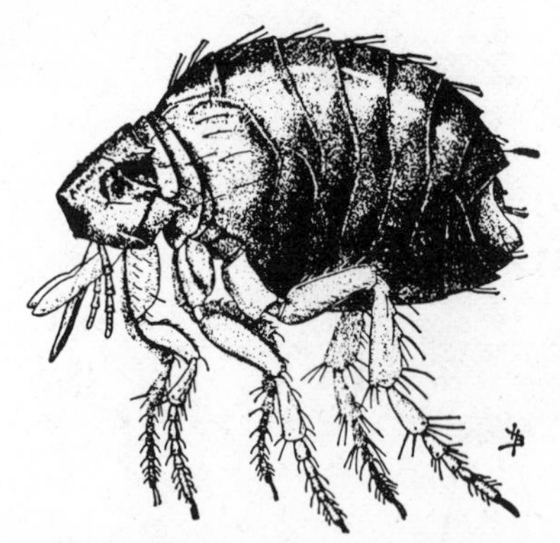

Figure 411. A female sticktight flea, *Echidnophaga gallinacea* (Westwood); 30×. (USDA)

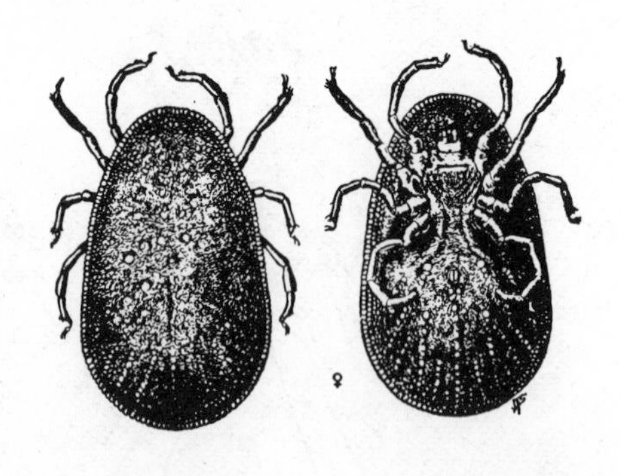

Figure 412. The fowl tick, *Argas persicus* (Oken); upper and lower surfaces of adult female; 2.5×. (USDA)

Sticktight Flea, *Echidnophaga gallinacea* (Westwood) (Fig. 411), is a pest well known in many parts of the South and is occasionally present in the northern sections of the country. It attaches itself in one place and remains there instead of hopping actively, as do other fleas.

When abundant they reduce egg production and often kill young chickens, sometimes even mature fowls, by their blood-sucking habit. They tend to congregate in clusters on the comb, wattles, and around the eye. This species also attacks dogs, cats, horses, man, and wild birds. The eggs are forcibly ejected from the female and drop to the ground where hatching occurs. The white larvae feed on excreta and filth in the litter or in cracks of the henhouse floor. After a period of 2 to 3 weeks they spin silken cocoons, transform to pupae, and emerge as adults. The life cycle may be completed in 1 to 2 months.

Effective control can be accomplished by area treatment of all grounds, as well as the litter, roosts, nests, and walls of the poultry house. Malathion is perhaps the best material to apply in all these places. Treat with 4% dust at the rate of 1 pound per 20 square feet, or use a spray at 0.5% concentration. Carbaryl, as suggested for poultry lice control, also checks flea populations. Area treatments are directed toward killing flea larvae. Infested birds should be given individual treatment by applying derris or cubé powder in vaseline to affected parts, being careful not to get the ointment in the eyes of the bird.

Other species of fleas that attack poultry and are known to be present in parts of this country are the European chicken flea, *Ceratophyllus gallinae* (Schr.), and the western chicken flea, *Ceratophyllus niger* Fox.

Fowl Tick, *Argas persicus* (Oken), is another ectoparasite of poultry which is common in the southwestern states but is little known elsewhere, except occasionally in Florida. It is red or brown, sometimes with a tinge of blue, which suggests the name "blue bug" by which it may be known locally (Fig. 412).

When fully grown this tick may be nearly 12 mm long. The females deposit many brown eggs in masses about poultry houses. Hatching occurs in 10 or more days, and the 6-legged nymphs or seed ticks attach to poultry at night, feed until engorged, which may require 3 to 10 days, and then drop off, molt, and gain the fourth pair of legs. Thereafter, the ticks attack poultry only at night after they have gone to roost. Their painful bites and blood-sucking habits greatly reduce the vitality and egg production of a poultry flock. Also, they are known to be carriers of the organisms causing fowl spirochaetosis, prevalent in other parts of the world.

Effective control depends on killing the ticks outside the poultry houses to prevent them from reaching the birds, and/or treating inside the houses to kill the ticks that leave the birds after engorgement. Chemicals recommended for use outside the houses are coumaphos, malathion, and ronnel; for inside use coumaphos, malathion, carbaryl, and naled. Do not contaminate water or feed with any of these chemicals.

REFERENCES: *USDA Farmers' Bul.* 1652, 1953; *Leaflets* 382, 1974; 383, 1964; *J. Econ. Ent.*, 45:748–749, 1952; 47:942–944, 1954; 54:1212–1214, 1961.

HOUSE FLY

Musca domestica, L., Family Muscidae

Found throughout most of the world, house flies are abundant wherever man makes his abode. They have been known during all periods of time for which we have records. Besides being annoying to man and other animals, house flies may serve as carriers of the organisms causing several important diseases: conjunctivitis, poliomyelitis, typhoid fever, tuberculosis, anthrax, leprosy, cholera, diarrhea, and dysentery. They also serve as intermediate hosts for the helminths, of which 3 species of tapeworms parasitic on poultry and 3 species of nematodes parasitic on horses, mules, and donkeys are the most important.

The adult is a typical fly with sucking or lapping mouthparts. The larvae or maggots are tapered and creamy white (Fig. 413). Female flies deposit 2 to 21 egg masses (Fig. 413), each containing about 130 white eggs, in manure or fermenting vegetable matter. Any kind of animal excrement may serve as a breeding medium if it is moist and of the proper temperature. Hatching takes place in 10 to 24 hours and the resulting maggots become fully grown in 3 to 7 days, crawl to the margins of the breeding material, and change to dark brown puparia, from which adults emerge 3 to 6 days later. In the laboratory, at a temperature of 80 F (27 C), the entire life cycle from egg to adult required 10 to 12 days. In the North overwintering is generally as larvae or puparia.

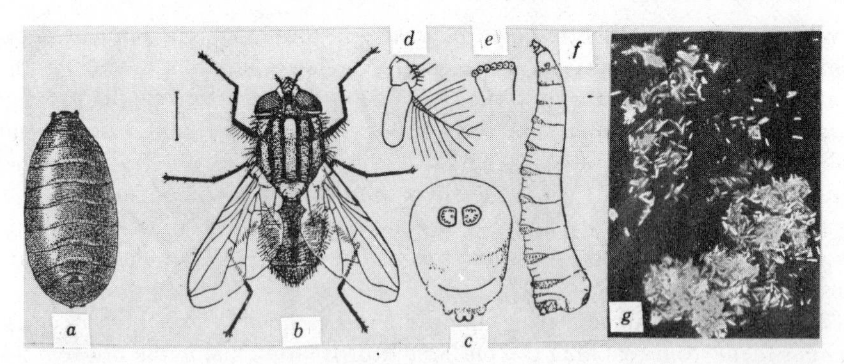

Figure 413. The house fly, *Musca domestica* L.: *a,* puparium; *b,* adult; *c,* enlarged posterior view of larva showing anal spiracles; *d,* enlarged antenna; *e,* enlarged thoracic spiracle; *f,* larva; *g,* eggs, natural size; *a, b,* and *f,* about 4×. (Howard, USDA)

House fly populations are checked or destroyed by sanitation, screening, trapping, baiting, aerosols, residual and space sprays of pesticides, parasites, and predators.

Elimination of breeding areas by daily removal and scattering of manure on the fields is recommended wherever practical. Proper wrapping, temporary storage, and disposal of garbage is also important in the prevention of fly breeding. Treating manure or other breeding materials with diazinon, malathion, ronnel, dichlorvos, or dimethoate will give good kill of larvae. Do not use diazinon in poultry houses.

Screening homes and buildings with 16-to-18-mesh screen wire is a means of preventing annoyance from flies and other insects. Traps of various types have been devised and are of value. Most of them employ a bait of ripened fruit, syrup, cheese, milk, or muscalure (a sex pheromone), which attracts the flies into an inescapable chamber or electrocutor. Tanglefoot ribbons are of value in reducing fly populations in screened buildings.

Poisoned baits, either liquid or dry, are very common for controlling flies, especially those that become resistant to certain pesticides. The most effective baits contain organic phosphorus compounds mixed with sugar or syrup as the attractant. Some preference for dry baits has been indicated because of convenience in handling. Many commercial formulations are available, and the directions given on the package should be followed. An effective bait formulation consists of 2 teaspoonfuls of 25% malathion wettable powder and 2 teaspoonfuls of sugar in 1 ounce of water. Mix this with ¾ cup of corn meal, allow to dry, and place in a shaker-type can. Other organic phosphorus compounds that have performed well as baits are dichlorvos, ronnel, trichlorfon, and naled. An old standard bait that will kill both normal and resistant strains of flies consists of 3 teaspoonfuls of commercial 40% formalin in a quart of milk, to which a little sugar or syrup has been added. All fly baits are highly poisonous and care must be exercised in handling them.

Atomizing commercial fly spray solutions containing dichlorvos or naled, plus synergized pyrethrins, is effective in enclosed places. Daily treatment is necessary. Residual sprays of crotoxyphos, dimethoate, ronnel, tetrachlorvinphos, dichlorvos, fenthion, naled, or chlorfenvinphos, may be applied to the walls and other areas inside and outside farm buildings where flies congregate. Animals must not be present during spraying and contamination of foods, feed troughs, water fountains, and milking equipment must be avoided. Surfaces should be wetted to the point of run-off. Places where flies rest at night are indicated by fly excrement; these areas should be thoroughly saturated with the pesticide. Cloth ribbons, strings, or strips of screen wire dipped in any of the residual pesticide solutions and then suspended in fly-infested buildings will serve as resting places and give remarkable control. Dichlorvos impregnated plastic strips are approved for use in milk processing rooms and other enclosures with little air movement.

Continuous use of the same pesticide results in populations of house flies that become resistant or can tolerate that chemical. Changing to an entirely different chemical and inauguration of bait sprays have helped to counteract the problem.

Introduction of parasites of the house fly, as well as other pest flies found breeding in manure, gives promise of successful biological control in some regions. By releasing parasites periodically, beginning in the spring, maintaining a manure height of 8 to 12 inches for proper fly breeding, and using poisoned baits to kill adults, fly populations were greatly reduced on poultry ranches in California. The relative merits of this management procedure over sanitation plus poisoned baits has not been assessed.

Many believe the best management operation for efficient and economical fly control is prompt disposal of all breeding materials.

REFERENCES: *USDA Leaflet* 390, 1975; EC–29, 1954; *Tech. Bul.* 1519, 1976; L. S. West, *The House Fly,* Comstock Publishing Assoc., Ithaca, N.Y., 1951; *J. Econ. Ent.* 46:172, 1953; *Ohio Ext. Bul.* 473, 1979; *Calif. Agr.* 29(5)8–10, 1975.

The cluster fly, *Pollenia rudis* (F.), is slightly larger than the house fly, more sluggish, with the larval stages parasitic on earthworms and possibly other hosts. The adults (Fig. 414) invade houses in autumn, often appearing in the attic or other places offering shelter, such as the space between windows and storm sash. Control measures are directed only against the adults when they become abundant in houses. Use aerosols of dichlorvos or propoxur plus synergized pyrethrins in enclosed places, then sweep up flies and discard.

Other flies found breeding where house flies abound are the little house fly, *Fannia canicularis* (L.) (Fig. 41), and the latrine fly, *F. scalaris* (F.).

Figure 414. The cluster fly, *Pollenia rudis* (F.); 4×. (Howard.)

MOSQUITOES

ORDER DIPTERA, FAMILY CULICIDAE

Many species of mosquitoes attack man and other animals; some of them are widespread, others rather local in their distribution. Common ones are the yellow-fever mosquito, *Aedes aegypti* (L.); salt-marsh mosquito, *A. sollicitans* (Wlkr.); brown salt-marsh mosquito, *A. cantator* (Coq.); California salt-marsh mosquito, *A. squamiger* (Coq.); black salt-marsh mosquito, *A. taeniorhynchus* (Wied.); tree-hole mosquito, *A. triseriatus* (Say); northwest coast mosquito, *A. aboriginis* Dyar; floodwater mosquito, *A. sticticus* (Meigen); western tree-hole mosquito, *A. sierrensis* (Ludlow); *A. dorsalis* (Meigen); *A. nigromaculis* (Ludlow); *A. vexans* (Meigen); the common malarial mosquitoes, *Anopheles quadrimaculatus* Say, *A. crucians* Wiedemann, *A. punctipennis* (Say), and *A. freeborni* Aitken; northern house mosquito, *Culex pipiens* L.; southern house mosquito, *C. quinquefasciatus* Say; *C. restuans* Theobald; *C. tarsalis* Coquillett; and species of *Mansonia* and *Psorophora*.

Mosquitos are important pests since they cause annoyance and discomfort; they may be responsible for reducing property values in heavily infested areas; they also transmit organisms causing malaria, encephalitis, yellow fever, dengue, and filariasis. The female mosquito pierces the skin and injects saliva at the time of feeding, which is responsible for the irritation that follows. Blood taken from man or other animals infected with disease-producing organisms in turn infects the mosquito, which transmits them to future hosts (Figs. 415 and 419).

The life cycles of mosquito species vary. Many of them overwinter as eggs, others as larvae or adults. In the North, the fertilized females of *Culex* and *Anopheles* species survive the winter, in warmer regions breeding occurs the

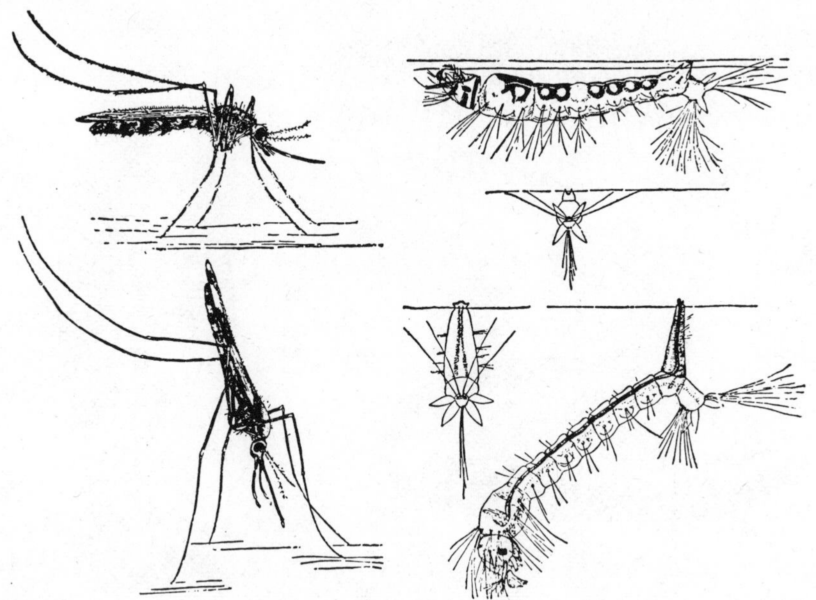

Figure 415. Feeding positions of mosquito adults and larvae: (upper left) adult *Culex,* (lower left) adult *Anopheles,* (upper right) *Anopheles* larva, (lower right) *Culex* larva; 8×. Side and posterior views of larvae. (USDA)

year around. *Aedes* and *Psorophora* normally pass the winter in the egg stage. *Culex* mosquitoes deposit masses of elongated eggs on end, which float on the water surface like a raft, whereas *Anopheles* and *Aedes* eggs are laid singly. Those of *Anopheles* float on the water, and those of *Aedes* are attached to moist objects near the water surface (Fig. 416). Eggs of *Aedes* are able to withstand long periods of drying and on being immersed in water hatch in several minutes. Usually a 2- or 3-day incubation period is required for *Culex* and *Anopheles* eggs. Among the blood-sucking mosquitoes, a blood meal is generally necessary for egg production.

The larvae of all mosquitoes are aquatic and, although possessing tracheal gills, must come to the surface for additional air obtained by means of a respiratory tube that projects through the surface film of water (Fig. 415). This tube varies in length according to the species; it is much shorter in *Anopheles* than in *Aedes* and *Culex*. The larvae move about by a series of spasmodic motions, feeding on protozoa, bacteria, algae, and other microorganisms. *Anopheles* larvae are primarily surface feeders and lie parallel to the water surface, whereas *Culex* and *Aedes* feed below the surface and lie at an angle of 45° when at rest (Fig. 415). The period of development requires several days and the larval skeleton is shed 4 times before the pupal stage appears.

The pupae have 2 short respiratory tubes on the thorax (Fig. 417), take no food, but can swim about in water. After a few days in this stage the adults

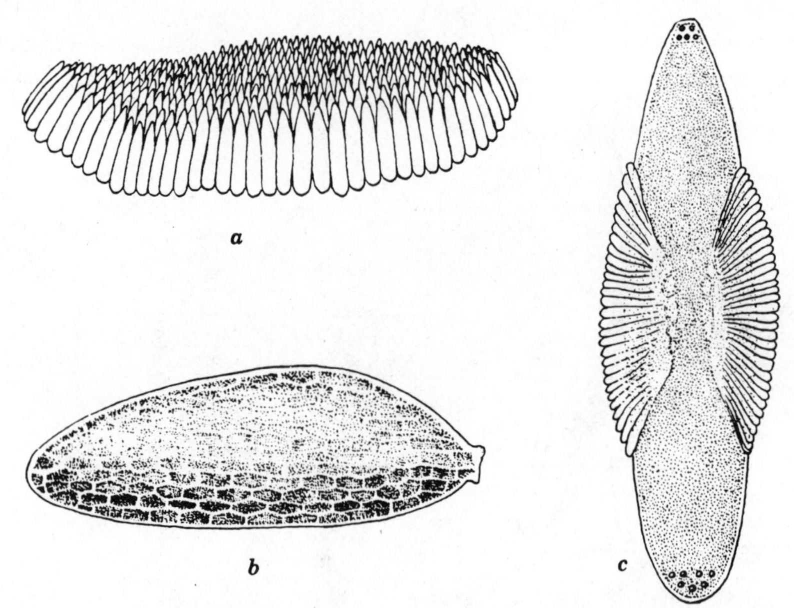

Figure 416. Eggs of mosquitoes: *a,* egg raft of *Culex restuans* Theob.; *b,* egg of *Aedes taeniorhynchus* (Wied.); *c,* egg of *Anopheles quadrimaculatus* Say, showing floats; (about 40×). (Howard, Dyar, and Knab.)

emerge, making a total period from egg to adult of 10 to 14 days at temperatures of 80 F (27 C). The length of the life cycle is influenced by climatic factors, usually a single to many generations developing each year depending on these factors and the species of mosquito.

Adult mosquitoes are small 2-winged flies, the body, legs, and wings covered with tiny scales. The feeding positions and head characters of adult females of *Culex* and *Anopheles,* are shown in Figs. 415, 418). Male mosquitoes do not

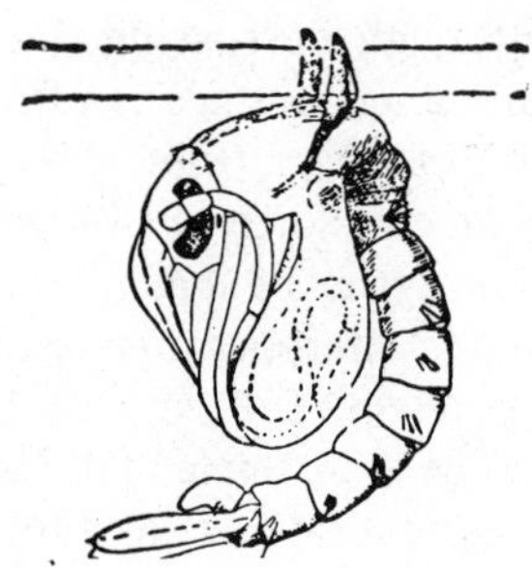

Figure 417. *Culex pupa;* 7×. (Howard, USDA)

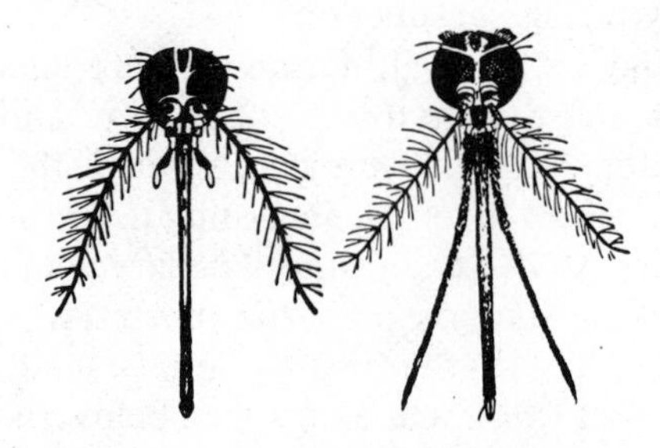

Figure 418. Head of female *Aedes* mosquito (left) compared with that of *Anopheles* (right); 25×. (Ky. Agr. Exp. Sta.)

suck blood but subsist on the nectar of flowers, fruit juices, and water. If none of these is available, death soon follows. The males are distinguished from females by their smaller size, more bushy antennae, and more prominent palpi.

Control of mosquitoes is accomplished in various ways. Where feasible, elimination of breeding places by drainage, filling, or sanitation is an effective measure. Frequent inspection of the premises should be made to see that bird baths, eave troughs, tin cans, bottles, jars, barrels, or other objects are not serving as breeding spots. The introduction of certain species of fishes (*Gambusia*) has proved of value in the control of mosquitoes in permanent ponds, pools, salt marshes, and irrigation ditches.

Successful biological control of mosquito larvae, without apparent disruption of the ecosystem, consists of treating breeding areas with hydra, the mermithid

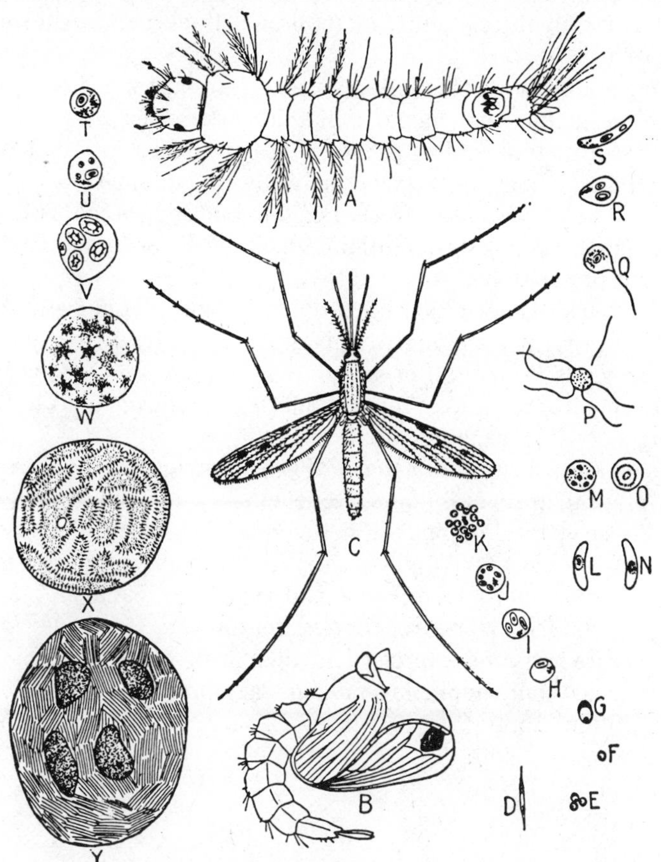

Figure 419. The malaria mosquito, *Anopheles* sp.: *A*, larva; *B*, pupa; *C*, adult; (*A*, *B*, *C*, about 6×) *D* to *Y*, stages in the development of the protozoan, *Plasmodium;* greatly magnified. (N. J. Agr. Exp. Sta.)

nematode, *Romanomermis culicivorax,* the bacteria, *Bacillus sphaericus,* and certain strains of *Bacillus thuringiensis.* These materials may be especially good for treating rice fields.

Application of larvicides by hand or power equipment has given excellent control, but care must be exercised because of the danger of killing fish and other aquatic animals in some habitats. Dust formulations are less toxic to wildlife than oil emulsions or solutions, but they are also less effective in control, except for the surface feeders. Paris green at the rate of 1 to 2 pounds per acre is effective in killing anopheline larvae. Kerosene or fuel oil (No. 2) alone or in combination with 0.007% pyrethrins at the rate of 20 to 30 gallons per acre has proved to be a good larvicide. Chlorpyrifos, methoxychlor, fenthion, malathion, temephos, carbaryl, and propoxur have also given excellent control of larvae. The growth regulator, methoprene (trade name Altosid), is a new effective larvicide.

For killing adults in enclosed places use aerosols or space sprays containing methoxychlor, dichlorvos, naled, malathion, or synergized pyrethrins. Residual sprays of methoxychlor, ronnel, fenthion, malathion, dichlorvos, stirofos, or dimethoate may be applied to walls of buildings, catch basins and other places where mosquitoes rest. Fogging out-door areas with solutions of propoxur, naled, dichlorvos, fenthion, malathion, or synergized pyrethrins reduces adult populations.

Screening houses is necessary for protection against adults and 18-mesh wire is required to prevent some individuals of *Aedes aegypti* from entering. Mosquito netting gives protection in camp areas or places not properly screened.

Where available, the smoke from smoldering pyrethrum coils has been useful in reducing as well as repelling adult mosquitoes.

There is no known substance that will give complete freedom from mosquito attack for more than several hours. Repellents are applied to the exposed skin and, if necessary, to clothing, the better materials being dimethyl phthalate, 2-ethyl-1,3-hexanediol, dimethyl carbate, diethyl toluamide, or a mixture containing six parts dimethyl phthalate and two parts each of 2-ethyl-1,3-hexanediol and indalone. These repellents are solvents for paints, varnish, and plastics, so care must be exercised in their use. Most of them cause some irritation if accidentally applied to mucous membranes. For relief from itching after being bitten, apply 5 to 10% benzocaine in ethyl alcohol, or Thephorin ointment.

REFERENCES: *USDA Misc. Pub.* 336, 1939; *Agr. Handbooks* 152, 1959; 173, 1960; 120, 1964; *Home and Garden Bul.* 84, 1972; *Conn. Agr. Exp. Sta. Bul.* 632, 1960; *Cir.* 224, 1970; *J. Econ. Ent.,* 42:586–590, 1949; 43:350–353, 1950; 44:428–429, 1951; 45:712–716, 1952; 46:164, 1953; 47:818–824, 1954; 56:58–60, 834–835, 1963; *Calif. Agr.* 30(9):11, 1976; 31(8):4–5, 1977; *Proc. Acad. Paris,* Vol. 286 (series D); 797–800; 1175–1178, 1978.

CHIGGER MITES

ORDER ACARI, FAMILY TROMBICULIDAE

A number of species of mites are troublesome to man and other animals, among them the chiggers or redbugs. The most numerous and widely distributed species is *Trombicula alfreddugèsi* (Oudemans), which occurs throughout the United States east of the Rocky Mountains. It has also been reported from Canada, California, Arizona, Mexico, Central America, South America, and neighboring islands. Related species are *T. splendens* Ewing, which has a similar but more restricted range, and *T. batatas* (Linn.), a tropical species found only in the warmer, southeastern part of the United States and Central and South American countries. Distinguishing characteristics of these three species of chiggers are shown in Fig. 420.

Contrary to popular belief, chiggers do not burrow into the skin; they only insert their mouthparts, usually in a skin pore or hair follicle, and begin feeding. On man they are found especially in regions of the body with tight-fitting clothing. The salivary juices secreted by the chiggers cause severe irritation accompanied by intense itching. This results in scattered red blotches of various sizes, with frequent secondary infection. Persons exposed repeatedly may develop immunity to the irritation.

Chiggers are found where vegetation is abundant, such as shaded areas, high grass or weeds, fruit orchards or berry patches. However, they may become serious pests in relatively dry areas, such as lawns, golf courses, and parks. They are tiny and not easily seen, and since their bites may not be felt for several hours after exposure it is difficult to know the exact location of infestation.

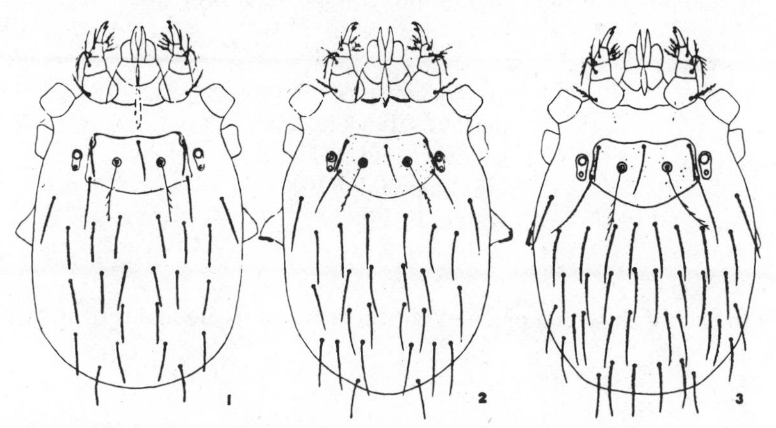

Figure 420. Dorsal views (legs absent) showing distinguishing characters of three species of chiggers attacking human beings: 1, *Trombicula alfreddugèsi;* 2, *T. splendens;* 3, *T. batatas;* about 100×. (Courtesy of Dale W. Jenkins.)

To determine such an area, place a piece of black cardboard edgewise on the ground where chiggers are suspected. If present, in a few minutes tiny yellow or pink mites will be seen moving rapidly over the cardboard and accumulating on the upper edge. They may also be detected on black polished shoes.

Chiggers pass the winter as adults near or slightly below the surface of the soil. In the spring they become active and deposit their eggs, varying from a few to as many as 15 per day. These hatch into 6-legged individuals called "larvae," the only stage that attacks man. After engorgement, which may require from one to several days, the larvae drop off, transform into 8-legged nymphs, and finally into the adult stage. Both nymphs and adults are said to feed on eggs of springtails, isopods, and mosquitoes. Adult female chiggers have lived over 1 year and have produced offspring throughout that period. Repeat generations occur in warmer climates, whereas only 2 or 3 develop each season in the North.

Treatment of infested areas with toxaphene, diazinon, or lindane provides a high degree of control for 1 to 2 months. Emulsion sprays usually give the best results, but wettable powder sprays and dusts have been quite satisfactory. Do not treat areas where livestock graze.

To prevent chigger attachment when going into probably infested areas, apply diethyl toluamide, dimethyl phthalate, dibutyl phthalate, dimethyl carbate, 2-ethyl-1,3-hexanediol, benzyl benzoate or sulfur dust to the clothing and also to the body, particularly to the legs, ankles, cuffs, waist, and sleeves. A thorough soapy bath soon after exposure will also be of help. Plastics, paints, and certain synthetic fibers are harmed by some repellents.

After itching begins some relief can be secured by treating the affected spots with the following formula which your druggist can prepare:

benzocaine, 5%
methyl salicylate, 2%
salicylic acid, 0.5%
ethyl alcohol, 73%
water, 19.5%

Apply to each welt with a piece of cotton or facial tissue and repeat as often as necessary.

REFERENCES: *USDA Bul.* 986, 1921; *Leaflet* 403, 1963; *Home and Garden Bul.* 137, 1976; *Am. J. Hygiene*, 48:22–35, 1948; *Ann. Ent. Soc. Amer.*, 42:289–318, 1949; *J. Econ. Ent.*, 55:22–23, 1962.

BED BUGS

ORDER HEMIPTERA, FAMILY CIMICIDAE

Nearly world-wide in distribution the bed bug, *Cimex lectularius* L., is found primarily in human habitations, but it also attacks poultry, birds, mice, rats, guinea pigs, and rabbits. A related species, *Cimex hemipterus* (F.), is found in Florida and other tropical areas. The poultry bug, *Haematosiphon inodorus* (Dugès), is another species that commonly infests birds and poultry and is found in some southwestern states and in Mexico. The swallow bug, *Oeciacus vicarius* Horvath, the bat bug, *Cimex pilosellus* Horvath, and the European pigeon bug, *Cimex columbarius* Jenyns, are other related species.

The bed bug *C. lectularius,* does most of its blood-sucking at night, and the bites are usually painless when inflicted. However, inflamed welts develop afterwards, along with severe itching caused by the injection of salivary juices during the feeding. The adult (Fig. 421) is a red-brown wingless (except for vestiges) insect, slightly more than 6 mm long, with a more or less flattened body when unfed but swollen and elongated when engorged with blood. Under favorable conditions the adult female lives from 2 to 10 months and deposits about 200 white eggs at the rate of 3 or 4 per day. Because of a sticky substance these adhere to the bedding or other objects in the infested area. At tempera-

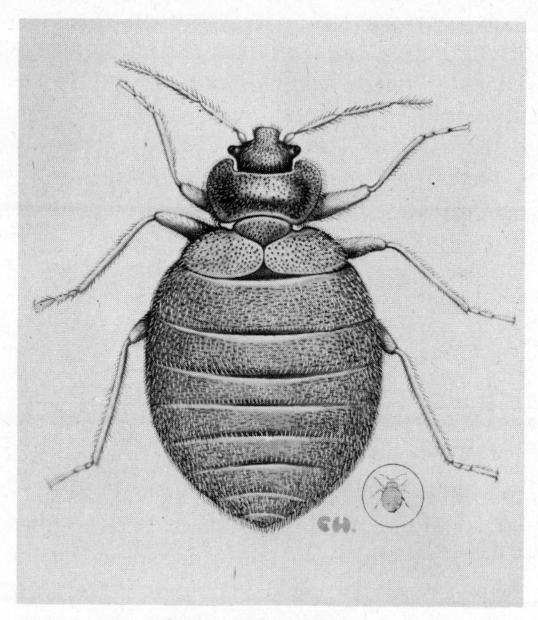

Figure 421. The bed bug, *Cimex lectularius* L.; 10×. (Courtesy of Can. Dept. Agr.)

tures of 70 F (21 C), or above, hatching occurs in 6 to 17 days, and the tiny nymphs molt 5 times before reaching the adult stage. Four to 6 weeks are required for completion of the life cycle, and there may be several generations in a year depending on the temperature and food supply. Development is greatly retarded as the temperature drops below 70 F (21 C), and no eggs are laid at 50 F (10 C), or lower. Newly hatched bugs may live several weeks without feeding during warm weather and for longer periods in cool weather. It is believed that adult bugs can live a year or longer without food. A blood meal is necessary before females produce fertile eggs.

It is not always necessary to see the bugs to detect their presence because of an offensive odor which is associated with them. At the beginning of an infestation the bugs are more likely to be found in the beds; as their numbers increase they spread to other parts of the room.

Bed bugs can be controlled by treating all the places where they hide. This includes bed frames, mattress and springs and, under heavy infestation, baseboards, closets and other furniture in the room.

Satisfactory control has resulted from residual sprays containing premium grade malathion, lindane, ronnel, or dichlorvos. In most instances where treatment of other areas is thorough there will be little need for treating the mattress. Mattresses may be treated lightly with any of these chemicals at reduced concentrations where adults sleep, whereas only synergized pyrethrins should be applied to mattresses where babies or children sleep. It is best to make all treatments in the morning so the solvent from the pesticides will be dissipated before nightfall. Dusts of these chemicals can be used for treating baseboards and closets. The common aerosol containing pyrethrins and piperonyl butoxide is effective if sufficient residue is deposited. Where sprays, dusts, or aerosols are not feasible, infestations may be eliminated by fumigating with methyl bromide or hydrogen cyanide by trained licensed persons only.

REFERENCES: *Thomas Say Foundation,* Vol. VII, 1966; *USDA Leaflet* 453, 1972; *Home and Garden Bul.* 96, 1976; *Conn. Agr. Exp. Sta. Cir.* 213, 1960; *Cir.* 224, 1970.

FLEAS

ORDER SIPHONAPTERA, FAMILY PULICIDAE

Fleas are widely distributed, blood-sucking, wingless insects whose bites cause irritation not only to man but also to cats, dogs, hogs, rabbits, rats, mice, and many other animals. They are important in the spread of the bacterial organism, *Pasteurella pestis* (L. and N.), causing bubonic plague, and serve as intermediate hosts for the dog and rodent tapeworms that occasionally parasitize man.

The oriental rat flea, *Xenopsylla cheopis* (Roth.), is a vector in the spread of endemic or murine typhus from rats to man and is prevalent in the United States. Besides the oriental rat flea, common species of importance are the cat flea, *Ctenocephalides felis* (Bouché), the dog flea, *Ctenocephalides canis* (Curtis) (Fig. 422), and the human flea, *Pulex irritans* Linn. (Fig. 423).

All irritation to the hosts is caused by the piercing-sucking adult stage. Adults are 2 to 4 mm long, very narrow-bodied, dark brown and spiny (Fig. 424), with well-developed jumping legs. Their white eggs may be laid either on the animal or on the floor of infested buildings, but development of the apodal white larvae takes place only in dry organic material on the ground or floor. In 2 weeks or more the chewing larvae become fully grown and then spin cocoons in which they pupate, the adults emerging in the course of one or more weeks. A complete generation may be produced in a month during warm weather; the life cycle is greatly extended by low temperatures. In colder climates the winter may be passed in the pupal stage inside the cocoon. Recently emerged adult fleas may live for several months without food. This accounts for the fact that

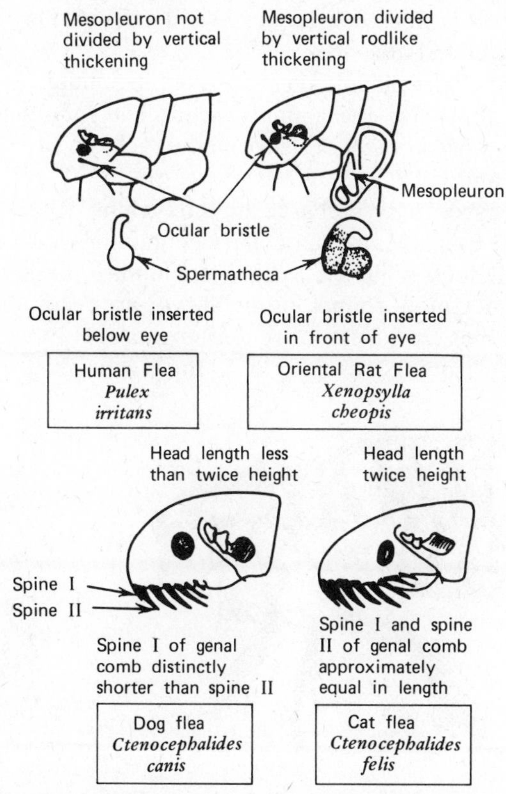

Figure 422. Characters used in separating common species of fleas. (Courtesy of USDHEW, Pub. Health Service)

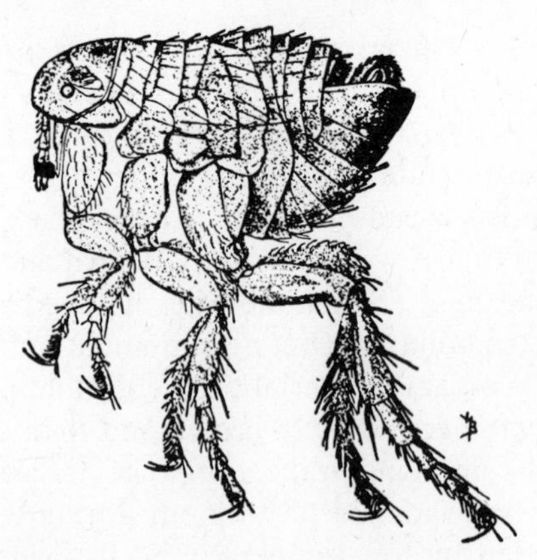

Figure 423. A male human flea, *Pulex irritans* Linn; 10×. (Bishopp, USDA)

people with pets often find their homes heavily infested with fleas when they return after an extended absence.

Successful control of fleas involves treating the animals as well as the premises, particularly the place the animals sleep. The following pesticidal dusts have given satisfactory control of fleas; methoxychlor, carbaryl, ronnel, malathion, rotenone, and 0.2% synergized pyrethrins. Cats groom themselves by licking; for this reason only carbaryl, methoxychlor, rotenone, or pyrethrum should be used to treat them. Flea collars containing dichlorvos are useful in treating cats and dogs. Oil-base sprays containing methoxychlor, premium grade malathion, diazinon, ronnel, or dichlorvos are recommended for treating inside infested homes. It is best to treat after thoroughly vacuum cleaning the house and then destroying the material in the cleaner bag which usually con-

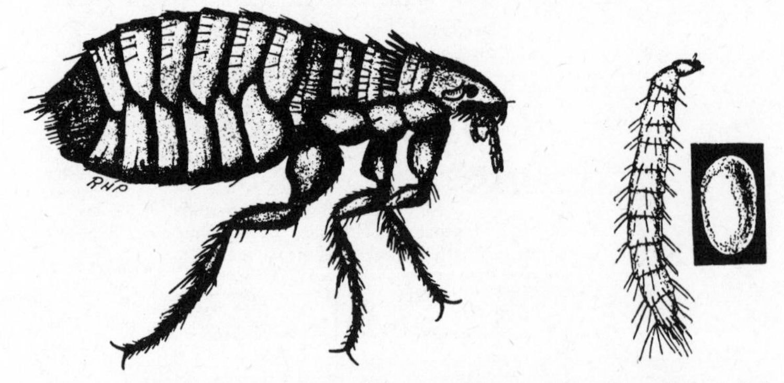

Figure 424. The cat flea, *Ctenocephalides felis* (Bouché); adult, larva, and egg; 12×. (Pettit, Mich. Agr. Exp. Sta.)

tains all life stages of fleas. Infested lawns, hog barns, or other places may be treated with the same chemicals. Elimination of rats, mice, and other rodents with poisoned baits is also necessary in a good flea control program.

REFERENCES: *USDA Leaflet* 392, 1964; *Home and Garden Bul.* 96, 1976; *Misc. Pub.* 500, 1943; *Am. J. Trop. Med.*, 31:252–256, 1951; *J. Econ. Ent.*, 46:598–601, 1953; 62:656–660, 1969.

LICE THAT ATTACK HUMANS

ORDER ANOPLURA, FAMILY PEDICULIDAE

Three species of lice, the body louse, *Pediculus humanus humanus* L., the head louse, *Pediculus humanus capitis* DeG., and the crab louse, *Phthirus pubis* (L.), attack humans the world over. All of them pierce the skin and suck blood, causing irritation, itching, and the development of objectionable welts. The body louse was the "grayback" of the Civil War and the "cootie" of World War I. It is a vector of the organisms causing epidemic typhus, trench, and relapsing fevers.

There is little morphological difference in the appearance of body and head lice; they are therefore normally differentiated on the basis of habits. Body lice are closely associated with clothing, laying their eggs or nits in the seams and other protected places, and moving to the skin only to feed but still maintaining a hold on the cloth. Head lice (Fig. 425) live among the hairs and skin on the head and glue their eggs to the hairs. Head and body lice can produce fertile offspring if crossbred and are therefore considered subspecies of a single species.

Crab lice (Fig. 425) are characteristic in appearance and frequent the pubic regions, the armpits, and sometimes the eyebrows. Their eggs adhere to the hairs of the host.

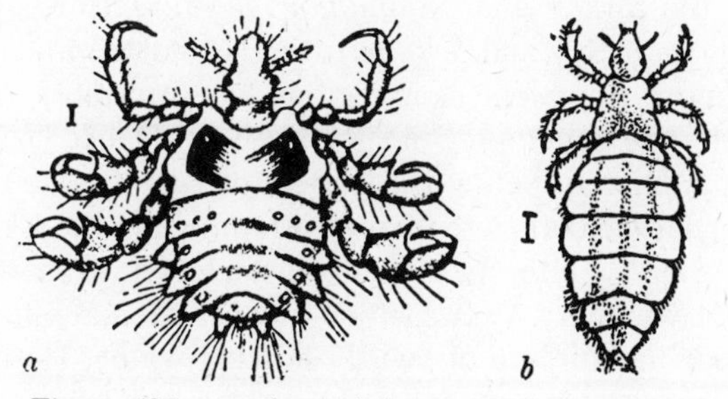

Figure 425. *a,* the crab louse, *Phthirus pubis* (L.); 12×; *b,* the human head louse, *Pediculus humanus capitis* DeGeer; 10×. (USDA)

All human lice eggs hatch in about a week, and after 3 nymphal molts the adult stage is reached, the total period from egg to adult requiring 3 or 4 weeks.

Shampoos containing lindane or DDT are available by prescription to control head and crab lice. Synergized pyrethrins are available without prescription. Powders containing lindane, DDT, or pyrethrins are useful in treating infested parts of the body and clothing. A hot water tub bath will kill crab lice or cause them to detach from the body. Clipping and destroying hairs from the body will eliminate many lice and eggs.

For body louse control additional measures are advocated, such as treatment of all clothing and bedding by steam sterilization, by dry heat at 150 F (66 C) for several hours, by washing in hot water, by exposure to deepfreeze temperatures, or by fumigation with methyl bromide or hydrogen cyanide. A 1% rotenone dust is of some merit in control, but repeated applications are required. Even when washed weekly, garments treated with a solution of 2% DDT in a volatile solvent will kill body lice for several weeks.

REFERENCES: *USDA Misc. Pub.* 606, 1946; *Yearbook of Agriculture,* pp. 487–491, 1952; *J. Econ. Ent.,* 46:524, 1953; 62:568–570, 1969.

BLACK FLIES

ORDER DIPTERA, FAMILY SIMULIIDAE

The females of many species of tiny, hump-backed, black-bodied flies with transparent wings suck the blood of humans as well as domesticated and wild animals. They vary in size up to 5 mm long, depending on the species. Besides the annoyance and irritation of their bites, some species are carriers of organisms causing disease. In Africa, Mexico, and Central American countries they are vectors of the filarial parasite causing onchocerciasis, a disease characterized by subcutaneous swellings.

A few common forms are the southern buffalo gnat, *Cnephia pecuarum* (Riley) (Fig. 426), the turkey gnat, *Simulium meridionale* Riley, *S. vittatum,* Zetterstedt, *S. venustum* Say and *S. tuberosum* (Lundström).

All species are more prevalent near regions where rapidly flowing streams occur, the aquatic larvae developing in such habitats. Dense larval populations are often found immediately below the outlets of lakes, ponds, or bogs (Fig. 427). Overwintering larvae complete their development in the spring, and adults emerge several days later. Outbreaks usually last a few days to a month or more but disappear in hot weather. After a blood meal the female flies deposit their eggs on objects near the surface of swiftly flowing water. Hatching occurs in approximately a week, the larvae feeding on aquatic microscopic plants and animals. The number of generations per year varies from one to several, depending on the species.

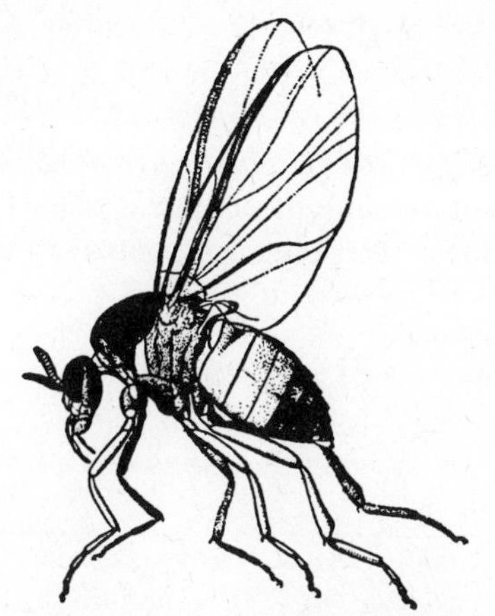

Figure 426. A black fly, the southern buffalo gnat, *Cnephia pecuarum* (Riley); 6×. (Garman, Ky. Agr. Exp. Sta.)

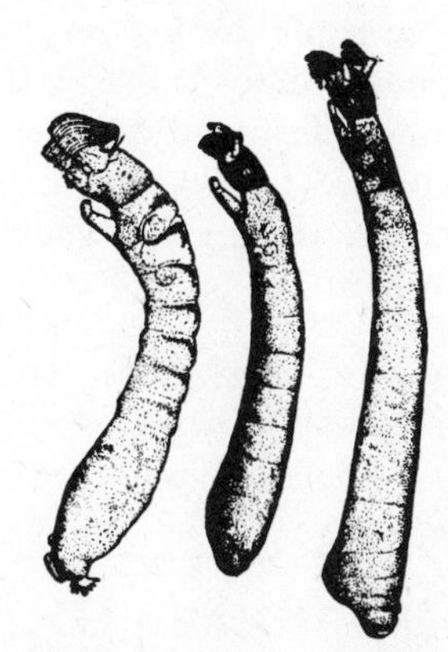

Figure 427. Black fly larvae; *Simulium pictipes* Hagen, *S. venustum* Say, and *Cnephia pecuarum* (Riley); 6×. (Knowlton, Utah Agr. Exp. Sta.)

Treatment of the breeding areas has shown promise for killing adults and larvae. Airplane application has proved to be advantageous where large inaccessible acreages are involved. Easily accessible areas can be treated by using hand or ground equipment. Control of adults has been done primarily with mist blowers or fogging devices on an area-wide basis. Some useful chemicals are dichlorvos, synergized pyrethrins, malathion, methoxychlor, ronnel, naled, fenthion, and chlorpyrifos. The choice of chemical and dosage depends on possible effects on fishes and other wildlife in the area to be treated.

Repellents such as dimethyl carbate, diethyl toluamide, dimethyl phthalate, 2-ethyl-1,3-hexandiol (6–12), indalone, or the 6-2-2 mixture of the last three materials, have given 2 to 7 hours' protection, depending on the species of black fly. When used alone 6–12 gave longer protection.

REFERENCES: *J. Econ. Ent.*, 38:694–699, 1945; 43:696–697, 702–705, 1950; 44:813–814, 1951; 47:135–141, 1954; 54:607–608, 1961; 55:636–638, 1962; 61:1072–1083, 1968.

EYE GNATS

ORDER DIPTERA, FAMILY CHLOROPIDAE

These tiny flies in the genus *Hippelates* are annoying to man and other animals in areas having moist soils, especially those of sand or high in organic matter and under cultivation. The life cycle requires a period of 3 or 4 weeks, and breeding may continue throughout the year. They are attracted to the secretions of the mucous membranes of animals and are thought to be carriers of the organisms causing conjunctivitis in man. Although widely distributed, annoyance is probably greater in California and in the southern coastal states of the United States.

Some widespread troublesome species are *Hippelates collusor* (Townsend), *H. pallipes* (Loew), and *H. pusio* Loew. Using attractant chemical baits for adults and treating soil where eye gnat larvae are developing are control methods of merit.

REFERENCES: *J. Econ. Ent.* 53:367–372, 1960; 54:130–131, 1961; *Ann. Ent. Soc. Amer.* 61: 368–372, 1968; *Calif. Agr.* 24(5):4–6, 1970.

SPIDERS

ORDER ARANEAE

Some 30,000 species of spiders occur throughout the world. As a group they are considered to be beneficial because of their predatory habits, mainly on insects. All spiders inject venom when biting to paralyze their prey. Perhaps the most feared species is the black widow or hourglass spider, *Latrodectus mactans* (F.) (Family Theridiidae), so-named because the female kills her mate following copulation, and because of the prominent hourglass-shaped spot on the ventral side of her abdomen (Fig. 428). Although black widow bites are likely to prove painful and even fatal, owing to the paralytic action of the venom on the nervous system, they are not often inflicted on man.

Another species said to be about as poisonous as the black widow spider is the brown recluse spider, *Loxosceles reclusa* Gertsch and Malaik (Family Loxoscelidae). Both sexes are senocular occurring in 3 pairs; the cephalothorax bears the dorsal violin-shaped spot which makes recognition easy (Fig. 429). Initially its bite is usually not painful, but within 8 to 12 hours pain intensifies

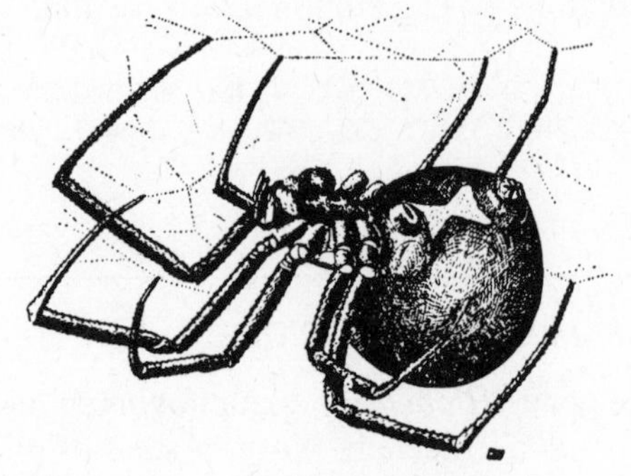

Figure 428. The black widow spider, *Latrodectus mactans* (F.); about 2×. (Knowlton, Utah Agr. Exp. Sta.)

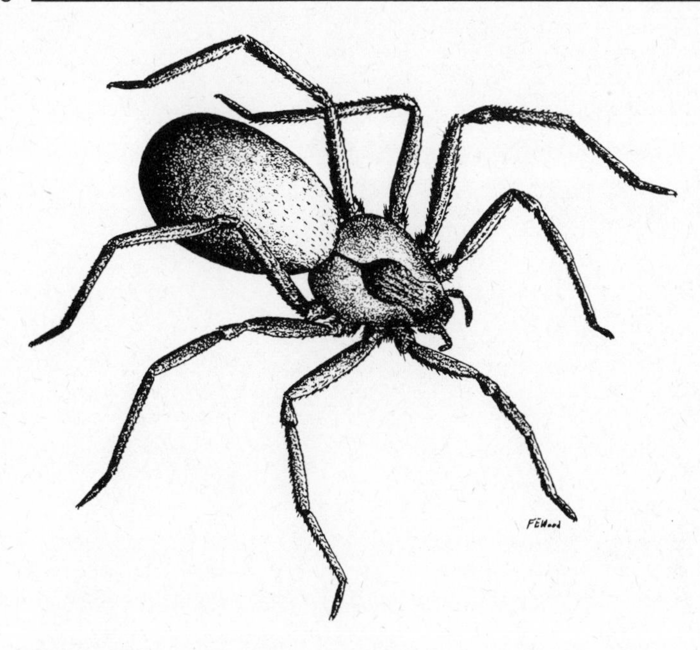

Figure 429. The brown recluse spider; 5×. (Courtesy of F. E. Wood)

and during the next few days the bitten area becomes ulcerous. The resulting sore heals slowly and often leaves a disfiguring scar. This species is found mainly in Central and South Central United States.

Control of these species as well as other spiders can be accomplished by treating outbuildings, under porches, around foundations, in basements, or other places likely to harbor spiders with one of the following pesticides: chlorpyrifos, propoxur, malathion, ronnel, dichlorvos, or lindane.

REFERENCES: *Calif. Agr. Exp. Sta. Bul.* 591, 1935; *Proc. North Central Branch, Ent. Soc. Amer.*, 19:115–118, 1964; *USDA Agr. Handbook* 290, 1965; *Home and Garden Bul.* 96, 1976; *Leaflet* 556, 1972; *Mo. Agr. Exp. Sta. Bul.* 738, 1968.

Index